ELEMENTARY DIFFERENTIAL EQUATIONS WITH LINEAR ALGEBRA

FOURTH EDITION

ELEMENTARY DIFFERENTIAL EQUATIONS WITH LINEAR ALGEBRA

FOURTH EDITION

ALBERT L. RABENSTEIN
WASHINGTON AND JEFFERSON COLLEGE

SAUNDERS COLLEGE PUBLISHING
HARCOURT BRACE JOVANOVICH COLLEGE PUBLISHERS
Fort Worth Philadelphia San Diego New York Orlando Austin San Antonio
Toronto Montreal London Sydney Tokyo

Associate Editor: Pamela Whiting
Manuscript Editor: Cindy Simpson
Production Editor: Michael Biskup
Designer: Lori McThomas
Art Editor: Avery Hallowell
Production Manager: Mary Kay Yearin

Cover: Copyright © 1991 by Anthony Grant

ISBN: 0-15-520984-1

Library of Congress Catalog Card Number: 91-72173

Printed in the United States of America

Preface

Differential equations and linear algebra are important in many fields of application. The purpose of this book is to present an integrated introduction to the two subjects that takes advantage of the relationship between them. Linear algebra is a useful tool in the study of linear (and nonlinear) differential equations, especially linear systems. Likewise, differential equations provide examples and applications of many concepts of linear algebra.

The main prerequisite for this book is two semesters of calculus. (Partial derivatives are used in Section 1.4 and are mentioned in a few other places.) Although there are no prerequisite courses for the study of linear algebra, some mathematical maturity is needed. The ideas of vector space, linear independence, and dimension are not easy for beginners to grasp. However, these concepts are basic, and an understanding of them pays off in the study of linear differential equations.

Chapters 2, 3, and 4 deal mainly with concepts of linear algebra. These ideas are applied to differential equations in Chapter 5 (Linear Differential Equations), Chapter 6 (Systems of Differential Equations), and Chapter 10 (Stability and the Phase Plane). Systems of differential equations with constant coefficients are treated by three techniques in Chapter 6. The first, the elimination method (Section 6.3), does not depend on linear algebra and can be utilized in a course where time does not permit full development of the material on characteristic values in Chapter 4. However the method of characteristic values (Sections 6.6 and 6.7) and the method that utilizes the exponential matrix (Sections 6.8, 6.9, and 6.10) require a knowledge of Chapter 4. Chapter 10 also depends on Chapter 4. Chapters 7, 8, 9, and 10 are independent of one another and may be taken up in any order.

One important addition to this edition is a new chapter about stability and the phase plane (Chapter 10). This edition also contains approximately 300 new exercises, many of which appear in the additional exercise sets at the ends of chapters. Other new material includes a section about linear transformations of the plane, an introduction to quadratic forms, and a section on numerical methods for systems of equations. Some computer programs in BASIC have also been added.

Answers to about half of the computational exercises are in an appendix at the end of the book. Answers to other exercises are available in the Instructor's Manual.

I would like to thank the following reviewers for their constructive criticism and helpful suggestions during the writing of this new edition: Gerald M. Armstrong, Brigham Young University; James Herod, Georgia Institute of Technology; Pao-sheng Hsu, University of Maine; David Johnson, Lehigh University; Yasutaka Sibuya, University of Minnesota; Vencil Skarda, Brigham Young University; William Snyder, University of Maine; and Raymond Terry, California Polytechnic State University.

I am also grateful for the efforts of Pamela Whiting, Associate Editor; Cindy Simpson, Manuscript Editor; Michael Biskup, Production Editor; Lori McThomas, Designer; Avery Hallowell, Art Editor; and Mary Kay Yearin, Production Manager.

Contents

C H A P T E R

Introduction to Differential Equations

1

1.1   Introduction   1
1.2   Separable Equations   8
1.3   Homogeneous Equations   12
1.4   Exact Equations   17
1.5   First-Order Linear Equations   23
1.6   Orthogonal Trajectories   28
1.7   Radioactive Decay   32
1.8   Mixing Problems   35
1.9   Population Growth   37
1.10 Cooling; The Rate of a Chemical Reaction   39
1.11 Two Special Types of Second-Order Equations   44
1.12 Falling Bodies   48
1.13 Some Theoretical Matters   55

C H A P T E R

Matrices and Determinants

2

2.1   Systems of Linear Equations   63
2.2   Homogeneous Systems   75
2.3   Applications   79
2.4   Matrices and Vectors   82
2.5   Matrix Multiplication   87
2.6   Inner Product and Length   93
2.7   Some Special Matrices   97
2.8   Determinants   103
2.9   Properties of Determinants   107
2.10 Cofactors   112

2.11 Cramer's Rule   115
2.12 The Inverse of a Matrix   121

C H A P T E R

3

Vector Spaces and Linear Transformations

3.1   Vector Spaces   129
3.2   Subspaces   133
3.3   Linear Dependence   137
3.4   Wronskians   142
3.5   Dimension   146
3.6   Orthogonal Bases   150
3.7   Linear Transformations   155
3.8   Properties of Linear Transformations   160
3.9   Transformations of the Plane   163
3.10 Differential Operators   169

C H A P T E R

4

Characteristic Values

4.1 Characteristic Values   177
4.2 An Application   182
4.3 Diagonalization   184
4.4 Real Symmetric Matrices   192
4.5 Functions of Matrices   197

C H A P T E R

5

Linear Differential Equations

5.1   Introduction   203
5.2   Polynomial Operators   207
5.3   Complex Solutions   211
5.4   Equations with Constant Coefficients   218
5.5   Cauchy-Euler Equations   225
5.6   Nonhomogeneous Equations   229
5.7   The Method of Undetermined Coefficients   232
5.8   Variation of Parameters   241
5.9   Simple Harmonic Motion   249
5.10 Electric Circuits   257

CHAPTER

Systems of Differential Equations

6

6.1   Introduction   263
6.2   First-Order Systems   267
6.3   Linear Systems with Constant Coefficients   271
6.4   Matrix Formulation of Linear Systems   279
6.5   Fundamental Sets of Solutions   283
6.6   Solutions by Characteristic Values   288
6.7   Repeated Characteristic Values   292
6.8   Series of Matrices   297
6.9   The Exponential Matrix Function   302
6.10 A Matrix Method   306
6.11 Nonhomogeneous Linear Systems   313
6.12 Mechanical Systems   317
6.13 The Two Body Problem   322
6.14 Electric Circuits   327

CHAPTER

Series Solutions

7

7.1 Power Series   333
7.2 Taylor Series   338
7.3 Ordinary Points   341
7.4 Singular Points   347
7.5 The Case of Equal Exponents   355
7.6 The Case when the Exponents Differ by an Integer   360
7.7 The Point at Infinity   365
7.8 Legendre Polynomials   367
7.9 Bessel Functions   371

CHAPTER

Numerical Methods

8

8.1 The Euler Method   381
8.2 Taylor Series Methods   386
8.3 Runge-Kutta Methods   389
8.4 A Multi-Step Method   393
8.5 Systems of Equations   398

C H A P T E R

9

Laplace Transforms

9.1 The Laplace Transform   405
9.2 Functions of Exponential Order   409
9.3 Properties of Laplace Transforms   413
9.4 Inverse Transforms   417
9.5 Applications to Differential Equations   422
9.6 Functions with Discontinuities   427

C H A P T E R

10

Stability and the Phase Plane

10.1 Basic Ideas   437
10.2 The Phase Plane   444
10.3 Nonlinear Systems   453
10.4 Competition Between Two Species   460
10.5 Periodic Solutions   472

References   479

Answers to Selected Exercises   481

Index   507

ELEMENTARY DIFFERENTIAL EQUATIONS WITH LINEAR ALGEBRA

FOURTH EDITION

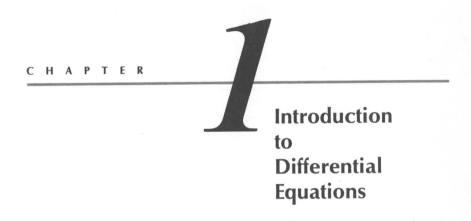

C H A P T E R

Introduction to Differential Equations

1.1
INTRODUCTION

An ordinary differential equation may be defined as an equation that involves a single unknown function of a single variable and some finite number of its derivatives. For example, a simple problem from calculus is that of finding all functions f for which

$$f'(x) = 3x^2 - 4x + 5 \tag{1.1}$$

for all x. Clearly a function f satisfies the condition (1.1) if and only if it is of the form

$$f(x) = x^3 - 2x^2 + 5x + c,$$

where c is an arbitrary number. A more difficult problem is that of finding all functions g for which

$$g'(x) + 2[g(x)]^2 = 3x^2 - 4x + 5. \tag{1.2}$$

Another difficult problem is that of finding all functions y for which (we use the abbreviation y for $y(x)$)

$$x^2 \frac{d^2y}{dx^2} - 3x\left(\frac{dy}{dx}\right)^2 + 4y = \sin x. \tag{1.3}$$

In each of the problems (1.1), (1.2), and (1.3) we are asked to find all functions that satisfy a certain condition, where the condition involves one or more *derivatives* of the function. We can reformulate our definition of a differential equation as follows. Let F be a function of $n + 2$ variables. Then the equation

$$F[x, y, y', y'', \dots, y^{(n)}] = 0 \tag{1.4}$$

is called an ordinary differential equation of order n for the unknown function y. The *order* of the equation is the order of the highest order derivative that appears in the equation. Thus, Eqs. (1.1) and (1.2) are first-order equations, while Eq. (1.3) is of second order.

A *partial* differential equation (as distinguished from an *ordinary* differential equation) is an equation that involves an unknown function of more than one independent variable, together with partial derivatives of the function. An example of a partial differential equation for an unknown function $u(x, t)$ of two variables is

$$\frac{\partial^2 u}{\partial x^2} = \frac{\partial u}{\partial t} + u.$$

Almost all the differential equations that we shall consider will be ordinary.

By a *solution* of an ordinary differential equation of order n, we mean a function that, on some interval,[1] possesses at least n derivatives and satisfies the equation. For example, a solution of the equation

$$\frac{dy}{dx} - 2y = 6$$

is given by the formula

$$y = e^{2x} - 3, \qquad \text{for all } x,$$

because

$$\frac{d}{dx}(e^{2x} - 3) - 2(e^{2x} - 3) = 2e^{2x} - 2e^{2x} + 6 = 6$$

for all x. The set of all solutions of a differential equation is called the *general solution* of the equation. For instance, the general solution of the equation

$$\frac{dy}{dx} = 3x^2 - 4x$$

---

[1] We shall use the notations (a, b), $[a, b]$, $(a, b]$, $[a, b)$, (a, ∞), $[a, \infty)$, $(-\infty, a]$, $(-\infty, a)$, $(-\infty, \infty)$ for intervals. Here (a, b) is the set of all real numbers x such that $a < x < b$, $[a, b]$ is the set of all real numbers x such that $a \le x \le b$, $[a, b)$ is the set of all real numbers x such that $a \le x < b$, and so on.

consists of all functions that are of the form

$$y = x^3 - 2x^2 + c, \qquad x \text{ in } \mathscr{I},$$

where c is an arbitrary constant and $\mathscr{I}$ is an arbitrary interval. To *solve* a differential equation is to find its general solution.

Let us now solve the second-order equation

$$\frac{d^2 y}{dx^2} = 12x + 8.$$

Integrating, we find that

$$\frac{dy}{dx} = 6x^2 + 8x + c_1,$$

where c_1 is an arbitrary constant. A second integration yields

$$y = 2x^3 + 4x^2 + c_1 x + c_2$$

for the general solution. Here c_2 is a second arbitrary constant.

The general solution of the third-order equation

$$y''' = 16e^{-2x}$$

can be found by three successive integrations. We find easily that

$$y'' = -8e^{-2x} + c_1',$$
$$y' = 4e^{-2x} + c_1' x + c_2,$$

and

$$y = -2e^{-2x} + \tfrac{1}{2}c_1' x^2 + c_2 x + c_3,$$

where c_1', c_2, and c_3 are arbitrary constants. If we replace the constant c_1' in the last formula by $2c_1$, it becomes

$$y = -2e^{-2x} + c_1 x^2 + c_2 x + c_3.$$

This last formula is slightly simpler in appearance. The two formulas describe the same set of functions since the coefficient of x^2 is completely arbitrary in both cases. Since $c_1' = 2c_1$, we see that to any arbitrarily assigned value for c_1, there corresponds a value for c_1' and vice versa.

If a formula can be found that describes the general solution of an *n*th-order equation, it usually involves *n* arbitrary constants. We note that this principle has been borne out in the last three examples, which admittedly are rather simple. Actually it is possible to find a simple formula that describes the general solution only for relatively few types of differential equations. Several such classes of first-order equations are discussed in the following three sections. In cases where it is not possible to find explicit formulas for the solutions, it still may be possible to discover certain properties of the solutions. For instance, it may be possible to show that a solution is bounded (or unbounded), to find its limiting value as the independent variable becomes infinite, or to establish that it is a periodic function. Much advanced work in differential equations is concerned with such matters.

Perhaps some reasons should now be given as to why we want to solve differential equations. Briefly, many experimentally discovered laws of science can be formulated as relations that involve not only magnitudes of quantities but also rates of change (usually with respect to time) of these magnitudes. Thus, the laws can be formulated as differential equations. A number of examples of problems that give rise to differential equations are presented in this book. Some applications will be described in Sections 1.6–1.10 after we have learned how to solve several kinds of first-order equations.

We have seen that ordinary differential equations can be classified as to order. We shall also categorize them in one more way. An equation of order n is said to be a *linear* equation if it is of the special form

$$a_0(x)y^{(n)} + a_1(x)y^{(n-1)} + \cdots + a_{n-1}(x)y' + a_n(x)y = f(x),$$

where $a_0, a_1, \ldots, a_n$ and f are given functions that are defined on an interval $\mathscr{I}$. Thus the general nth-order equation (1.4) is linear if the function F is a first-degree polynomial in $y, y', \ldots, y^{(n)}$. An equation that is not linear is said to be a *nonlinear* equation. For example, each of the equations

$$y' + (\cos x)y = e^x,$$

$$xy'' + y' = x^2,$$

$$xy''' - e^x y' + (\sin x)y = 0,$$

is linear, while each of the equations

$$y' + y^2 = 1,$$

$$y'' + (\cos x)yy' = \sin x,$$

$$y''' - x(y')^3 + y = 0,$$

is nonlinear. Because linear equations possess special properties, they will be treated in a separate chapter, Chapter 5.

In most applications that involve differential equations, the unknown function is required not only to satisfy the differential equation but also to satisfy certain other auxiliary conditions. These auxiliary conditions often specify the values of the function and some of its derivatives at one or more points. As an example, suppose we are asked to find a solution of the equation

$$\frac{dy}{dx} = 3x^2$$

that satisfies the auxiliary condition $y = 1$ when $x = 2$, or

$$y(2) = 1.$$

Thus, we require the graph of our solution (which is called a *solution curve* or *integral curve*) to pass through the point $(2, 1)$ in the xy plane. The general solution of the equation is

$$y = x^3 + c,$$

where c is an arbitrary constant. In order to find a specific solution that satisfies the initial condition, we set $x = 2$ and $y = 1$ in the last formula, finding that $1 = 8 + c$ or $c = -7$. Thus, there is only one value of c for which the condition is satisfied. The equation possesses one and only one solution (defined for all x) that satisfies the condition, namely,

$$y = x^3 - 7.$$

For an nth-order equation of the form

$$y^{(n)} = G[x, y, y', y'', \ldots, y^{(n-1)}], \tag{1.5}$$

auxiliary conditions of the type

$$y(x_0) = k_0, \quad y'(x_0) = k_1, \quad y''(x_0) = k_2, \ldots, \quad y^{(n-1)}(x_0) = k_{n-1}, \tag{1.6}$$

where the k_i are given numbers, are common. We note that there are n conditions for the nth-order equation. These conditions specify the values of the unknown function and its first $n - 1$ derivatives at a single point x_0. For a first-order equation

$$y' = H(x, y),$$

we would have only one condition

$$y(x_0) = k_0$$

specifying the value of the unknown function itself at x_0. In the case of a second-order equation

$$y'' = K(x, y, y'),$$

we would have two conditions

$$y(x_0) = k_0, \qquad y'(x_0) = k_1.$$

A set of auxiliary conditions of the form (1.6) is called a set of *initial conditions* for the Eq. (1.5). The equation (1.5) together with the conditions (1.6) constitute an *initial value problem*. The reason for this terminology is that in many applications the independent variable x represents time and the conditions are specified at the instant x_0 at which some process begins.

In specifying the values of the first $n - 1$ derivatives of a solution of Eq. (1.5) at x_0, we have essentially specified the values of any higher derivatives that might exist. The values of these higher derivatives can be found from the differential equation itself. For example, let us consider the initial value problem

$$y'' = x^2 - y^3$$
$$y(1) = 2, \qquad y'(1) = -1.$$

From the differential equation we see that

$$y''(1) = 1 - 8 = -7.$$

By differentiating through in the differential equation, we find that

$$y''' = 2x - 3y^2 y'$$

and hence

$$y'''(1) = 2 - (3)(4)(-1) = 14.$$

The values of higher derivatives at $x = 1$ can be found by repeated differentiation.

If a function can be expanded in a power series about a point x_0, a knowledge of the values of the function and its derivatives at x_0 completely determines the function. This discussion suggests that the initial value problem (1.5) and (1.6) can have but one solution if the function G is infinitely differentiable with respect to all variables. Actually it can be shown that, under rather mild restrictions on G, the initial value problem possesses a solution and that it has only one solution. (See the final section of this chapter.) In most of the problems and examples of this chapter, it is possible to actually find all the solutions of the differential equation at hand. In cases where this is impossible, it is comforting to know that the problem being considered actually has a solution and that there is only one solution. An initial value problem purporting to describe some physical process would not be very valuable without these two properties.

Also, numerical methods, some of which are described in Chapter 8, can be used in connection with computers to produce tables of approximate values of solutions. Computer software is available to draw graphs of solutions. (One such program is mentioned in the references at the end of the book.)

―――――――――――――――― **Exercises for Section 1.1** ――――――――――――――――

1. Find the order of the differential equation and determine whether it is linear or nonlinear.

    (a) $y' = e^x$

    (b) $y'' + xy = \sin x$

    (c) $y' + e^y = 0$

    (d) $y'' + 2y' + y = \cos x$

    (e) $y'' + xyy' + y = 2$

    (f) $y^{(4)} + 3(\cos x)y''' + y' = 0$

    (g) $y''' = 0$

    (h) $yy''' + y' = 0$

2. Find the general solution of the differential equation.

    (a) $y' = 2x - 3$ (b) $y' = 3x^2 \sin x^3$

    (c) $y' = \dfrac{4}{x(x-4)}$ (d) $y'' = 12e^{-2x} + 4$

    (e) $y'' = \sec^2 x$ (f) $y'' = 8e^{-2x} + e^x$

    (g) $y''' = 24x - 6$ (h) $y^{(4)} = 32 \sin 2x$

3. Find a solution of the differential equation that satisfies the specified conditions.

    (a) $y' = 0, y(2) = -5$

    (b) $y' = x, y(2) = 9$

    (c) $y' = 4x - 3, y(4) = 3$

    (d) $y' = 3x^2 - 6x + 1, y(-2) = 0$

    (e) $y'' = 0, y(2) = 1, y'(2) = -1$

    (f) $y'' = 9e^{-3x}, y(0) = 1, y'(0) = 2$

    (g) $y'' = \cos x, y(\pi) = 2, y'(\pi) = 0$

    (h) $y''' = e^{-x}, y(0) = -1, y'(0) = 1, y''(0) = 3$

4. Show that a function is a solution of the equation $y' + ay = 0$, where a is a constant, if, and only if, it is a solution of the equation $(e^{ax}y)' = 0$. Hence show that the general solution of the equation is described by the formula $y = ce^{-ax}$, where c is an arbitrary constant.

5. Use the result of Exercise 4 to find the general solution of the given differential equation.

    (a) $y' + 3y = 0$ (b) $y' - 3y = 0$

    (c) $3y' - y = 0$ (d) $3y' + 2y = 0$

6. Verify that the differential equation has the given function as a solution.

    (a) $xy' + y = 3x^2, y = x^2$, all x.

    (b) $xy' + y = 0, y = 1/x, x > 0$.

    (c) $y' + 2xy = 0, y = \exp(-x^2)$, all x.

    (d) $y'' + 4y = 0, y = \cos 2x$, all x.

    (e) $y'' + y' - 2y = 0, y = e^{-2x}$, all x.

    (f) $2x^2y'' + 3xy' - y = 0, y = \sqrt{x}, x > 0$.

7. Verify that each of the functions $y = e^{-x}$ and $y = e^{3x}$ is a solution of the equation $y'' - 2y' - 3y = 0$ on any interval. Then show that $c_1 e^{-x} + c_2 e^{3x}$ is a solution for every choice of the constants c_1 and c_2.

8. Suppose that a function f is a solution of the initial value problem $y' = x^2 + y^2, y(1) = 2$. Find $f'(1), f''(1),$ and $f'''(1)$.

9. If the function g is a solution of the initial value problem

    $$y'' + yy' - x^3 = 0,$$

    $$y(-1) = 1, \qquad y'(-1) = 2,$$

    find $g''(-1)$ and $g'''(-1)$.

10. Show that the problem $y' = 2x, y(0) = 0, y(1) = 100$, has no solution. Is this an initial value problem?

1.2

SEPARABLE EQUATIONS

A first-order differential equation that can be written in the form

$$p(y)\frac{dy}{dx} = q(x),\qquad\qquad (1.7)$$

or

$$p(y)\,dy = q(x)\,dx$$

where p and q are given functions, is called a *separable* equation. Examples of such equations are

$$y^{-2}\frac{dy}{dx} = 2x, \qquad y^{-1}\frac{dy}{dx} = (x+1)^{-1}, \qquad (3y^2 + e^y)\frac{dy}{dx} = \cos x.$$

If a function f is a solution of Eq. (1.7) on an interval $\mathcal{I}$, then

$$p[f(x)]f'(x) = q(x)$$

for x in $\mathcal{I}$. Taking antiderivatives, we have

$$\int p[f(x)]f'(x)\,dx = \int q(x)\,dx + c$$

or

$$\int p(y)\,dy = \int q(x)\,dx + c.$$

If P and Q are functions such that $P'(y) = p(y)$ and $Q'(x) = q(x)$, then the solution f must satisfy the equation

$$P(y) = Q(x) + c,\qquad\qquad (1.8)$$

where c is a constant. That is,

$$P[f(x)] = Q(x) + c$$

for x in $\mathcal{I}$. Conversely, if y is any differentiable function that satisfies Eq. (1.8), we see by implicit differentiation that

$$P'(y)\frac{dy}{dx} = Q'(x)$$

or

$$p(y)\frac{dy}{dx} = q(x).$$

Thus, a function is a solution of Eq. (1.7) if and only if it satisfies an equation of the form (1.8) for some choice of the constant c. It may not be possible to find an explicit formula for y in terms of x from Eq. (1.8). However, we say that Eq. (1.8) determines the solutions of the differential equation *implicitly*. Let us now consider some examples of separable equations.

Example 1

$$\frac{dy}{dx} = 2xy^2. \qquad (1.9)$$

"Separating the variables," we have

$$y^{-2}\frac{dy}{dx} = 2x$$

or

$$y^{-2}\,dy = 2x\,dx.$$

Taking antiderivatives, we have

$$\int y^{-2}\,dy = \int 2x\,dx + c$$

or

$$-\frac{1}{y} = x^2 + c.$$

Thus, the functions defined by the formula

$$y = \frac{-1}{x^2 + c} \qquad (1.10)$$

are solutions of Eq. (1.9). Note, however, that the identically zero function $(y = 0)$ is also a solution. In arriving at formula (1.10), we started out by dividing both sides of the original equation by y^2, and this procedure is not valid when $y = 0$.

Suppose that it is desired to find the solution curve that passes through the point $(2, -1)$ in the xy plane. Then our initial condition is

$$y(2) = -1.$$

Setting $x = 2$ and $y = -1$ in formula (1.10), we see that

$$-1 = \frac{-1}{4 + c}$$

or $c = -3$. Then the desired solution is given by the formula

$$y = \frac{1}{3 - x^2}, \qquad \sqrt{3} < x < \infty.$$

The reader should note the domain of the solution. There is nothing about the differential equation (1.9) or the initial condition that indicates anything special about the number $\sqrt{3}$.

Another approach to finding the specific solution for which $y(2) = -1$ is to incorporate the initial conditions into the limits of integration. Going back to the relation

$$y^{-2}\, dy = 2x\, dx$$

we write

$$\int_{-1}^{y} y^{-2}\, dy = \int_{2}^{x} 2x\, dx.$$

Then

$$\left[-\frac{1}{y} \right]_{-1}^{y} = [x^2]_{2}^{x}$$

and

$$-\frac{1}{y} - 1 = x^2 - 4$$

so

$$y = \frac{1}{3 - x^2}.$$

In general, to solve the equation

$$f(y)\, dy = g(x)\, dx$$

with initial condition $y(x_0) = y_0$ we would write

$$\int_{y_0}^{y} f(y)\, dy = \int_{x_0}^{x} g(x)\, dx.$$

Finally, suppose we seek the solution of Eq. (1.9) for which $y(3) = 0$. If we set $x = 3$ in formula (1.10), we obtain the equation

$$0 = \frac{-1}{9 + c}$$

for the constant c, but no solution exists. We must remember that the zero function $(y = 0)$ is a solution of the equation, and it certainly satisfies the initial condition $y(3) = 0$.

Example 2

$$(x + 1)\frac{dy}{dx} = 2y.$$

Here we have

$$\frac{dy}{y} = 2\frac{dx}{x + 1}$$

or

$$\ln|y| = \ln(x + 1)^2 + c'.$$

Then

$$|y| = e^{c'}(x + 1)^2$$

and

$$y = \pm e^{c'}(x + 1)^2, \qquad (1.11)$$

where c' is an arbitrary constant. But $\pm e^{c'}$ can have any value except zero, so the set of functions described by formula (1.11) is also described by the simpler formula

$$y = c(x + 1)^2. \qquad (1.12)$$

where c is a constant different from zero but otherwise is arbitrary. However, since $y = 0$ is obviously a solution of the differential equation, formula (1.12) also describes a solution when $c = 0$. This formula, with c completely arbitrary, gives the general solution of the equation.

Example 3 Consider the initial value problem

$$\frac{dy}{dx} = \frac{\cos x}{3y^2 + e^y}, \qquad y(0) = 2.$$

From the differential equation we have

$$(3y^2 + e^y)\, dy = \cos x \, dx$$

or

$$y^3 + e^y = \sin x + c.$$

Setting $x = 0$ and $y = 2$ (these values come from the initial condition) we find that

$$8 + e^2 = c.$$

Hence, the desired solution (if such a solution exists) is implicitly determined by the equation

$$y^3 + e^y = \sin x + 8 + e^2.$$

Exercises for Section 1.2

In Exercises 1–20, find the general solution, if possible. Otherwise find a relation that defines the solutions implicitly. If an initial condition is specified, also find the particular solution that satisfies the condition.

1. $yy' = 4x, \quad y(1) = -3$

2. $xy' = 4y, \quad y(1) = -3$

3. $(x^2 + 4)y' = xy, \quad y(0) = 6$

4. $y' = e^x(1 - y^2)^{1/2}, \quad y(0) = \frac{1}{2}$

5. $y' = \dfrac{1 + y^2}{1 + x^2}, \quad y(2) = 3$

6. $e^y y' = 4, \quad y(0) = 2$

7. $2(y - 1)y' = e^x, \quad y(0) = -2$

8. $2y' = y(y - 2)$

9. $3y^2 y' = (1 + y^3)\cos x$

10. $(\cos^2 x)y' = y^2(y - 1)\sin x$

11. $(\cos y)y' = 1$

12. $(\cos^2 x)y' = (1 + y^2)^{1/2}$

13. $y' = e^{x+y}$

14. $y' = y \tan x$

15. $y' = 2xy \ln y$

16. $x(x + 1)y' = y(y - 1)$

17. $y' \tan^{-1} y = x(1 + y^2)$

18. $axy' + by = 0, (a \neq 0)$,

19. $y' = \tan y \cot x$

20. $y' = xye^x$

21. Solve the initial value problem; it is not necessary to find all solutions of the equation.
    (a) $y' = e^x(\sin x)(y + 1), y(2) = -1$
    (b) $xy' = y(y - 2), y(3) = 2$
    (c) $e^x y' = \sin y, y(0) = \pi$

22. Show that an equation of the form $y' = F(ay + bx + c), \ a \neq 0$, becomes separable under the change of dependent variable $v = ay + bx + k$, where k is any number.

23. Use the result of Exercise 22 to solve the differential equation.
    (a) $y' = (y + 4x - 1)^2$
    (b) $(y - x + 1)y' = y - x$
    (c) $(y - 3x)y' = 3(y - 3x + 2)$
    (d) $(y - 2x)y' = 3y - 6x + 1$

1.3 HOMOGENEOUS EQUATIONS

Some differential equations that are not separable as they stand become separable after a change of variable. One such class of equations consists of those that can be written in the form

$$\frac{dy}{dx} = F\left(\frac{y}{x}\right). \qquad (1.13)$$

An example of such an equation is

$$(x^4 + y^4)\frac{dy}{dx} = x^3 y,$$

which may be rewritten as

$$\frac{dy}{dx} = \frac{x^3 y}{x^4 + y^4}$$

or

$$\frac{dy}{dx} = \frac{y/x}{1 + (y/x)^4}.$$

An equation of the form (1.13) can be made separable by introducing a new dependent variable, v where

$$v = \frac{y}{x}.$$

For then

$$y = vx, \qquad \frac{dy}{dx} = x\frac{dv}{dx} + v,$$

and Eq. (1.13) becomes

$$x\frac{dv}{dx} + v = F(v)$$

or

$$\frac{1}{F(v) - v}\frac{dv}{dx} = \frac{1}{x},$$

which is separable. An equation of the form (1.13) is called a *homogeneous*[2] differential equation.

We shall presently describe a technique for determining whether a first-order equation is homogeneous. First, we make the definition that a function g of two variables is said to be *homogeneous of degree m* if

$$g(tx, ty) = t^m g(x, y)$$

---

[2] The term *homogeneous* has several different meanings, in mathematics generally, and in the field of differential equations in particular. In this book it will be used in Chapters 2 and 5 to describe entirely different concepts.

for all t in some interval. For example, if $g(x, y) = x^3y^2 - 3x^5$ then

$$g(tx, ty) = (tx)^3(ty)^2 - 3(tx)^5$$
$$= t^5(x^3y^2 - 3x^5)$$
$$= t^5 g(x, y)$$

for all t, so g is homogeneous of degree 5. However, the function h, where $h(x, y) = 2x^3y - 5x^2$, is not homogeneous. As another example, let us consider $q(x, y) = \sqrt{2x^3 - xy^2}$. Then

$$q(tx, ty) = \sqrt{2(tx)^3 - (tx)(ty)^2}$$
$$= \sqrt{t^3(2x^3 - xy^2)}$$
$$= t^{3/2} q(x, y)$$

for $t \geq 0$, so q is homogeneous of degree 3/2.

Given a differential equation

$$N(x, y)\frac{dy}{dx} = M(x, y), \qquad (1.14)$$

which may be written as

$$\frac{dy}{dx} = \frac{M(x, y)}{N(x, y)},$$

we claim that if M and N are both homogeneous of the same degree, then the equation is homogeneous. To see this, suppose that M and N are both homogeneous of degree m. Then

$$M(x, y) = t^{-m}M(tx, ty), \qquad N(x, y) = t^{-m}N(tx, ty).$$

Setting $t = x^{-1}$, we have

$$M(x, y) = x^m M(1, y/x), \qquad N(x, y) = x^m N(1, y/x)$$

and

$$\frac{dy}{dx} = \frac{M(1, y/x)}{N(1, y/x)}.$$

This equation is of the form (1.13).

Example   We consider the equation

$$x^2\frac{dy}{dx} = y(3x + 2y).$$

The functions $f(x, y) = x^2$ and $g(x, y) = y(3x + 2y)$ are both homogeneous of degree two, because

$$f(tx, ty) = (tx)^2 = t^2x^2 = t^2f(x, y)$$
$$g(tx, ty) = ty(3tx + 2ty) = t^2y(3x + 2y) = t^2g(x, y)$$

Hence the equation is homogeneous and the change of variable (from y to v) $y = vx$ will produce a separable equation. We have

$$x^2\left(x\frac{dv}{dx} + v\right) = vx(3x + 2vx)$$

or

$$x\frac{dv}{dx} = 2(v^2 + v).$$

Separating the variables, we have

$$\frac{dv}{v^2 + v} = 2\frac{dx}{x}$$

or

$$\left(\frac{1}{v} - \frac{1}{v + 1}\right)dv = 2\frac{dx}{x}.$$

Integrating yields

$$\ln\left|\frac{v}{v + 1}\right| = \ln x^2 + \ln c', \qquad c' > 0.$$

Solving for v, we have

$$\left|\frac{v}{v + 1}\right| = c'x^2$$

or

$$\frac{v}{v + 1} = cx^2, \qquad c = \pm c'.$$

Then

$$v = \frac{cx^2}{1 - cx^2}$$

and, since $v = y/x$,

$$y = \frac{cx^3}{1 - cx^2}.$$

(Here c may be zero, since $y = 0$ is a solution of the original equation.)

──────────────── **Exercises for Section 1.3** ────────────────

1. Determine whether the function is homogeneous. If it is, state the degree.

   (a) $f(x, y) = 5xy^2 - 4y^3$

   (b) $f(x, y) = 5x^2 + 2xy$

   (c) $g(x, y) = 2x^2 + 3y^2 + 4$

   (d) $g(x, y) = 3x - 4y - 2$

   (e) $h(x, y) = e^{xy}$

   (f) $h(x, y) = 2ye^{x/y} - 3x$

   (g) $M(x, y) = 5x^{1/3}y^{2/3}$

   (h) $M(x, y) = \sin xy$

   (i) $N(x, y) = xy \cos \dfrac{y}{x} + x^2$

   (j) $N(x, y) = 2x + 3y \sin \dfrac{x}{y}$

   (k) $P(x, y) = \ln 2x - \ln 3y$

   (l) $P(x, y) = 2 \ln x - 3 \ln y$

2. Which of the following equations are separable? Which are homogeneous? (Do not solve the equations.)

   (a) $x^2 \dfrac{dy}{dx} = 3y^2$

   (b) $(x^2 + y^2) \dfrac{dy}{dx} = 5xy$

   (c) $(x^2 + y^2) \dfrac{dy}{dx} = 5y$

   (d) $2xy \dfrac{dy}{dx} = 3$

   (e) $\dfrac{dy}{dx} = \sin \dfrac{y}{x}$

   (f) $x \dfrac{dy}{dx} = y \cos \dfrac{x}{y}$

   (g) $y \dfrac{dy}{dx} = x \cos y$

   (h) $x^2 \dfrac{dy}{dx} = y\sqrt{x^2 + y^2}$

In Exercises 3–16, find the general solution, if possible. Otherwise find a relation that defines the solutions implicitly.

3. $x^2 y' = xy - y^2$

4. $x^2 y' = y^2 + 2xy$

5. $xyy' = 2y^2 - x^2$

6. $xy' = y - xe^{y/x}$

7. $e^{y/x}y' = 2(e^{y/x} - 1) + \dfrac{y}{x} e^{y/x}$

8. $y' = \dfrac{y}{x} - 3\left(\dfrac{y}{x}\right)^{4/3}$

9. $xy' = y + (x^2 + y^2)^{1/2}$

10. $3xy^2 y' = 4y^3 - x^3$

11. $xy' - y = x \tan \dfrac{y}{x}$

12. $xy' - y = 2y(\ln y - \ln x)$

13. $x^2 y' - xy = (x^2 + y^2) \tan^{-1} \dfrac{y}{x}$

14. $xy' - y = 2\sqrt{xy}$

15. $xy' - y = \sqrt{x^2 + y^2}$

16. $xy' - y = x(1 + e^{-y/x})$

17. Consider a first-order differential equation of the form

$$(a_1 x + b_1 y + c_1)y' = a_2 x + b_2 y + c_2.$$

   (a) If $a_1 b_2 - b_1 a_2 = 0$, show that the equation is of the type considered in Exercise 22, Section 1.2.

   (b) If $a_1 b_2 - b_1 a_2 \neq 0$ introduce new variables u and v, where

$$u = x + p,$$

$$v = y + q.$$

   Show that the constants p and q can be chosen in such a way that the equation takes on the form

$$(a_1 u + b_1 v)\frac{dv}{du} = a_2 u + b_2 v.$$

Hence

$$\frac{dv}{du} = F\left(\frac{v}{u}\right).$$

18. Use the results of Exercise 17 to solve the equation.

(a) $(x + y + 1)y' = y + 2$

(b) $(3x - y + 1)y' = -x + 3y + 5$

19. Given the equation $y' = F(x, y)$, suppose that there is a number n such that

$$F(tx, t^n y) = t^{n-1}F(x, y).$$

Show that the equation can be written

$$\frac{dy}{dx} = x^{n-1}F\left(1, \frac{y}{x^n}\right)$$

and that the change of variable $y = x^n v$ makes it separable. The next two exercises provide examples.

20. Show that the equation

$$(x^2 + y)\frac{dy}{dx} = -2xy$$

is neither separable nor homogeneous. Then show that the change of variable $y = vx^2$ makes the equation separable, and solve it.

21. Do as in the previous exercise, but make the change of variable $y = vx^3$.

$$(x^3 + y)\frac{dy}{dx} = 3x^2 y$$

1.4 EXACT EQUATIONS

The first-order equation

$$M(x, y) + N(x, y)\frac{dy}{dx} = 0 \qquad (1.15)$$

is said to be *exact* (in some region of the xy-plane) if there exists a function ϕ with continuous first partial derivatives such that

$$\frac{\partial \phi(x, y)}{\partial x} = M(x, y), \qquad \frac{\partial \phi(x, y)}{\partial y} = N(x, y). \qquad (1.16)$$

The relationship between the function ϕ and the solutions of the differential equation is described in the following theorem.

Theorem 1.1   If the differential equation (1.15) is exact and if the function ϕ has the properties (1.16) then a function f, with $y = f(x)$, is a solution of the differential equation if and only if it satisfies an equation of the form

$$\phi(x, y) = c,$$

where c is a constant.

Proof   Suppose that the function f is a solution of Eq. (1.15). If $y = f(x)$ we have

$$\frac{\partial \phi(x, y)}{\partial x} + \frac{\partial \phi(x, y)}{\partial y} \frac{dy}{dx} = 0$$

or

$$\frac{d\phi(x, y)}{dx} = 0.$$

Hence $\phi(x, y) = c$. Conversely, suppose that a (differentiable) function f satisfies the equation $\phi(x, y) = c$. Then by implicit differentiation we have

$$\frac{\partial \phi(x, y)}{\partial x} + \frac{\partial \phi(x, y)}{\partial y} \frac{dy}{dx} = 0$$

or

$$M(x, y) + N(x, y) \frac{dy}{dx} = 0.$$

Hence the function is a solution of the differential equation.

Notice that if Eq. (1.15) is exact, then the total differential of ϕ is

$$d\phi = \frac{\partial \phi}{\partial x} dx + \frac{\partial \phi}{\partial y} dy = M \, dx + N \, dy.$$

Along any solution curve, $d\phi = 0$, and the solutions satisfy equations of the form $\phi(x, y) = c$.

We need a criterion for determining whether or not an equation is exact. We also need a method for finding the function ϕ when it is exact. In what follows we assume that the functions M and N and their first partial derivatives are continuous in some region.

Suppose that Eq. (1.15) is exact. Then there exists a function ϕ such that $M = \partial \phi / \partial x$ and $N = \partial \phi / \partial y$. Hence

$$\frac{\partial M}{\partial y} = \frac{\partial^2 \phi}{\partial y \, \partial x}, \qquad \frac{\partial N}{\partial x} = \frac{\partial^2 \phi}{\partial x \, \partial y},$$

and because the mixed second partial derivatives of ϕ are equal (since ϕ, ϕ_x, ϕ_y, ϕ_{xy}, and ϕ_{yx} are continuous), we have

$$\frac{\partial M}{\partial y} = \frac{\partial N}{\partial x}. \tag{1.17}$$

Thus if the equation is exact, the condition (1.17) is satisfied.

It can be shown that if M and N satisfy the condition (1.17) in a *simply connected region* then the differential equation is exact. A simply connected region is such that every simple closed curve[3] in the region contains only points of the region inside it. The interior of an ellipse or a rectangle is a simply connected region but the region bounded by two concentric circles is not simply connected. We shall prove that the condition (1.17) is sufficient for exactness in the special case of a rectangle. Let D be the rectangle

$$\{(x, y): a < x < b,\ c < y < d\},$$

where any or all of a, b, c, and d may be infinite. Such a region is simply connected.

Theorem 1.2   Let M and N satisfy the condition (1.17) in the rectangle D. Then the equation $M + Ny' = 0$ is exact.

Proof   Let (x_0, y_0) be any fixed point in D. We define a function ϕ of two variables by means of the formula

$$\phi(x, y) = \int_{x_0}^{x} M(s, y_0)\, ds + \int_{y_0}^{y} N(x, t)\, dt \tag{1.18}$$

for (x, y) in D. See Figure 1.1. We need to verify that $\partial\phi/\partial x = M$ and $\partial\phi/\partial y = N$. Differentiating with respect to x, we have[4]

$$\frac{\partial\phi(x, y)}{\partial x} = M(x, y_0) + \int_{y_0}^{y} \frac{\partial N(x, t)}{\partial x}\, dt.$$

Since the condition (1.17) is satisfied, $\partial N(x, t)/\partial x = \partial M(x, t)/\partial t$. Hence

$$\frac{\partial\phi(x, y)}{\partial x} = M(x, y_0) + \int_{y_0}^{y} \frac{\partial M(x, t)}{\partial t}\, dt$$

$$= M(x, y_0) + M(x, y) - M(x, y_0)$$

$$= M(x, y).$$

In similar fashion it can be shown (Exercise 1) that $\partial\phi/\partial y = N$. Since the function ϕ has the property (1.16) the differential equation is exact.

Formula (1.18) can be used to find a function ϕ when the equation (1.15) is exact. However, the function can usually be found by means of a simpler

---

[3] A *simple* closed curve does not cross itself. A circle is a simple closed curve. A figure eight is closed but not simple.

[4] We have differentiated with respect to x under the integral sign, a procedure that requires justification. This situation is covered by *Leibniz's rule*, which is discussed in most advanced calculus books.

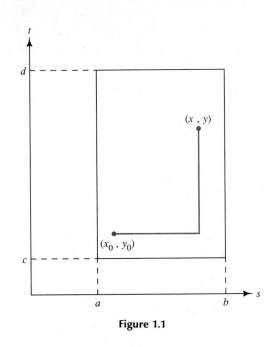

Figure 1.1

procedure which does not necessitate memorizing the formula. We shall illustrate the method with an example.

Example 1   The equation

$$3x^2 - 2y^2 + (1 - 4xy)\frac{dy}{dx} = 0 \tag{1.19}$$

is exact, since

$$M(x, y) = 3x^2 - 2y^2, \qquad N(x, y) = 1 - 4xy,$$

and

$$\frac{\partial M(x, y)}{\partial y} = -4y = \frac{\partial N(x, y)}{\partial x}$$

for all (x, y). Hence, by Theorem 1.2, there exists a function ϕ such that

$$\frac{\partial \phi(x, y)}{\partial x} = 3x^2 - 2y^2, \qquad \frac{\partial \phi(x, y)}{\partial y} = 1 - 4xy. \tag{1.20}$$

Integrating with respect to x in the first of these relations, we see that ϕ is of the form

$$\phi(x, y) = x^3 - 2xy^2 + f(y), \tag{1.21}$$

where f can be any function of y only, since $\partial f(y)/\partial x = 0$. We must choose f so that the second of the conditions (1.20) is satisfied. We require that

$$\frac{\partial \phi(x, y)}{\partial y} = -4xy + f'(y) = 1 - 4xy$$

or

$$f'(y) = 1.$$

One possible choice for f is $f(y) = y$. Then from Eq. (1.21) we have

$$\phi(x, y) = x^3 - 2xy^2 + y.$$

The solutions of Eq. (1.19) are those differentiable functions that satisfy equations of the form

$$x^3 - 2xy^2 + y = c.$$

If the equation $M + Ny' = 0$ is not exact, it may be possible to make it exact by multiplying through by some function. That is, we may be able to find a function μ such that

$$\mu(x, y)M(x, y) + \mu(x, y)N(x, y)y' = 0$$

is an exact equation. If such a function μ exists, it is called an *integrating factor* for the original equation.

Example 2   The equation

$$(xy^2 + 4x^2y) + (3x^2y + 4x^3)y' = 0 \qquad (1.22)$$

is not exact, since

$$\frac{\partial M(x, y)}{\partial y} = 2xy + 4x^2, \qquad \frac{\partial N(x, y)}{\partial x} = 6xy + 12x^2.$$

Let us see if there is an integrating factor of the form $\mu(x, y) = x^m y^n$. (There may not be one.) Multiplying through in the equation by $x^m y^n$, we have

$$(x^{m+1}y^{n+2} + 4x^{m+2}y^{n+1}) + (3x^{m+2}y^{n+1} + 4x^{m+3}y^n)y' = 0. \quad (1.23)$$

Then

$$\frac{\partial M(x, y)}{\partial y} = (n + 2)x^{m+1}y^{n+1} + 4(n + 1)x^{m+2}y^n,$$

$$\frac{\partial N(x, y)}{\partial x} = 3(m + 2)x^{m+1}y^{n+1} + 4(m + 3)x^{m+2}y^n.$$

Comparing the terms with the same exponents in these two quantities, we see that they are equal if

$$n + 2 = 3(m + 2),$$
$$4(n + 1) = 4(m + 3).$$

This is a system of two equations for m and n. We find that $m = -1$ and $n = 1$. Consequently an integrating factor is $\mu(x, y) = y/x$. Upon multiplying through in the original equation by this factor, it becomes

$$(y^3 + 4xy^2) + (3xy^2 + 4x^2 y)y' = 0.$$

As a check on our calculations, we find $\partial M/\partial y$ and $\partial N/\partial x$. The result is

$$\frac{\partial M(x, y)}{\partial y} = 3y^2 + 8xy = \frac{\partial N(x, y)}{\partial x}.$$

We know that there is a function ϕ such that

$$\frac{\partial \phi(x, y)}{\partial x} = y^3 + 4xy^2, \qquad \frac{\partial \phi(x, y)}{\partial y} = 3xy^2 + 4x^2 y.$$

Proceeding as in the previous example, we find that

$$\phi(x, y) = xy^3 + 2x^2 y^2.$$

The solutions of the differential equation are determined by the equation

$$xy^3 + 2x^2 y^2 = c.$$

Exercises for Section 1.4

1. Show that $\partial \phi(x, y)/\partial y = N(x, y)$, where ϕ is defined as in Eq. (1.18).

In Exercises 2–13, first determine if the equation is exact. If it is exact, find the general solution, or at least a relation that defines the solutions implicitly.

2. $3x^3 y^2 y' + 3x^2 y^3 - 5x^4 = 0$

3. $(3x^2 y^2 - 4xy)y' + 2xy^3 - 2y^2 = 0$

4. $xe^{xy}y' + ye^{xy} - 4x^3 = 0$

5. $(x + y^2)y' + 2x^2 - y = 0$

6. $[\cos(x^2 + y) - 3xy^2]y' + 2x \cos(x^2 + y) - y^3 = 0$

7. $(x^2 - y)y' + 2x^3 + 2xy = 0$

8. $(x + y \sin x)y' + y + x \sin y = 0$

9. $(y^3 - x^2 y)y' - xy^2 = 0$

10. $(y^{-1/3} - y^{-2/3}e^x)y' - 3(e^x y^{1/3} + e^{2x}) = 0$

11. $(e^{2y} - xe^y)y' - e^y - x = 0$

12. $y^2(x^6 + y^3)^{1/3}y' + 2x^5[(x^6 + y^3)^{1/3} - x^2] = 0$

13. $(y^{-3} - y^{-2}\sin x)y' + y^{-1}\cos x = 0$

14. Show that the separable equation $p(y)y' - q(x) = 0$ is exact.

15. Show that the function μ is an integrating factor for the equation $M + Ny' = 0$ if it satisfies the partial differential equation

$$N\frac{\partial\mu}{\partial x} - M\frac{\partial\mu}{\partial y} = \mu\left(\frac{\partial M}{\partial y} - \frac{\partial N}{\partial x}\right).$$

16. Show that an integrating factor for the equation $y' - F(y/x) = 0$ is

$$\mu(x, y) = \frac{1}{xF(y/x) - y}.$$

In Exercises 17–20, determine if the equation has an integrating factor of the form $\mu(x, y) = x^m y^n$. If it does, solve the equation.

17. $(1 - xy)y' + y^2 + 3xy^3 = 0$

18. $(3x^2 + 5xy^2)y' + 3xy + 2y^3 = 0$

19. $(x^2 + xy^2)y' - 3xy + 2y^3 = 0$

20. $3xy' + xy^3 + y^2 = 0$

21. Let $P(x) = \int p(x)\,dx$. Show that $e^{P(x)}$ is an integrating factor for the linear equation $y' + p(x)y - q(x) = 0$.

22. Consider the equation

$$3x^2y - x^3\frac{dy}{dx} = 0.$$

(a) Show that the equation is not exact.
(b) Show that $f(x, y) = y^{-2}$ is an integrating factor.
(c) Show that $g(x, y) = y^{-1}x^{-3}$ is an integrating factor.
(d) Show that $h(x, y) = x^{-6}$ is an integrating factor.
(e) Solve the equation.

23. Verify that

$$d\tan^{-1}\frac{x}{y} = \frac{y\,dx - x\,dy}{x^2 + y^2},$$

$$\frac{1}{2}d\ln(x^2 + y^2) = \frac{x\,dx + y\,dy}{x^2 + y^2}.$$

Then use these formulas to find integrating factors for, and to solve, the following equations:

(a) $x\frac{dy}{dx} + 3x^4 + 3x^2y^2 - y = 0$

(b) $(2x^2y + 2y^3 - x)\frac{dy}{dx} + y = 0$

(c) $y\frac{dy}{dx} + (3x^4 + 3x^2y^2 + x) = 0$

(d) $(2x^2 + 3y^2)\frac{dy}{dx} + xy = 0$

1.5 FIRST-ORDER LINEAR EQUATIONS

As defined in Section 1.1, a *linear* differential equation of order n has the form

$$a_0(x)y^{(n)} + a_1(x)y^{(n-1)} + \cdots + a_{n-1}(x)y' + a_n(x)y = f(x),$$

where the functions a_i and f are specified on some interval. We assume that $a_0(x) \neq 0$ for all x in this interval. (A solution may not exist throughout an interval on which a_0 vanishes.)

A first-order linear equation is of the form

$$a_0(x)y' + a_1(x)y = f(x).$$

Since $a_0(x)$ is never zero, we can divide through by a_0 and write this equation in the form

$$y' + p(x)y = q(x), \tag{1.24}$$

where $p = a_1/a_0$ and $q = f/a_0$. A formula for the solutions of Eq. (1.24) is given in the following theorem.

Theorem 1.3   Let $\mu(x) = \exp \int^x p(s)\, ds$, where $\int^x p(s)\, ds$ is any antiderivative of $p(x)$[5]. Then the solutions of Eq. (1.24) are given by

$$\mu(x)y(x) = \int \mu(x)q(x)\, dx, \tag{1.25}$$

where the indefinite integral on the right represents *all* of the antiderivatives of $\mu(x)q(x)$.

Proof   If both sides of Eq. (1.24) are multiplied by $\mu(x)$, it becomes

$$\mu(x)y'(x) + p(x)\mu(x)y(x) = \mu(x)q(x).$$

Noting that $\mu'(x) = p(x)\mu(x)$, we see that this last relation may be written as

$$\mu(x)y'(x) + \mu'(x)y(x) = \mu(x)q(x)$$

or

$$\frac{d}{dx}\left[\mu(x)y(x)\right] = \mu(x)q(x).$$

Taking antiderivatives, we have

$$\mu(x)y(x) = \int \mu(x)q(x)\, dx, \tag{1.26}$$

which is the same as Eq. (1.25). Thus, if a solution of Eq. (1.24) exists, it must be of the form (1.26). Conversely, any function defined by Eq. (1.26) is a solution of Eq. (1.24), as can be verified by retracing steps.

The formula (1.25) may be written as

$$y(x)e^{\int^x p(s)\, ds} = \int q(x)e^{\int^x p(s)\, ds}\, dx. \tag{1.27}$$

The reader who does not wish to memorize this formula can simply remember to multiply through in the differential equation by $\exp \int^x p(s)\, ds$ after it has been put in the form (1.24).

---

[5] The symbol exp (a) means the same as e^a.

The trick in solving Eq. (1.24) was to multiply through by $\exp \int^x p(s)\, ds$. Some motivation for doing this is provided by the following reasoning. Suppose we attempt to find an integrating factor for Eq. (1.24) that depends on x only. Multiplying through in the equation by $\mu(x)$ and collecting all terms on one side of the equals sign, we have

$$[p(x)y - q(x)]\mu(x) + \mu(x)y' = 0.$$

For this equation to be exact, we must have

$$\frac{\partial}{\partial y}\{[p(x)y - q(x)]\mu(x)\} = \frac{\partial}{\partial x}\mu(x)$$

or

$$p(x)\mu(x) = \mu'(x).$$

This is a separable equation for μ. We find that

$$\frac{\mu'(x)}{\mu(x)} = p(x), \qquad \ln|\mu(x)| = \int p(x)\, dx + k$$

and

$$\mu(x) = ce^{\int p(x)\, dx},$$

where $c = \pm e^k$. Choosing $c = 1$ leads to our choice for $\mu(x)$.

Example 1   An example of a first-order linear equation is

$$(x + 1)y' - y = x, \qquad x > -1. \tag{1.28}$$

Dividing through by $x + 1$ to put it in the form (1.24), we have

$$y' - \frac{1}{x + 1}y = \frac{x}{x + 1}, \qquad x > -1.$$

Note that the restriction $x > -1$ confines x to an interval where

$$p(x) = \frac{-1}{x + 1}, \qquad q(x) = \frac{x}{x + 1}$$

are both defined. We find that

$$\int p(x)\, dx = -\ln(x + 1) = \ln(x + 1)^{-1},$$

so

$$\mu(x) = e^{\ln(x+1)^{-1}} = (x+1)^{-1} = \frac{1}{x+1}.$$

From formula (1.26) or (1.27) we have

$$\frac{y}{x+1} = \int \frac{x}{(x+1)^2}\,dx.$$

Partial fractions can be used to evaluate the integral. We find that

$$\frac{y}{x+1} = \int \left[\frac{1}{x+1} - \frac{1}{(x+1)^2} \right] dx,$$

so

$$\frac{y}{x+1} = \ln(x+1) + \frac{1}{x+1} + c$$

and

$$y = c(x+1) + 1 + (x+1)\ln(x+1), \qquad x > -1.$$

Sometimes a nonlinear equation can be put in the form (1.24) by means of a change of variable. One set of equations for which this can always be accomplished is the class of *Bernoulli equations*. These are of the form

$$y' + p(x)y = q(x)y^n,$$

where n is any number other than 0 or 1. Division by y^n yields the equation

$$y^{-n}y' + p(x)y^{1-n} = q(x).$$

If we let $u = y^{1-n}$, then $u' = (1-n)y^{-n}y'$ and the equation becomes

$$\frac{1}{1-n}u' + p(x)u = q(x).$$

This is a linear equation that can be solved by the method described earlier in this section.

Example 2   An example of a Bernoulli equation is

$$y' + \frac{3}{x}y = x^2 y^2, \qquad x > 0. \tag{1.29}$$

Dividing through by y^2, we have

$$y^{-2}y' + \frac{3}{x}y^{-1} = x^2.$$

If we set $u = y^{-1}$, then $u' = -y^{-2}y'$ and the equation becomes

$$u' - \frac{3}{x}u = -x^2.$$

An integrating factor is

$$\mu(x) = \exp\left(-3 \int x^{-1} \, dx\right) = x^{-3}.$$

Using formula (1.26), we find that

$$ux^{-3} = -\int x^{-1} \, dx = -\ln x + c$$

and

$$u = x^3(c - \ln x).$$

Since $u = y^{-1}$ we have

$$y = x^{-3}(c - \ln x)^{-1}.$$

It should be noted that $y = 0$ is also a solution of the original equation. In dividing through by y^2 we tacitly assumed that y was never zero.

Exercises for Section 1.5

In Exercises 1–12, find the general solution of the equation. If an initial condition is given, also find the solution that satisfies the condition.

1. $xy' + 2y = 4x^2$, $y(1) = 4$

2. $xy' - 3y = x^3$, $y(1) = 0$

3. $xy' + (x - 2)y = 3x^3e^{-x}$

4. $y' - 2y = 4x$, $y(0) = 1$

5. $y' - 2xy = 1$, $y(a) = b$

6. $y' + (\cos x)y = \cos x$, $y(\pi) = 0$

7. $x(\ln x)y' + y = 2 \ln x$

8. $(x^2 + 1)y' - 2xy = x^2 + 1$, $y(1) = \pi$

9. $y' + 2xy = 2x$

10. $y' + (\cot x)y = 3 \sin x \cos x$

11. $x(x + 1)y' - y = 2x^2(x + 1)$

12. $xy' - y = x \sin x$

13. Show that the solution of the initial value problem

$$y' + p(x)y = q(x), \qquad y(a) = b$$

is given by the formula

$$y = be^{-P(x)} + \int_a^x e^{-[P(x)-P(t)]}q(t)\,dt,$$

where

$$P(x) = \int_a^x p(t)\,dt.$$

Suggestion: integrate from a to x in the equation preceding (1.26).

14. Use the result of Exercise 13 to solve the following initial value problems.
    (a) Exercise 1.   (b) Exercise 2.
    (c) Exercise 5.   (d) Exercise 6.

In Exercises 15–20, solve the differential equation.

15. $xy' + y + x^2y^2e^x = 0$

16. $xy' - (3x + 6)y = -9xe^{-x}y^{4/3}$

17. $3xy^2y' - 3y^3 = x^4 \cos x$

18. $xyy' = y^2 - x^2$

19. $y' - 2(\sin x)y = -2y^{3/2}\sin x$

20. $2y' + \dfrac{1}{x+1}y + 2(x^2 - 1)y^3 = 0$

In Exercises 21–23, find a new dependent variable such that the equation becomes linear in that variable. Then solve the equation.

21. $xe^yy' - e^y = 3x^2$ (Suggestion: let $u = e^y$.)

22. $\dfrac{1}{y^2 + 1}y' + \dfrac{2}{x}\tan^{-1}y = \dfrac{2}{x}$

23. $y' - \dfrac{1}{x+1}y \ln y = (x + 1)y$

24. An equation of the form

$$\frac{dy}{dx} = p(x)y^2 + q(x)y + r(x)$$

is called a *Riccati equation*.
    (a) If $p(x) \equiv 0$ show that the equation is linear. If $r(x) \equiv 0$ show that it is a Bernoulli equation.
    (b) If $y = y_1(x)$ is some particular solution, show that the change of variable $y = y_1(x) + 1/u$ leads to a linear equation for u.

25. Use the results of the previous exercise to solve the following equations. A particular solution of each equation is given.
    (a) $y' = y^2 + 2xy + (x^2 - 1);\quad y_1 = -x$
    (b) $y' = y^2 - 4xy + (4x^2 + 2);\quad y_1 = 2x$
    (c) $y' = y^2 + \left(\dfrac{2}{x} - 2x^2\right)y + x^4;\quad y_1 = x^2$
    (d) $y' = y^2 + \left(\dfrac{1}{x} - 2x\right)y + x^2;\quad y_1 = x$

26. A function f is said to be *bounded* on an interval I if there exists a number M such that $|f(x)| \le M$ for x in I. Let the function q be continuous and bounded on the interval $[0, \infty)$. Let k be a positive constant. Show that every solution of the equation $y' + ky = q(x)$ is bounded on the interval $[0, \infty)$. (Use Exercise 13, with $a = 0$.)

1.6 ORTHOGONAL TRAJECTORIES

If c is an arbitrary constant, the equation

$$y = cx^2 \tag{1.30}$$

describes a family of parabolas. Some of these are shown by the solid curves in Figure 1.2. Through every point (x_0, y_0) in the plane, except those points on the y-axis, there passes exactly one curve of the family. For if we specify (x_0, y_0) with $x_0 \ne 0$, then c is determined by the condition $y_0 = cx_0^2$ or $c = y_0/x_0^2$.

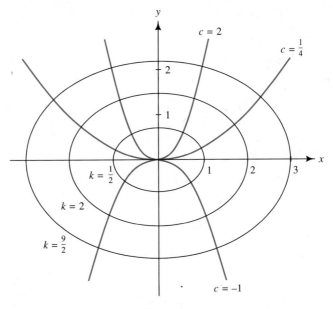

Figure 1.2

The slope of the curve through the point (x, y) is

$$y' = 2cx.$$

But since $c = y/x^2$, we have the formula

$$y' = 2\frac{y}{x} \qquad (1.31)$$

for the slope at the point (x, y) of the curve of the family (1.30) that passes through that point.

   Suppose that we wish to find a second family of curves, with exactly one curve of the family passing through each point (x, y) and such that at each point the curve of this second family is orthogonal or perpendicular to the curve of the original family (1.30) that passes through the point. The slope of the curve of the second family must be the negative reciprocal of the slope of the curve of the first family. In view of formula (1.31), we must find a family of curves for which

$$y' = -\frac{x}{2y}.$$

This is a separable equation. We have

$$2y\,dy = -x\,dx$$

and hence

$$y^2 = -\tfrac{1}{2}x^2 + k$$

or

$$\frac{x^2}{2k} + \frac{y^2}{k} = 1.$$

Here k must be a positive constant; otherwise there is no curve.) This is a family of ellipses, a few of which are shown by the black curves in Figure 1.2.

To consider a slightly different problem, suppose we wish to find a third family of curves, with one curve of the family through each point, such that at (x, y) the curve of the third family makes an angle of $\pi/4$ with the curve of the first family (1.30). The angle is to be measured counterclockwise from the curve of the third family to the curve of the first family, as shown in Figure 1.3. Using the notation shown in the figure, we must have

$$\tan(\phi_2 - \phi_1) = \frac{\tan \phi_2 - \tan \phi_1}{1 + \tan \phi_2 \tan \phi_1} = \tan\frac{\pi}{4}.$$

But $\tan \phi_2 = 2y/x$, and if we set $\tan \phi_1 = y'$, we require that

$$\frac{(2y/x) - y'}{1 + (2y/x)y'} = 1.$$

Simplification yields the differential equation

$$(2y + x)y' = 2y - x.$$

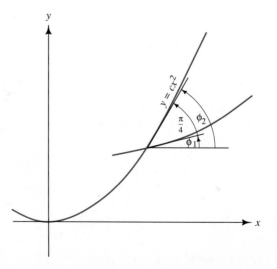

Figure 1.3

This equation is homogeneous. The change of variable $y = vx$ leads to the separable equation

$$\frac{4v + 2}{2v^2 - v + 1} v' = -\frac{2}{x}.$$

Integration yields the relation

$$\ln(2v^2 - v + 1) + \frac{6}{\sqrt{7}} \tan^{-1}\left(\frac{4v - 1}{\sqrt{7}}\right) = -\ln x^2 + k.$$

In terms of x and y this relation is

$$\ln(2y^2 - xy + x^2) + \frac{6}{\sqrt{7}} \tan^{-1}\left(\frac{4y - x}{\sqrt{7}\,x}\right) = k.$$

Exercises for Section 1.6

In Exercises 1–10 find the orthogonal trajectories of the family of curves. In Exercises 1–6, also sketch several curves of each family.

1. $y = e^x + c$

2. $y = ce^x$

3. $y = cx$

4. $y = \tan^{-1} x + c$

5. $x^2 + \dfrac{y^2}{4} = c^2$

6. $x^2 + (y - c)^2 = c^2$

7. $x^2 - y^2 = 2cx$

8. $y = x \ln cx$

9. $y = \dfrac{x}{cx + 1}$

10. $y = \dfrac{1 + cx}{1 - cx}$

In Exercises 11–14, find a family of curves making an angle of $\pi/4$ with the curves of the given family. The angle is to be measured counterclockwise toward the curve of the given family.

11. $y = cx^{-1}$

12. $y = \dfrac{1}{x + c}$

13. $y = \dfrac{x^3}{3} + c$

14. $y = \sin x + c$

15. Find a family of curves that cuts the family $y = x^2 + c$ at an angle of $\pi/6$. The angle is to be measured counterclockwise toward the curve of the given family.

16. Find a family of curves that cuts the family $y = x^3 + c$ at an angle of $\pi/3$. The angle is to be measured counterclockwise toward the curve of the given family.

17. Let (r, θ) be polar coordinates. If ψ is the positive angle measured counterclockwise from the radius vector to the tangent line to a curve $r = f(\theta)$, then it is shown in calculus that

$$\tan \psi = \frac{r}{dr/d\theta}.$$

Show that two curves $r = f_1(\theta)$ and $r = f_2(\theta)$ are orthogonal at a point if the corresponding values of $r/(dr/d\theta)$ are negative reciprocals, provided that the same coordinates (r, θ) are used to describe the same point on both curves.

Use the results of Exercise 17 to find the orthogonal trajectories of the family of curves in Exercises 18–21. Sketch a few curves of each family.

18. $r = c(1 + \cos \theta), \quad c > 0$

19. $r = c \sin \theta$

20. $r = c \sin 2\theta$

21. $r^2 = c \cos 2\theta$

22. If the first-order differential equation for a family of curves is of the specified type, what can be said about the type of equation for the family of orthogonal trajectories?
     (a) Separable   (b) Homogeneous
     (c) Linear   (d) Exact

23. For a three-dimensional surface $z = f(x, y)$ the curves of the family $f(x, y) = c$ are called

the *level curves* for the surface. It is shown in multivariable calculus that the curves orthogonal to the level curves are curves of steepest descent; that is, at each point (x, y) the direction in which z decreases most rapidly (or, in the opposite direction increases most rapidly) is along such a curve. Find the curves of steepest descent for the following surfaces.
     (a) $z = xy$
     (b) $z = x^2 y$
     (c) $z = e^{-y} \cos x$
     (d) $z = \sqrt{8 - 2x^2 - 8y^2}$

1.7 RADIOACTIVE DECAY

In Section 1.1, we stated that many natural laws of science could be formulated as differential equations. In this and the next several sections we consider some phenomena whose mathematical descriptions lead to first-order equations.

First we shall consider the decay of a radioactive substance. If $x(t)$ is the amount of substance present at time t, then x is a decreasing function since atoms of the substance decay into another element or elements. In a given very short time interval, it seems reasonable that the number of atoms that decay should be in proportion to the total number of undecayed atoms present at the beginning of the time interval. In other words, $-\Delta x/\Delta t \approx kx(t)$, where $\Delta x = x(t + \Delta t) - x(t)$ and k is a positive constant. Letting Δt shrink to zero, we find that

$$\frac{dx}{dt} = -kx. \tag{1.32}$$

In words this law says that the radioactive substance decays at a rate proportional to the amount of undecayed substance remaining.

Instead of specifying the value of k for a particular substance, most tables of physical constants specify the *half-life*; that is, the time required for half of an amount of the substance to decay. As we shall see, the half-life, which we denote by T, does not depend on the amount $x(0)$ present at time $t = 0$.

Separating the variables in Eq. (1.32) we have

$$\frac{dx}{x} = -k \, dt. \tag{1.33}$$

Integrating, we find that

$$\ln x = -kt + c' = -kt + \ln c.$$

(Here $x > 0$ and $\ln c$ is an arbitrary constant with $c > 0$.) Thus

$$x = ce^{-kt}. \tag{1.34}$$

The constants c and k can now be obtained from the conditions

$$x(0) = x_0, \qquad x(T) = \tfrac{1}{2}x_0. \qquad (1.35)$$

Here x_0 is the amount of undecayed substance present at $t = 0$ and T is the half-life. Putting $t = 0$ in Eq. (1.34), we see that

$$x_0 = ce^0 \quad \text{or} \quad c = x_0.$$

Thus

$$x(t) = x_0 e^{-kt}. \qquad (1.36)$$

Putting $t = T$ in this relation yields

$$\tfrac{1}{2}x_0 = x_0 e^{-kT},$$

from which it follows that

$$e^{kT} = 2 \quad \text{and} \quad kT = \ln 2.$$

Thus T and k are related as follows:

$$T = \frac{1}{k}\ln 2, \qquad k = \frac{1}{T}\ln 2.$$

Notice that T does not depend on x_0, the amount of substance present initially. If T is known from experimentation, then k is found from the last formula. Substitution of this value into Eq. (1.36) gives

$$x(t) = x_0 e^{-t(\ln 2)/T}. \qquad (1.37)$$

The half-lives of some radioactive substances are displayed in Table 1.1.

Table 1.1
Half-lives of Radioactive Substances

Substance	Half-life
Carbon-14	5730 years
Strontium-90	28.1 years
Radium-226	1600 years
Uranium-238	4.51×10^9 years
Uranium-235	7.07×10^8 years
Plutonium-244	8×10^7 years

The number associated with the name of the element in the first column of the table indicates its atomic weight. For instance, carbon-14 is a radioactive isotope of carbon. Most carbon found in nature is carbon-12, which is not radioactive.

Example   Every living plant or animal contains minute traces of carbon-14, a radioactive isotope of carbon whose half-life is 5730 years. While the plant or animal is alive, the ratio of carbon-14 to "ordinary" nonradioactive carbon in it remains constant, since plants absorb carbon dioxide from the air and animals eat plants. (The carbon dioxide in the air remains radioactive because of cosmic radiation.) When an organism dies, the ratio of radioactive carbon to ordinary carbon decreases because of the radioactive decay of the carbon-14. Given a piece of wood, it is possible to determine the proportion of carbon-14 that remains from the time the wood was part of a living tree. (This is done by counting the number of disintegrations per unit time per unit mass of carbon, both for the piece of wood and for a living tree.)

Suppose that in 1980 a piece of wood is analyzed and the amount of carbon-14 present is found to be 0.256 of the amount that was present when the tree was alive. To determine when the tree was cut down, we use the fact (from the table) that the half-life of carbon-14 is 5730 years. Then from Eq. (1.37) we have

$$x(t) = x_0 e^{-t(\ln 2)/5730},$$

with t in years. If the tree was cut down at time $t = 0$, then $x(t)/x_0 = 0.256$. Hence

$$0.256 = e^{-t(\ln 2)/5730}$$

and so

$$t = -5730 \, \frac{\ln 0.256}{\ln 2} = 11{,}264.$$

The approximate date on which the tree was cut down is

$$1980 - 11264 = -9284 \text{ or } 9285 \text{ BC.}$$

Willard F. Libby was awarded a Nobel prize for his work on carbon dating. The reader who is interested in learning more about the subject is referred to the *Encyclopaedia Britanica*, 11th ed., s.v. "Dating, Relative and Absolute."

Exercises for Section 1.7
---

1. At a certain instant 100 gm of a radioactive substance are present. After 4 yr, 20 gm remain. How much of the substance remains after 8 yr?

2. Show that formula (1.37) can be written as

$$x(t) = x_0 2^{-t/T}.$$

3. After 6 hr, 60 gm of a radioactive substance are present. After 8 hr (2 hr later) 50 gm are present. How much of the substance was present initially?

4. At a certain instant, 10 gm of a radioactive substance are present. After 20 min, 3 gm are present. Determine the half-life.

5. If the half-life of a radioactive substance is 10 yr, when does 25% of the substance remain?

6. If the half-life of a radioactive substance is 28.1 yr and if 40% of the substance remains, how long ago did decay begin?

7. If a radioactive substance decays according to the formula

$$\frac{dx}{dt} = -0.0837x,$$

where t is in years, what is the half-life?

8. The remains of a basket found in a cave were analyzed and the proportion of carbon-14 remaining was determined to be 0.324. How long ago was the cave used by humans?

9. The proportion of carbon-14 remaining in wood used for construction was determined to be 0.590. How long ago was the construction carried out?

10. If some charcoal from a wood fire was determined to be 10,000 years old, what is the proportion of carbon-14 remaining?

1.8
MIXING
PROBLEMS

We consider another type of problem (called a mixing problem) whose mathematical formulation yields a first-order differential equation. An example will illustrate the main features.

Example   Suppose that a tank contains 20 gal of a solution of a certain chemical and that 5 lb of the chemical are in the solution. Starting at a certain instant, a solution of the same chemical, with a concentration of 2 lb/gal, is allowed to flow into the tank at the rate of 3 gal/min. The mixture is drained off at the rate of 2 gal/min, so the volume of the solution in the tank is $20 + t$ gallons after t minutes. (We make the simplifying assumption that the concentration of the solution in the tank is kept uniform, perhaps by stirring.) The problem we wish to solve is this: after 10 minutes, how many pounds of solute are in the tank, and what is the concentration?

It is true that each gallon of the solution coming in brings with it 2 lb of the chemical, but the mixture leaving takes some of the chemical with it. We can solve our problem if we can obtain a formula for the amount of chemical in the tank at time t. If $x(t)$ is the amount in the tank at time t, then the rate of change of x, dx/dt, is given by the following rule: dx/dt is equal to the rate at which the chemical enters the tank minus the rate at which the chemical leaves the tank. This rule expresses the principle of *conservation of mass*: the chemical is neither created nor destroyed in the tank. The rate at which the chemical enters is $2 \times 3 = 6$ lb/min, since 3 gal of solution flow in per minute and each gallon contains 2 lb of the chemical. At time t the concentration of the solution in the tank is $x(t)/(t + 20)$, since the volume is $t + 20$. Hence the rate at which the chemical is leaving the tank is $2 \times x/(t + 20)$ lb/min. Thus we arrive at the differential equation

$$\frac{dx}{dt} = 6 - \frac{2x}{t + 20}.$$

This is a linear equation. Its general solution is

$$x = 2(t + 20) + \frac{c}{(t + 20)^2}.$$

Using the fact that $x = 5$ when $t = 0$, we have $5 = 2(0 + 20) + c/20^2$ or $c = -14000$. Then

$$x = 2(t + 20) - \frac{14000}{(t + 20)^2}.$$

Setting $t = 10$ to find the amount of solute after 10 minutes, we find that

$$x = 2(10 + 20) - \frac{14000}{30^2} = 44.444 \text{ lb}.$$

The concentration is

$$\frac{x}{t + 20} = \frac{44.444}{30} = 1.481 \text{ lb/gal}.$$

--------- Exercises for Section 1.8 ---------

1. A tank initially contains 100 gal of a solution that holds 30 lb of a chemical. Water runs into the tank at the rate of 2 gal/min and the solution runs out at the same rate. How much of the chemical remains in the tank after 20 min?

2. A tank initially contains 50 gal of a solution that holds 30 lb of a chemical. Water runs into the tank at the rate of 3 gal/min and the mixture runs out at the rate of 2 gal/min. After how long will there be 25 lb of the chemical in the tank?

3. A tank initially contains 100 gal of a solution that holds 40 lb of a chemical. A solution containing 2 lb/gal of the chemical runs into the tank at the rate of 2 gal/min and the mixture runs out at the rate of 3 gal/min. How much chemical is in the tank after 50 min?

4. A tank initially contains 50 gal of water. Alcohol enters at the rate of 2 gal/min and the mixture leaves at the same rate. When will the concentration of alcohol be 25 percent?

5. A tank initially holds 25 gal of water. Alcohol enters at the rate of 2 gal/min and the mixture leaves at the rate of 1 gal/min. What will be the concentration of alcohol when 50 gal of fluid is in the tank?

6. A tank initially contains 100 gal of a solution that holds 10 lb of a chemical. Water runs in at a constant rate and the mixture runs out at the same rate. What should be the rate of flow if 1 lb is to remain after 20 min?

7. A tank initially contains 100 gal of a solution that holds 30 lb of a chemical. A solution con-taining 2 lb/gal of the chemical runs in at a constant rate and the mixture runs out at the same rate. What should be the rate of flow if the tank is to contain 70 lb of the chemical after 40 min?

8. A tank initially holds 100 liters of a solution in which is dissolved 200 gm of a radioactive substance. The half-life is 10 minutes. If water flows in at the rate of 2 liters/min and the solution flows out at the same rate, find a formula for the amount of radioactive substance in the tank after t min.

1.9 POPULATION GROWTH

The applications of this section involve mathematical models for the population growth of a biological species. We shall consider models in which the rate of change of the population at any time depends only on the population at that time. Thus, if $N(t)$ is the population at time t, our mathematical formulation of the problem is a differential equation of the form

$$\frac{dN}{dt} = f(N). \tag{1.38}$$

In general, such a model will exclude the effects of immigration, emigration, competition with other species, changes in climate, and environmental changes other than those caused by the change of the population itself. The time interval over which the model is valid may be short.

If the birth rate is higher than the death rate, dN/dt should increase with N. Then, in Eq. (1.38), f should be an increasing function. The simplest model is

$$\frac{dN}{dt} = kN, \tag{1.39}$$

where k is a positive constant. The solution,

$$N(t) = N_0 e^{kt},$$

where $N_0 = N(0)$, is an increasing function of time. In fact, as t becomes infinite, so does $N(t)$.

We might consider models in which $f(N) = kN^\alpha$, where α is a positive constant. Then f will be an increasing function and $f(0) = 0$. This last condition is desirable because the growth rate dN/dt should be zero when the population is zero. We leave a discussion of these models to the exercises.

The model (1.39) cannot be valid over a long time interval because a real population cannot increase indefinitely. Instead, we expect the population to level off at a certain value, called the *saturation value*. At this point the growth rate will be zero. We therefore consider the equation

$$\frac{dN}{dt} = kN(a - N) \tag{1.40}$$

in which k and a are positive constants. If $N_0 = N(0)$ is less than a, dN/dt is positive so long as N remains less than a. If N should ever exceed the value a, then dN/dt will be negative and N will decrease. The solution of this separable equation is found to be

$$N(t) = \frac{a}{1 + \left(\dfrac{a}{N_0} - 1\right)e^{-akt}}. \tag{1.41}$$

Notice that as t increases, $N(t)$ approaches the limiting value a, regardless of the value of N_0.

Equation (1.40) can be written as

$$\frac{1}{N}\frac{dN}{dt} = ka - kN. \tag{1.42}$$

The left-hand side, $\dfrac{1}{N}\dfrac{dN}{dt}$, is the per capita growth rate. The right-hand side can be regarded as the difference between the birth rate ka and the death rate kN. In this model, the birth rate is constant but the death rate increases with N, perhaps because of overcrowding, food shortage, etc. A variety of models can be constructed by making different assumptions about the birth and death rate.

Problems involving the growth of the populations of two species in competition will be considered in Chapter 10.

--- **Exercises for Section 1.9** ---

1. The population of a certain species is initially 2000. After two hours it is 2500. Find a formula for the population as a function of time, assuming that the model (1.39) applies.

2. Using the data of Exercise 1, find the population as a function of time, assuming that the model $dN/dt = kN^{1/2}$ applies.

3. (a) Find a formula for $N(t)$ if $N(0) = N_0$ and

$$\frac{dN}{dt} = kN^{\alpha},$$

where α is a positive constant, $\alpha \neq 1$.

(b) If $0 < \alpha < 1$ in part (a), what happens to $N(t)$ as t becomes infinite?

(c)   If $\alpha > 1$ in part (a), show that $N(t)$ becomes infinite after a finite time.

4.   Derive the solution (1.41) of Eq. (1.40).

5.   Initially the population of a species is 5000. After 10 days it is 8000. After a very long time the population stabilizes at 15000. Find a formula for the population as a function of time, assuming that a model of the form (1.40) applies.

6.   What is $N(t)$ if $N_0 = 0$ in the equation (1.40)?

7.   The U.S. population in 1850 and 1900 was 23.192 and 75.995 million, respectively. Use these facts and the model (1.39) to predict the population in the year (a) 1950 and (b) 2000. The actual 1950 population was 150.697 million. *Suggestion*: Use $t = 0, 1, 2, 3$ instead of $t = 1850$, etc.

8.   The U.S. population in 1800, 1850, and 1900 was 5.308, 23.192, and 75.995 million, respectively. Use these facts and the model (1.40) to

predict the population in the year (a) 1950 and (b) 2000. Compare these results with those of Exercise 7. What is the limiting population predicted by this model?

9.   From the formula (1.40) show that

$$\frac{d^2N}{dt^2} = k^2 N(a - N)(a - 2N).$$

Then show that a solution curve is concave up when $0 < N < a/2$ or when $N > a$, but concave down when $a/2 < N < a$.

10.   Suppose that we have a constant death rate and a birth rate proportional to the population N. Show that Eq. (1.42) must be replaced by the equation

$$\frac{1}{N} \frac{dN}{dt} = kN - ka.$$

Find the solution for which $N(0) = N_0$. Show that if $N_0 > a$ then $N(t) \to \infty$, but that if $N_0 < a$ then $N(t) \to 0$ as $t \to \infty$.

1.10 COOLING; THE RATE OF A CHEMICAL REACTION

The first problem that we consider in this section has to do with the change in temperature in a cooling body. If a body cools in a surrounding medium (such as air or water) it might be expected that the rate of change of the temperature of the body would depend on the difference between the temperature of the body and that of the surrounding medium. *Newton's law of cooling* asserts that the rate of change is directly proportional to the difference of the temperatures. Thus if $u(t)$ is the temperature of the body at time t and if u_0 is the (constant) temperature of the surrounding medium, we have

$$\frac{du}{dt} = -k(u - u_0),$$

where k is a positive constant. The minus sign occurs because du/dt will be negative when $u > u_0$.

Example 1   As an example, suppose that an object is heated to 300°F and allowed to cool in a room whose air temperature is 80°F. If after 10 min the temperature of the body is 250°, what will be its temperature after 20 min? The differential equation becomes

$$\frac{du}{dt} = -k(u - 80)$$

and the initial condition is $u(0) = 300$. We also know that $u(10) = 250$.

This equation is linear; it is also separable. Treating it as a separable equation, we write

$$\frac{du}{u - 80} = -k \, dt. \tag{1.43}$$

Using the specified conditions to determine k, we have

$$\int_{300}^{250} \frac{du}{u - 80} = -k \int_{0}^{10} dt.$$

From this equation we find that

$$k = \frac{1}{10} \ln \frac{220}{170} = 0.0258.$$

To determine u when $t = 20$, we go back to Eq. (1.43) and write

$$\int_{300}^{u} \frac{du}{u - 80} = -k \int_{0}^{20} dt.$$

We find that

$$\ln \frac{u - 80}{220} = -20k$$

or

$$u = 220e^{-20k} + 80.$$

Since $e^{-20k} = e^{-0.516} = 0.5971$, we have

$$u(20) = 211°.$$

Our next application of differential equations concerns the rate of a chemical reaction. Suppose that A and B are two chemicals that react in solution. Let $x(t)$ and $y(t)$ denote the concentrations (in moles[6] per liter) at time t of A and B, respectively.

---

[6] If the molecular weight of a chemical is w, then one mole of that chemical consists of w grams.

Example 2   We shall assume that one molecule of A combines with one molecule of B to form a new product or products. Then x and y decrease at the same rate, so that $dx/dt = dy/dt$. Let $z(t)$ be the amount by which x and y have decreased in time t. If a and b are the initial concentrations of A and B, respectively, then

$$x(t) = a - z(t), \qquad y(t) = b - z(t) \tag{1.44}$$

and

$$\frac{dz}{dt} = -\frac{dx}{dt} = -\frac{dy}{dt}.$$

The quantity dz/dt is called the *rate of reaction*. For many reactions of this type (one molecule of A combining with one molecule of B), it is found that[7] under conditions of constant temperature

$$\frac{dz}{dt} = kxy \tag{1.45}$$

or

$$\frac{dz}{dt} = k(a - z)(b - z), \tag{1.46}$$

where k is a positive constant of proportionality. Thus the rate of reaction is directly proportional to the concentration of each reactant.

The differential equation (1.46) is nonlinear, but separable. Separating the variables and using the fact that $z(0) = 0$, we have

$$\int_0^z \frac{dz}{(a - z)(b - z)} = k \int_0^t dt.$$

The integral on the left can be evaluated by the use of partial fractions. We find that

$$\frac{1}{a - b} \int_0^z \left(\frac{-1}{a - z} + \frac{1}{b - z} \right) dz = kt$$

or

$$\frac{1}{a - b} \ln \frac{b(a - z)}{a(b - z)} = kt.$$

---

[7] The formula (1.46) does not always apply. In some cases nonreacting substances (catalysts) influence the rate of reaction. In any case the rate of reaction must ultimately be determined by experiment.

The value of k can be determined by experiment, in which z is measured for various values of t. From the relation just shown we obtain the formula

$$z = ab \, \frac{e^{k(a-b)t} - 1}{ae^{k(a-b)t} - b}$$

for z. The concentrations of A and B can be found from Eq. (1.44).

Example 3   Let us next consider a reaction in which two molecules of chemical B combine with one molecule of chemical A to form new products. Then two moles of B are used up for every mole of A, so that

$$\frac{dy}{dt} = 2 \frac{dx}{dt}.$$

If $z(t)$ is the decrease in chemical A in time t, then

$$x = a - z, \qquad y = b - 2z.$$

In this case (two molecules of B combining with one of A), it is found that[8]

$$\frac{dz}{dt} = kxy^2$$

or

$$\frac{dz}{dt} = k(a - z)(b - 2z)^2. \tag{1.47}$$

Here dz/dt is directly proportional to x and to the square of y. The integration of Eq. (1.47) is left to the exercises.

In the general case where m molecules of A combine with n molecules of B, we have

$$n \frac{dx}{dt} = m \frac{dy}{dt}.$$

If we set

$$z = \frac{1}{m}(a - x) = \frac{1}{n}(b - y),$$

then $z(0) = 0$ and

$$\frac{dz}{dt} = -\frac{1}{m} \frac{dx}{dt} = -\frac{1}{n} \frac{dy}{dt}.$$

---

[8] See footnote 7.

The equation for z is found to be

$$\frac{dz}{dt} = kx^m y^n$$

or

$$\frac{dz}{dt} = k(a - mz)^m (b - nz)^n.$$

The exponents m and n are often called the *orders* of the reaction with respect to the concentrations of A and B.

───────────────── **Exercises for Section 1.10** ─────────────────

1. An object whose initial temperature is 150°F is allowed to cool in a room where the temperature of the air is 75°F. After 10 min the temperature of the object is 125°. When will its temperature be 100°?

2. A heated object is allowed to cool in air whose temperature is 20°C. After 5 min its temperature is 200°C. After 10 min (5 min later) its temperature is 160°. What was the temperature of the object initially?

3. An object whose temperature is 220°F is placed in a room where the temperature is 60°F. After 10 min the temperature of the object is 200°. At this point refrigeration equipment, which lowers the temperature of the room at the rate of 1°F/min, is turned on. What is the temperature of the object t min after the equipment is turned on?

4. An object with a temperature of 10°F is placed in a room where the temperature is 80°F. After 10 min the temperature of the object is 30°. What will be the temperature of the object after it has been in the room for 30 min?

5. An object at room temperature of 20°C is put in a pan of boiling water (100°C). Four minutes later the temperature of the object is 70°C. Three minutes after that, the object is removed from the pan. How long after removal will the temperature be 21°C?

6. An object at room temperature of 20°C is put in a pan of boiling water (100°C) at noon. Four minutes later the temperature of the object is 70°C. The object is later removed from the pan, and at 12:10 P.M. its temperature is 50°C. When was the object removed from the pan?

7. Suppose that in a chemical reaction where one molecule of A combines with one molecule of B, the rule (1.45) applies. Assume that A and B have the same initial concentration a.
    (a)   Find a formula for the concentrations of A and B at time t.
    (b)   Find a formula for the half-life of the reaction, which is the time required for the concentrations of the reactants to be halved.

8. Suppose that in a chemical reaction where two molecules of B combine with one of A, the rule (1.47) applies. Find a formula that expresses k in terms of z and t in the case where
    (a) $b \neq 2a$ (b) $b = 2a$

9. A chemical A breaks down when heated, with n molecules of A reacting to form new products. The law of reaction is

    $$\frac{dx}{dt} = -kx^n,$$

    where $x(t)$ is the amount of the chemical remaining at time t. If half the chemical decomposes after T min, find a formula for x in terms of t. Let a denote the initial amount of the chemical.

1.11
TWO SPECIAL
TYPES OF
SECOND-ORDER
EQUATIONS

A second-order differential equation is of the form

$$F\left(t, x, \frac{dx}{dt}, \frac{d^2x}{dt^2}\right) = 0.$$

In this section we shall consider two classes of second-order equations that can be solved by successively solving two first-order equations. Thus the methods of solution for first-order equations that were presented earlier in this chapter may be used.

We consider first the class of second-order equations in which the dependent variable x is absent. Such an equation is of the form

$$G\left(t, \frac{dx}{dt}, \frac{d^2x}{dt^2}\right) = 0.$$

Suppose that x is a solution of this equation. If we set $v = dx/dt$, then v must be a solution of the first-order equation

$$G\left(t, v, \frac{dv}{dt}\right) = 0.$$

If we can solve this equation for v, then the solutions of the original equation can be found from the relation

$$\frac{dx}{dt} = v(t)$$

by integration.

Example 1   Let us consider the equation

$$t\frac{d^2x}{dt^2} = 2\left[\left(\frac{dx}{dt}\right)^2 - \frac{dx}{dt}\right]. \tag{1.48}$$

Note that x itself is absent. Setting $v = dx/dt$, we obtain the first-order equation

$$t\frac{dv}{dt} = 2(v^2 - v) \tag{1.49}$$

for v. This equation is separable and we have

$$\frac{dv}{v^2 - v} = 2\frac{dt}{t}$$

or

$$\left(\frac{1}{v-1} - \frac{1}{v}\right) dv = 2\frac{dt}{t}.$$

Integrating, we find that

$$\ln\left|\frac{v-1}{v}\right| = 2\ln|t| + c_1'$$

or

$$\frac{v-1}{v} = c_1 t^2.$$

Then

$$v = \frac{dx}{dt} = \frac{1}{1 - c_1 t^2}.$$

If c_1 is positive, say $c_1 = a^2$, we have

$$\frac{dx}{dt} = \frac{1}{1 - a^2 t^2},$$

so that

$$x = \frac{1}{2a} \ln\left|\frac{1 + at}{1 - at}\right| + c_2.$$

If c_1 is negative, say $c_1 = -b^2$, then

$$\frac{dx}{dt} = \frac{1}{1 + b^2 t^2}$$

and

$$x = \frac{1}{b} \tan^{-1} bt + c_2.$$

Finally, we observe that since the constant functions $v = 0$ and $v = 1$ are solutions of Eq. (1.49), the functions

$$x = c,$$

$$x = t + c$$

are solutions of Eq. (1.48).

The second class of second-order equations that we shall consider are those in which the independent variable t is missing. Such equations are of the form

$$H\left(x, \frac{dx}{dt}, \frac{d^2x}{dt^2}\right) = 0.$$

Suppose that x is a solution and let $v = dx/dt$. On an interval where x is a strictly increasing, or decreasing, function, t can be regarded as a function of x and we can write

$$\frac{d^2x}{dt^2} = \frac{dv}{dt} = \frac{dv}{dx}\frac{dx}{dt} = v\frac{dv}{dx}.$$

Then the equation becomes

$$H\left(x, v, v\frac{dv}{dx}\right) = 0$$

and this is a first-order equation for v. If we can solve it, finding a solution v, then a solution of the original equation can be found by solving the first-order equation

$$\frac{dx}{dt} = v(x).$$

Example 2   As an illustration, we consider the equation

$$x\frac{d^2x}{dt^2} = \left(\frac{dx}{dt}\right)^2 + 2\frac{dx}{dt}, \tag{1.50}$$

in which t is missing. Setting

$$\frac{dx}{dt} = v, \qquad \frac{d^2x}{dt^2} = v\frac{dv}{dx},$$

we have

$$xv\frac{dv}{dx} = v^2 + 2v. \tag{1.51}$$

By inspection we see that $v = 0$ and $v = -2$ are solutions of this equation, so that

$$x = c, \qquad x = -2t + c$$

are solutions of Eq. (1.50). To find the remaining solutions, we divide through by v in Eq. (1.51), obtaining the separable equation

$$x \frac{dv}{dx} = v + 2.$$

We easily find that

$$v = c_1 x - 2.$$

Now we must solve the equation

$$\frac{dx}{dt} = c_1 x - 2.$$

For $c_1 \neq 0$ we find that

$$x = \frac{1}{c_1} (c_2 e^{c_1 t} + 2).$$

(When $c_1 = 0$ we have $v = -2$, which was considered previously.)

Some applications that give rise to the types of differential equations discussed here are presented in the next section.

──────────────────── **Exercises for Section 1.11** ────────────────────

Solve the following differential equations.

1. $t \dfrac{d^2 x}{dt^2} = 2 \dfrac{dx}{dt} + 2$

2. $\dfrac{d^2 x}{dt^2} = \dfrac{dx}{dt} + 2t$

3. $2t \dfrac{dx}{dt} \dfrac{d^2 x}{dt^2} = \left(\dfrac{dx}{dt} \right)^2 + 1$

4. $\dfrac{d^2 x}{dt^2} = -2t \left(\dfrac{dx}{dt} \right)^2$

5. $2t \dfrac{d^2 x}{dt^2} = \left(\dfrac{dx}{dt} \right)^2 - 1$

6. $t^2 \dfrac{d^2 x}{dt^2} + \left(\dfrac{dx}{dt} \right)^2 = 2t \dfrac{dx}{dt}$

7. $\left(\dfrac{dx}{dt} - t \right) \dfrac{d^2 x}{dt^2} - \dfrac{dx}{dt} = 0$

8. $t \exp\left(\dfrac{dx}{dt} \right) \dfrac{d^2 x}{dt^2} = \exp\left(\dfrac{dx}{dt} \right) - 1$

9. $t \dfrac{d^2 x}{dt^2} = \dfrac{dx}{dt} + 2 \sqrt{ t^2 + \left(\dfrac{dx}{dt} \right)^2 }$

10. $\dfrac{d^2 x}{dt^2} = \dfrac{1}{t} \dfrac{dx}{dt} + \tanh\left(\dfrac{dx}{dt} \Big/ t \right)$

11. $\dfrac{d^2 x}{dt^2} + x^{-3} = 0$

12. $x \dfrac{d^2 x}{dt^2} = \left(\dfrac{dx}{dt} \right)^2$

13. $\dfrac{d^2 x}{dt^2} + \left(\dfrac{dx}{dt} \right)^3 = 0$

14. $3x \dfrac{dx}{dt} \dfrac{d^2x}{dt^2} = \left(\dfrac{dx}{dt}\right)^3 - 1$

15. $\dfrac{d^2x}{dt^2} + e^{-x} \dfrac{dx}{dt} = 0$

16. $(x^2 + 1)\dfrac{d^2x}{dt^2} = 2x\left(\dfrac{dx}{dt}\right)^2$

17. $x^3 \dfrac{d^2x}{dt^2} = 2\left(\dfrac{dx}{dt}\right)^3$

18. $\dfrac{d^2x}{dt^2} = \left(\dfrac{dx}{dt}\right)^2 \tanh x$

19. $x \dfrac{d^2x}{dt^2} = \dfrac{dx}{dt}\left(\dfrac{dx}{dt} + 2\right)$

20. $\dfrac{d^2x}{dt^2} + 2\left(\dfrac{dx}{dt}\right)^2 \tan x = 0$

21. $\dfrac{d^2x}{dt^2} \exp\left(\dfrac{dx}{dt}\right) = 1$

22. $t \dfrac{d^3x}{dt^3} = 2\dfrac{d^2x}{dt^2}$

1.12
FALLING BODIES

The applications in this section involve the motion of a solid body whose center of mass moves in a straight line. Let us denote by x the directed distance of the center of mass from some fixed point on the line of motion. Then x depends on time t. The velocity and acceleration of the center of mass are dx/dt and d^2x/dt^2, respectively. The notations

$$\dot{x} = \frac{dx}{dt}, \qquad \ddot{x} = \frac{d^2x}{dt^2}$$

are commonly used.

According to *Newton's second law of motion*, the mass of the body times the acceleration of the center of mass is proportional to the force acting on the body. Actually the commonly used systems of units for measuring mass, distance, time, and force are arranged so that the constant of proportionality may be taken as unity. Thus

$$m\frac{d^2x}{dt^2} = F, \tag{1.52}$$

where m is the mass of the body and F is the force. When F depends on t, x, and dx/dt, Eq. (1.52) is a second-order differential equation for x. The units in Eq. (1.52) must be chosen appropriately. Equation (1.52) is sometimes called the *equation of motion* of the body.

In the *centimeter-gram-second* system of units (c.g.s.) distance is in *centimeters*, time is in *seconds*, mass is in *grams*, and force is in *dynes*. In the *British* system of units, distance is in *feet*, time is in *seconds*, mass is in *slugs*, and force is in *pounds*. In the MKS system, distance is in *meters*, time in *seconds*, mass in *kilograms*, and force in *newtons*.

In the examples of this section we shall examine the motion of falling (and rising) bodies. One of the forces present is that of gravity. The *weight* of a body is very nearly[9] the force exerted on the body by the earth's gravitational field. Near the surface of the earth the force due to gravity is mg, where g is approximately 980 cm/sec^2 or 32 ft/sec^2. Actually the value of g (the acceleration due to gravity) varies slightly over the surface of the earth, being slightly larger at the poles than at the equator. If the weight of a body is w lb, then the mass m of the body in slugs is given by the formula

$$m = \frac{w}{g}.$$

In our first example let us suppose that an object is thrown directly upward from the surface of the earth, with an initial velocity v_0. Let x denote the directed distance upward of the object from the surface of the earth. If we assume that the only force acting on the object is that due to gravity, then Eq. (1.52) becomes

$$m\ddot{x} = -mg. \tag{1.53}$$

The minus sign occurs because the force acts in the direction of decreasing x. The initial conditions are

$$x(0) = 0, \qquad \dot{x}(0) = v_0,$$

assuming that $t = 0$ is the time at which the object is thrown.

A first integration of Eq. (1.53) yields the relation

$$\dot{x} = -gt + c_1,$$

where c_1 is a constant. The condition $\dot{x}(0) = v_0$ tells us that $c_1 = v_0$, so we have

$$\dot{x} = -gt + v_0. \tag{1.54}$$

Integrating again, we have

$$x = -\frac{1}{2} gt^2 + v_0 t + c_2.$$

Since $x(0) = 0$ we must have $c_2 = 0$ and

$$x = -\frac{1}{2} gt^2 + v_0 t. \tag{1.55}$$

---

[9] The rotation of the earth complicates an exact definition of weight, but for most practical considerations the weight of a body may be taken to be the force due to gravity.

Formulas (1.54) and (1.55) describe the velocity and position of the object at time t, $t \geq 0$. From formula (1.54) we see that the velocity is positive until $t = v_0/g$, after which time it becomes negative, with the object descending. Thus the time required for the object to reach its maximum height is v_0/g and the maximum height, as found from formula (1.55), is

$$h = \frac{v_0^2}{2g}.$$

The time when the object returns to earth can be found by setting $x = 0$ in Eq. (1.55) and solving for t. We find that the time is $2v_0/g$ so that the time that it takes the object to fall back to earth is the same as the time going up. The velocity with which the object strikes the earth is found by setting $t = 2v_0/g$ in formula (1.54). This velocity is found to be $-v_0$. The magnitude of the final velocity is therefore the same as that of the initial velocity. The symmetry that occurs in this problem (time going up equals time coming down, and final velocity equals initial velocity) does not always arise when forces other than gravity are considered. Examples are presented in the exercises.

If an object is dropped from a specified height h above the earth, it is probably more convenient to let x denote the directed distance *downward* from the point of release. Then the equation of motion becomes

$$m\ddot{x} = mg. \tag{1.56}$$

Here the force acts in the direction of increasing x, and there is no minus sign. The initial conditions are

$$x(0) = 0, \qquad \dot{x}(0) = 0.$$

Two integrations of Eq. (1.56) yield the formula

$$x = \frac{1}{2} g t^2 \tag{1.57}$$

for the distance through which the object falls in time t. If a stone is dropped from a bridge and 3 sec elapse before it hits the water below, we can estimate the height of the bridge above the water from formula (1.57). Using the value $g = 32$ ft/sec^2, we have

$$h = \frac{1}{2} (32)(3^2) = 144 \text{ ft.}$$

Actually, when a body moves through the air (or another surrounding medium) the air exerts a damping force F_d on the body. This force depends on the velocity of the body (and on the shape of the body and the nature of the surrounding medium). In some situations the damping force is proportional to the velocity. In others it is more nearly proportional to the square or cube of the velocity.

Let us now reexamine the problem of a falling object that is released from a height h above the earth. Let x be the directed distance downward from the point of release. If the damping force F_d is proportional to the velocity, then

$$F_d = -c\dot{x},$$

where c is a positive constant of proportionality called the *damping constant*. The minus sign indicates that the force acts in a direction opposite to that of the velocity vector. The resultant force F is the sum of the forces acting, and in this case is

$$F = -c\dot{x} + mg.$$

The equation of motion becomes

$$m\ddot{x} = -c\dot{x} + mg \tag{1.58}$$

and the initial conditions are

$$x(0) = 0, \qquad \dot{x}(0) = 0.$$

In the second-order equation (1.58) both x and t are absent. Setting $\dot{x} = v$ and $\ddot{x} = \dot{v}$, we obtain the first-order linear equation

$$m\dot{v} + cv = mg \tag{1.59}$$

for v. Solving, and using the initial condition $v(0) = 0$, we find that

$$v = \dot{x} = \frac{mg}{c}(1 - e^{-ct/m}). \tag{1.60}$$

An integration yields the formula

$$x = \frac{mg}{c}\left(t + \frac{m}{c}e^{-ct/m}\right) - g\left(\frac{m}{c}\right)^2 \tag{1.61}$$

for x.

We notice from formula (1.60) that as t becomes infinite the velocity tends to the limiting value

$$v_\infty = \frac{mg}{c}.$$

Given that there is a limiting velocity, its value could have been found from the equation of motion (1.59). As the velocity of the falling object increases, the damping force increases in magnitude until it balances the force due to gravity. The acceleration thus tends to zero. Setting $\dot{v} = 0$ in Eq. (1.59), we find that $cv = mg$ or $v = mg/c$, which is the value given in the last formula.

When the damping constant c is small, we might expect the formula (1.61) to agree closely with the formula (1.57), at least when the time interval is short. The latter formula was derived in the absence of a damping force ($c = 0$). To make a comparison, we expand the exponential function that appears in formula (1.61) in a Maclaurin series,

$$e^{-ct/m} = 1 - \frac{ct/m}{1!} + \frac{(ct/m)^2}{2!} - \frac{(ct/m)^3}{3!} + \cdots .$$

Formula (1.60) becomes

$$x = g\left(\frac{m}{c}\right)^2 \left[\frac{1}{2}\left(\frac{ct}{m}\right)^2 - \frac{1}{6}\left(\frac{ct}{m}\right)^3 + \cdots\right]$$

or

$$x = \frac{1}{2} gt^2 - \frac{1}{6}\frac{cg}{m} t^3 + \cdots .$$

When ct/m is small compared with unity, we see that formula (1.57) is a good approximation to the more complicated formula (1.61).

In the case of a body falling from a great height we can no longer assume that the force due to gravity is constant. Instead we must use the more accurate "inverse square law" of Newton. This implies that the force of attraction between any two spherically symmetric bodies is directly proportional to the product of their masses and inversely proportional to the square of the distance between their centers of mass. Consider the case of a body of mass m falling toward the earth. We assume that the earth is a sphere of mass M and radius R. If r is the distance from the center of the earth to the center of mass of the falling body, then the force is

$$F = k\frac{mM}{r^2}, \tag{1.62}$$

where k is a constant of proportionality. To determine k we use the fact that, when $r = R$, F has the value mg. Setting $r = R$ in formula (1.62) we have

$$mg = k\frac{mM}{R^2} \quad \text{or} \quad k = \frac{gR^2}{M} .$$

Substituting this value for k into Eq. (1.63), we obtain the formula

$$F = \frac{mgR^2}{r^2}.$$

As a check we observe that $F = mg$ when $r = R$.

Let us denote by x the directed distance from the center of the earth to the center of mass of the falling body. Then the equation of motion of the body is

$$m\ddot{x} = -\frac{mgR^2}{x^2}, \qquad (1.63)$$

neglecting air resistance and other forces.

As an example, suppose that a projectile is fired directly upward from the surface of the earth with a velocity v_0. Then the initial conditions are

$$x(0) = R, \qquad \dot{x}(0) = v_0.$$

In the second-order equation (1.63), the independent variable t is missing. Setting $\dot{x} = v$ we have

$$\ddot{x} = \frac{dv}{dt} = \frac{dv}{dx}\frac{dx}{dt} = v\frac{dv}{dx}$$

and the equation becomes

$$v\frac{dv}{dx} = -\frac{gR^2}{x^2}.$$

Then

$$\int_{v_0}^{v} v\,dv = -gR^2 \int_{R}^{x} \frac{dx}{x^2},$$

$$\frac{1}{2}(v^2 - v_0^2) = gR^2\left(\frac{1}{x} - \frac{1}{R}\right),$$

or

$$v^2 = v_0^2 + 2gR^2\left(\frac{1}{x} - \frac{1}{R}\right).$$

Initially v is positive and must be the positive square root of the right-hand side of this equation. Thus

$$v = \frac{dx}{dt} = \left[v_0^2 - 2gR + \frac{2gR^2}{x}\right]^{1/2}. \qquad (1.64)$$

As x increases we see that the velocity decreases. If the quantity in brackets in Eq. (1.64) vanishes, the velocity becomes negative and the body starts to fall back to earth. However, if $v_0^2 - 2gR \geq 0$, v can never become negative no matter how large x becomes. The critical value $v_0 = \sqrt{2gR}$ is called the *escape velocity* of the earth. Unless v_0 is greater than, or equal to, this value, the projectile will ultimately fall back to earth. The value of the escape velocity is approximately 7 miles/sec. It should be pointed out that we have ignored a number of forces acting on the body, such as air resistance and the attractive forces of other celestial bodies.

Exercises for Section 1.12
---

1. A stone weighing 8 lb is dropped from a bridge 1000 ft above the surface of the water below. If there is no air resistance, when does the stone hit the water and what is its velocity on impact?

2. An object that weighs 176 lb falls from rest through the air, which resists with a force proportional to the velocity. If the limiting velocity is 24 ft/sec, what is the velocity after 2 sec?

3. An object weighing 64 lb falls from rest through the air, which resists with a force proportional to the velocity. When the velocity is 10 ft/sec, the resistance is 30 lb. Find formulas for the velocity and position as functions of time.

4. A projectile weighing 64 lb is fired directly upward with velocity 2000 ft/sec. Assume that the acceleration due to gravity is constant, and that the air resists the motion with a force proportional to the velocity. When the velocity is 1000 ft/sec, the resisting force is 20 lb. Find the maximum height attained by the projectile, and the time required for it to reach this height.

5. A projectile of mass 5 grams is fired directly upward with a velocity of 10^4 cm/sec through air that resists with a force proportional to the square of the velocity. When the velocity is 100 cm/sec, the resistance is 1 dyne. Assume a constant force mg due to gravity ($g = 980$ cm/sec^2). Find the maximum height

attained by the projectile, and the time required to reach this height.

6. An object that weighs 128 lb falls from rest through the air which resists with a force proportional to the square of the velocity. When the velocity is 40 ft/sec, the resistance is 32 lb. Find (a) the velocity when the object has travelled 50 ft. (b) the velocity after 5 sec.

7. An object of mass m is thrown directly upward from the surface of the earth with velocity v_0, where $v_0 < (2gR)^{1/2}$, R being the radius of the earth. Assume that Newton's inverse square law applies and neglect air resistance.
   (a) Find the maximum height h attained by the object.
   (b) Find the time t_1 required for the object to attain its maximum height.
   (c) Show that the time required for the object to fall back to earth from its maximum height is the same as the time required for it to reach this height. Also show that the velocity of impact is the same in magnitude as the initial velocity.

8. A ship of mass m traveling in a straight line with speed v_0 shuts off its engines and coasts. Assume that the water resistance is equal to c times the α power of the velocity.
   (a) Show that the ship travels a finite distance before coming to rest if $0 < \alpha < 2$ and find this distance.
   (b) What is the time required for the ship to come to rest?

9. An object of mass m is thrown directly upward from the surface of the earth with velocity v_0. Assume that the acceleration due to gravity has the constant value g and that the force due to air resistance is equal to c times the fourth power of the velocity.
    (a) Find the maximum height attained by the object.
    (b) From what height must an object of mass m be dropped in order to achieve an impact velocity of v_0?

10. A particle of mass m moves along a straight line, with $x(t)$ its directed distance from a fixed point on the line of motion at time t. The particle is repelled from the point $x = 0$ by a force of magnitude m/x^2. It starts at $x = 1$ with a speed v_0 toward $x = 0$.
    (a) How close does the particle come to the origin?
    (b) What is the time required for the particle to reach the position where it is closest to the origin?

11. A simple pendulum consists of an object of mass m attached to the end of a massless rod of length L. The other end of the rod is connected to a frictionless pivot, as in Figure 1.4. Assume that the acceleration due to gravity has the constant value g and that air resistance is negligible.
    (a) By equating $m\ddot{s}$ to the tangential component of force, derive the equation of

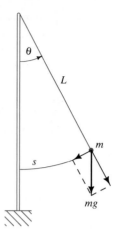

Figure 1.4

motion

$$\ddot{\theta} + \frac{g}{L}\sin\theta = 0.$$

    (b) To what physical situations do the constant solutions $\theta = n\pi$, n an integer, of the equation in part (a) correspond?
    (c) Suppose that the pendulum released from rest at time $t = 0$ from the position $\theta = -\alpha$, where $0 < \alpha < \pi$. Show that on the first half-swing of the pendulum (θ increasing)

$$t = \left(\frac{L}{2g}\right)^{1/2} \int_{-\alpha}^{\theta} (\cos\theta - \cos\alpha)^{-1/2}\, d\theta.$$

1.13
SOME
THEORETICAL
MATTERS

In the first section of this chapter, we said that the initial value problem

$$y'(x) = f(x, y(x)), \qquad y(x_0) = y_0 \tag{1.65}$$

had a solution if the function f satisfied certain mild restrictions. The reader may well wonder how we can make this claim, since for many specific problems, such as

$$y' = x^2 + y^2, \qquad y(0) = 0,$$

a formula for a solution cannot be expressed simply in terms of elementary functions. In this section we shall show that, under certain conditions, the problem (1.65) has a solution and that it has only one solution. We begin with a discussion of the restrictions on the function f.

A function $f(x, y)$ is said to satisfy a *Lipschitz condition* (in the second argument) in a region D if there exists a positive constant K such that

$$|f(x, y_1) - f(x, y_2)| \leq K|y_1 - y_2|$$

whenever the points (x, y_1) and (x, y_2) both lie in D. The constant K is called a *Lipschitz constant* for the function $f(x, y)$.

As an example, let us consider the function

$$f(x, y) = a(x)y + b(x)$$

(which is linear in y), where $a(x)$ and $b(x)$ are continuous on a closed interval $\alpha \leq x \leq \beta$. We shall show that this function satisfies a Lipschitz condition in the region $\alpha \leq x \leq \beta$, $-\infty < y < +\infty$. Let K be the maximum value of $|a(x)|$ on the interval $[\alpha, \beta]$. Then

$$|f(x, y_1) - f(x, y_2)| = |a(x)(y_1 - y_2)| \leq K|y_1 - y_2|.$$

As a second example, let $f(x, y)$ be continuous, along with its partial derivative $f_y(x, y)$ on a rectangle R of the form

$$|x - x_0| \leq a, \qquad |y - y_0| \leq b. \tag{1.66}$$

Then $f(x, y)$ satisfies a Lipschitz condition on R, and a Lipschitz constant K is given by the maximum value of $|f_y(x, y)|$ on R. For if (x, y_1) and (x, y_2) lie in R, we have by the mean-value theorem that

$$f(x, y_1) - f(x, y_2) = (y_1 - y_2)f_y(x, y_3),$$

where y_3 is between y_1 and y_2. Since $|f_y(x, y)| \leq K$ for all (x, y) in R, we have

$$|f(x, y_1) - f(x, y_2)| \leq K|y_1 - y_2|.$$

We are now in a position to state and discuss the following basic theorem.

Theorem 1.4   Let $f(x, y)$ and $f_y(x, y)$ be continuous on the rectangle R, as defined by the inequalities (1.66). Let M be the maximum value of $|f(x, y)|$ on R, and let $\alpha = \min(a, b/M)$. Then the equation $y' = f(x, y)$ possesses a solution $y(x)$ on the interval $|x - x_0| \leq \alpha$ which satisfies the initial condition $y(x_0) = y_0$. If $y_1(x)$ and $y_2(x)$ are both solutions of the initial value problem on an interval that contains x_0, then $y_1(x) = y_2(x)$; that is, the solution of the initial-value problem is unique.

Although we shall not give a detailed proof of this theorem, we shall describe generally the method employed. To begin with, we reformulate our initial-value problem as an integral equation. If $y(x)$ is a solution of the initial-value problem, we have $y(x_0) = y_0$ and $y'(x) = f(x, y(x))$ on an interval that contains x_0. Integrating both members of this last equation from x_0 to x, we see that $y(x)$ satisfies the integral equation

$$y(x) - y_0 = \int_{x_0}^{x} f(t, y(t)) \, dt. \tag{1.67}$$

The existence of the integral is guaranteed by the continuity of f and y. Conversely, if $y(x)$ is any continuous function that satisfies the integral equation (1.67) on an interval that contains x_0, we see that $y(x_0) = y_0$ and, upon differentiating, that $y'(x) = f(x, y(x))$. Hence the initial-value problem possesses a solution if, and only if, the integral equation (1.67) possesses a solution.

To prove the existence of a solution of the equation (1.67), we first define a sequence of functions $\{y_n(x)\}$, $n \geq 0$, by setting

$$y_0(x) = y_0 \tag{1.68}$$

and

$$y_{k+1}(x) = y_0 + \int_{x_0}^{x} f(t, y_k(t)) \, dt, \qquad k \geq 0. \tag{1.69}$$

Let x be restricted to the interval $|x - x_0| \leq \alpha$, where $\alpha = \min(a, b/M)$. Then

$$\left| y_1(x) - y_0 \right| = \left| \int_{x_0}^{x} f(t, y_0) \, dt \right| \leq M \left| \int_{x_0}^{x} dt \right| \leq M\alpha \leq b.$$

Consequently, the points $(x, y_1(x))$, for $|x - x_0| \leq \alpha$, lie in the rectangle R, and this ensures that the function $y_2(x)$ is well defined. It can be shown by induction that each of the functions $y_n(x)$ is well defined. It can also be shown that the sequence $\{y_n(x)\}$ converges to a function $y(x)$ that is a solution of the integral equation (1.67). The fact that $f(x, y)$ satisfies a Lipschitz condition in R is used in establishing the convergence of the sequence (Exercises 6 and 7). The method of proof described above, sometimes called the *method of successive approximations*, is due to Picard.

We now consider the uniqueness part of the theorem. Suppose that $y_1(x)$ and $y_2(x)$ are both solutions of the initial-value problem (and hence of the integral equation (1.67)) on an interval I. Then we have

$$y_1(x) - y_2(x) = \int_{x_0}^{x} [f(t, y_1(t)) - f(t, y_2(t))] \, dt.$$

Since $f(x, y)$ satisfies a Lipschitz condition on the rectangle R, we have

$$\left| y_1(x) - y_2(x) \right| \leq K \left| \int_{x_0}^{x} \left| y_1(t) - y_2(t) \right| \, dt \right|.$$

It follows (by Exercise 8) that $y_1(x) - y_2(x) \equiv 0$.

The interval $|x - x_0| \leq \alpha$ may be small even when the rectangle R is large. In the example

$$y' = 2xy^2, \qquad y(0) = 1,$$

the functions $f(x, y) = 2xy^2$ and $f_y(x, y) = 4xy$ are continuous everywhere, and hence on any rectangle of the form $|x| \leq a, |y - 1| \leq b$. But the solution of the initial-value problem, as found by elementary methods, is

$$y = \frac{1}{1 - x^2};$$

it exists only on the interval $|x| < 1$, and so it is clear that $\alpha < 1$.

In practice, the function $f(x, y)$ and its derivative $f_y(x, y)$ will be continuous in a region D of the xy plane that is not a rectangle. They may even be continuous for all x and y, as in the example above. However, Theorem 1.4 can be applied by considering a rectangle contained in the region D. Theorem 1.4 assures the existence of a solution only on an interval $|x - x_0| \leq \alpha$, which may be small. However, it may be possible to continue, or extend, the solution to the right of the point $x_0 + \alpha$ (or to the left of the point $x_0 - \alpha$). Suppose that $y(x)$ is a solution on the interval $|x - x_0| \leq \alpha$, and that the point $P: (x_0 + \alpha, y(x_0 + \alpha))$ lies in the interior of the region D. Then there exists a rectangle, with center at P and contained in D. According to Theorem 1.4, a solution $\tilde{y}(x)$, satisfying $\tilde{y}(x_0 + \alpha) = y(x_0 + \alpha)$, exists on some interval $|x - (x_0 + \alpha)| \leq \alpha_1$. But by the uniqueness part of Theorem 1.4, the functions $\tilde{y}(x)$ and $y(x)$ must coincide on the interval on which both are defined. In this way the solution $y(x)$ is continued to the right of the point $x_0 + \alpha$, in fact up to the point $x_0 + \alpha + \alpha_1$. If the point $(x_0 + \alpha + \alpha_1, y(x_0 + \alpha + \alpha_1))$ lies in D, this process can be repeated.

It may happen that the solution can be continued for all x greater than x_0. If not, a deeper analysis shows that the solution can be continued up to a point x_1, and that as $x \to x_1 -$, either $y(x)$ becomes infinite or else the integral curve approaches the boundary of the region D.

If, in Theorem 1.4, we drop the hypothesis that $f_y(x, y)$ exists and is continuous, and assume only that $f(x, y)$ is continuous on the rectangle R, it is still possible to prove that a solution to the initial-value problem exists. However, a different method of proof must be employed. Also, the solution may not be unique. Consider, for example, the problem

$$y' = 3y^{2/3}, \qquad y(0) = 0.$$

One solution is found to be $y = x^3$. But it is evident that the zero function, $y = 0$, is also a solution. It should be noted that although $f(x, y) = 3y^{2/3}$ is continuous for all x and y, the function $f_y(x, y) = 2y^{-1/3}$ is not continuous at $(0, 0)$, or at any point on the x axis. Thus Theorem 1.4 cannot be applied to this initial-value problem.

We conclude this section with an example to illustrate the method of successive approximations.

Example Let us calculate a few terms of the sequence of functions that converges to the solution of the problem

$$y' = xy + y^2, \qquad y(0) = 1.$$

From Eq. (1.68) we see that $y_0(x) = 1$. Formula (1.69) becomes, for this example,

$$y_{k+1}(x) = 1 + \int_0^x \{ty_k(t) + [y_k(t)]^2\} \, dt.$$

For $k = 1$ we have

$$y_1(x) = 1 + \int_0^x (t+1) \, dt = 1 + x + \frac{1}{2}x^2.$$

For $k = 2$ we have

$$y_2(x) = 1 + \int_0^x \left[t\left(1 + t + \frac{1}{2}t^2\right) + \left(1 + t + \frac{1}{2}t^2\right)^2 \right] dt$$

$$= 1 + x + \frac{3}{2}x^2 + x^3 + \frac{3}{8}x^4 + \frac{1}{20}x^5.$$

The computations have become difficult. The method of successive approximations is important for proving the existence of a solution, but it is not always useful in finding an explicit formula for the solution.

Exercises for Section 1.13
---

1. Use Theorem 1.4 to show that the initial-value problem possesses a unique solution. In parts (a) and (b), actually find the solution.

   (a) $y' = \dfrac{3\sqrt{x}}{4y}$, $y(1) = -2$

   (b) $(x + y) + (x - y)y' = 0$, $y(0) = 1$

   (c) $y' = x^2 + y^2$, $y(0) = 0$

2. Find at least two solutions of the given initial-value problem. Show that the hypotheses of Theorem 1.4 are not satisfied in any rectangle of the form $|x - x_0| \le a$, $|y - y_0| \le b$.

   (a) $y' = \sqrt{1 - y^2}$, $y(0) = 1$

   (b) $y' = \dfrac{3}{2}y^{1/3}$, $y(0) = 0$

3. Find, for the problem $y' = x - y^2$, $y(0) = 1$, the functions $y_0(x)$, $y_1(x)$, and $y_2(x)$ in the sequence of successive approximation defined by the relations (1.68) and (1.69).

4. Do as in Exercise 3 for the initial-value problem $y' = x^2 + y^2$, $y(0) = -1$.

5. Prove by induction that each of the functions $y_n(x)$ in the sequence (1.68), (1.69) is well defined and satisfies $|y_n(x) - y_0| \le b$ for $|x - x_0| \le a$.

6. Prove by induction that the functions $y_n(x)$ in the sequence (1.68), (1.69) satisfy the inequalities

$$|y_n(x) - y_{n-1}(x)| \le \frac{MK^{n-1}}{n!}|x - x_0|^n, \quad n \ge 1,$$

and hence that

$$|y_n(x) - y_{n-1}(x)| \le \frac{MK^n \alpha^n}{Kn!},$$

$$|x - x_0| \le \alpha, \quad n \ge 1.$$

Suggestion: Use the integral equation (1.67), and the fact that $f(x, y)$ satisfies a Lipschitz condition.

7. Observing that

$$y_n(x) = \sum_{k=1}^{n} [y_k(x) - y_{k-1}(x)] + y_0(x),$$

prove that the sequence $\{y_n(x)\}$ converges for $|x - x_0| \le \alpha$ by proving that the series

$$\sum_{k=1}^{\infty} [y_k(x) - y_{k-1}(x)]$$

converges. Suggestion: Use the result of Exercise 6.

8. On an interval I, which contains the point x_0, let the function $w(x)$ be defined and continuous, and satisfy an inequality of the form

$$|w(x)| \le M\left|\int_{x_0}^{x} |w(t)|\, dt\right|, \quad (1)$$

where M is a positive constant. Prove that

$w(x)$ is identically zero on I. Suggestion: for $x \ge x_0$, let

$$W(x) = \int_{x_0}^{x} |w(t)|\, dt.$$

Then $W(x_0) = 0$ and $W'(x) - MW(x) \le 0$ for $x \ge x_0$. If both sides of the inequality are multiplied by the quantity $e^{-M(x-x_0)}$, it becomes

$$\frac{d}{dx}[W(x)e^{-M(x-x_0)}] \le 0.$$

An integration from x_0 to x yields the inequality

$$W(x) \le W(x_0)e^{M(x-x_0)}.$$

Hence $W(x) = 0$, and from the original inequality (1) we have $w(x) = 0$. In order to treat the case where $x \le x_0$, let

$$W(x) = -\int_{x_0}^{x} |w(t)|\, dt,$$

and proceed as before.

9. Let $f(x, y)$ be continuous and satisfy a Lipschitz condition in a region D. If $y_1(x)$ and $y_2(x)$ are solutions of the equation $y' = f(x, y)$ on an interval I, and if $y_1(x_0) = a_1$ and $y_2(x_0) = a_2$, show that

$$|y_1(x) - y_2(x)| \le |a_1 - a_2| + K\left|\int_{x_0}^{x} |y_1(t) - y_2(t)|\, dt\right|.$$

From this inequality, show that

$$|y_1(x) - y_2(x)| \le |a_1 - a_2|e^{K|x-x_0|}.$$

Additional Exercises for Chapter 1

In Exercises 1–16, find the general solution. If an initial condition is specified, also find the solution that satisfies it.

1. $xy' - 2y - x^3 \sin x = 0, \quad x > 0$

2. $y^3 - 4x + (3xy^2 + 1)y' = 0$

3. $e^{2x+y}y' = 4x, \quad y(0) = 1$

4. $xy' - y + e^x y^2 = 0, \quad x > 0$

5. $xy^2y' = x^3 + y^3, \quad x > 0, \quad y(1) = 2$

6. $xy' = y(y - 2)$

7. $(x^2 + 1)y' = 2xy + 2x(x^2 + 1)$

8. $(x - 2x^2y)y' + (3y - 4xy^2) = 0$

9. $3xy^2y' = 3y^3 + 2x^{3/2}\sqrt{x^3 + y^3}$

10. $2xy' + 3y + x^4y^3 = 0$

11. $y' = \dfrac{3x - 2y}{y + 2x}$

12. $(x^3e^y - e^x)y' + (3x^2e^y - ye^x) = 0$

13. $\dfrac{d^2x}{dt^2} + 2t\dfrac{dx}{dt} = 2t$

14. $-3\left(\dfrac{dx}{dt}\right)^3 - 2t^5 + 3t\left(\dfrac{dx}{dt}\right)^2\dfrac{d^2x}{dt^2} = 0$

15. $x\dfrac{d^2x}{dt^2} = \left(\dfrac{dx}{dt}\right)^2 - x^2\dfrac{dx}{dt}$

16. $2x\dfrac{d^2x}{dt^2} = \left(\dfrac{dx}{dt}\right)^2 + 1$

17. Find all curves with the property that the angle between the tangent and normal lines at each point is bisected by the line through the point and the origin.

18. Find the orthogonal trajectories of the family

$$y^2 = \dfrac{x^2}{1 + cx^2}.$$

19. A tank initially contains 100 gal of a solution that holds 40 lb of a chemical. Water enters at the rate of 5 gal/min and the mixture leaves at the same rate. After how long a time will 1 percent of the chemical remain in the tank?

20. An object whose temperature is 0°C is brought into a room where the air temperature is 20°C. After 30 minutes, the temperature of the object is 8°C. When will the

temperature be 18°C? Assume Newton's law of cooling applies.

21. An object with temperature 100°C is placed in a refrigerator. After two minutes its temperature is 80°. After two more minutes its temperature is 64.5°. What is the temperature inside the refrigerator?

22. Let P, S, and D be the price, supply, and demand of a commodity; all are functions of time t. Suppose that the rate of change of price is proportional to the difference between supply and demand, so that

$$\dfrac{dP}{dt} = k(D - S).$$

Solve the equation under the assumptions that $D = a - bP$ and $S = c + dP$ are linear decreasing and increasing functions of price, respectively.

23. A buoy weighing 64 lb has the shape of a right circular cone of height 4 ft and radius 2 ft. The buoy bobs up and down, with pointed end downward. The density of water is 62.4 lb/ft³, and the buoyant force is equal to the weight of the water displaced. Find the differential equation of motion for x, where x is the downward displacement of the buoy from its equilibrium position.

24. Suppose that a colony of bacteria has a constant immigration of h organisms per unit time. Assuming a constant birth rate and a death rate proportional to the size of the colony, derive the model

$$\dfrac{dN}{dt} = kN(a - N) + h.$$

Find the one constant solution.

25. The banks of a river are represented by the lines $x = 0$ and $x = a$, and the river flows in the positive y-direction with constant speed v_1. A boat leaves the point $(a, 0)$ and travels with constant speed v_2 in such a way that it always points toward the point $(0, 0)$ on the

opposite bank. Let $(x(t), y(t))$ be the coordinates of the boat at time t.

(a)   Show that

$$\frac{dx}{dt} = -v_2 \frac{x}{r} \quad \text{and} \quad \frac{dy}{dt} = -v_2 \frac{y}{r} + v_1,$$

where r is the distance from $(0, 0)$.

(b)   Find an expression for $\frac{dy}{dx}$, thus obtaining the differential equation for the path of the boat.

(c)   Find the xy equation of the path. What happens if $v_2 < v_1$?

26.   The first-order equation $\frac{dx}{dt} = f(x, t)$ is said to be *autonomous* if f depends only on x, so that $\frac{dx}{dt} = f(x)$. If $x_1 = g(t)$ is a solution, show that $x_2 = g(t + c)$ is also a solution, for every value of c.

27.   Let q be continuous on $[0, \infty)$ and let $\lim_{x \to \infty} q(x) = L$. If k is a positive number,

show that every solution of the equation $y' + ky = q(x)$ tends to the limit L/k as x becomes infinite. Suggestion: given $\varepsilon > 0$ there is a positive number x_0 such that $|q(x) - L| < \varepsilon$ if $x \geq x_0$. Let $h(x) = q(x) - L$.

28.   Let the function q be continuous and bounded on $[0, \infty)$ and let k be a positive constant. Show that the equation $y' - ky = q(x)$ possesses solutions that are not bounded on $[0, \infty)$.

29.   Show that the solution of the initial value problem

$$y' = (1 + x^2 + y^2)^{-1}, \qquad y(0) = 1$$

satisfies the inequalities

$$1 \leq y(x) \leq x + 1 \quad \text{and} \quad 1 \leq y(x) \leq 1 + \frac{\pi}{2}$$

for $x \geq 0$. Suggestion: show that $0 \leq y'(x) \leq 1/(1 + x^2) \leq 1$.

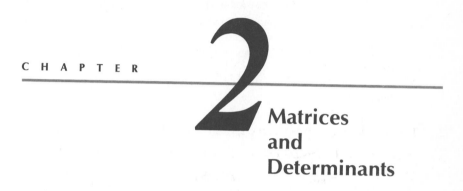

C H A P T E R

2

Matrices
and
Determinants

<table>
<tr><td rowspan="4">

2.1

SYSTEMS OF

LINEAR

EQUATIONS

</td><td>

Let us begin by considering a linear system of two equations with two unknowns. Such a system is of the form

</td></tr>
</table>

$$a_{11}x_1 + a_{12}x_2 = b_1,$$

$$a_{21}x_1 + a_{22}x_2 = b_2,$$

where a_{11}, a_{12}, a_{21}, a_{22}, b_1, and b_2 are given numbers.[1] A *solution* of the system is an ordered pair of numbers (x_1, x_2) that satisfies all the equations. For example, a solution of the system

$$2x_1 - 3x_2 = 5,$$

$$-x_1 + 2x_2 = -2,$$

is the ordered pair (4, 1), because

$$2(4) - 3(1) = 5 \qquad \text{and} \qquad -(4) + 2(1) = -2.$$

Note, however, that the ordered pair (1, 4) is *not* a solution.

---

[1] We assume for the moment that all numbers are real. This restriction will be removed shortly.

The system possesses a simple geometrical interpretation. Each equation of the system can be regarded as that of a straight line in a rectangular $x_1 x_2$ coordinate system. A solution of the system represents the coordinates of a point of intersection of the two lines. In general, two lines intersect in one point and there will be one, and only one, solution of the system. However, other possibilities are apparent. The two lines may be parallel and noncoincident. In this case, no solution of the system will exist. If the two lines coincide, the system will possess infinitely many solutions. In fact, the coordinates of every point on the common line will constitute a solution.

A system need not have the same number of equations as unknowns. For example, let us consider a system of the form

$$a_{11}x_1 + a_{12}x_2 = b_1,$$

$$a_{21}x_1 + a_{22}x_2 = b_2,$$

$$a_{31}x_1 + a_{32}x_2 = b_3,$$

$$a_{41}x_1 + a_{42}x_2 = b_4.$$

It is likely that this system will have no solution, because four arbitrary lines are unlikely to pass through a common point. But if the four lines do have a common point a solution will exist. If all four lines coincide, the system will possess infinitely many solutions.

Real systems of equations that involve three unknowns, such as

$$a_{11}x_1 + a_{12}x_2 + a_{13}x_3 = b_1,$$

$$a_{21}x_1 + a_{22}x_2 + a_{23}x_3 = b_2,$$

$$\cdots\cdots\cdots\cdots\cdots\cdots\cdots\cdots\cdots$$

$$a_{m1}x_1 + a_{m2}x_2 + a_{m3}x_3 = b_m,$$

also permit a geometric interpretation. Each equation can be regarded as that of a plane in a three dimensional $x_1 x_2 x_3$ rectangular coordinate system. A solution of the system is an ordered triple (x_1, x_2, x_3) of numbers that satisfies the system. If the m planes pass through a common point P, then the coordinates of P constitute a solution of the system. A little reflection should convince the reader that the system may have exactly one solution, no solution, or infinitely many solutions (the latter when all the planes pass through a line, or are coincident).

The most general linear system is of the form

$$a_{11}x_1 + a_{12}x_2 + a_{13}x_3 + \cdots + a_{1n}x_n = b_1,$$

$$a_{21}x_1 + a_{22}x_2 + a_{23}x_3 + \cdots + a_{2n}x_n = b_2,$$

$$a_{31}x_1 + a_{32}x_2 + a_{33}x_3 + \cdots + a_{3n}x_n = b_3, \qquad (2.1)$$

$$\cdots\cdots\cdots\cdots\cdots\cdots\cdots\cdots\cdots\cdots\cdots$$

$$a_{m1}x_1 + a_{m2}x_2 + a_{m3}x_3 + \cdots + a_{mn}x_n = b_m.$$

This system involves m equations and n unknowns. The mn numbers[2] a_{ij}, $1 \leq i \leq m$, $1 \leq j \leq n$, are called the *coefficients* of the system. The pattern of the subscripts should be noted carefully. The first subscript indicates the number of the equation to which the coefficient belongs and the second subscript indicates the number of the unknown by which the coefficient is multiplied. Thus all the coefficients a_{2j}, $1 \leq j \leq n$, belong to the second equation (or row). Each coefficient of the form a_{i3}, $1 \leq i \leq m$, is multiplied by x_3 (and belongs to the third column, so to speak, of the system). The m numbers $b_1, b_2, \ldots, b_m$ are called the *constants* of the system. When these numbers are all zero, the system is said to be a *homogeneous* system. Otherwise the system is said to be *nonhomogeneous*.

The coefficients and constants of the linear system (2.1) can be displayed in rectangular arrays:

$$\begin{bmatrix} a_{11} & a_{12} & \cdots & a_{1n} \\ a_{21} & a_{22} & \cdots & a_{2n} \\ \vdots & \vdots & & \vdots \\ a_{m1} & a_{m2} & \cdots & a_{mn} \end{bmatrix}, \qquad \begin{bmatrix} a_{11} & a_{12} & \cdots & a_{1n} & b_1 \\ a_{21} & a_{22} & \cdots & a_{2n} & b_2 \\ \vdots & \vdots & & \vdots & \vdots \\ a_{m1} & a_{m2} & \cdots & a_{mn} & b_m \end{bmatrix}.$$

The first, with m rows and n columns, is called the *coefficient matrix* of the system (2.1). The second, with m rows and $n + 1$ columns, is called the *augmented matrix*. The numbers a_{ij} and b_i appearing in a matrix are called its *elements*. For example, the augmented matrix of the system

$$2x_1 - 3x_2 = 5,$$
$$-x_1 + 2x_2 = -2$$

is

$$\begin{bmatrix} 2 & -3 & 5 \\ -1 & 2 & -2 \end{bmatrix}.$$

Knowing the augmented matrix of a system is tantamount to knowing the system. Matrices are discussed in detail in Section 2.4.

A solution of the system (2.1) is simply an ordered n-tuple of numbers[3] $(x_1, x_2, \ldots, x_n)$ that satisfies the system. If the system possesses at least one solution it is said to be *consistent*. If no solution exists, the system is said to be *inconsistent*. The set of all solutions of a system of equations is called the *general solution* or *complete solution* of the system. In regard to the number

---

[2] We now allow all numbers to be complex. Although we shall present examples that involve only real numbers, all the theory in this chapter is valid when complex numbers are permitted, except where noted. We have made the restriction to real numbers until now in order to obtain geometric interpretations in the cases $n = 2$ and $n = 3$.

[3] Usually when the coefficients and constants of the system are real we are interested only in real solutions.

of solutions of the system, it turns out that there are three possibilities:

1. No solution exists.

2. Exactly one solution exists.

3. Infinitely many solutions exist.

Thus it can never happen that the system possesses exactly two or exactly 50 solutions. Either there is none, one, or infinitely many. In the cases $n = 2$ and $n = 3$, when all numbers are restricted to be real, these facts are fairly apparent from the geometric interpretation of the systems. We shall elaborate on the general case below. When infinitely many solutions exist, it is still possible to describe the "number" of solutions, in a sense to be explained in Chapter 3.

One important property of the systems of equations that we have been considering should not escape notice. Each equation of the system (2.1) is *linear*. That is, each equation is a *polynomial equation of first degree* in the unknowns $x_1, x_2, \ldots, x_n$. Systems of linear equations are called *linear systems*. It is possible to consider nonlinear systems, such as

$$x_1^2 x_2 - 2x_1^2 = 3, \qquad 2x_1 + 4x_2 = 5$$

or

$$e^{x_1} + x_2 \sin x_1 = 0, \qquad \cos(x_1 x_2) - x_2 = 3;$$

however, we shall not be concerned with such systems in this chapter.

Two systems of equations are said to be *equivalent* if they have exactly the same solutions. For example, the two systems

$$2x_1 - 3x_2 = 1, \qquad x_1 - 7x_2 = -5,$$
$$x_1 + 4x_2 = 6, \qquad 2x_1 + 8x_2 = 12$$

are equivalent because each possesses the single solution (2, 1). Many of the standard procedures for solving a linear system involve the finding of an equivalent but simpler system that can be solved easily. Let us write our system as

$$f_1(x_1, x_2, \ldots, x_n) = b_1,$$
$$f_2(x_1, x_2, \ldots, x_n) = b_2,$$
$$\cdots\cdots\cdots\cdots\cdots\cdots\cdots$$
$$f_m(x_1, x_2, \ldots, x_n) = b_m,$$

where

$$f_i(x_1, x_2, \ldots, x_n) = a_{i1}x_1 + a_{i2}x_2 + \cdots + a_{in}x_n, \qquad 1 \le i \le m.$$

In finding an equivalent system, three types of operations are used. These are

1. Interchanging two equations.

2. Multiplying both sides of an equation by a number that is not zero.

3. Adding to one equation an equation that is formed by multiplying both sides of another equation by a number.

It is apparent that the first type of operation leads to an equivalent system. To justify the second type of operation, we merely observe that an ordered n-tuple $(x_1, x_2, \ldots, x_n)$ that satisfies either of the equations

$$f_i(x_1, x_2, \ldots, x_n) = b_i,$$
$$cf_i(x_1, x_2, \ldots, x_n) = cb_i,$$

must also satisfy the other if c is a number other than zero. The restriction $c \neq 0$ is necessary; see Exercise 17 at the end of this section.

We now consider the third type of operation. Taking the jth and kth equations

$$f_j(x_1, x_2, \ldots, x_n) = b_j,$$
$$f_k(x_1, x_2, \ldots, x_n) = b_k,$$

of the system, let us add c times the jth equation to the kth. The resulting pair of equations is

$$f_j(x_1, x_2, \ldots, x_n) = b_j,$$
$$f_k(x_1, x_2, \ldots, x_n) + cf_j(x_1, x_2, \ldots, x_n) = b_k + cb_j.$$

The reader can easily verify that any ordered n-tuple that satisfies either pair of equations also satisfies the other. Hence the third type of operation always yields an equivalent system.

The method of solving linear systems that we shall now describe is based on a process of elimination. Step by step, one unknown is eliminated from all equations but one. We consider some examples before discussing the general case.

Example 1

$$x_1 - 2x_2 = 1,$$
$$-2x_1 + x_2 = 4,$$
$$x_1 + 3x_2 = -9.$$

We eliminate x_1 from the second and third equations by adding appropriate multiples of the first equation to the others. Thus we add 2 times the first equation to the second and -1 times the first equation to the third. The result is the equivalent system

$$x_1 - 2x_2 = 1,$$
$$-3x_2 = 6,$$
$$5x_2 = -10.$$

We next multiply through in the second equation by $-\frac{1}{3}$ to make the coefficient of x_2 in that equation equal to 1. The resulting system is

$$x_1 - 2x_2 = 1,$$
$$x_2 = -2,$$
$$5x_2 = -10.$$

We now use the second equation to eliminate x_2 from all other equations. Addition of 2 times the second equation to the first and -5 times the second to the third yields the system

$$x_1 = -3,$$
$$x_2 = -2,$$
$$0 = 0.$$

It is now evident that the ordered pair $(-3, -2)$ is a solution, and the only solution.

We now solve the system of Example 1 again, this time using matrix notation. The augmented matrix of the system is

$$\begin{bmatrix} 1 & -2 & 1 \\ -2 & 1 & 4 \\ 1 & 3 & -9 \end{bmatrix}.$$

If we add twice the first row to the second and -1 times the first row to the third, we obtain the matrix

$$\begin{bmatrix} 1 & -2 & 1 \\ 0 & -3 & 6 \\ 0 & 5 & -10 \end{bmatrix}.$$

This is the matrix of the system

$$x_1 - 2x_2 = 1,$$
$$-3x_2 = 6,$$
$$5x_2 = -10$$

obtained previously. Then we multiply through in the second row of the last matrix by $-\frac{1}{3}$. The new matrix is

$$\begin{bmatrix} 1 & -2 & 1 \\ 0 & 1 & -2 \\ 0 & 5 & -10 \end{bmatrix}.$$

Adding twice the second row to the first and -5 times the second row to the third (this step corresponds to eliminating x_2 from all equations other than the second) gives

$$\begin{bmatrix} 1 & 0 & -3 \\ 0 & 1 & -2 \\ 0 & 0 & 0 \end{bmatrix}.$$

This is the matrix of the system

$$x_1 = -3,$$
$$x_2 = -2,$$
$$0 = 0,$$

whose solution is $(-3, -2)$.

Example 2

$$2x_1 - 4x_2 = -4,$$
$$3x_1 - 4x_2 = -2,$$
$$-4x_1 + 3x_2 = 6.$$

We immediately write down the augmented matrix of the system,

$$\begin{bmatrix} 2 & -4 & -4 \\ 3 & -4 & -2 \\ -4 & 3 & 6 \end{bmatrix}.$$

We multiply through in the first row by $\frac{1}{2}$, to put a 1 in the first column, first row position. The resulting matrix is

$$\begin{bmatrix} 1 & -2 & -2 \\ 3 & -4 & -2 \\ -4 & 3 & 6 \end{bmatrix}.$$

By adding -3 times the first row to the second, and 4 times the first row to the third, we obtain the matrix of a system in which x_1 has been eliminated from all equations but the first. The matrix is

$$\begin{bmatrix} 1 & -2 & -2 \\ 0 & 2 & 4 \\ 0 & -5 & -2 \end{bmatrix}.$$

Next we use the second equation to eliminate x_2 from all other equations. To do this, we multiply through in the second row of the last matrix by $\frac{1}{2}$ and then add appropriate multiples of the new second row to the first and third rows. The result is

$$\begin{bmatrix} 1 & 0 & 2 \\ 0 & 1 & 2 \\ 0 & 0 & 8 \end{bmatrix}.$$

The corresponding system is

$$x_1 = 2,$$

$$x_2 = 2,$$

$$0x_1 + 0x_2 = 8.$$

No solution exists, since the last equation cannot be satisfied for any choice of x_1 and x_2.

Example 3

$$x_1 - x_2 + x_3 - 2x_4 = -1,$$

$$2x_1 - 2x_2 + x_3 - 2x_4 = -3,$$

$$-x_1 + x_2 - 2x_3 + 4x_4 = 0,$$

$$-3x_1 + 3x_2 - x_3 + 2x_4 = 5.$$

The augmented matrix of the system is

$$\begin{bmatrix} 1 & -1 & 1 & -2 & -1 \\ 2 & -2 & 1 & -2 & -3 \\ -1 & 1 & -2 & 4 & 0 \\ -3 & 3 & -1 & 2 & 5 \end{bmatrix}.$$

We add appropriate multiples of the first row to other rows to introduce zeros in the first column below the first row. The result is

$$\begin{bmatrix} 1 & -1 & 1 & -2 & -1 \\ 0 & 0 & -1 & 2 & -1 \\ 0 & 0 & -1 & 2 & -1 \\ 0 & 0 & 2 & -4 & 2 \end{bmatrix}.$$

As it happens, only zeros appear in the second column below the first row. However, there are nonzero elements in the third column below the first row, and we can use the second equation to eliminate x_3 from all other equations. We multiply through in the second row by -1 and then add appropriate multiples of the second row to the other rows, so as to introduce zeros in the third column. The resulting matrix is

$$\begin{bmatrix} 1 & -1 & 0 & 0 & -2 \\ 0 & 0 & 1 & -2 & 1 \\ 0 & 0 & 0 & 0 & 0 \\ 0 & 0 & 0 & 0 & 0 \end{bmatrix}.$$

The corresponding system is

$$x_1 - x_2 \qquad\qquad = -2,$$
$$x_3 - 2x_4 = 1.$$

If we assign any values whatsoever to the unknowns x_2 and x_4, the remaining unknowns x_1 and x_3 are completely determined by the equations of the system. Setting $x_2 = a$ and $x_4 = b$, where a and b are arbitrary constants, we see that $x_1 = a - 2$ and $x_3 = 2b + 1$. Hence the general solution is described by the ordered 4-tuple

$$(a - 2, a, 2b + 1, b).$$

We now give a systematic description of the elimination procedure. The first step is to write down the matrix of the system. Then we perform a sequence of operations on the rows of the matrix. These operations, called *elementary row operations*,[4] are of three types: interchange of two rows, multiplication of a row by a nonzero constant, and addition of a multiple of one row to another row.

Assuming that the unknown x_1 actually appears in one of the equations, there will be at least one nonzero element in the first column of the matrix. By interchanging rows if necessary, we place a nonzero element in the position of row 1 and column 1. We then multiply through in the first row by the reciprocal of this element, thus placing a 1 in the upper left-hand corner. The resulting matrix has the form

$$\begin{bmatrix} 1 & a_{12} & a_{13} & \cdots & a_{1n} & b_1 \\ a_{21} & a_{22} & a_{23} & \cdots & a_{2n} & b_2 \\ \vdots & \vdots & \vdots & & \vdots & \vdots \\ a_{m1} & a_{m2} & a_{m3} & \cdots & a_{mn} & b_m \end{bmatrix}.$$

By adding appropriate multiples of the first row to the other rows, we introduce zeros in the first column below the first position. This amounts to

---

[4] Elementary column operations are defined similarly, but we will make no use of them.

eliminating x_1 from all equations of the system other than the first. It may happen that x_2 and other unknowns are also eliminated. Suppose that in the process of eliminating x_1 we also eliminate $x_2, x_3, \ldots, x_{r-1}$ but not x_r. Then the matrix of the resulting system has the form

$$
\begin{bmatrix}
1 & a_{12} & a_{13} & \cdots & a_{1r} & \cdots & a_{1n} & b_1 \\
0 & 0 & 0 & \cdots & a'_{2r} & \cdots & a'_{2n} & b'_2 \\
\vdots & \vdots & \vdots & & \vdots & & \vdots & \vdots \\
0 & 0 & 0 & \cdots & a'_{mr} & \cdots & a'_{mn} & b'_m
\end{bmatrix},
$$

where not all of $a'_{2r}, a'_{3r}, \ldots, a'_{mr}$ are zero. By interchanging rows if necessary, we place a nonzero element at the junction of row 2 and column r. We make this element equal to 1 by multiplying through in row 2 by the appropriate number. Then we eliminate x_r from all other equations, including the first equation. This amounts to introducing zeros in all places of column r of the matrix, except the second. In other words, we apply the same process to the matrix

$$
\begin{bmatrix}
a'_{2r} & \cdots & a'_{2n} & b'_2 \\
\vdots & & \vdots & \vdots \\
a'_{mr} & \cdots & a'_{mn} & b'_m
\end{bmatrix}
$$

that we applied to the original matrix, but we also include the step of eliminating x_r from the first equation of the system.

This procedure is continued, either until we run out of equations or until we run out of unknowns. The final coefficient matrix (consisting of the first n columns of the augmented matrix) is said to be in *row-echelon form*. The augment matrix is of the form.

$$
\begin{bmatrix}
1 & * & * & \cdots & * & 0 & * & \cdots & * & 0 & * & \cdots & * & b''_1 \\
0 & 0 & 0 & \cdots & 0 & 1 & * & \cdots & * & 0 & * & \cdots & * & b''_2 \\
\hdashline
0 & 0 & 0 & \cdots & 0 & 0 & 0 & \cdots & 0 & 1 & * & \cdots & * & b''_k \\
0 & 0 & 0 & \cdots & 0 & 0 & 0 & \cdots & 0 & 0 & 0 & \cdots & 0 & b''_{k+1} \\
\hdashline
0 & 0 & 0 & \cdots & 0 & 0 & 0 & \cdots & 0 & 1 & 0 & \cdots & 0 & b''_m
\end{bmatrix}.
$$

Here the asterisks represent elements that are not necessarily zero or one. The elements b''_i in the right-hand column may be different from the elements b_i in the right-hand column of the original matrix. If we run out of equations before unknowns, the final rows, which correspond to equations with zero coefficients, will not be present, but if we run out of unknowns before equations, one or more rows corresponding to an equation of the form $0 = b''_s$ will be present. In this case, unless the numbers $b''_{k+1}, b''_{k+2}, \ldots,$ b''_m are all zero, the system will have no solutions. If $b''_{k+1}, b''_{k+2}, \ldots, b''_m$ are all zero (or if $m = k$) then the k unknowns whose subscripts are those of the columns in which the 1's appear can be expressed in terms of the remaining unknowns, and these remaining $n - k$ unknowns can be assigned arbitrary values.

We present a final example to illustrate some of these ideas.

Example 4   The augmented matrix of the system

$$x_1 - 2x_2 - x_3 + x_4 = 5$$
$$-2x_1 + 4x_2 + 3x_3 = -13$$
$$3x_1 - 6x_2 + 2x_3 + 13x_4 + x_5 = 2$$
$$2x_1 - 4x_2 + x_3 + 8x_4 + x_5 = 3$$

is

$$\begin{bmatrix} 1 & -2 & -1 & 1 & 0 & 5 \\ -2 & 4 & 3 & 0 & 0 & -13 \\ 3 & -6 & 2 & 13 & 1 & 2 \\ 2 & -4 & 1 & 8 & 1 & 3 \end{bmatrix}.$$

Adding appropriate multiples of the first row to the other rows (to introduce zeros in the first column), we obtain the matrix

$$\begin{bmatrix} 1 & -2 & -1 & 1 & 0 & 5 \\ 0 & 0 & 1 & 2 & 0 & -3 \\ 0 & 0 & 5 & 10 & 1 & -13 \\ 0 & 0 & 3 & 6 & 1 & -7 \end{bmatrix}.$$

Notice that all elements in the *second* column below the first row are also zero, but the third column has a nonzero element below the first row. In particular, there is a nonzero element in the second row of the third column, so we can use the second row to introduce zeros in all positions of the third column (including the first) except the second. The resulting matrix is

$$\begin{bmatrix} 1 & -2 & 0 & 3 & 0 & 2 \\ 0 & 0 & 1 & 2 & 0 & -3 \\ 0 & 0 & 0 & 0 & 1 & 2 \\ 0 & 0 & 0 & 0 & 1 & 2 \end{bmatrix}.$$

All elements in the fourth column below the second row are also zero. The fifth column has a nonzero element in the third row, so we can use the third row to introduce zeros in all positions of the fifth column other than the third. The resulting matrix, which is in row-echelon form, is

$$\begin{bmatrix} 1 & -2 & 0 & 3 & 0 & 2 \\ 0 & 0 & 1 & 2 & 0 & -3 \\ 0 & 0 & 0 & 0 & 1 & 2 \\ 0 & 0 & 0 & 0 & 0 & 0 \end{bmatrix}.$$

The first nonzero elements (reading from left to right) in the rows are in columns 1, 3, and 5. Then x_1, x_3, and x_5 can be expressed in terms of x_2 and x_4. These last mentioned unknowns may be assigned arbitrary values. The equations of the system are

$$
\begin{aligned}
x_1 - 2x_2 + 3x_4 &= 2, \\
x_3 + 2x_4 &= -3, \\
x_5 &= 2.
\end{aligned}
$$

Setting $x_2 = a$, $x_4 = b$, where a and b are arbitrary constants, we find that the general solution is

$$
x_1 = 2a - 3b + 2, \qquad x_2 = a, \qquad x_3 = -2b - 3, \qquad x_4 = b, \qquad x_5 = 2.
$$

In the special case where the system is consistent and $k = n$, there is exactly one solution. For the matrix has the form

$$
\begin{bmatrix}
1 & 0 & 0 & \cdots & 0 & b_1'' \\
0 & 1 & 0 & \cdots & 0 & b_2'' \\
0 & 0 & 1 & \cdots & 0 & b_3'' \\
\cdots & \cdots & \cdots & \cdots & \cdots & \cdots \\
0 & 0 & 0 & \cdots & 1 & b_n'' \\
0 & 0 & 0 & \cdots & 0 & 0 \\
0 & 0 & 0 & \cdots & 0 & 0
\end{bmatrix}
$$

and we see that necessarily $x_i = b_i''$, $i = 1, 2, \ldots, n$. If $k < n$ and the system is consistent, then there are infinitely many solutions, since $n - k$ of the unknowns may be assigned arbitrary values.

Exercises for Section 2.1

In Exercises 1–16, find all real solutions of the system by reducing the matrix of the system to row-echelon form.

1. $\begin{aligned} x_1 + 2x_2 &= 3 \\ 3x_1 - x_2 &= -5 \end{aligned}$

2. $\begin{aligned} -2x_1 + 6x_2 &= -8 \\ x_1 - 3x_2 &= 4 \end{aligned}$

3. $\begin{aligned} -2x_1 + x_2 &= 5 \\ 4x_1 - 2x_2 &= -1 \end{aligned}$

4. $\begin{aligned} 2x_1 + 8x_2 &= 14 \\ x_1 - 3x_2 &= 0 \\ 4x_1 + 2x_2 &= 14 \end{aligned}$

5. $\begin{aligned} 3x_1 - x_2 &= 4 \\ 6x_1 - 2x_2 &= 8 \\ -9x_1 + 3x_2 &= -12 \end{aligned}$

6. $\begin{aligned} x_1 - 2x_2 + x_3 &= 5 \\ -x_1 + x_2 - 4x_3 &= -7 \\ 3x_1 + 3x_2 + x_3 &= 4 \end{aligned}$

7. $2x_1 - x_2 + x_3 = 5$
    $x_1 - x_2 - x_3 = 4$
    $-2x_1 + 2x_2 + x_3 = -6$

8. $3x_1 - 5x_2 + x_3 = 0$
    $-x_1 - x_2 + x_3 = -4$
    $2x_1 - 4x_2 + x_3 = -1$

9. $x_1 + 2x_2 - x_3 = -3$
    $3x_1 - x_2 - 2x_3 = 13$
    $x_1 - 5x_2 = 19$

10. $-2x_1 + 2x_2 - 2x_3 = -8$
     $x_1 - x_2 + x_3 = 4$
     $2x_1 - 2x_2 + 2x_3 = 8$

11. $-3x_1 - x_2 + x_3 = 5$
     $2x_1 + x_2 = -3$
     $-5x_1 - x_2 + x_3 = 9$

12. $x_1 + x_2 - x_3 = 5$
     $2x_1 + x_3 = -2$
     $x_1 - x_2 + 2x_3 = 0$

13. $2x_1 - x_2 + x_3 = -1$
     $x_1 + x_2 + x_3 = 3$

14. $x_1 + x_2 - x_3 + x_4 = 4$
     $2x_1 + x_2 - x_3 + x_4 = 5$
     $x_2 + x_3 = -1$
     $x_2 + 2x_4 = 4$

15. $2x_1 + x_2 - 2x_3 = 4$
     $x_1 + x_2 - 3x_3 + x_4 = 7$
     $3x_1 + x_2 - 6x_3 + x_4 = 3$

16. $x_1 + 2x_2 - x_3 = 1$
     $2x_1 + 2x_3 + x_4 = 4$
     $x_1 - 2x_2 + 3x_3 + x_4 = 3$
     $3x_1 - 2x_2 + 5x_3 + 2x_4 = 7$

17. Suppose that the first equation of the system $2x_1 - x_2 = -1$, $x_1 + x_2 = 7$, is multiplied through by zero, yielding the system $0x_1 + 0x_2 = 0$, $x_1 + x_2 = 7$. Show that the two systems are not equivalent.

18. Show that the system of equations $x_1 - x_2 = -7$, $x_1^2 + x_2 = 9$, possesses the two solutions $(1, 8)$ and $(-2, 5)$, and no others. Does this fact contradict the assertions made about systems of the form (2.1)? Explain.

19. For what values of a and b does the system have a solution?
     (a) $x_1 - 2x_2 = a$
          $-3x_1 + 6x_2 = b$
     (b) $-x_1 - 2x_3 = a$
          $2x_1 + x_2 + x_3 = 0$
          $x_1 + x_2 - x_3 = b$

20. Show all the 3×4 matrix configurations that can result when a system of 3 equations in 3 unknowns is reduced to row-echelon form.

2.2 HOMOGENEOUS SYSTEMS

The procedure for solving a homogeneous system

$$a_{11}x_1 + a_{12}x_2 + \cdots + a_{1n}x_n = 0,$$
$$\cdots\cdots\cdots\cdots\cdots\cdots\cdots\cdots\cdots\cdots$$
$$a_{m1}x_1 + a_{m2}x_2 + \cdots + a_{mn}x_n = 0 \tag{2.2}$$

is no different from that for solving any other linear system. One fact peculiar to homogeneous systems should be pointed out, however. A homogeneous system always has a solution, since it is satisfied if $x_1, x_2, \ldots, x_n$ are all zero. This solution is called the *trivial solution*. The system may or may not have nontrivial solutions in addition. If the matrix of the system is reduced to row-echelon form, it will be of the type

$$
\begin{bmatrix}
1 & \cdots & 0 & \cdots & 0 & \cdots\cdots & 0 \\
0 & \cdots & 1 & \cdots & 0 & \cdots\cdots & 0 \\
\multicolumn{7}{c}{\cdots\cdots\cdots\cdots\cdots\cdots\cdots\cdots\cdots\cdots\cdots\cdots} \\
0 & \cdots\cdots\cdots\cdots & 1 & \cdots\cdots & 0 & 0 \\
0 & \cdots\cdots\cdots\cdots\cdots\cdots & 0 & 0 \\
\multicolumn{7}{c}{\cdots\cdots\cdots\cdots\cdots\cdots\cdots\cdots\cdots\cdots\cdots} \\
0 & \cdots\cdots\cdots\cdots\cdots\cdots\cdots & 0 & 0
\end{bmatrix}
\quad
\begin{array}{l}
\\[12pt]
\text{(row } k) \\[24pt]
\text{(row } m)
\end{array}
$$

(The last rows whose elements are all zero will not be present if $k = m$.) The k unknowns whose subscripts are those of the columns in which the 1's appear can be expressed in terms of the remaining unknowns. The remaining unknowns can be assigned arbitrary values. If $k < n$, the system has infinitely many solutions. But if $k = n$, then the only solution is the trivial solution. Since necessarily $k \leq m$, there being m equations in the system, we always have a nontrivial solution if $n > m$. This important fact we state as a theorem.

Theorem 2.1   The homogeneous linear system (2.2) possesses nontrivial solutions if it has more unknowns than equations ($n > m$).

We now consider some examples.

Example 1

$$
\begin{aligned}
x_1 - x_2 \qquad &= 0, \\
2x_1 - 3x_2 + x_3 &= 0, \\
x_1 + x_2 - 5x_3 &= 0.
\end{aligned}
$$

The augmented matrix of the system is

$$
\begin{bmatrix}
1 & -1 & 0 & 0 \\
2 & -3 & 1 & 0 \\
1 & 1 & -5 & 0
\end{bmatrix}.
$$

Adding appropriate multiples of the first row to the other rows, we have

$$
\begin{bmatrix}
1 & -1 & 0 & 0 \\
0 & -1 & 1 & 0 \\
0 & 2 & -5 & 0
\end{bmatrix}.
$$

We multiply through in the second row by -1 and then add multiples of this row to the other rows so as to introduce zeros in the second column. The result is

$$
\begin{bmatrix}
1 & 0 & -1 & 0 \\
0 & 1 & -1 & 0 \\
0 & 0 & -3 & 0
\end{bmatrix}.
$$

Multiplying through in the last equation by $-\frac{1}{3}$ and then using this last row to introduce zeros in the third column, we arrive at the row-echelon matrix

$$\begin{bmatrix} 1 & 0 & 0 & 0 \\ 0 & 1 & 0 & 0 \\ 0 & 0 & 1 & 0 \end{bmatrix}.$$

The corresponding system,

$$x_1 = 0, \qquad x_2 = 0, \qquad x_3 = 0,$$

has only the trivial solution.

Example 2

$$x_1 - x_2 - 2x_3 + x_4 = 0,$$
$$-3x_1 + 3x_2 + x_3 - x_4 = 0,$$
$$2x_1 - 2x_2 + x_3 = 0.$$

Since the number of unknowns (4) is greater than the number of equations (3), we are guaranteed that the system has nontrivial solutions. In what follows, we have written down the matrix of the system and then reduced it to row-echelon form by using elementary row operations.

$$\begin{bmatrix} 1 & -1 & -2 & 1 & 0 \\ -3 & 3 & 1 & -1 & 0 \\ 2 & -2 & 1 & 0 & 0 \end{bmatrix}, \quad \begin{bmatrix} 1 & -1 & -2 & 1 & 0 \\ 0 & 0 & -5 & 2 & 0 \\ 0 & 0 & 5 & -2 & 0 \end{bmatrix},$$

$$\begin{bmatrix} 1 & -1 & -2 & 1 & 0 \\ 0 & 0 & 1 & -\frac{2}{5} & 0 \\ 0 & 0 & 5 & -2 & 0 \end{bmatrix}, \quad \begin{bmatrix} 1 & -1 & 0 & \frac{1}{5} & 0 \\ 0 & 0 & 1 & -\frac{2}{5} & 0 \\ 0 & 0 & 0 & 0 & 0 \end{bmatrix}.$$

The system corresponding to the last matrix is

$$x_1 - x_2 \qquad + \frac{1}{5}x_4 = 0,$$
$$x_3 - \frac{2}{5}x_4 = 0,$$
$$0 = 0.$$

The unknowns x_2 and x_4 may be assigned arbitrary values. Setting $x_2 = a$ and $x_4 = b$, we have

$$x_1 = a - \frac{1}{5}b, \qquad x_2 = a, \qquad x_3 = \frac{2}{5}b, \qquad x_4 = b.$$

The form of the solution becomes simpler if we replace the arbitrary constant b by a new symbol, $b = 5b'$. Then

$$x_1 = a - b', \qquad x_2 = a, \qquad x_3 = 2b', \qquad x_4 = 5b'.$$

Exercises for Section 2.2

1. In Exercises 3–12, which systems have more unknowns than equations? Without solving these systems, what can be said about the existence of nontrivial solutions?

2. If a homogeneous system has more equations than unknowns, is it possible for the system to have a nontrivial solution? Justify your answer.

In Exercises 3–12, find the general solution.

3. $2x_1 - x_2 = 0,$
   $3x_1 + 4x_2 = 0.$

4. $2x_1 - 3x_2 = 0,$
   $-4x_1 + 6x_2 = 0,$
   $6x_1 - 9x_2 = 0.$

5. $3x_1 - x_2 + x_3 = 0,$
   $x_1 - x_2 - x_3 = 0,$
   $x_1 + x_2 + x_3 = 0.$

6. $3x_1 - 3x_2 - x_3 = 0,$
   $2x_1 - x_2 - x_3 = 0.$

7. $3x_1 + x_2 - 5x_3 - x_4 = 0,$
   $2x_1 + x_2 - 3x_3 - 2x_4 = 0,$
   $x_1 + x_2 - x_3 - 3x_4 = 0.$

8. $x_1 - x_2 - 3x_3 = 0,$
   $x_1 + x_2 + x_3 = 0,$
   $2x_1 + 2x_2 + x_3 = 0.$

9. $3x_1 + x_2 + 2x_3 = 0,$
   $-9x_1 + 2x_2 - 2x_3 - 8x_4 = 0,$
   $-6x_1 + x_2 - 4x_4 = 0,$
   $6x_1 - x_2 + 2x_3 + 5x_4 = 0.$

10. $2x_1 - 2x_2 - x_3 + x_4 = 0,$
    $-x_1 + x_2 + x_3 - 2x_4 = 0,$
    $3x_1 - 3x_2 + x_3 - 6x_4 = 0,$
    $2x_1 - 2x_2 - 2x_4 = 0.$

11. $x_1 - 2x_2 + x_3 - x_4 + 2x_5 = 0,$
    $2x_1 - 4x_2 + 2x_3 - x_4 + x_5 = 0,$
    $x_1 - 2x_2 + x_3 + 2x_4 - 7x_5 = 0.$

12. $2x_1 - x_2 + x_3 + x_4 = 0,$
    $2x_1 + x_2 - x_3 + 2x_4 = 0,$
    $2x_1 + x_2 + 2x_3 + 4x_4 = 0,$
    $2x_1 - x_2 + 4x_3 + 4x_4 = 0.$

13. For what values of a does the system have nontrivial solutions?
    (a) $x_1 + ax_2 = 0$
        $-3x_1 + 2x_2 = 0$
    (b) $x_1 + ax_2 - 2x_3 = 0$
        $2x_1 - x_2 - x_3 = 0$
        $-x_1 - x_2 + x_3 = 0$

14. Consider a linear system of two equations with two unknowns.
    (a) Suppose that the ordered pair (a_1, a_2) is a solution. If the system is homogeneous, show that (ka_1, ka_2) is also a solution for every constant k. Is it a solution if the system is not homogeneous? Justify your answer.
    (b) Suppose that (u_1, u_2) and (v_1, v_2) are both solutions. If the system is homogeneous, show that $(u_1 + v_1, u_2 + v_2)$ is also a solution. Is it a solution if the system is not homogeneous? Justify your answer.

2.3
APPLICATIONS

Example 1 Concrete (as used for sidewalks, foundations, etc.) is usually a mixture of portland cement, sand, and gravel. A contractor has three previously mixed batches. Batch 1 contains cement, sand, and gravel in the proportions 1:4:4 by volume, batch 2 in the proportions 1:2:3, and batch 3 in the proportions 1:2:0. The contractor desires a mixture with proportions 1:3:3. Is it possible to obtain such a mixture from the three batches? If so, in what proportions should the three batches be mixed?

Let x_1, x_2, and x_3 be the numbers of units (say cubic feet) to be taken from the three batches. Then the amounts of the various ingredients are

$$\text{cement:}\qquad \frac{1}{9}x_1 + \frac{1}{6}x_2 + \frac{1}{3}x_3$$

$$\text{sand:}\qquad \frac{4}{9}x_1 + \frac{2}{6}x_2 + \frac{2}{3}x_3$$

$$\text{gravel:}\qquad \frac{4}{9}x_1 + \frac{3}{6}x_2.$$

To obtain the proportions 1:3:3 we require

$$3\left(\frac{1}{9}x_1 + \frac{1}{6}x_2 + \frac{1}{3}x_3\right) = \frac{4}{9}x_1 + \frac{2}{6}x_2 + \frac{2}{3}x_3$$

$$3\left(\frac{1}{9}x_1 + \frac{1}{6}x_2 + \frac{1}{3}x_3\right) = \frac{4}{9}x_1 + \frac{3}{6}x_2.$$

These equations simplify to

$$2x_1 - 3x_2 - 6x_3 = 0$$

$$x_1 \qquad\quad - 9x_3 = 0.$$

We are interested only in nontrivial solutions whose components are all nonnegative. If we let $x_3 = c$, we see from the second equation that $x_1 = 9c$. Then, from the first equation, we have $3x_2 = 18c - 6c = 12c$ or $x_2 = 4c$. (We could have reduced the system to row-echelon form, of course.) Thus the three batches should be mixed in the proportions 9:4:1.

Example 2 Consider the electrical network of Figure 2.1, in which 5 resistors and 2 batteries are wired so as to form three loops. The currents in the loops are denoted by I_1, I_2, and I_3. The current is considered positive when the flow is in the direction of the arrow. (The direction of the arrow that orients the loop can be chosen arbitrarily.) The current through the resistor of R_3 ohms is $\pm(I_1 - I_2)$, the plus sign depending on which direction is considered positive. The current in the resistor of R_5 ohms is $\pm(I_2 - I_3)$. The voltage drop across any resistor is RI, where R is the resistance in ohms and I is the

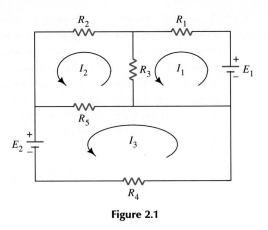

Figure 2.1

current in amperes. According to a law of Kirchhoff, the sum of the voltage drops around each loop must be equal to the applied voltage. Applying this law to each of the three loops in the figure, we arrive at the system of equations

$$R_1 I_1 + R_3(I_1 - I_2) = E_1,$$

$$R_2 I_2 + R_3(I_2 - I_1) + R_5(I_2 - I_3) = 0,$$

$$R_4 I_3 + R_5(I_3 - I_2) = -E_2$$

for the loop currents I_1, I_2, and I_3.

Example 3   Consider a square slab, of side L, as shown in Figure 2.2. Let $h = L/n$, where n is a positive integer, and let $x_i = ih$, $y_i = ih$, for $i = 0, 1, \ldots, n$. The points

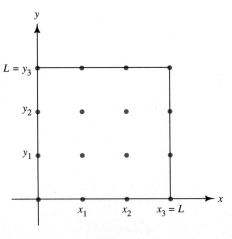

Figure 2.2

(x_j, y_k) are called *lattice points*. In Figure 2.2 we have illustrated the case $n = 3$. We denote the temperature at a point (x, y) in the slab by $u(x, y)$. The abbreviation $u(x_i, y_j) = u_{ij}$ will be used for the temperatures at the lattice points. We assume that the temperature at each point on the boundary of the slab is known. It can be shown that the temperature at each lattice point in the interior of the slab is approximately equal to the average of the temperatures at the four nearest lattice points if the spacing h is small. We therefore have, approximately,

$$u_{11} = \frac{1}{4}(u_{01} + u_{21} + u_{10} + u_{12}),$$

$$u_{12} = \frac{1}{4}(u_{02} + u_{22} + u_{11} + u_{13}),$$

$$u_{21} = \frac{1}{4}(u_{11} + u_{31} + u_{20} + u_{22}),$$

$$u_{22} = \frac{1}{4}(u_{12} + u_{32} + u_{21} + u_{23}).$$

This constitutes a system of equations for u_{11}, u_{12}, u_{21}, and u_{22}, since the temperatures u_{01}, u_{02}, u_{31}, u_{32}, u_{10}, u_{20}, u_{13}, and u_{23} are assumed to be known.

The smaller the spacing h, the better the approximation in this method. Therefore, in practice, systems with a large number of unknowns and equations are encountered and a computer must be used for their solution. The method we have described here can be generalized so as to apply to slabs that are not square shaped.

Exercises for Section 2.3

1. In Example 1, can a mixture of cement, sand, and gravel in the proportions 1:3:4 be obtained from the three batches? If so, in what proportions should they be mixed?

2. A coffee dealer has 3 kinds of beans costing $3.00, $3.75, and $4.00 per pound. If the dealer wants 100 lb of a blend that costs $3.60 per lb, what amounts of the 3 beans can he use?

3. A gardener has 3 bags of fertilizer labeled 5-5-10, 2-10-5, and 10-5-5. (The 3 numbers on a bag indicate the percentages, by weight, of nitrogen, phosphate, and potash. Thus 5-5-10 fertilizer contains 5 percent nitrogen, 5 percent phosphate, and 10 percent potash.) The gardener really wants 5-10-10 fertilizer. Can the 3 fertilizers be mixed so that the 3 active ingredients are in the proportions 1:2:2? If so, how many lb of the mixture are equivalent to 1 lb of 5-10-10 fertilizer?

4. A company has on hand 60 lb of chemical A and 50 lb of chemical B that it wants to use up. Each unit of product 1 uses 3 lb of A and 2 lb of B, while each unit of product 2 uses 1 lb of A and 1 lb of B. Can the company use up all of A and B in making the two products, and if so, how many units of each should it make? (The number of units of each product need not be an integer.)

5. Proceed as in Exercise 4 if there are 40 lb of A, 80 lb of B, each unit of product 1 uses 2 lb of A and 3 lb of B, and each unit of product 2 uses 2 lb of A and 1 lb of B.

6. A factory uses 3 ingredients to produce 4 products. The number of lb of each ingredient needed to produce one unit of each product is shown in the table below. If 200 lb of each ingredient are available, how many units of each product should be produced if all ingredients are to be used up? Assume fractional units are permissible.

Ingredient	Product			
	1	2	3	4
1	1	1	2	3
2	0	1	4	0
3	2	7	0	0

7. Proceed as in Exercise 6, but assume that 200 lb of ingredients 1 and 2 are available, and 100 lb of ingredient 3.

Ingredient	Product			
	1	2	3	4
1	1	0	1	4
2	2	5	1	2
3	0	0	1	1

8. Find the loop currents I_1, I_2, and I_3 in the network of Figure 2.3.

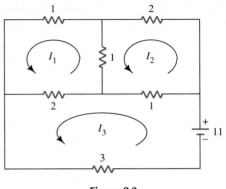

Figure 2.3

9. In Example 3, find the temperatures u_{11}, u_{12}, u_{21}, u_{22} if $u = 0$ on the sides $x = 0$, $x = L$, $y = 0$, and $u_{13} = 1$, $u_{23} = 2$.

10. A rectangular plate is twice as long as it is wide. The temperatures at the (equally spaced) points on the edges are as shown in Figure 2.4. Find the steady-state temperatures at the 3 interior points.

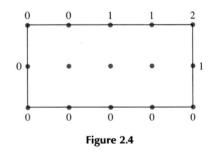

Figure 2.4

2.4 MATRICES AND VECTORS

The coefficients of a linear system

$$a_{11}x_1 + a_{12}x_2 + \cdots + a_{1n}x_n = b_1,$$

$$a_{21}x_1 + a_{22}x_2 + \cdots + a_{2n}x_n = b_2,$$

$$\cdots\cdots\cdots\cdots\cdots\cdots\cdots\cdots\cdots$$

$$a_{m1}x_1 + a_{m2}x_2 + \cdots + a_{mn}x_n = b_m,$$

(2.3)

can be displayed in a rectangular array of the form

$$
\begin{bmatrix}
a_{11} & a_{12} & \cdots & a_{1n} \\
a_{21} & a_{22} & \cdots & a_{2n} \\
\vdots & \vdots & & \vdots \\
a_{m1} & a_{m2} & \cdots & a_{mn}
\end{bmatrix}
\tag{2.4}
$$

with m rows and n columns. Such an array is called a *matrix of size $m \times n$*. To be more precise, a matrix of size $m \times n$ is a function of two variables whose domain is the set of ordered pairs of integers (i, j), where $1 \le i \le m$, $1 \le j \le n$. For example, in the above arrangement a_{23} is the value of the function that corresponds to the ordered pair $(2, 3)$. The numbers a_{ij} are called the *elements* of the matrix. A matrix with the same number of rows as columns is called a *square matrix*. A square $n \times n$ matrix is said to be a matrix of *order n*. With the exceptions noted later in this section, we shall denote matrices by capital letters A, B, C, and so on. Usually the elements of a matrix will be denoted by corresponding lowercase letters with subscripts. Thus b_{ij} will be the element in the ith row and jth column of a matrix B.

The matrix (2.4) is called the *coefficient matrix* of the linear system (2.3).

Two matrices A and B are said to be *equal*, written $A = B$, if they are of the same size and have equal corresponding elements. Thus two matrices A and B of size $m \times n$ are equal if $a_{ij} = b_{ij}$ for $1 \le i \le m$, $1 \le j \le n$.

If A is a matrix and c a number, we define the *product* $cA = Ac$ to be the matrix obtained from A by multiplying each element of A by C. Thus if

$$
A = \begin{bmatrix} 3 & -1 & 1 \\ 4 & 2 & -3 \end{bmatrix},
$$

then

$$
2A = \begin{bmatrix} 6 & -2 & 2 \\ 8 & 4 & -6 \end{bmatrix}, \qquad -A = (-1)A = \begin{bmatrix} -3 & 1 & -1 \\ -4 & -2 & 3 \end{bmatrix}.
$$

If A and B are matrices of the same size, we define the *sum* of A and B, written $A + B$, to be the matrix of the same size as A and B whose elements are the sums of the corresponding elements of A and B. Thus if A and B are $m \times n$ matrices and $C = A + B$, then C is an $m \times n$ matrix with elements $c_{ij} = a_{ij} + b_{ij}$. For example, if

$$
A = \begin{bmatrix} 1 & 2 & 3 \\ 4 & 5 & 6 \end{bmatrix}, \qquad B = \begin{bmatrix} 3 & -7 & 1 \\ 0 & 1 & -8 \end{bmatrix},
$$

then

$$
A + B = \begin{bmatrix} 4 & -5 & 4 \\ 4 & 6 & -2 \end{bmatrix}.
$$

Note that the sum of matrices of different sizes is not defined. We define *subtraction* of matrices of the same size by means of the formula

$$A - B = A + (-1)B.$$

We state a number of easily verifiable properties as a theorem. The proofs are left as an exercise at the end of this section.

Theorem 2.2   If A, B, and C are matrices of the same size and if a, b, and c are numbers, then

$$B + A = A + B,$$

$$(A + B) + C = A + (B + C),$$

$$a(bA) = (ab)A = (ba)A = b(aA),$$

$$(a + b)A = aA + bA,$$

$$c(A + B) = cA + cB.$$

A matrix with one column and n rows, such as

$$\begin{bmatrix} x_1 \\ x_2 \\ \vdots \\ x_n \end{bmatrix}$$

is also called an n-dimensional *column vector*. A matrix with one row and n columns, such as

$$\begin{bmatrix} y_1 & y_2 & \cdots & y_n \end{bmatrix}$$

is called an n-dimensional *row vector*. The elements of a row or column vector are sometimes called the *components* of the vector. We shall designate row and column vectors by lowercase letters in boldface type. Thus we might write

$$\mathbf{u} = \begin{bmatrix} u_1 \\ u_2 \\ u_3 \end{bmatrix}, \qquad \mathbf{v} = \begin{bmatrix} v_1 & v_2 & v_3 \end{bmatrix}.$$

Associated with the $m \times n$ matrix (2.4) are the n column vectors

$$\begin{bmatrix} a_{11} \\ a_{21} \\ \vdots \\ a_{m1} \end{bmatrix}, \begin{bmatrix} a_{12} \\ a_{22} \\ \vdots \\ a_{m2} \end{bmatrix}, \dots, \begin{bmatrix} a_{1n} \\ a_{2n} \\ \vdots \\ a_{mn} \end{bmatrix},$$

called the *column vectors of the matrix*. The m row vectors

$$[a_{11} \quad a_{12} \cdots a_{1n}], \quad [a_{21} \quad a_{22} \cdots a_{2n}], \ldots, \quad [a_{m1} \quad a_{m2} \cdots a_{mn}]$$

are called the *row vectors of the matrix*.

If $\mathbf{u}_1, \mathbf{u}_2, \ldots, \mathbf{u}_n$ are each m-dimensional column vectors, we write

$$[\mathbf{u}_1, \mathbf{u}_2, \ldots, \mathbf{u}_n]$$

to denote the $m \times n$ matrix whose column vectors are $\mathbf{u}_1, \mathbf{u}_2, \ldots, \mathbf{u}_n$. If $\mathbf{v}_1, \mathbf{v}_2, \ldots, \mathbf{v}_m$ are n-dimensional row vectors, we write

$$\begin{bmatrix} \mathbf{v}_1 \\ \mathbf{v}_2 \\ \vdots \\ \mathbf{v}_m \end{bmatrix}$$

for the $m \times n$ matrix whose row vectors are $\mathbf{v}_1, \mathbf{v}_2, \ldots, \mathbf{v}_m$. For example, if

$$\mathbf{u}_1 = \begin{bmatrix} 1 \\ 0 \end{bmatrix}, \qquad \mathbf{u}_2 = \begin{bmatrix} -2 \\ 3 \end{bmatrix}, \qquad \mathbf{u}_3 = \begin{bmatrix} 3 \\ 1 \end{bmatrix},$$

then

$$[\mathbf{u}_1, \mathbf{u}_2, \mathbf{u}_3] = \begin{bmatrix} 1 & -2 & 3 \\ 0 & 3 & 1 \end{bmatrix}.$$

We denote by R^n the set of all ordered n-tuples of real numbers. Thus R^1 is the set of all real numbers, and R^2 is the set of all ordered pairs (a, b) of real numbers a and b. A particular ordered n-tuple $(x_1, x_2, \ldots, x_n)$ is called a *vector* in R^n, or an *n-dimensional vector*. We also denote n-dimensional vectors by boldface symbols. If

$$\mathbf{x} = (x_1, x_2, \ldots, x_n), \qquad \mathbf{y} = (y_1, y_2, \ldots, y_n)$$

are n-dimensional vectors, we define the *product* of a real number c and the vector $\mathbf{x}$ to be

$$c\mathbf{x} = \mathbf{x}c = (cx_1, cx_2, \ldots, cx_n).$$

We also define the *sum*

$$\mathbf{x} + \mathbf{y} = (x_1 + y_1, x_2 + y_2, \ldots, x_n + y_n).$$

The particular vector $(0, 0, \ldots, 0)$ is called the n-dimensional *zero vector* and is denoted by $\mathbf{0}$.

Associated with a vector in R^n are a column vector and a row vector. We write

$$\begin{bmatrix} x_1 \\ x_2 \\ \vdots \\ x_n \end{bmatrix} \leftrightarrow (x_1, x_2, \ldots, x_n) \leftrightarrow \begin{bmatrix} x_1 & x_2 \cdots x_n \end{bmatrix}$$

to indicate the correspondence. This correspondence is preserved under addition and multiplication by a number. To see this for column vectors, we observe that

$$c \begin{bmatrix} x_1 \\ x_2 \\ \vdots \\ x_n \end{bmatrix} = \begin{bmatrix} cx_1 \\ cx_2 \\ \vdots \\ cx_n \end{bmatrix} \leftrightarrow (cx_1, cx_2, \ldots, cx_n) = c(x_1, x_2, \ldots, x_n)$$

and

$$\begin{bmatrix} x_1 \\ x_2 \\ \vdots \\ x_n \end{bmatrix} + \begin{bmatrix} y_1 \\ y_2 \\ \vdots \\ y_n \end{bmatrix} = \begin{bmatrix} x_1 + y_1 \\ x_2 + y_2 \\ \vdots \\ x_n + y_n \end{bmatrix} \leftrightarrow (x_1 + y_1, x_2 + y_2, \ldots, x_n + y_n)$$

$$= (x_1, x_2, \ldots, x_n) + (y_1, y_2, \ldots, y_n).$$

For convenience, we shall sometimes refer to the column (or row) vector $\mathbf{u} = (u_1, u_2, \ldots, u_n)$, when actually we mean the column (or row) vector associated with the n-tuple $(u_1, u_2, \ldots, u_n)$.

Exercises for Section 2.4

1. Find the coefficient matrix of the given linear system.

   (a) $2x_1 - x_2 = 4$
   $-x_1 + 3x_2 = 0$

   (b) $x_1 + 2x_2 - x_3 = 0$
   $3x_1 - 3x_2 \quad = 1$

   (c) $\quad x_2 + 2x_3 = 4$
   $x_1 + x_2 + x_3 = 0$
   $2x_1 \quad - x_3 = -1$

   (d) $\quad x_1 - x_2 = 0$
   $3x_1 + 2x_2 = 5$
   $-x_1 + x_2 = -1$
   $-2x_1 + 2x_2 = -2$

2. Find the linear *homogeneous* system of equations that has the given matrix as its coefficient matrix.

   (a) $\begin{bmatrix} 3 & -5 \\ -1 & 2 \end{bmatrix}$
   (b) $\begin{bmatrix} 2 & 0 & 1 \\ 0 & 1 & -1 \end{bmatrix}$

   (c) $\begin{bmatrix} 1 & 1 & 0 \\ 0 & 5 & -1 \\ 0 & 4 & 2 \end{bmatrix}$
   (d) $\begin{bmatrix} 4 & -2 \\ -2 & 1 \\ -6 & 3 \\ 2 & -1 \end{bmatrix}$

3. If the given matrix is denoted by A, find $3A$, $-A$, $-2A$, and $0A$.

   (a) $\begin{bmatrix} 2 & -5 \\ 1 & 0 \end{bmatrix}$
   (b) $\begin{bmatrix} -1 & 2 & 0 \\ 3 & 0 & 1 \end{bmatrix}$

4. Given the matrices A and B below, find (a) $A + B$, (b) $A - B$, (c) $2A - 3B$.

$$A = \begin{bmatrix} 1 & 0 & 2 \\ -3 & 1 & 4 \end{bmatrix},$$

$$B = \begin{bmatrix} -2 & 2 & 4 \\ 1 & 1 & 3 \end{bmatrix}$$

5. If A and B are as in Exercise 4, find a matrix X such that $2X + A = B$.

6. Find the sum of any two of the matrices below for which the sum is defined.

$$A = \begin{bmatrix} 2 & 0 \\ 1 & 4 \\ 1 & 1 \end{bmatrix}, \quad B = \begin{bmatrix} -2 & 1 & 0 \\ 3 & 0 & 0 \\ 0 & 0 & 0 \end{bmatrix}$$

$$C = \begin{bmatrix} 3 & 4 \\ 5 & 6 \end{bmatrix}, \quad D = \begin{bmatrix} 1 & 2 \\ 3 & 4 \end{bmatrix}$$

7. Are any two of the following matrices equal?

$$A = \begin{bmatrix} 0 & 0 \\ 0 & 0 \end{bmatrix}, \quad B = \begin{bmatrix} 0 & 0 \\ 0 & 0 \\ 0 & 0 \end{bmatrix}$$

$$C = \begin{bmatrix} 2 & 2 \\ 2 & 2 \end{bmatrix}, \quad D = \begin{bmatrix} 2 & 2 & 2 \\ 2 & 2 & 2 \end{bmatrix}$$

8. Find the column vectors of the given matrix.

(a) $\begin{bmatrix} 2 & 1 & -3 & 0 \\ 5 & 0 & 0 & 2 \\ 0 & 1 & 4 & 0 \end{bmatrix}$,

(b) $\begin{bmatrix} 6 & -1 \\ 3 & 0 \\ 2 & 2 \end{bmatrix}$

9. Find the row vectors for each of the matrices of Exercise 8.

10. Display the elements of the matrix whose *column vectors* are as indicated.
    (a) $\mathbf{u}_1 = (2, 3), \quad \mathbf{u}_2 = (-4, 1), \quad \mathbf{u}_3 = (3, 5)$
    (b) $\mathbf{u}_1 = (1, 2, 3), \quad \mathbf{u}_2 = (3, -1, 0),$ $\mathbf{u}_3 = (-1, 4, 5)$
    (c) $\mathbf{u}_1 = (1, 0, 1, 4), \quad \mathbf{u}_2 = (2, 1, 5, -1),$ $\mathbf{u}_3 = (1, 2, 3, 4)$

11. Display the elements of the matrix whose *row vectors* are as indicated.
    (a) $\mathbf{v}_1 = (-2, 3), \quad \mathbf{v}_2 = (1, 0), \quad \mathbf{v}_3 = (4, 5)$
    (b) $\mathbf{v}_1 = (1, 2, 3), \quad \mathbf{v}_2 = (0, 4, -1)$
    (c) $\mathbf{v}_1 = (2, 3, 0, -1), \quad \mathbf{v}_2 = (0, 1, 5, 0),$ $\mathbf{v}_3 = (4, 3, 2, 1)$

12. Prove Theorem 2.2.

2.5 MATRIX MULTIPLICATION

Let us again consider the system of equations

$$a_{11}x_1 + a_{12}x_2 + \cdots + a_{1n}x_n = b_1,$$
$$\cdots\cdots\cdots\cdots\cdots\cdots\cdots\cdots\cdots$$
$$a_{i1}x_1 + a_{i2}x_2 + \cdots + a_{in}x_n = b_i,$$
$$\cdots\cdots\cdots\cdots\cdots\cdots\cdots\cdots\cdots$$
$$a_{m1}x_1 + a_{m2}x_2 + \cdots + a_{mn}x_n = b_m.$$

We form the matrices

$$A = \begin{bmatrix} a_{11} & a_{12} & \cdots & a_{1n} \\ & & & \\ a_{i1} & a_{i2} & \cdots & a_{in} \\ & & & \\ a_{m1} & a_{m2} & \cdots & a_{mn} \end{bmatrix}, \quad \mathbf{x} = \begin{bmatrix} x_1 \\ x_2 \\ \vdots \\ x_n \end{bmatrix}, \quad \mathbf{b} = \begin{bmatrix} b_1 \\ b_2 \\ \vdots \\ b_m \end{bmatrix}.$$

Notice that the left-hand member of the ith equation of the system is formed by taking the products of the n elements in the ith row of A with the corresponding components of the column vector $\mathbf{x}$ and summing. This sum can be written as

$$\sum_{k=1}^{n} a_{ik} x_k.$$

The number of elements in each row of A is the same as the number of components of $\mathbf{x}$. We now define the *product* $A\mathbf{x}$ of an $m \times n$ matrix A and an n-dimensional column vector $\mathbf{x}$ to be the m-dimensional column vector with components

$$\sum_{k=1}^{n} a_{ik} x_k, \qquad 1 \le i \le m.$$

Then the system can be written as

$$A\mathbf{x} = \mathbf{b}.$$

This equation asserts the equality of two m-dimensional column vectors. It is satisfied if and only if each equation of the system is satisfied.

As an example, let

$$A = \begin{bmatrix} 2 & -1 & 3 \\ 1 & 4 & -2 \end{bmatrix}, \qquad \mathbf{x} = \begin{bmatrix} x_1 \\ x_2 \\ x_3 \end{bmatrix}, \qquad \mathbf{b} = \begin{bmatrix} 0 \\ 8 \end{bmatrix}.$$

Since the number of elements in each row of A is the same as the number of components of $\mathbf{x}$, the product $A\mathbf{x}$ is defined and

$$A\mathbf{x} = \begin{bmatrix} 2 & -1 & 3 \\ 1 & 4 & -2 \end{bmatrix} \begin{bmatrix} x_1 \\ x_2 \\ x_3 \end{bmatrix} = \begin{bmatrix} 2x_1 - x_2 + 3x_3 \\ x_1 + 4x_2 - 2x_3 \end{bmatrix}.$$

The vector equation $A\mathbf{x} = \mathbf{b}$ corresponds to the system

$$2x_1 - x_2 + 3x_3 = 0,$$
$$x_1 + 4x_2 - 2x_3 = 8.$$

The particular column vector $\mathbf{c} = (2, 1, -1)$ is a *solution* of the equation $A\mathbf{x} = \mathbf{b}$ because

$$A\mathbf{c} = \begin{bmatrix} 2 & -1 & 3 \\ 1 & 4 & -2 \end{bmatrix} \begin{bmatrix} 2 \\ 1 \\ -1 \end{bmatrix} = \begin{bmatrix} 4 - 1 - 3 \\ 2 + 4 + 2 \end{bmatrix} = \begin{bmatrix} 0 \\ 8 \end{bmatrix} = \mathbf{b}.$$

Notice that if

$$B = \begin{bmatrix} 3 & -1 & 1 \\ 0 & 2 & 4 \end{bmatrix}, \qquad \mathbf{u} = \begin{bmatrix} 2 \\ 3 \end{bmatrix},$$

the product *Bu* is *not* defined because the number of elements in a row of *B* is not the same as the number of components of **u**.

Having defined the product of a matrix and a column vector, we now attempt a more general definition of the product of two matrices. Let *A* be an $m \times n$ matrix and let *B* be an $n \times p$ matrix. Let us write

$$B = [\mathbf{b}_1, \mathbf{b}_2, \ldots, \mathbf{b}_p],$$

where each column vector $\mathbf{b}_j$ of *B* has *n* components. Since each row of *A* has *n* elements (*A* has *n* columns) each of the products

$$A\mathbf{b}_1, A\mathbf{b}_2, \ldots, A\mathbf{b}_p$$

is defined and is an *m*-dimensional column vector. We now define the *product AB* to be the $m \times p$ matrix with these column vectors; that is,

$$AB = [A\mathbf{b}_1, A\mathbf{b}_2, \ldots, A\mathbf{b}_p].$$

Let $C = AB$. Then *C* is an $m \times p$ matrix. The element c_{ij} in the *i*th row and *j*th column of *C* is the *i*th component of the vector $A\mathbf{b}_j$. This *i*th component is formed by multiplying the elements in the *i*th row of *A* by the corresponding elements in the *j*th column of *B*. Thus

$$c_{ij} = \sum_{k=1}^{n} a_{ik}b_{kj}, \qquad 1 \le i \le m, \qquad 1 \le j \le p. \tag{2.5}$$

Thus if *A* is an $m \times n$ matrix and *B* is an $n \times p$ matrix (the number of columns of *A* must be the same as the number of rows of *B*) the product *AB* is an $m \times p$ matrix *C* whose elements are given by formula (2.5). To illustrate, let

$$A = \begin{bmatrix} 1 & -1 \\ 3 & 0 \\ 2 & 4 \end{bmatrix}, \qquad B = \begin{bmatrix} 2 & 1 \\ -3 & -2 \end{bmatrix}.$$

Since *A* is of size 3×2 and *B* is of size 2×2, the product *AB* is defined and is a 3×2 matrix. We have

$$AB = \begin{bmatrix} (1)(2) + (-1)(-3) & (1)(1) + (-1)(-2) \\ (3)(2) + (0)(-3) & (3)(1) + (0)(-2) \\ (2)(2) + (4)(-3) & (2)(1) + (4)(-2) \end{bmatrix} = \begin{bmatrix} 5 & 3 \\ 6 & 3 \\ -8 & -6 \end{bmatrix}.$$

Notice that the product BA is not defined in this example, since the number of columns of B is not the same as the number of rows of A. Even when AB and BA are both defined, the products are not necessarily the same. For instance, if

$$A = \begin{bmatrix} 1 & 2 \\ 3 & 1 \end{bmatrix}, \qquad B = \begin{bmatrix} 1 & -1 \\ 2 & 0 \end{bmatrix},$$

the reader can verify that

$$AB = \begin{bmatrix} 5 & -1 \\ 5 & -3 \end{bmatrix}, \qquad BA = \begin{bmatrix} -2 & 1 \\ 2 & 4 \end{bmatrix},$$

so that $AB \neq BA$. Thus matrix multiplication is not in general a commutative operation. If it happens that $AB = BA$, we say that A and B *commute*.

Some properties of matrices that involve multiplication are given in the following theorem.

Theorem 2.3 If c is a number and if A, B, C, and D are matrices such that the indicated sums and products are defined, then

$$(cA)B = A(cB) = c(AB)$$
$$A(B + C) = AB + AC$$
$$(B + C)D = BD + CD$$
$$A(BD) = (AB)D.$$

We shall prove the second and fourth properties only, leaving the proofs of the remaining properties as exercises. First, consider the second property. Let the size of A be $m \times n$ and let that of B and C be $n \times p$. Then AB, AC, and $A(B + C)$ are defined and have size $m \times p$. Let $E = A(B + C)$, $F = AB$, and $G = AC$. We want to show that $E = F + G$ or that $e_{ij} = f_{ij} + g_{ij}$ for all i and j. We have

$$e_{ij} = \sum_{k=1}^{n} a_{ik}(b_{kj} + c_{kj})$$

$$= \sum_{k=1}^{n} a_{ik}b_{kj} + \sum_{k=1}^{n} a_{ik}c_{kj}$$

$$= f_{ij} + g_{ij},$$

as we wished to prove.

In order to prove the fourth property, we need a preliminary result. Let S be an $m \times n$ matrix,

$$S = \begin{bmatrix} s_{11} & s_{12} & \cdots & s_{1n} \\ s_{21} & s_{22} & \cdots & s_{2n} \\ \vdots & \vdots & & \vdots \\ s_{m1} & s_{m2} & \cdots & s_{mn} \end{bmatrix},$$

and let s denote the sum of all the elements of S. The sum of the elements in the ith row of S is

$$\sum_{j=1}^{n} s_{ij}, \qquad 1 \le i \le m.$$

Consequently

$$s = \sum_{i=1}^{m} \sum_{j=1}^{n} s_{ij}.$$

But the sum of the elements in the jth column of S is

$$\sum_{i=1}^{m} s_{ij}, \qquad 1 \le j \le n.$$

Hence we also have

$$s = \sum_{j=1}^{n} \sum_{i=1}^{m} s_{ij}.$$

Comparing the formulas for s, we see that

$$\sum_{i=1}^{m} \sum_{j=1}^{n} s_{ij} = \sum_{j=1}^{n} \sum_{i=1}^{m} s_{ij}. \qquad (2.6)$$

This relation says that the order of summation in a repeated sum can be reversed without changing its value.

We now turn to a proof of the fourth property. Let the size of A be $m \times n$, that of B be $n \times p$, and that of D be $p \times q$. Then BD is of size $n \times q$ and AB is of size $m \times p$. Both $A(BD)$ and $(AB)D$ are of size $m \times q$. Let $E = BD$, $F = AB$, $G = AE$, and $H = FD$. Now

$$g_{ij} = \sum_{k=1}^{n} a_{ik} e_{kj} = \sum_{k=1}^{n} a_{ik} \sum_{r=1}^{p} b_{kr} d_{rj}.$$

Since a_{ik} does not depend on r, we may write

$$g_{ij} = \sum_{k=1}^{n} \sum_{r=1}^{p} a_{ik} b_{kr} d_{rj}.$$

Next,

$$h_{ij} = \sum_{r=1}^{p} f_{ir}d_{rj} = \sum_{r=1}^{p}\left(\sum_{k=1}^{n} a_{ik}b_{kr}\right)d_{rj} = \sum_{r=1}^{p}\sum_{k=1}^{n} a_{ik}b_{kr}d_{rj}.$$

But, according to the relation (2.6), this is the same as

$$\sum_{k=1}^{n}\sum_{r=1}^{p} a_{ik}b_{kr}d_{rj},$$

which is g_{ij}. Thus $G = H$ or $A(BD) = (AB)D$.

One additional feature of matrix multiplication should be noted.

Theorem 2.4   Let

$$A = [\mathbf{a}_1, \mathbf{a}_2, \ldots, \mathbf{a}_n]$$

be an $m \times n$ matrix and let

$$\mathbf{c} = \begin{bmatrix} c_1 \\ c_2 \\ \vdots \\ c_n \end{bmatrix}$$

be an n-dimensional column vector. Then the matrix product $A\mathbf{c}$ of the $m \times n$ matrix A and the $n \times 1$ matrix $\mathbf{c}$ is defined and

$$A\mathbf{c} = c_1\mathbf{a}_1 + c_2\mathbf{a}_2 + \cdots + c_n\mathbf{a}_n.$$

We leave the verification to the reader.

--- Exercises for Section 2.5 ---

In Exercises 1–10, find all the products AB and BA that are defined.

1. $A = \begin{bmatrix} 4 & 2 \\ 1 & 3 \end{bmatrix}$, $B = \begin{bmatrix} 1 & -1 \\ -2 & 2 \end{bmatrix}$

2. $A = \begin{bmatrix} -2 & 5 \\ 1 & 4 \end{bmatrix}$, $B = \begin{bmatrix} 3 & 2 \\ -3 & 4 \end{bmatrix}$

3. $A = \begin{bmatrix} 1 & 2 & 0 \\ 2 & -1 & 3 \\ 0 & 1 & 2 \end{bmatrix}$, $B = \begin{bmatrix} 3 & -3 & 2 \\ 2 & 1 & -1 \\ -3 & 0 & 0 \end{bmatrix}$

4. $A = \begin{bmatrix} 1 & 4 & -3 \\ 2 & 0 & 1 \\ -3 & -2 & 0 \end{bmatrix}$, $B = \begin{bmatrix} 2 & 3 & 4 \\ -3 & -2 & 3 \\ 0 & 1 & 0 \end{bmatrix}$

5. $A = \begin{bmatrix} 1 & 2 \\ -1 & 3 \end{bmatrix}$, $B = \begin{bmatrix} 3 & 2 & 1 \\ 2 & 1 & 4 \end{bmatrix}$

6. $A = \begin{bmatrix} 2 & 1 & 0 \\ -1 & -2 & 2 \end{bmatrix}$, $B = \begin{bmatrix} 2 & 4 \\ 1 & -1 \\ 3 & 1 \end{bmatrix}$

7. $A = \begin{bmatrix} -4 & 3 \\ 2 & 0 \end{bmatrix}$, $\quad B = \begin{bmatrix} 1 & 2 \\ -1 & 1 \\ 0 & 0 \end{bmatrix}$

8. $A = \begin{bmatrix} 2 & 3 & -1 \end{bmatrix}$, $\quad B = \begin{bmatrix} 2 & 0 \\ 1 & 4 \\ -2 & 1 \end{bmatrix}$

9. $A = \begin{bmatrix} 4 & 2 & 3 \end{bmatrix}$, $\quad B = \begin{bmatrix} 2 & 1 & -5 \end{bmatrix}$

10. $A = \begin{bmatrix} 6 & -2 & 1 \end{bmatrix}$, $\quad B = \begin{bmatrix} 1 \\ 3 \\ 2 \end{bmatrix}$

11. Find a matrix A and column vectors $\mathbf{x}$ and $\mathbf{b}$ such that the vector equation $A\mathbf{x} = \mathbf{b}$ corresponds to the given system.

(a) $3x_1 + x_2 = 7$
$\quad 4x_1 - 2x_2 = -3$

(b) $2x_1 + 5x_2 = -1$
$\quad 3x_1 - 2x_2 = 0$
$\quad x_1 + x_2 = 3$

(c) $2x_1 - x_2 + x_3 = 4$
$\quad -x_1 + x_2 + 5x_3 = -2$
$\quad 2x_1 + x_2 \qquad = 3$

(d) $3x_1 + x_2 - x_3 = 2$
$\quad 2x_1 \qquad + x_3 = -4$

12. Find the system of equations that corresponds to the vector equation $A\mathbf{x} = \mathbf{b}$, where A and $\mathbf{b}$ are as given.

(a) $A = \begin{bmatrix} 2 & -3 \\ 1 & 4 \end{bmatrix}$, $\quad \mathbf{b} = \begin{bmatrix} 2 \\ -5 \end{bmatrix}$

(b) $A = \begin{bmatrix} -2 & 1 & 3 \\ 3 & 0 & 1 \end{bmatrix}$, $\quad \mathbf{b} = \begin{bmatrix} 0 \\ 0 \end{bmatrix}$

(c) $A = \begin{bmatrix} 2 & -1 & 3 \\ 0 & 3 & -2 \\ 1 & 0 & 4 \end{bmatrix}$, $\quad \mathbf{b} = \begin{bmatrix} 1 \\ -1 \\ 0 \end{bmatrix}$

13. Show that the matrices commute.

(a) $\begin{bmatrix} 1 & 2 \\ -1 & 0 \end{bmatrix}$, $\begin{bmatrix} -1 & 2 \\ -1 & -2 \end{bmatrix}$

(b) $\begin{bmatrix} 1 & 0 & 2 \\ 0 & 1 & 1 \\ 3 & 0 & -1 \end{bmatrix}$, $\begin{bmatrix} 7 & 0 & 0 \\ 3 & 1 & 0 \\ 0 & 0 & 7 \end{bmatrix}$

14. Show that for any square matrix A, A and A^2 commute. (Here $A^2 = A \cdot A$. Use Theorem 2.3.)

15. Prove the first and third properties of Theorem 2.3.

16. (a) If $\mathbf{u}$ and $\mathbf{v}$ are solutions of the equation $A\mathbf{x} = \mathbf{0}$, show that $c\mathbf{u}$ (where c is any number) and $\mathbf{u} + \mathbf{v}$ are also solutions.

(b) If $\mathbf{y}$ is a solution of the equation $A\mathbf{x} = \mathbf{b}$, and $\mathbf{u}$ is a solution of the equation $A\mathbf{x} = \mathbf{0}$, verify that $c\mathbf{u} + \mathbf{y}$ is a solution of the equation $A\mathbf{x} = \mathbf{b}$.

17. Let A be an $m \times n$ matrix. Let $\mathbf{x}_1, \mathbf{x}_2, \ldots,$ $\mathbf{x}_k$ be n-dimensional column vectors, and let $X = [\mathbf{x}_1, \mathbf{x}_2, \ldots, \mathbf{x}_k]$. Show that $AX = 0$ if and only if each of the vectors $\mathbf{x}_1, \mathbf{x}_2, \ldots, \mathbf{x}_k$ is a solution of the equation $A\mathbf{x} = \mathbf{0}$.

18. Prove Theorem 2.4.

19. If A and B are square matrices of the same size, show that in general $(A + B)^2 \neq A^2 + 2AB + B^2$. Show, however, that equality always holds if A and B commute.

20. A vector $\mathbf{v}$ is said to be a *linear combination* of the vectors $\mathbf{v}_1, \mathbf{v}_2, \ldots, \mathbf{v}_m$ if there are numbers $c_1, c_2, \ldots, c_m$ such that $\mathbf{v} = c_1\mathbf{v}_1 + c_2\mathbf{v}_2 + \cdots + c_m\mathbf{v}_m$. Use Theorem 2.4 to show that the system of equations $A\mathbf{x} = \mathbf{b}$ has a solution if, and only if, $\mathbf{b}$ is a linear combination of the column vectors of the matrix A.

2.6
INNER PRODUCT
AND LENGTH

If we think of the numbers x_1, x_2, and x_3 as the rectangular coordinates of a point in space, then there is associated with each point a vector (x_1, x_2, x_3) in R^3. A *geometric vector* is something characterized by length and direction. It can be represented by an arrow (directed line segment) with tail at the origin and tip at a point P. Thus to each geometric vector there corresponds a point P in space, and hence an element of R^3. The length of the geometric vector represented by the arrow from the origin O to the point P is denoted by $|OP|$. Thus

$$|OP| = (x_1^2 + x_2^2 + x_3^2)^{1/2}.$$

Let us now consider two elements of R^3, (x_1, x_2, x_3) and (y_1, y_2, y_3), and their corresponding points P and Q. These points and their corresponding geometric vectors are shown in Figure 2.5. By the law of cosines,

$$|PQ|^2 = |OP|^2 + |OQ|^2 - 2|OP||OQ| \cos \theta,$$

where θ is the angle between the vectors. But

$$|OP|^2 = x_1^2 + x_2^2 + x_3^2, \qquad |OQ|^2 = y_1^2 + y_2^2 + y_3^2,$$
$$|PQ|^2 = (y_1 - x_1)^2 + (y_2 - x_2)^2 + (y_3 - x_3)^2,$$

and a little algebra shows that

$$x_1 y_1 + x_2 y_2 + x_3 y_3 = |OP||OQ| \cos \theta.$$

The quantity $x_1 y_1 + x_2 y_2 + x_3 y_3$ evidently depends on the lengths of the two vectors and the angle between them. It is called the inner product (or scalar product) of the vectors.

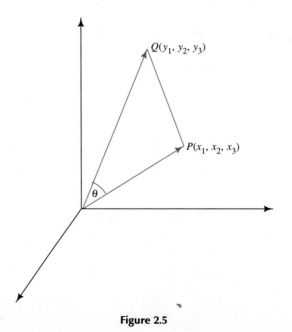

Figure 2.5

We now consider vectors in R^n. If $\mathbf{x} = (x_1, x_2, \ldots, x_n)$ and $\mathbf{y} = (y_1, y_2, \ldots, y_n)$, we define the *inner product* (or *scalar product*) $\mathbf{x} \cdot \mathbf{y}$ of $\mathbf{x}$ and $\mathbf{y}$ to be

$$\mathbf{x} \cdot \mathbf{y} = x_1 y_1 + x_2 y_2 + \cdots + x_n y_n.$$

The *length* $\|\mathbf{x}\|$ of $\mathbf{x}$ is defined as

$$\|\mathbf{x}\| = (\mathbf{x} \cdot \mathbf{x})^{1/2} = (x_1^2 + x_2^2 + \cdots + x_n^2)^{1/2}.$$

The *angle*[5] θ between $\mathbf{x}$ and $\mathbf{y}$ is defined to be the angle in the interval $[0, \pi]$ for which

$$\cos \theta = \frac{\mathbf{x} \cdot \mathbf{y}}{\|\mathbf{x}\| \, \|\mathbf{y}\|}.$$

When n is 2 or 3, these quantities have the geometric interpretations described at the beginning of this section. For $n > 3$ there are no such geometrical interpretations. Nevertheless, the above definitions are useful. We leave it to the reader to verify the following properties of inner products in R^n. (In the equations, c stands for a real number.)

Property 1 $\mathbf{x} \cdot \mathbf{x} \geq 0.$

Property 2 $\mathbf{y} \cdot \mathbf{x} = \mathbf{x} \cdot \mathbf{y}.$

Property 3 $(c\mathbf{x}) \cdot \mathbf{y} = c(\mathbf{x} \cdot \mathbf{y}) = \mathbf{x} \cdot (c\mathbf{y}).$

Property 4 $(\mathbf{x} + \mathbf{y}) \cdot \mathbf{z} = \mathbf{x} \cdot \mathbf{z} + \mathbf{y} \cdot \mathbf{z}.$

In connection with Property 1, it should be noted that $\mathbf{x} \cdot \mathbf{x} = 0$ if and only if $\mathbf{x} = \mathbf{0}$.

Let us denote the set of all ordered n-tuples of *complex* numbers by C^n. If

$$\mathbf{u} = (u_1, u_2, \ldots, u_n), \qquad \mathbf{v} = (v_1, v_2, \ldots, v_n)$$

are elements of C^n, we define

$$c\mathbf{u} = (cu_1, cu_2, \ldots, cu_n)$$

for every complex number c and

$$\mathbf{u} + \mathbf{v} = (u_1 + v_1, u_2 + v_2, \ldots, u_n + v_n).$$

---

[5] It can be shown that the quantity here is in the interval $[0, 1]$ for all n. See Exercise 9.

We define the *inner product* $\mathbf{u} \cdot \mathbf{v}$ of $\mathbf{u}$ and $\mathbf{v}$ as

$$\mathbf{u} \cdot \mathbf{v} = u_1 \bar{v}_1 + u_2 \bar{v}_2 + \cdots + u_n \bar{v}_n,$$

where the bar indicates the complex conjugate. The *length* $\|\mathbf{u}\|$ of $\mathbf{u}$ is defined to be

$$\|\mathbf{u}\| = (\mathbf{u} \cdot \mathbf{u})^{1/2} = \left(|u_1|^2 + |u_2|^2 + \cdots + |u_n|^2 \right)^{1/2}.$$

Note that $\|\mathbf{u}\|$ is a nonnegative real number. The reader may verify the following properties (where c stands for a complex number):

Property 1 $\mathbf{u} \cdot \mathbf{u} \geq 0$.

Property 2 $\mathbf{v} \cdot \mathbf{u} = \overline{\mathbf{u} \cdot \mathbf{v}}$.

Property 3 $(c\mathbf{u}) \cdot \mathbf{v} = c(\mathbf{u} \cdot \mathbf{v}) = \mathbf{u} \cdot (\bar{c}\mathbf{v})$.

Property 4 $(\mathbf{u} + \mathbf{v}) \cdot \mathbf{w} = \mathbf{u} \cdot \mathbf{w} + \mathbf{v} \cdot \mathbf{w}$.

Two vectors $\mathbf{x}$ and $\mathbf{y}$ (in R^n or in C^n) are said to be *orthogonal* if

$$\mathbf{x} \cdot \mathbf{y} = 0.$$

A set of nonzero vectors $\mathbf{x}_1, \mathbf{x}_2, \ldots, \mathbf{x}_m$ is called an *orthogonal set* of vectors if $\mathbf{x}_i \cdot \mathbf{x}_j = 0$, whenever $i \neq j$. For example, the vectors

$$\mathbf{x}_1 = (1, -1, 2, 0), \qquad \mathbf{x}_2 = (3, 1, -1, 0),$$
$$\mathbf{x}_3 = (0, 0, 0, 1), \qquad \mathbf{x}_4 = (-1, 7, 4, 0)$$

constitute an orthogonal set in R^4.

A vector $\mathbf{x}$ such that $\|\mathbf{x}\| = 1$ is called a *unit vector*. An orthogonal set of unit vectors is called an *orthonormal set* of vectors. If $\mathbf{u}$ is any nonzero vector, then $\mathbf{v} = \mathbf{u}/\|\mathbf{u}\|$ is a unit vector. To see this, we observe that

$$\|\mathbf{v}\|^2 = \mathbf{v} \cdot \mathbf{v} = \left(\frac{1}{\|\mathbf{u}\|} \mathbf{u} \right) \cdot \left(\frac{1}{\|\mathbf{u}\|} \mathbf{u} \right) = 1.$$

Thus $\mathbf{v}$ has unit length. If

$$\mathbf{y}_1 = \frac{1}{\sqrt{6}} \mathbf{x}_1, \qquad \mathbf{y}_2 = \frac{1}{\sqrt{11}} \mathbf{x}_2, \qquad \mathbf{y}_3 = \mathbf{x}_3, \qquad \mathbf{y}_4 = \frac{1}{\sqrt{66}} \mathbf{x}_4,$$

where the $\mathbf{x}_i$ are as in the preceding example, then $\mathbf{y}_1, \mathbf{y}_2, \mathbf{y}_3, \mathbf{y}_4$ form an orthonormal set.

––––––––––––––––––– **Exercises for Section 2.6** –––––––––––––––––

1. If $\mathbf{x}$ and $\mathbf{y}$ are as given, find $\mathbf{x} \cdot \mathbf{y}$, $\|\mathbf{x}\|$, and $\|\mathbf{y}\|$.
   (a) $\mathbf{x} = (1, -2, 0)$, $\mathbf{y} = (2, 5, -4)$
   (b) $\mathbf{x} = (1, 1, -2, 0)$, $\mathbf{y} = (0, 2, -2, 1)$
   (c) $\mathbf{x} = (2, 0, 1, -3)$, $\mathbf{y} = (0, 0, -1, 5)$
   (d) $\mathbf{x} = (-1, 1, 2, 3, 1)$, $\mathbf{y} = (2, 1, 1, -1, 2)$

2. Find the angle between the given vectors.
   (a) $\mathbf{x} = (1, -2, 0)$, $\mathbf{y} = (2, 5, -4)$
   (b) $\mathbf{x} = (3, 6, 2)$, $\mathbf{y} = (-1, 2, 2)$

3. If $\mathbf{u}$ and $\mathbf{v}$ are as given, find $\mathbf{u} \cdot \mathbf{v}$, $\mathbf{v} \cdot \mathbf{u}$, $\|\mathbf{u}\|$, and $\|\mathbf{v}\|$.
   (a) $\mathbf{u} = (2 - i, 1)$, $\mathbf{v} = (-i, -2i)$
   (b) $\mathbf{u} = (1, i, 2 + i)$, $\mathbf{v} = (1 - 3i, 0, 1 + i)$
   (c) $\mathbf{u} = (1, i, -i)$, $\mathbf{v} = (3, -i, 2i)$
   (d) $\mathbf{u} = (1, -i, 1 + i, 1)$, $\mathbf{v} = (-i, 1, -i, 2 - i)$

4. Verify that the given set of vectors is orthogonal. Is it also orthonormal?
   (a) $\mathbf{x} = \left(\dfrac{3}{7}, \dfrac{6}{7}, \dfrac{2}{7}\right)$, $\mathbf{y} = \left(\dfrac{6}{7}, -\dfrac{2}{7}, -\dfrac{3}{7}\right)$,
        $\mathbf{z} = \left(\dfrac{2}{7}, -\dfrac{3}{7}, \dfrac{6}{7}\right)$
   (b) $\mathbf{x} = \left(\dfrac{2}{3}, \dfrac{2}{3}, -\dfrac{1}{3}\right)$, $\mathbf{y} = \left(-\dfrac{1}{3}, \dfrac{2}{3}, \dfrac{2}{3}\right)$,
        $\mathbf{z} = \left(\dfrac{2}{3}, -\dfrac{1}{3}, \dfrac{2}{3}\right)$
   (c) $\mathbf{x} = (3, 6, 2)$, $\mathbf{y} = (6, -2, -3)$,
        $\mathbf{z} = (2, -3, 6)$
   (d) $\mathbf{x} = (2, 1, -1, 3)$, $\mathbf{y} = (-1, 3, 1, 0)$,
        $\mathbf{z} = (5, 1, 2, -3)$

5. Verify the four properties of the inner product for R^n.

6. Verify the four properties of the inner product for C^n.

7. If $\mathbf{x}$ and $\mathbf{y}$ are in R^n, verify that
$$\|\mathbf{x} + \mathbf{y}\|^2 = \|\mathbf{x}\|^2 + 2\mathbf{x} \cdot \mathbf{y} + \|\mathbf{y}\|^2.$$

8. If $\mathbf{u}$ and $\mathbf{v}$ are in C^n, verify that
$$\|\mathbf{u} + \mathbf{v}\|^2 = \|\mathbf{u}\|^2 + 2\,\mathrm{Re}(\mathbf{u} \cdot \mathbf{v}) + \|\mathbf{v}\|^2.$$
Here Re denotes the real part.

9. Let $\mathbf{x}$ and $\mathbf{y}$ be elements of R^n, with $\mathbf{x} \neq \mathbf{0}$. Let $P(\lambda) = \|\lambda \mathbf{x} + \mathbf{y}\|^2$.
   (a) Verify that P is a second degree polynomial.
   (b) Show that $P(\lambda) \geq 0$ for all λ, and then deduce from this fact the *Schwarz inequality*
$$|\mathbf{x} \cdot \mathbf{y}| \leq \|\mathbf{x}\| \,\|\mathbf{y}\|.$$

10. Use the results of Exercises 7 and 9 to show that
$$\|\mathbf{x} + \mathbf{y}\| \leq \|\mathbf{x}\| + \|\mathbf{y}\|.$$

11. Find all unit vectors orthogonal to all the given vectors.
    (a) $(1, -2, 0)$, $(2, 1, -1)$
    (b) $(2, -3)$
    (c) $(1, 0, 1, 0)$, $(0, 2, -1, 0)$, $(-1, 1, 0, 2)$
    (d) $(2, 1, 1)$, $(-1, 0, 3)$

2.7
SOME SPECIAL
MATRICES

The $m \times n$ matrix whose elements are all zero is called the *zero matrix* of size $m \times n$. We denote it by $0_{m \times n}$, or simply by 0 when the size is clear. If A is any $m \times n$ matrix, then

$$A + 0_{m \times n} = A$$

and

$$A + (-1)A = 0_{m \times n}.$$

We also see that

$$A0_{n \times p} = 0_{m \times p}$$

and

$$0_{q \times m} A = 0_{q \times n}.$$

Let us define the useful symbol δ_{ij}, where i and j are positive integers, by means of the formula

$$\delta_{ij} = \begin{cases} 0 & \text{if } i \neq j, \\ 1 & \text{if } i = j. \end{cases}$$

The symbol δ_{ij} is called the *Kronecker delta*. We have, for example, $\delta_{11} = \delta_{22} = \delta_{33} = 1$ and $\delta_{12} = \delta_{13} = \delta_{23} = 0$. The matrix of size $n \times n$ (same number of rows as columns) whose element in the ith row and jth column is δ_{ij}, $1 \leq i \leq n$, $1 \leq j \leq n$, is called the $n \times n$ *identity matrix*. We denote it by I_n, or simply by I when the size is evident. Thus

$$I_3 = \begin{bmatrix} 1 & 0 & 0 \\ 0 & 1 & 0 \\ 0 & 0 & 1 \end{bmatrix}.$$

The identity matrix has the following important properties.

Theorem 2.5   If A is any $m \times n$ matrix, then

$$AI_n = A$$

and

$$I_m A = A.$$

Proof   We shall prove the first of these properties. The establishment of the second is left as an exercise. Let $B = AI_n$. Then B is an $m \times n$ matrix with elements

$$b_{ij} = \sum_{k=1}^{n} a_{ik} \delta_{kj}.$$

Each term in the sum is zero except the one where k is equal to j. Hence

$$b_{ij} = a_{ij} \delta_{jj} = a_{ij}$$

so $B = A$ or $AI_n = A$.

If A is a square $n \times n$ matrix, the elements $a_{11}, a_{22}, \ldots, a_{nn}$ are called the *diagonal* elements of A. If all the elements of a square matrix D that are not diagonal elements are zero, then D is called a *diagonal matrix*. We write $D = \text{diag}(d_1, d_2, \ldots, d_n)$ to indicate the diagonal matrix D with diagonal elements $d_1, d_2, \ldots, d_n$. For example, if $D = \operatorname{diag}(2, -1, 0)$ then

$$D = \begin{bmatrix} 2 & 0 & 0 \\ 0 & -1 & 0 \\ 0 & 0 & 0 \end{bmatrix}.$$

If $D = \operatorname{diag}(d_1, d_2, \ldots, d_n)$ then the element in the ith row and jth column of D is $d_{ij} = d_i \delta_{ij} = d_j \delta_{ij}$. A diagonal matrix with identical diagonal elements is called a *scalar matrix*. Thus a 3×3 scalar matrix has the form

$$\begin{bmatrix} d & 0 & 0 \\ 0 & d & 0 \\ 0 & 0 & d \end{bmatrix}.$$

The $n \times n$ identity matrix and the $n \times n$ zero matrix are examples of scalar matrices. Note that each $n \times n$ scalar matrix is of the form cI, where I is the $n \times n$ identity matrix.

Theorem 2.6   Let A be an $m \times n$ matrix, let $C = \operatorname{diag}(c_1, c_2, \ldots, c_m)$ and let $D = \operatorname{diag}(d_1, d_2, \ldots, d_n)$. Then CA is the matrix obtained from A by multiplying every element in the ith row of A by c_i, and AD is the matrix obtained from A by multiplying every element in the jth column of A by d_j.

Proof   We shall prove the part for CA, leaving the proof of the remaining part as an exercise. Let $E = CA$. Then E is an $m \times n$ matrix, and

$$e_{ij} = \sum_{k=1}^{n} c_{ik} a_{kj} = \sum_{k=1}^{n} c_i \delta_{ik} a_{kj} = c_i a_{ij}.$$

Thus every element e_{ij} in the ith row of E, $1 \leq j \leq n$, is formed by multiplying a_{ij} by c_i. The proof of the following corollary is left as an exercise.

Corollary   Let A be an $m \times n$ matrix. Let C be an $m \times m$ scalar matrix with diagonal elements equal to c, and let D be an $n \times n$ scalar matrix with diagonal elements equal to d. Then

$$CA = cA$$
$$AD = dA.$$

If A is a square matrix, it is said to be *upper triangular* if $a_{ij} = 0$ whenever $i > j$. It is said to be *lower triangular* if $a_{ij} = 0$ whenever $i < j$. For example, the first of the matrices

$$\begin{bmatrix} 2 & 1 & -6 \\ 0 & 0 & 2 \\ 0 & 0 & 5 \end{bmatrix}, \qquad \begin{bmatrix} 4 & 0 & 0 \\ 0 & 1 & 0 \\ 2 & 3 & 2 \end{bmatrix}$$

is upper triangular, and the second is lower triangular.

If A is an $m \times n$ matrix, we define the *transpose* of A, written A^T, to be the $n \times m$ matrix whose elements a_{ij}^T are given by the relation

$$a_{ij}^T = a_{ji}, \qquad 1 \le i \le n, \qquad 1 \le j \le m.$$

The elements in the ith row of A^T are the corresponding elements in the ith column of A. The elements in the jth column of A^T are the corresponding elements in the jth row of A. For example, if

$$B = \begin{bmatrix} 1 & 2 & 3 \\ 4 & 5 & 6 \end{bmatrix},$$

then

$$B^T = \begin{bmatrix} 1 & 4 \\ 2 & 5 \\ 3 & 6 \end{bmatrix}.$$

The derivations of the properties

$$(A^T)^T = A$$

$$(A + B)^T = A^T + B^T$$

$$(AB)^T = B^T A^T$$

are left as exercises.

A matrix A with the property that $A^T = A$ is said to be *symmetric*. If $A^T = -A$, then A is said to be *skew-symmetric*. The matrices

$$\begin{bmatrix} 5 & 2 & -1 \\ 2 & 0 & 3 \\ -1 & 3 & -2 \end{bmatrix}, \qquad \begin{bmatrix} 0 & -2 & 4 \\ 2 & 0 & -3 \\ -4 & 3 & 0 \end{bmatrix}$$

are symmetric and skew-symmetric, respectively.

The transpose of a row vector is a column vector and the transpose of a column vector is a row vector. If $\mathbf{u}$ and $\mathbf{v}$ are both n-dimensional column vectors; then $\mathbf{u}^T$ is a row vector and we have

$$\mathbf{u}^T \mathbf{v} = \begin{bmatrix} u_1 & u_2 & \cdots & u_n \end{bmatrix} \begin{bmatrix} v_1 \\ v_2 \\ \vdots \\ v_n \end{bmatrix} = \begin{bmatrix} u_1 v_1 + u_2 v_2 + \cdots + u_n v_n \end{bmatrix}.$$

If $\mathbf{u}$ and $\mathbf{v}$ are both real then

$$\mathbf{u}^T \mathbf{v} = \begin{bmatrix} \mathbf{u} \cdot \mathbf{v} \end{bmatrix}.$$

If A is an $n \times n$ matrix and $\mathbf{x}$ is an n-dimensional column vector, then $A\mathbf{x}$ is an $n \times 1$ matrix (or column vector) whose ith component is

$$\sum_{j=1}^{n} a_{ij} x_j.$$

Hence $\mathbf{x}^T A \mathbf{x}$ is a 1×1 matrix whose single element is

$$f(\mathbf{x}) = \sum_{i=1}^{n} \sum_{j=1}^{n} x_i a_{ij} x_j.$$

This quantity is called a *quadratic form* in the n variables $x_1, x_2, \ldots, x_n$. Such forms arise in the physical sciences, and in theoretical economics. For $n = 2$ we have

$$f(\mathbf{x}) = a_{11} x_1^2 + (a_{12} + a_{21}) x_1 x_2 + a_{22} x_2^2$$

while for $n = 3$ we have

$$f(\mathbf{x}) = a_{11} x_1^2 + a_{22} x_2^2 + a_{33} x_3^2 + (a_{23} + a_{32}) x_2 x_3$$
$$+ (a_{31} + a_{13}) x_3 x_1 + (a_{12} + a_{21}) x_1 x_2$$

We may always assume that $a_{ij} = a_{ji}$ if $i \neq j$ because the term

$$(a_{ij} + a_{ji}) x_i x_j$$

can be replaced by

$$(a'_{ij} + a'_{ji})$$

where

$$a'_{ij} = a'_{ji} = \frac{1}{2}(a_{ij} + a_{ji}).$$

Thus associated with a quadratic form is a real symmetric matrix. For example, suppose that $n = 3$ and that

$$f(\mathbf{x}) = 2x_1^2 - 5x_3^2 + 2x_1 x_2 - 6x_2 x_3.$$

Writing

$$f(\mathbf{x}) = 2x_1^2 - 5x_3^2 + (1 + 1)x_1 x_2 - (3 + 3)x_2 x_3,$$

we see that $f(\mathbf{x}) = \mathbf{x}^T A \mathbf{x}$ where A is the symmetric matrix

$$\begin{bmatrix} 2 & 1 & 0 \\ 1 & 0 & -3 \\ 0 & -3 & -5 \end{bmatrix}.$$

—————————— Exercises for Section 2.7 ——————————

In Exercises 1–6, determine whether or not the given matrix is (a) diagonal, (b) scalar, (c) upper triangular, or (d) lower triangular.

1. $\begin{bmatrix} 2 & 0 & 0 \\ 0 & -1 & 0 \\ 0 & 0 & 0 \end{bmatrix}$ *diagonal.*

2. $\begin{bmatrix} 0 & 3 & 1 & 4 \\ 0 & 0 & 0 & 0 \\ 0 & 0 & 2 & 5 \\ 0 & 0 & 0 & -1 \end{bmatrix}$

3. $\begin{bmatrix} 5 & 0 & 0 & 0 \\ 0 & -1 & 0 & 0 \\ 0 & 0 & 0 & 0 \\ 6 & 1 & 4 & 0 \end{bmatrix}$

4. $\begin{bmatrix} 5 & 0 & 0 \\ 0 & 5 & 0 \\ 0 & 0 & 5 \end{bmatrix}$ *scalar*

5. $\begin{bmatrix} 2 & 0 & 0 & 0 \\ 0 & 0 & 0 & 0 \\ 0 & 0 & 0 & 0 \end{bmatrix}$

6. $\begin{bmatrix} 0 & 0 \\ 0 & 0 \end{bmatrix}$

7. If $D = \text{diag}(2, -1, 0)$ find DA, where A is the given matrix

(a) $\begin{bmatrix} 1 & -1 & 2 \\ 0 & 3 & -2 \\ 4 & 4 & -1 \end{bmatrix}$ (b) $\begin{bmatrix} 4 & -1 \\ 5 & 2 \\ 3 & 6 \end{bmatrix}$

8. If $D = \text{diag}(-2, 0, 3)$ find BD, where B is the given matrix

(a) $\begin{bmatrix} 3 & 5 & -1 \\ 0 & 1 & 3 \\ -4 & -4 & 2 \end{bmatrix}$

(b) $\begin{bmatrix} 4 & 1 & 2 \\ -3 & 6 & -1 \end{bmatrix}$

9. Prove the second part of Theorem 2.5.

10. Prove the corollary to Theorem 2.6.

11. If A and B are both diagonal matrices of the same size, show that A and B commute.

12. Find the transpose of the given matrix.

(a) $\begin{bmatrix} 2 & -1 \\ 4 & 5 \end{bmatrix}$

(b) $\begin{bmatrix} 3 & 0 & -2 \\ -1 & 4 & 0 \end{bmatrix}$

(c) $\begin{bmatrix} 6 & 2 & 4 \\ 1 & 0 & 3 \\ 2 & 1 & 1 \end{bmatrix}$

(d) $\begin{bmatrix} 5 \\ 1 \\ 3 \end{bmatrix}$

13. Prove the three facts about transposes.

14. (a) Give an example of a 3×3 symmetric matrix, none of whose elements is zero.
(b) Show that if A is any square matrix then AA^T and A^TA are symmetric.
(c) If A and B are both symmetric matrices of the same size, is AB necessarily symmetric?
(d) Given an example of a 3×3 skew-symmetric matrix that is not the zero matrix.

15. If A and B are upper triangular matrices of the same size, is AB upper triangular? Justify your answer.

16. If A is a square matrix, show that $A^TA = I$ if and only if the column vectors of A are mutually orthogonal unit vectors.

17. Find the real symmetric matrix associated with the quadratic form.
(a) $-2x_1^2 + x_1x_2 + x_2^2$, $n = 2$
(b) $-2x_1^2 + x_1x_2 + x_2^2$, $n = 3$
(c) $3x_1^2 + x_2^2 - 4x_3^2 + 3x_2x_3 - 2x_1x_2$, $n = 3$
(d) $x_2^2 - 2x_3^2 + 4x_2x_3 - x_1x_3 + 6x_1x_2$, $n = 3$

18. Find the quadratic form associated with the matrix.

(a) $\begin{bmatrix} 2 & -1 \\ -1 & 0 \end{bmatrix}$

(b) $\begin{bmatrix} -1 & \frac{1}{2} \\ \frac{1}{2} & 3 \end{bmatrix}$

(c) $\begin{bmatrix} -2 & \frac{3}{2} & 0 \\ \frac{3}{2} & 4 & 1 \\ 0 & 1 & 1 \end{bmatrix}$

(d) $\begin{bmatrix} 0 & -1 & -\frac{3}{2} \\ -1 & 2 & \frac{5}{2} \\ -\frac{3}{2} & \frac{5}{2} & -5 \end{bmatrix}$

2.8 DETERMINANTS

Let A be a square matrix of order n. Associated with such a matrix is a number, called the *determinant* of A and denoted by "det A." If

$$A = \begin{bmatrix} a_{11} & a_{12} & \cdots & a_{1n} \\ a_{21} & a_{22} & \cdots & a_{2n} \\ \vdots & \vdots & & \vdots \\ a_{n1} & a_{n2} & \cdots & a_{nn} \end{bmatrix}, \tag{2.7}$$

we write

$$\det A = \begin{vmatrix} a_{11} & a_{12} & \cdots & a_{1n} \\ a_{21} & a_{22} & \cdots & a_{2n} \\ \vdots & \vdots & & \vdots \\ a_{n1} & a_{n2} & \cdots & a_{nn} \end{vmatrix} \tag{2.8}$$

when we wish to display the elements of A. If A is of order n, we say that det A is a determinant of order n.

If A is of order 1, with a single element a_{11}, we define det $A = a_{11}$. If A is of order 2,

$$A = \begin{bmatrix} a_{11} & a_{12} \\ a_{21} & a_{22} \end{bmatrix},$$

we define

$$\det A = a_{11}a_{22} - a_{12}a_{21}.$$

Thus

$$\begin{vmatrix} 2 & -3 \\ 1 & 5 \end{vmatrix} = (2)(5) - (-3)(1) = 13.$$

We shall presently define the determinant of a square matrix of arbitrary order n. First, however, we must develop some preliminary ideas. Consider a set[6] $\{j_1, j_2, \ldots, j_n\}$ whose distinct elements $j_1, j_2, \ldots, j_n$ are positive

---

[6] One way to describe a set is to list the members of the set, enclosed in braces.

integers. Each possible ordering of the elements of the set is called a *permutation* of the set. We use parentheses to denote an *ordered* set. For example, the possible permutations of the set $\{1, 2, 5\}$ are $(1, 2, 5)$, $(1, 5, 2)$, $(2, 1, 5)$, $(2, 5, 1)$, $(5, 1, 2)$, and $(5, 2, 1)$. The number of possible permutations of n integers (or of any n objects) is $n!$.

Let $(j_1, j_2, \ldots, j_n)$ be a permutation of a set of n positive integers. Let α_1 be the number of integers following j_1 that are smaller than j_1, let α_2 be the number of integers following j_2 that are smaller than j_2, and so on. Note that α_n is always zero. The sum $\alpha_1 + \alpha_2 + \cdots + \alpha_{n-1}$ is called the number of *inversions* in the permutation $(j_1, j_2, \ldots, j_n)$. For example, in the permutation $(2, 4, 1, 3)$ of $\{1, 2, 3, 4\}$, we have $\alpha_1 = 1$, $\alpha_2 = 2$, and $\alpha_3 = 0$, so the number of inversions is three. A permutation is said to have *even or odd parity* according to whether the number of its inversions is even or odd. We define

$$\delta(j_1, j_2, \ldots, j_n)$$

to be one if the parity of $(j_1, j_2, \ldots, j_n)$ is even and minus one if the parity is odd.

We are now in a position to define the determinant of the nth-order matrix (2.7). We form all possible products of the form

$$a_{1j_1} a_{2j_2} a_{3j_3} \cdots a_{nj_n} \qquad (2.9)$$

in which there occurs exactly one element from each row and each column of A. Thus $(j_1, j_2, j_3, \ldots, j_n)$ is a permutation of $\{1, 2, 3, \ldots, n\}$. The determinant of A is defined by the formula

$$\det A = \sum \delta(j_1, j_2, \ldots, j_n) a_{1j_1} a_{2j_2} \cdots a_{nj_n}, \qquad (2.10)$$

where the sum is taken over all possible permutations $(j_1, j_2, \ldots, j_n)$ of $\{1, 2, \ldots, n\}$.

It can be verified that this definition agrees with those previously given for the cases $n = 1$ and $n = 2$. Let us apply the definition to find the determinant of a 3×3 matrix,

$$A = \begin{bmatrix} a_{11} & a_{12} & a_{13} \\ a_{21} & a_{22} & a_{23} \\ a_{31} & a_{32} & a_{33} \end{bmatrix}.$$

The products of the form (2.9) are

$$a_{11}a_{22}a_{33}, \quad a_{11}a_{23}a_{32}, \quad a_{12}a_{21}a_{33},$$

$$a_{12}a_{23}a_{31}, \quad a_{13}a_{21}a_{32}, \quad a_{13}a_{22}a_{31}.$$

Finding the proper signs and summing, we have

$$\det A = a_{11}a_{22}a_{33} + a_{13}a_{21}a_{32} + a_{12}a_{23}a_{31}$$
$$- a_{13}a_{22}a_{31} - a_{12}a_{21}a_{33} - a_{11}a_{23}a_{32}.$$

This formula is not easy to remember. However the device

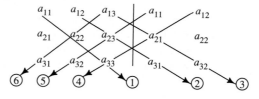

is useful. We have written down the elements of the matrix A in array form and then repeated the first two columns. We form the six products associated with the six arrows, each product having as factors the three elements pierced by that arrow. We assign a plus sign to the products determined by arrows ①, ②, and ③, and a minus sign to the products associated with arrows ④, ⑤, and ⑥. The sum of the six signed products is $\det A$. As an example, we evaluate the determinant

$$\begin{vmatrix} 1 & 2 & 3 \\ -1 & 3 & 0 \\ 2 & -4 & 5 \end{vmatrix}.$$

Forming the array

$$\begin{array}{ccc|cc} 1 & 2 & 3 & 1 & 2 \\ -1 & 3 & 0 & -1 & 3 \\ 2 & -4 & 5 & 2 & -4 \end{array}$$

and using the procedure described above, we find that the value of the determinant is

$$(1)(3)(5) + (2)(0)(2) + (3)(-1)(-4)$$
$$-(2)(-1)(5) - (1)(0)(-4) - (3)(3)(2) = 19.$$

The reader should be warned that the scheme just described works only for third-order determinants and not for higher order ones. A practical method for evaluating a determinant of any order will be developed in the next section.

A square matrix whose determinant is zero is said to be *singular*. A square matrix that is not singular is called *nonsingular*.

We consider one more topic, the derivative of the determinant of a matrix function. Consider an ordered set of real differentiable functions a_{ij}, $1 \le i \le n$, $1 \le j \le n$, with a common domain. Associated with each number x in the domain is the matrix

$$A(x) = \begin{bmatrix} a_{11}(x) & a_{12}(x) & \cdots & a_{1n}(x) \\ a_{21}(x) & a_{22}(x) & \cdots & a_{2n}(x) \\ \vdots & \vdots & & \vdots \\ a_{n1}(x) & a_{n2}(x) & \cdots & a_{nn}(x) \end{bmatrix}.$$

For each x we can inquire about $(d/dx) \det A(x)$ Recalling that

$$\det A(x) = \sum \pm a_{1j_1}(x) a_{2j_2}(x) \cdots a_{nj_n}(x),$$

we see that

$$(\det A)' = \sum \pm a'_{1j_1} a_{2j_2} \cdots a_{nj_n} + \sum \pm a_{1j_1} a'_{2j_2} \cdots a_{nj_n}$$
$$+ \cdots + \sum \pm a_{1j_1} a_{2j_2} \cdots a'_{nj_n}$$

or

$$(\det A)' = \begin{vmatrix} a'_{11} & a'_{12} & \cdots & a'_{1n} \\ a_{21} & a_{22} & \cdots & a_{2n} \\ \vdots & \vdots & & \vdots \\ a_{n1} & a_{n2} & \cdots & a_{nn} \end{vmatrix} + \begin{vmatrix} a_{11} & a_{12} & \cdots & a_{1n} \\ a'_{21} & a'_{22} & \cdots & a'_{2n} \\ \vdots & \vdots & & \vdots \\ a_{n1} & a_{n2} & \cdots & a_{nn} \end{vmatrix}$$

$$+ \cdots + \begin{vmatrix} a_{11} & a_{12} & \cdots & a_{1n} \\ a_{21} & a_{22} & \cdots & a_{2n} \\ \vdots & \vdots & & \vdots \\ a'_{n1} & a'_{n2} & \cdots & a'_{nn} \end{vmatrix}. \tag{2.11}$$

Thus the derivative of $\det A$ is the sum of n determinants that are obtained by successively differentiating the rows of A. As will be shown in the next section the derivative of $\det A$ can also be expressed as the sum of n determinants that are obtained by successively differentiating the columns of A.

For purposes of illustration, we observe that

$$\frac{d}{dx} \begin{vmatrix} x & x^2 & x^3 \\ e^x & 1 & 0 \\ \sin x & 0 & 0 \end{vmatrix}$$

$$= \begin{vmatrix} 1 & 2x & 3x^2 \\ e^x & 1 & 0 \\ \sin x & 0 & 0 \end{vmatrix} + \begin{vmatrix} x & x^2 & x^3 \\ e^x & 0 & 0 \\ \sin x & 0 & 0 \end{vmatrix}$$

$$+ \begin{vmatrix} x & x^2 & x^3 \\ e^x & 1 & 0 \\ \cos x & 0 & 0 \end{vmatrix}.$$

--- Exercises for Section 2.8 ---

1. Find all the permutations of the set $\{1, 2, 3, 4\}$.

2. Find the number of inversions in each of the following permutations:
   (a) (3, 1, 2)   (b) (3, 2, 1)
   (c) (1, 2, 3)   (d) (1, 3, 2, 4)
   (e) (4, 2, 3, 1)   (f) (2, 3, 4, 1)
   (g) (1, 3, 5, 2, 4) (h) (4, 1, 5, 3, 2)

3. With $\delta(j_1, j_2, \ldots, j_n)$ defined as in the text, find the value of each of the following:
   (a) $\delta(2, 1, 3)$ (b) $\delta(1, 2, 3)$
   (c) $\delta(2, 4, 1, 3)$ (d) $\delta(4, 3, 2, 1)$
   (e) $\delta(1, 3, 5, 2, 4)$

4. (a) If two adjacent elements j_k and j_{k+1} in the permutation $(j_1, \ldots, j_k, j_{k+1}, \ldots, j_n)$ are interchanged, show that the parity changes. (Consider the two cases $j_k < j_{k+1}$ and $j_k > j_{k+1}$.)
   (b) If any two elements in a permutation $(j_1, j_2, \ldots, j_n)$ are interchanged, show that the parity changes. (Think of successively interchanging adjacent elements.)

5. If every element in a row of A is zero, show that $\det A = 0$.

6. Derive the formula for the value of a second-order determinant from the general formula (2.10).

7. Find the value of each of the second-order determinants.
   (a) $\begin{vmatrix} 2 & -3 \\ 4 & 5 \end{vmatrix}$ (b) $\begin{vmatrix} 2 & -3 \\ -4 & 6 \end{vmatrix}$
   (c) $\begin{vmatrix} -5 & 2 \\ 1 & 3 \end{vmatrix}$ (d) $\begin{vmatrix} 0 & 3 \\ 2 & 4 \end{vmatrix}$

8. Find the value of each of the third-order determinants.
   (a) $\begin{vmatrix} 1 & 3 & 2 \\ 3 & 1 & -1 \\ -2 & 4 & 5 \end{vmatrix}$
   (b) $\begin{vmatrix} 2 & 0 & -2 \\ 1 & 1 & 5 \\ 3 & 4 & 5 \end{vmatrix}$
   (c) $\begin{vmatrix} -4 & -1 & 3 \\ 2 & 2 & 1 \\ 3 & 5 & 0 \end{vmatrix}$
   (d) $\begin{vmatrix} 1 & 2 & 3 \\ 3 & 2 & 1 \\ 1 & 1 & 1 \end{vmatrix}$

9. If A is a square matrix of order n, prove that $\det(-A) = (-1)^n \det A$.

10. If A is a square matrix of order n and c is a number, prove that
$$\det(cA) = c^n \det A.$$

11. Show, by means of an example, that $\det(A + B) \neq \det A + \det B$ in general.

12. If $D = \text{diag}(d_1, d_2, \ldots, d_n)$, show that $\det D = d_1 d_2 \ldots d_n$.

13. Evaluate the derivative of the determinant
$$\begin{vmatrix} x^3 + 1 & x^2 \\ x^2 & 2x \end{vmatrix}$$
in two ways: first, by evaluating the determinant and then taking the derivative; second, by applying formula (2.11).

2.9
PROPERTIES OF DETERMINANTS

Listed here are some elementary properties of determinants. Each property is illustrated by an example. A proof of only the first property is given in the text. Proofs of others are left as exercises. The reader who encounters difficulty with these will find proofs in most college algebra textbooks and in some books on calculus. Property 8 is more difficult to prove, and we ask for a proof only for 2×2 matrices.

Property 1 $\det A^T = \det A$:

$$\begin{vmatrix} 1 & -1 & 2 \\ 3 & 0 & 1 \\ 2 & 1 & 5 \end{vmatrix} = \begin{vmatrix} 1 & 3 & 2 \\ -1 & 0 & 1 \\ 2 & 1 & 5 \end{vmatrix}.$$

Property 2   If every element in a row (column) of A is zero, then $\det A = 0$:

$$\begin{vmatrix} 2 & 0 & 3 \\ 1 & 0 & 1 \\ -1 & 0 & 4 \end{vmatrix} = 0.$$

Property 3   If every element in one row (column) of A is multiplied by the number c, the determinant of the resulting matrix is equal to $c \det A$:

$$\begin{vmatrix} -6 & 3 & -9 \\ 2 & 1 & 0 \\ 1 & 1 & 2 \end{vmatrix} = -3 \begin{vmatrix} 2 & -1 & 3 \\ 2 & 1 & 0 \\ 1 & 1 & 2 \end{vmatrix}.$$

Property 4   If two rows (columns) of A are interchanged, the determinant of the resulting matrix is equal to $-\det A$:

$$\begin{vmatrix} 1 & -1 & 3 \\ 0 & 4 & 1 \\ 2 & 2 & 5 \end{vmatrix} = - \begin{vmatrix} 2 & 2 & 5 \\ 0 & 4 & 1 \\ 1 & -1 & 3 \end{vmatrix}.$$

Property 5   If two rows (columns) of A are identical, then $\det A = 0$:

$$\begin{vmatrix} 2 & 1 & 2 \\ 1 & -4 & 1 \\ 3 & 5 & 3 \end{vmatrix} = 0.$$

Property 6   If any column vector of A, say the ith, is the sum of two column vectors, so that $\mathbf{a}_i = \mathbf{a}_i' + \mathbf{a}_i''$, then $\det A = \det[\mathbf{a}_1, \ldots, \mathbf{a}_i', \ldots, \mathbf{a}_n] + \det[\mathbf{a}_1, \ldots, \mathbf{a}_i'', \ldots, \mathbf{a}_n]$. An analogous property holds for rows:

$$\begin{vmatrix} 1 & 2+3 & 3 \\ 0 & 1-4 & 5 \\ 2 & -2+0 & 6 \end{vmatrix} = \begin{vmatrix} 1 & 2 & 3 \\ 0 & 1 & 5 \\ 2 & -2 & 6 \end{vmatrix} + \begin{vmatrix} 1 & 3 & 3 \\ 0 & -4 & 5 \\ 2 & 0 & 6 \end{vmatrix}.$$

Property 7   If to every element in a row (column) of A is added c times the corresponding element of a different row (column), the determinant of the resulting matrix is equal to det A:

$$\begin{vmatrix} 2 & -1 & 0 \\ 1 & 2 & -3 \\ 4 & 5 & 6 \end{vmatrix} = \begin{vmatrix} 4 & 3 & -6 \\ 1 & 2 & -3 \\ 4 & 5 & 6 \end{vmatrix}.$$

(Here, twice the second row has been added to the first row.)

Property 8   If A and B are both matrices of order n, then $\det(AB) = \det A \cdot \det B$:

$$\begin{vmatrix} 1 & -1 \\ 2 & 1 \end{vmatrix} \cdot \begin{vmatrix} 2 & 1 \\ -1 & 3 \end{vmatrix} = \begin{vmatrix} 3 & -2 \\ 3 & 5 \end{vmatrix}.$$

The following lemma is needed in the derivations of some of the properties described above.

Lemma   If two adjacent elements in a permutation are interchanged, the parity of the permutation is changed.

Proof   Suppose that j_α and j_β are interchanged in the permutation $(j_1, j_2, \ldots, j_\alpha, j_\beta, \ldots, j_n)$. If $j_\alpha < j_\beta$, the interchange of j_α and j_β introduces one new inversion. If $j_\alpha > j_\beta$, the number of inversions is decreased by one. In each case the parity is changed.

Proof of Property 1   Let a_{ij}^T be the element in the ith row and jth column of A^T. Then $a_{ij}^T = a_{ji}$. The determinant of A^T is the sum of terms of the form

$$\delta(j_1, j_2, \ldots, j_n) a_{1j_1}^T a_{2j_2}^T \cdots a_{nj_n}^T = \delta(j_1, j_2, \ldots, j_n) a_{j_1 1} a_{j_2 2} \cdots a_{j_n n}.$$

Now

$$a_{j_1 1} a_{j_2 2} \cdots a_{j_n n} = a_{1k_1} a_{2k_2} \cdots a_{nk_n},$$

where $(k_1, k_2, \ldots, k_n)$ is some permutation of $\{1, 2, \ldots, n\}$. We can think of the product on the right-hand side as being formed from the product on the left-hand side by successively interchanging adjacent factors. Each interchange changes the parity of the ordered set of first subscripts and simultaneously changes the parity of the ordered set of second subscripts. Consequently the parity of $(k_1, k_2, \ldots, k_n)$ must be the same as that of $(j_1, j_2, \ldots, j_n)$. Thus $\delta(k_1, k_2, \ldots, k_n) = \delta(j_1, j_2, \ldots, j_n)$, so that

$$\det A^T = \sum \delta(k_1, k_2, \ldots, k_n) a_{1k_1} a_{2k_2} \cdots a_{nk_n} = \det A.$$

The rows of A^T are the same as the columns of A. Consequently, Property 1 allows us to convert theorems about rows of a determinant into corresponding theorems about columns. For example, the derivative of det A can be found by differentiating det A^T by rows; this amounts to differentiating det A by columns.

We shall presently give a practical method for evaluating a determinant of any order. Our method is based on the following result.

Theorem 2.7   If A is an $n \times n$ triangular (upper or lower) matrix, then

$$\det A = a_{11}a_{22} \cdots a_{nn}. \qquad (2.12)$$

Proof   The determinant of A is the sum of terms of the form

$$\pm a_{1j_1}a_{2j_2} \cdots a_{nj_n}. \qquad (2.13)$$

If one factor, say a_{pj_p}, is such that $p < j_p$, there must be another factor, say a_{qj_q}, such that $q > j_q$, and conversely. This is because

$$1 + 2 + \cdots + n = j_1 + j_2 + \cdots + j_n$$

for every permutation $(j_1, j_2, \ldots, j_n)$. In a triangular matrix, one of these two factors must be zero. Thus, if A is triangular, every term of the form (2.13) is zero except the one where $j_1 = 1, j_2 = 2, \ldots, j_n = n$. Hence $\det A$ is given by formula (2.12).

By using the elementary properties listed at the beginning of this section, we can reduce the problem of evaluating any determinant to one of evaluating the determinant of a triangular matrix. The essential features of the reduction are shown in the following example. Consider the determinant

$$\det A = \begin{vmatrix} 0 & 2 & 1 & -1 \\ 2 & -2 & 4 & 0 \\ -1 & 2 & 0 & 1 \\ -2 & 1 & 1 & 3 \end{vmatrix}.$$

We wish to place a nonzero element in the first row and column. Interchanging the first and third rows, we have

$$\det A = - \begin{vmatrix} -1 & 2 & 0 & 1 \\ 2 & -2 & 4 & 0 \\ 0 & 2 & 1 & -1 \\ -2 & 1 & 1 & 3 \end{vmatrix}.$$

We place zeros in every position of the first column below the first by adding appropriate multiples of the first row to the second and fourth rows. Thus

$$\det A = - \begin{vmatrix} -1 & 2 & 0 & 1 \\ 0 & 2 & 4 & 2 \\ 0 & 2 & 1 & -1 \\ 0 & -3 & 1 & 1 \end{vmatrix} = -2 \begin{vmatrix} -1 & 2 & 0 & 1 \\ 0 & 1 & 2 & 1 \\ 0 & 2 & 1 & -1 \\ 0 & -3 & 1 & 1 \end{vmatrix}.$$

We place zeros in every position of the second column below the second by adding appropriate multiples of the second row to the third and fourth rows. We find that

$$\det A = -2 \begin{vmatrix} -1 & 2 & 0 & 1 \\ 0 & 1 & 2 & 1 \\ 0 & 0 & -3 & -3 \\ 0 & 0 & 7 & 4 \end{vmatrix} = 6 \begin{vmatrix} -1 & 2 & 0 & 1 \\ 0 & 1 & 2 & 1 \\ 0 & 0 & 1 & 1 \\ 0 & 0 & 7 & 4 \end{vmatrix}.$$

We add (-7) times the third row to the fourth to obtain the triangular form

$$\det A = 6 \begin{vmatrix} -1 & 2 & 0 & 1 \\ 0 & 1 & 2 & 1 \\ 0 & 0 & 1 & 1 \\ 0 & 0 & 0 & -3 \end{vmatrix}.$$

Using Theorem 2.7, we have

$$\det A = (6)(-1)(1)(1)(-3) = 18.$$

─────────────── **Exercises for Section 2.9** ───────────────

1. Derive Properties 2 and 3 of determinants that are given at the beginning of this section.

2. Prove Property 4 (Suggestion: use the lemma of this section). Then use Property 4 to prove Property 5.

3. Derive Properties 6 and 7.

4. Prove that $\det(AB) = (\det A) \cdot (\det B)$ when A and B are matrices of order two.

In Exercises 5–9, evaluate the determinant by using Theorem 2.7 and elementary properties of determinants.

5. (a) $\begin{vmatrix} 1 & 3 & 2 \\ 2 & -1 & 5 \\ -2 & 4 & -4 \end{vmatrix}$

   (b) $\begin{vmatrix} 5 & 2 & 3 \\ 2 & -1 & 0 \\ 3 & 4 & 7 \end{vmatrix}$

6. (a) $\begin{vmatrix} -3 & 5 & 7 \\ -5 & -4 & 3 \\ 2 & 5 & 6 \end{vmatrix}$

   (b) $\begin{vmatrix} 6 & 5 & 2 \\ 4 & 3 & 2 \\ 7 & 3 & 7 \end{vmatrix}$

7. (a) $\begin{vmatrix} 2 & 1 & 4 & 7 \\ 3 & 0 & 1 & 5 \\ -4 & -3 & 3 & 4 \\ 2 & 2 & -1 & 0 \end{vmatrix}$

   (b) $\begin{vmatrix} -2 & 1 & 4 & 2 \\ -3 & 0 & 1 & 6 \\ 1 & 2 & 3 & 4 \\ -4 & 3 & -2 & 1 \end{vmatrix}$

8. (a) $\begin{vmatrix} 4 & 0 & 2 & 0 \\ 0 & 1 & 0 & 3 \\ 5 & 0 & 7 & 0 \\ 0 & 8 & 0 & 6 \end{vmatrix}$

   (b) $\begin{vmatrix} 6 & 2 & 8 & 0 \\ 1 & 3 & 5 & 2 \\ 2 & 1 & 0 & -3 \\ 2 & -5 & -2 & -7 \end{vmatrix}$

9.
$$\begin{vmatrix} 2 & -1 & 0 & 4 & 1 \\ 1 & 5 & 2 & 0 & -2 \\ -1 & 3 & -3 & 1 & 0 \\ 0 & 1 & 1 & 2 & -2 \\ 2 & 2 & 1 & 0 & -1 \end{vmatrix}$$

10. Let $C = AB$, where A and B are square matrices of the same order. If $\det C = 0$, show that either $\det A$ or $\det B$, or both, is zero.

11. (a) If $A^2 = I$, show that $\det A = \pm 1$.
    (b) If $A^T A = I$, show that $\det A = \pm 1$.
    (c) If $A^2 = A$, what can be said about $\det A$?

12. Let $f_1, f_2, \ldots, f_n$ be functions that are defined and possess at least n derivatives on an interval. If

$$A(x) = \begin{bmatrix} f_1(x) & f_2(x) & \cdots & f_n(x) \\ f_1'(x) & f_2'(x) & \cdots & f_n'(x) \\ \vdots & \vdots & & \vdots \\ f_1^{(n-1)}(x) & f_2^{(n-1)}(x) & \cdots & f_n^{(n-1)}(x) \end{bmatrix},$$

show that

$$\frac{d}{dx} \det A(x) = \begin{vmatrix} f_1(x) & f_2(x) & \cdots & f_n(x) \\ f_1'(x) & f_2'(x) & \cdots & f_n'(x) \\ \vdots & \vdots & & \vdots \\ f_1^{(n-2)}(x) & f_2^{(n-2)}(x) & \cdots & f_n^{(n-2)}(x) \\ f_1^{(n)}(x) & f_2^{(n)}(x) & \cdots & f_n^{(n)}(x) \end{vmatrix}$$

2.10 COFACTORS

Let A be an $m \times n$ matrix. Any matrix that is formed from A by deleting rows of A or columns of A, or both, is called a *submatrix* of A. In addition, it is sometimes convenient to regard A as a submatrix of itself (deletion of no rows and no columns).

We now restrict attention to the case where A is a square $n \times n$ matrix. If we delete the ith row and kth column of A (the row and column containing the element a_{ik}), we obtain a square submatrix of order $n - 1$. The determinant of this submatrix is called the *minor* of the element a_{ik}; we denote the minor by M_{ik}. We may write

$$M_{ik} = \sum \delta(j_1, \ldots, j_{i-1}, j_{i+1}, \ldots, j_n) a_{1j_1} \cdots a_{i-1, j_{i-1}} a_{i+1, j_{i+1}} \cdots a_{nj_n},$$

where k is excluded from the column subscripts and the sum is over all permutations of $1, 2, \ldots, k-1, k+1, \ldots, n$. The quantity

$$A_{ik} = (-1)^{i+k} M_{ik}$$

is called the *cofactor* of the element a_{ik}. For example, if

$$A = \begin{bmatrix} 1 & -2 & 0 \\ 3 & 1 & 4 \\ 2 & 2 & 1 \end{bmatrix},$$

then

$$A_{11} = \begin{vmatrix} 1 & 4 \\ 2 & 1 \end{vmatrix} = -7, \qquad A_{12} = -\begin{vmatrix} 3 & 4 \\ 2 & 1 \end{vmatrix} = 5,$$

and so on.

The determinant of a matrix can be expressed in terms of the cofactors of the elements in any row or column, as is shown by the following theorem.

Theorem 2.8   If each element in any one row (column) of an $n \times n$ matrix A is multiplied by its cofactor, the sum of the n products so formed is equal to det A. Thus

$$\sum_{j=1}^{n} a_{ij}A_{ij} = \det A, \qquad 1 \le i \le n,$$

$$\sum_{i=1}^{n} a_{ij}A_{ij} = \det A, \qquad 1 \le j \le n.$$

Proof   We establish the first relation, starting with the formula

$$\det A = \sum \delta(j_1, j_2, \ldots, j_n)a_{1j_1}a_{2j_2}\cdots a_{nj_n}.$$

For fixed i, we collect the products that involve $a_{i1}, a_{i2}, \ldots, a_{in}$. Then

$$\det A = \sum_{j_i=1}^{n} a_{ij_i} \sum \delta(j_1, \ldots, j_i, \ldots, j_n)a_{1j_1}\cdots a_{i-1,j_{i-1}}a_{i+1,j_{i+1}}\cdots a_{nj_n}.$$

where the inner sum is taken over all permutations $(j_1, \ldots, j_i, \ldots, j_n)$ of $\{1, 2, \ldots, n\}$ with j_i fixed. Since j_i is in the ith position,

$$\delta(j_1, \ldots, j_i, \ldots, j_n) = (-1)^{i-1}\delta(j_i, j_1, \ldots, j_{i-1}, j_{i+1}, \ldots, j_n).$$

In the symbol on the right, j_i is followed by $j_i - 1$ smaller terms, so this is equal to

$$(-1)^{i-1}(-1)^{j_i-1}\delta(j_1, \ldots, j_{i-1}, j_{i+1}, \ldots, j_n).$$

Hence

$$\det A = \sum_{j_i=1}^{n} a_{ij_i}(-1)^{i+j_i}M_{ij_i} = \sum_{j_i=1}^{n} a_{ij_i}A_{ij_i}.$$

This establishes the first equation in the statement of the theorem.

The second equation can be derived by expressing det A in terms of the cofactors of the elements of the jth row of A^T, and observing that the cofactor of a_{ji}^T is the same as the cofactor of a_{ij}. We omit the details.

To illustrate the theorem, we evaluate the determinant of a third-order matrix using cofactors of the elements of the first row. We have

$$\begin{vmatrix} 1 & -2 & 0 \\ 3 & 1 & 4 \\ 2 & 2 & 1 \end{vmatrix} = (1)\begin{vmatrix} 1 & 4 \\ 2 & 1 \end{vmatrix} - (-2)\begin{vmatrix} 3 & 4 \\ 2 & 1 \end{vmatrix} + (0)\begin{vmatrix} 3 & 1 \\ 2 & 2 \end{vmatrix}$$

$$= (1)(-7) - (-2)(-5) + 0$$

$$= -17.$$

Theorem 2.9   If the elements of the ith row (column) of an $n \times n$ matrix A are multiplied by the cofactors of the corresponding elements of the jth row (column), the sum of the products is det A if $i = j$ and zero if $i \neq j$. In symbols,

$$\sum_{k=1}^{n} a_{ik} A_{jk} = \delta_{ij} \det A,$$

$$\sum_{k=1}^{n} a_{ki} A_{kj} = \delta_{ij} \det A,$$

where δ_{ij} is the Kronecker delta.

Proof   The validity of the formulas follows from Theorem 2.8 if $i = j$. For the case $i \neq j$, the sum on the left-hand side in the first equation can be regarded as the determinant of a matrix whose jth row is the same as its ith row. Hence its value must be zero. Similarly, the left-hand member of the second equation can be regarded as the determinant of a matrix whose jth column is the same as its ith column.

Theorem 2.9 will be used to derive some results of theoretical importance in the next section. Although it can be used to evaluate determinants, the method described in the previous section requires fewer arithmetic operations to be performed, and is more efficient when the order of the matrix is large.

──────────────── **Exercises for Section 2.10** ────────────────

1. Find the cofactor of each element of the given matrix.

    (a) $\begin{bmatrix} 2 & 3 \\ -1 & 0 \end{bmatrix}$ (b) $\begin{bmatrix} a & b \\ c & d \end{bmatrix}$

    (c) $\begin{bmatrix} 2 & -1 & 3 \\ 1 & 0 & -2 \\ 3 & 1 & 1 \end{bmatrix}$ (d) $\begin{bmatrix} 4 & 0 & 2 \\ 0 & 2 & -2 \\ 1 & 3 & 1 \end{bmatrix}$

2. Evaluate the given determinant by applying Theorem 2.8. Use any row or column.

    (a) $\begin{vmatrix} 1 & 2 & -1 \\ 4 & 1 & 2 \\ 1 & 1 & -3 \end{vmatrix}$

    (b) $\begin{vmatrix} 4 & 3 & -2 \\ -1 & 2 & 0 \\ 1 & -1 & 3 \end{vmatrix}$

    (c) $\begin{vmatrix} 0 & 2 & 1 & 0 \\ 1 & -1 & 3 & 1 \\ 2 & 0 & 1 & 4 \\ 0 & -1 & -2 & 1 \end{vmatrix}$

    (d) $\begin{vmatrix} 4 & 0 & 1 & 3 \\ -1 & 3 & 1 & 1 \\ 1 & 0 & -1 & 1 \\ 2 & 0 & 2 & 1 \end{vmatrix}$

3. Find det D if

    $$D = \begin{bmatrix} 0 & \cdots & 0 & 0 & 0 & d_1 \\ 0 & \cdots & 0 & 0 & d_2 & 0 \\ 0 & \cdots & 0 & d_3 & 0 & 0 \\ \vdots & & \vdots & \vdots & \vdots & \vdots \\ d_n & \cdots & 0 & 0 & 0 & 0 \end{bmatrix}$$

4. Let A be a third-order matrix with elements a_{ij}. Verify that the sum

    $$a_{11} A_{21} + a_{12} A_{22} + a_{13} A_{23}$$

(whose terms are the products of the elements in the first row of A with the cofactors of the corresponding elements of the second row) is the determinant of the matrix whose first and second rows are identical. (Hence the sum is equal to zero.)

5. Use Theorem 2.8 to prove Theorem 2.7.

6. Let B be the matrix obtained from A by setting $b_{ij} = A_{ji}$, where A_{ij} is the cofactor of a_{ij}. (In other words, B is obtained from A by replacing each element of A by its cofactor and then taking the transpose of the resulting matrix.)
   (a) Show that $AB = (\det)I$.
   (b) Show that $\det B = (\det A)^{n-1}$.

7. Let P_1 and P_2 be distinct points with rectangular coordinates (x_1, y_1) and (x_2, y_2) in a plane. Show that the equation

$$\begin{vmatrix} x & y & 1 \\ x_1 & y_1 & 1 \\ x_2 & y_2 & 1 \end{vmatrix} = 0$$

is that of the straight line through P_1 and P_2.

8. Let $x_1, x_2, \ldots, x_n$ be distinct numbers. Show that the formula

$$P(x) = \begin{vmatrix} 1 & x & x^2 & \cdots & x^n \\ 1 & x_1 & x_1^2 & \cdots & x_1^n \\ 1 & x_2 & x_2^2 & \cdots & x_2^n \\ \vdots & \vdots & \vdots & & \vdots \\ 1 & x_n & x_n^2 & \cdots & x_n^n \end{vmatrix}$$

defines a polynomial P whose zeros are $x_1, x_2, \ldots, x_n$.

2.11
CRAMER'S RULE

We consider a system of linear equations

$$\begin{aligned}
a_{11}x_1 + a_{12}x_2 + \cdots + a_{1n}x_n &= b_1, \\
a_{21}x_1 + a_{22}x_2 + \cdots + a_{2n}x_n &= b_2, \\
&\cdots \cdots \cdots \cdots, \\
a_{n1}x_1 + a_{n2}x_2 + \cdots + a_{nn}x_n &= b_n
\end{aligned} \tag{2.14}$$

with n equations and n unknowns. The $n \times n$ coefficient matrix of the system is denoted by A.

Theorem 2.10   If $\det A \neq 0$, the system (2.14) possesses exactly one solution.

Proof   We first write down the $n \times (n + 1)$ augmented matrix of the system,

$$\begin{bmatrix} a_{11} & a_{12} & \cdots & a_{1n} & b_1 \\ a_{21} & a_{22} & \cdots & a_{2n} & b_2 \\ \vdots & \vdots & & \vdots & \vdots \\ a_{n1} & a_{n2} & \cdots & a_{nn} & b_n \end{bmatrix}. \tag{2.15}$$

By performing elementary row operations (Section 2.1), we obtain the matrix of a system that is equivalent to the original system. Furthermore, the determinant of the coefficient matrix of the new system is a nonzero multiple of $\det A$. Hence, if $\det A \neq 0$ the coefficient matrix of the new system is non-singular.

Since A is nonsingular, there is at least one element in the first column of the matrix (2.15) that is not zero. By interchanging rows, if necessary, we place a nonzero element in the first row and first column. Then by adding appropriate multiples of the first row to the other rows we introduce zeros in all positions of the first column below the first. The result is an augmented matrix

$$\begin{bmatrix} a'_{11} & a'_{12} & \cdots & a'_{1n} & b'_1 \\ 0 & a'_{22} & \cdots & a'_{2n} & b'_2 \\ \vdots & \vdots & & \vdots & \vdots \\ 0 & a'_{n2} & \cdots & a'_{nn} & b'_n \end{bmatrix}$$

of a system that is equivalent to the system (2.14). Since the coefficient matrix of this system is nonsingular, at least one element in the second column below the first position must be different from zero. We place a nonzero element in the second position of the second column by interchanging rows if necessary. Then we introduce zeros in all positions of the second column below the second position by adding appropriate multiples of the second row to the lower rows. The result is an augmented matrix of the form

$$\begin{bmatrix} a''_{11} & a''_{12} & a''_{13} & \cdots & a''_{1n} & b''_1 \\ 0 & a''_{22} & a''_{23} & \cdots & a''_{2n} & b''_2 \\ 0 & 0 & a''_{33} & \cdots & a''_{3n} & b''_3 \\ \vdots & \vdots & \vdots & & \vdots & \vdots \\ 0 & 0 & a''_{n3} & \cdots & a''_{nn} & b''_n \end{bmatrix},$$

where a''_{11} and a''_{22} are not zero. Since $\det A \neq 0$, at least one element in the third column of this matrix below the second position must be different from zero, so we can continue this process. Finally, we arrive at an augmented matrix of the form

$$\begin{bmatrix} \tilde{a}_{11} & \tilde{a}_{12} & \tilde{a}_{13} & \cdots & \tilde{a}_{1n} & \tilde{b}_1 \\ 0 & \tilde{a}_{22} & \tilde{a}_{23} & \cdots & \tilde{a}_{2n} & \tilde{b}_2 \\ 0 & 0 & \tilde{a}_{33} & \cdots & \tilde{a}_{3n} & \tilde{b}_3 \\ \vdots & \vdots & \vdots & & \vdots & \vdots \\ 0 & 0 & 0 & \cdots & \tilde{a}_{nn} & \tilde{b}_n \end{bmatrix},$$

where none of the elements $\tilde{a}_{11}, \tilde{a}_{22}, \ldots, \tilde{a}_{nn}$ is zero. The last row of this matrix corresponds to the equation $\tilde{a}_{nn}x_n = \tilde{b}_n$, so we can solve for x_n, finding $x_n = \tilde{b}_n / \tilde{a}_{nn}$. Then $(n-1)$st row corresponds to the equation

$$\tilde{a}_{n-1,n-1}x_{n-1} + \tilde{a}_{n-1,n}x_n = \tilde{b}_{n-1}$$

and x_{n-1} is now found from this equation. By working upward, we find all the solution components, which are uniquely determined. This concludes our proof.

In establishing the existence of a solution of the system, we have also described a method for finding that solution. This method, in which the coefficient matrix is reduced to triangular form, is known as the *Gauss reduction method* or method of *Gaussian elimination*. It provides a practical, efficient method for solving a system with n equations and n unknowns. We could have solved the system by reducing its matrix to row-echelon form, the procedure used in Section 2.1. However, for large values of n the number of arithmetic operations required in the Gauss reduction method is considerably less than in the other method. It is for this reason the Gauss method is preferable, particularly in solving large systems.

Example   Let us solve the system

$$x_1 - x_2 - 3x_3 = -3,$$
$$x_1 + 2x_2 + 3x_3 = 3,$$
$$x_1 + 2x_2 + 6x_3 = 7.$$

The matrix of the system is

$$\begin{bmatrix} 1 & -1 & -3 & -3 \\ 1 & 2 & 3 & 3 \\ 1 & 2 & 6 & 7 \end{bmatrix}.$$

Elimination of x_1 from the last two equations of the system yields the matrix

$$\begin{bmatrix} 1 & -1 & -3 & -3 \\ 0 & 3 & 6 & 6 \\ 0 & 3 & 9 & 10 \end{bmatrix}.$$

Subtraction of the second row from the third gives

$$\begin{bmatrix} 1 & -1 & -3 & -3 \\ 0 & 3 & 6 & 6 \\ 0 & 0 & 3 & 4 \end{bmatrix}.$$

The corresponding system is

$$x_1 - x_2 - 3x_3 = -3,$$
$$3x_2 + 6x_3 = 6,$$
$$3x_3 = 4.$$

From the last equation we find that $x_3 = \dfrac{4}{3}$. The second equation gives

$$3x_2 = 6 - 6x_3 = 6 - 6\left(\frac{4}{3}\right) = -2$$

or $x_2 = -\dfrac{2}{3}$. We obtain x_1 from the first equation, finding that

$$x_1 = -3 + 3x_3 + x_2 = -3 + 3\left(\dfrac{4}{3}\right) - \dfrac{2}{3} = \dfrac{1}{3}.$$

Hence the solution is

$$x_1 = \dfrac{1}{3}, \qquad x_2 = -\dfrac{2}{3}, \qquad x_3 = \dfrac{4}{3}.$$

We shall now derive a formula, known as *Cramer's rule*, for the solution of the system (2.14). The use of this rule to compute the solution components is less efficient than the use of the Gauss reduction method and in practice is seldom used for that purpose. Nevertheless, it is important for some theoretical purposes, as we shall illustrate. Cramer's rule can be stated as follows.

Theorem 2.11   If $\det A \neq 0$, the components of the solution of the system (2.14) are given by the formula

$$x_k = \dfrac{\det B_k}{\det A}, \qquad 1 \leq k \leq n, \tag{2.16}$$

where the matrix B_k is the same as A except that the elements a_{ik}, $1 \leq i \leq n$, in the kth column of A have been replaced by the terms b_i, $1 \leq i \leq n$, respectively.

We shall look at an example before proving the theorem. Consider the system

$$2x_1 + 2x_2 - x_3 = 2,$$

$$-3x_1 - x_2 + 3x_3 = -2,$$

$$4x_1 + 2x_2 - 3x_3 = 0.$$

Calculation shows that the determinant of the coefficient matrix is 2. Since this matrix is nonsingular, Theorem 2.11 applies. Using formula (2.16), we have

$$x_1 = \dfrac{1}{2}\begin{vmatrix} 2 & 2 & -1 \\ -2 & -1 & 3 \\ 0 & 2 & -3 \end{vmatrix} = \dfrac{-14}{2} = -7,$$

$$x_2 = \dfrac{1}{2}\begin{vmatrix} 2 & 2 & -1 \\ -3 & -2 & 3 \\ 4 & 0 & -3 \end{vmatrix} = \dfrac{10}{2} = 5,$$

$$x_3 = \dfrac{1}{2}\begin{vmatrix} 2 & 2 & 2 \\ -3 & -1 & -2 \\ 4 & 2 & 0 \end{vmatrix} = \dfrac{-12}{2} = -6.$$

Proof Let $\mathbf{a}_1, \mathbf{a}_2, \ldots, \mathbf{a}_n$ be the column vectors of A. Then

$$B_k = [\mathbf{a}_1, \ldots, \mathbf{a}_{k-1}, \mathbf{b}, \mathbf{a}_{k+1}, \ldots, \mathbf{a}_n].$$

Since $\det A \neq 0$, we know by Theorem 2.10 that the system (2.14) has a solution $(x_1, x_2, \ldots, x_n)$. Consequently (see Theorem 2.4)

$$x_1 \mathbf{a}_1 + x_2 \mathbf{a}_2 + \cdots + x_n \mathbf{a}_n = \mathbf{b}.$$

Then

$$\det B_k = \sum_{j=1}^{n} \det[\mathbf{a}_1, \ldots, \mathbf{a}_{k-1}, x_j \mathbf{a}_j, \mathbf{a}_{k+1}, \ldots, \mathbf{a}_n].$$

But all the terms in the sum on the right are zero except the one with $j = k$. Hence

$$\det B_k = \det[\mathbf{a}_1, \ldots, \mathbf{a}_{k-1}, x_k \mathbf{a}_k, \mathbf{a}_{k+1}, \ldots, \mathbf{a}_n]$$
$$= x_k \det A.$$

Since $\det A \neq 0$, we have

$$x_k = \frac{\det B_k}{\det A}.$$

Notice that the equation $\det B_k = x_k \det A$ is valid regardless of whether or not $\det A = 0$. If $\det A \neq 0$ the system has a unique solution; if $\det A = 0$ the system can have a solution only if $\det B_k = 0$ for all k.

The case of a homogeneous system, $A\mathbf{x} = \mathbf{0}$, is of particular interest. Such a system always possesses the trivial solution, $\mathbf{x} = \mathbf{0}$. If $\det A \neq 0$, Theorem 2.10 says that this is the only solution. To see what happens when $\det A = 0$, we reduce the matrix of the system to row-echelon form. The reduced $n \times (n + 1)$ matrix has the form

$$\begin{bmatrix} 0 & \cdots & 1 & \cdots\cdots\cdots\cdots\cdots\cdots & 0 \\ \cdots\cdots\cdots\cdots\cdots\cdots\cdots\cdots\cdots \\ 0 & \cdots\cdots\cdots & 1 & \cdots\cdots\cdots & 0 \\ \cdots\cdots\cdots\cdots\cdots\cdots\cdots\cdots\cdots \\ 0 & \cdots\cdots\cdots\cdots\cdots & 1 & \cdots & 0 \\ 0 & \cdots\cdots\cdots\cdots\cdots\cdots\cdots & 0 \\ \cdots\cdots\cdots\cdots\cdots\cdots\cdots\cdots \\ 0 & \cdots\cdots\cdots\cdots\cdots\cdots\cdots\cdots & 0 \end{bmatrix} \begin{matrix} \\ \\ \\ \text{(row } k) \\ \\ \\ \\ \text{(row } n) \end{matrix}$$

We recall that elementary row operations transform a singular matrix into a singular matrix. If $k = n$, the coefficient matrix of the reduced system would be the nonsingular identity matrix. This is impossible, since A is singular. Consequently $k < n$ and $n - k$ of the unknowns may be assigned arbitrary values. We summarize as follows.

Theorem 2.12   The equation $A\mathbf{x} = \mathbf{0}$, where A is an $n \times n$ matrix, has a nontrivial solution if and only if $\det A = 0$.

As an application of Cramer's rule, let the quantities a_{ij} and b_i, $1 \le i \le n$, $1 \le j \le n$, be *continuous functions* defined on an interval $\mathscr{I}$. If $\det A(x)$ does not vanish for any x in $\mathscr{I}$, the system

$$a_{11}(x)f_1(x) + \cdots + a_{1n}(x)f_n(x) = b_1(x),$$

$$\cdots\cdots\cdots\cdots\cdots\cdots\cdots\cdots\cdots\cdots\cdots\cdots$$

$$a_{n1}(x)f_1(x) + \cdots + a_{nn}(x)f_n(x) = b_n(x),$$

determines a set of functions $f_1, f_2, \ldots, f_n$ each defined on $\mathscr{I}$. Cramer's rule allows us to conclude that each of these functions is *continuous*, since each can be expressed as the quotient of quantities that are the sums of products of continuous functions. It is not necessary to actually solve the system to determine this important property of the solution functions.

Exercises for Section 2.11

1. Let A be of size $m \times n$. Under what conditions can Cramer's rule be used to find the solutions of the equation $A\mathbf{x} = \mathbf{b}$?

In Exercises 2–9, if Cramer's rule applies, solve the system using the rule and also solve it by the Gauss reduction method. If Cramer's rule does not apply, solve the system by reducing the matrix of the system to row-echelon form.

2. $\begin{aligned} 3x_1 - 2x_2 &= 1 \\ -2x_1 + 2x_2 &= 5 \end{aligned}$

3. $\begin{aligned} 4x_1 + 5x_2 &= 8 \\ 2x_1 + x_2 &= -7 \end{aligned}$

4. $\begin{aligned} 2x_1 - 6x_2 &= 1 \\ -x_1 + 3x_2 &= 4 \end{aligned}$

5. $\begin{aligned} -3x_1 + x_2 &= 6 \\ 9x_1 - 3x_2 &= -18 \end{aligned}$

6. $\begin{aligned} 2x_1 - x_2 + 3x_3 &= 1 \\ x_2 + 2x_3 &= -3 \\ x_1 \phantom{{}+x_2} + x_3 &= 0 \end{aligned}$

7. $\begin{aligned} -4x_1 + x_2 \phantom{{}+ 4x_3} &= 3 \\ 2x_1 + 2x_2 + x_3 &= -2 \\ 3x_1 \phantom{{}+ 2x_2} + 4x_3 &= 2 \end{aligned}$

8. $\begin{aligned} 3x_1 \phantom{{}+ 2x_2} + x_3 &= -2 \\ x_1 + 2x_2 - x_3 &= 0 \\ x_1 - 4x_2 + 3x_3 &= 1 \end{aligned}$

9. $\begin{aligned} -2x_1 - x_2 \phantom{{}- x_3} &= 3 \\ x_1 + 3x_2 - x_3 &= 0 \\ 5x_2 - 2x_3 &= 3 \end{aligned}$

10. Let A be a singular matrix. Show that the equation $A\mathbf{x} = \mathbf{b}$ is either inconsistent or else possesses infinitely many solutions.

In Exercises 11 and 12, determine all values of c for which the system has nontrivial solutions, and then find all such solutions.

11. $\begin{aligned} x_1 + 2x_2 + cx_3 &= 0 \\ 3x_1 - x_2 \phantom{{}+ x_3} &= 0 \\ -2x_1 + x_2 + x_3 &= 0 \end{aligned}$

12. $\begin{aligned} 2x_1 - x_2 + 5x_3 &= 0 \\ cx_2 - 2x_3 &= 0 \\ x_1 + 2cx_2 \phantom{{}+ x_3} &= 0 \end{aligned}$

13. (a) If A is nonsingular and $AB = 0$, show that $B = 0$. (Look at the column vectors of B.)
    (b) If A is nonsingular and $BA = 0$, show that $B = 0$.

(c) If A is singular and $AB = 0$, is B necessarily 0?

14. If A is nonsingular and $AB = AC$, show that $B = C$.

15. Consider the system

$$a_{11}x_1 + a_{12}x_2 = b_1,$$

$$a_{21}x_1 + a_{22}x_2 = b_2,$$

where neither a_{11} nor a_{21} is zero. Clearly, we can eliminate x_1 from the second equation by

multiplying through in the first equation by a_{21}/a_{11} and subtracting the resulting equation from the second equation. Alternatively, we could eliminate x_1 from the first equation by multiplying through in the second equation by a_{11}/a_{21} and subtracting the resulting equation from the first equation. In a practical problem the numbers a_{ij} and b_i would probably be rounded and not exact. If $|a_{11}/a_{21}| < 1$, explain why it would be better to eliminate x_1 from the first equation rather than the second.

2.12
THE INVERSE
OF A MATRIX

Let us consider the equation

$$A\mathbf{x} = \mathbf{b},\tag{2.17}$$

where A is a nonsingular $n \times n$ matrix. If we can find an $n \times n$ matrix B with the property that

$$BA = I,$$

then we can solve the equation easily. Upon multiplying both sides of Eq. (2.17) by B we have

$$BA\mathbf{x} = B\mathbf{b}$$

or

$$\mathbf{x} = B\mathbf{b}.$$

If there does exist a matrix B such that $BA = I$, then A and B must both be nonsingular. For det $I = 1$ and hence det $(BA) = \det B \cdot \det A = 1$.

If $BA = I$ then it is also true that $AB = I$. To prove this, let $C = AB$. Then

$$BC = BAB = IB = B,$$

so that

$$B(C - I) = 0.$$

Hence every column vector of $C - I$ is a solution of the equation $B\mathbf{x} = \mathbf{0}$. Since B is nonsingular, Cramer's rule asserts that $C - I = 0$ or $C = AB = I$.

Suppose that A is nonsingular. Let $\mathbf{e}_1, \mathbf{e}_2, \ldots, \mathbf{e}_n$ be the column vectors of I and let B be an $n \times n$ matrix with column vectors $\mathbf{b}_1, \mathbf{b}_2, \ldots, \mathbf{b}_n$. Then B satisfies the equation $AB = I$ if and only if

$$A[\mathbf{b}_1, \mathbf{b}_2, \ldots, \mathbf{b}_n] = [\mathbf{e}_1, \mathbf{e}_2, \ldots, \mathbf{e}_n]$$

or

$$Ab_1 = e_1, \qquad Ab_2 = e_2, \ldots, \qquad Ab_n = e_n. \qquad (2.18)$$

Since A is nonsingular, there exists exactly one vector b_i such that $Ab_i = e_i$, $1 \le i \le n$. We summarize our results thus far as follows.

Theorem 2.13   If A is a nonsingular $n \times n$ matrix, there exists one and only one $n \times n$ matrix B such that

$$AB = BA = I.$$

The column vectors of B are the solutions of Eqs. (2.18).
  The matrix B is called the *inverse* of the matrix A and is denoted by A^{-1}.
  As an example, let us find the inverse of the (nonsingular) matrix

$$A = \begin{bmatrix} 0 & 1 & 0 \\ 2 & 0 & -1 \\ 1 & 3 & 1 \end{bmatrix}. \qquad (2.19)$$

To do this, we need to find vectors b_1, b_2, and b_3 such that

$$Ab_1 = \begin{bmatrix} 1 \\ 0 \\ 0 \end{bmatrix}, \qquad Ab_2 = \begin{bmatrix} 0 \\ 1 \\ 0 \end{bmatrix}, \qquad Ab_3 = \begin{bmatrix} 0 \\ 0 \\ 1 \end{bmatrix}.$$

Then the inverse of A will be given by the formula

$$A^{-1} = [b_1, b_2, b_3].$$

We can solve the three systems $Ab_i = e_i$, $i = 1, 2, 3$, by writing down the matrix of each system and reducing it to row-echelon form (or by performing a Gauss reduction). However, since the coefficient matrix is the same for each system, we can save some work by using the following procedure. First we write down the 3×6 matrix

$$\begin{bmatrix} 0 & 1 & 0 & 1 & 0 & 0 \\ 2 & 0 & -1 & 0 & 1 & 0 \\ 1 & 3 & 1 & 0 & 0 & 1 \end{bmatrix}$$

formed by adjoining the identity matrix to A. Then we reduce this matrix to row-echelon form. As the first step, we interchange rows 1 and 3. Then we place zeros in the positions of the first column below row 1. The result is

$$\begin{bmatrix} 1 & 3 & 1 & 0 & 0 & 1 \\ 0 & -6 & -3 & 0 & 1 & -2 \\ 0 & 1 & 0 & 1 & 0 & 0 \end{bmatrix}.$$

Next we multiply through in the second row by $-\frac{1}{6}$ and add multiples of this row to the other rows so as to introduce zeros in the second column. Thus we obtain the matrix

$$\begin{bmatrix} 1 & 0 & -\frac{1}{2} & 0 & \frac{1}{2} & 0 \\ 0 & 1 & \frac{1}{2} & 0 & -\frac{1}{6} & \frac{1}{3} \\ 0 & 0 & -\frac{1}{2} & 1 & \frac{1}{6} & -\frac{1}{3} \end{bmatrix}.$$

Finally we multiply through in the last row by -2 and add multiples of the last row to the other rows so as to arrive at the row-echelon matrix

$$\begin{bmatrix} 1 & 0 & 0 & -1 & \frac{1}{3} & \frac{1}{3} \\ 0 & 1 & 0 & 1 & 0 & 0 \\ 0 & 0 & 1 & -2 & -\frac{1}{3} & \frac{2}{3} \end{bmatrix}.$$

The three systems $Ab_i = e_i$, $i = 1, 2, 3$ are equivalent to the systems

$$I\mathbf{b}_1 = \begin{bmatrix} -1 \\ 1 \\ -2 \end{bmatrix}, \quad I\mathbf{b}_2 = \begin{bmatrix} \frac{1}{3} \\ 0 \\ -\frac{1}{3} \end{bmatrix}, \quad I\mathbf{b}_3 = \begin{bmatrix} \frac{1}{3} \\ 0 \\ \frac{2}{3} \end{bmatrix},$$

respectively. The last three columns of the row-echelon matrix are therefore the column vectors of A^{-1}. Hence

$$A^{-1} = \begin{bmatrix} -1 & \frac{1}{3} & \frac{1}{3} \\ 1 & 0 & 0 \\ -2 & -\frac{1}{3} & \frac{2}{3} \end{bmatrix}.$$

In general, if A is an $n \times n$ nonsingular matrix, we can form an $n \times 2n$ matrix by adjoining the $n \times n$ identity matrix to A. We write this matrix as

$$[A \quad I].$$

This matrix can be reduced to the form

$$[I \quad C]$$

by elementary row operations. Since the equation $A\mathbf{b}_k = \mathbf{e}_k$ is equivalent to the equation $I\mathbf{b}_k = \mathbf{c}_k$, $1 \le k \le n$, it is clear that $C = A^{-1}$. Thus this last matrix is

$$[I \quad A^{-1}].$$

Our next theorem gives a formula for the elements of A^{-1}. While not very practical for computational purposes, the formula is occasionally useful for theoretical purposes.

Theorem 2.14   Let the element in the ith row and jth column of A^{-1} be denoted by $\tilde{a}_{ij}$. Then

$$\tilde{a}_{ij} = \frac{A_{ji}}{\det A}, \tag{2.20}$$

where A_{ji} is the cofactor of the element a_{ji} of A. Thus to find A^{-1} we replace each element a_{ij} of A by its cofactor A_{ij}, form the transpose of the resulting matrix, and divide each element by $\det A$.

Proof   We must show that if $\tilde{a}_{ij}$ is given by formula (2.20) then

$$\sum_{k=1}^{n} a_{ik}\tilde{a}_{kj} = \delta_{ij}, \qquad 1 \leq i \leq n, \qquad 1 \leq j \leq n.$$

Using Theorem 2.9, we see that

$$\sum_{k=1}^{n} a_{ik}\tilde{a}_{kj} = \frac{1}{\det A} \sum_{k=1}^{n} a_{ik}A_{jk}$$

$$= \frac{1}{\det A} \delta_{ij} \det A$$

$$= \delta_{ij}.$$

Let us apply the theorem to find the inverse of a nonsingular 2×2 matrix

$$A = \begin{bmatrix} a & b \\ c & d \end{bmatrix}.$$

Calculating cofactors, we have

$$\begin{bmatrix} A_{11} & A_{12} \\ A_{21} & A_{22} \end{bmatrix} = \begin{bmatrix} d & -c \\ -b & a \end{bmatrix}.$$

Taking the transpose and dividing by $\det A = ad - bc$, we have

$$A^{-1} = \frac{1}{ad - bc} \begin{bmatrix} d & -b \\ -c & a \end{bmatrix}.$$

To illustrate this formula, we observe that the inverse of the matrix

$$A = \begin{bmatrix} 2 & -1 \\ 1 & 3 \end{bmatrix}$$

is

$$A^{-1} = \frac{1}{7} \begin{bmatrix} 3 & 1 \\ -1 & 2 \end{bmatrix}.$$

For matrices of higher order, Theorem 2.12 is not efficient for calculating the inverse matrix. A more practical use of the theorem is illustrated by the following result, which will be of theoretical importance in our study of linear differential equations.

Theorem 2.15   Let each of the functions a_{ij}, $1 \le i \le n$, $1 \le j \le n$, be defined and continuous on an interval $\mathscr{I}$. Suppose that the matrix $A(x)$ with elements $a_{ij}(x)$ is nonsingular for each x in $\mathscr{I}$. If the elements of $[A(x)]^{-1}$ are denoted by $\tilde{a}_{ij}(x)$ then the functions $\tilde{a}_{ij}$ are continuous on $\mathscr{I}$.

Proof   The functions A_{ij} and det A are continuous, being the sums of products of continuous functions. According to Theorem 2.14, the functions $\tilde{a}_{ij}$ are the quotients of continuous functions and therefore are continuous.

Yet another method for finding the inverse of a matrix is described in Section 4.3.

──────────── **Exercises for Section 2.12** ────────────

1. If A is singular, why can there exist no matrix B such that $AB = I$ or $BA = I$?

In Exercises 2–11, find the inverse of each nonsingular matrix (a) by using the method of the example (2.19); (b) by using Theorem 2.14.

2. $\begin{bmatrix} -2 & 2 \\ -4 & 3 \end{bmatrix}$

3. $\begin{bmatrix} 2 & 1 \\ -5 & -4 \end{bmatrix}$

4. $\begin{bmatrix} 2 & -1 \\ -4 & 2 \end{bmatrix}$

5. $\begin{bmatrix} 3 & 1 \\ 1 & 1 \end{bmatrix}$

6. $\begin{bmatrix} 1 & -2 & 0 \\ 2 & -1 & 1 \\ 0 & 4 & 2 \end{bmatrix}$

7. $\begin{bmatrix} 1 & 0 & 2 \\ 2 & -3 & 4 \\ 0 & 2 & 1 \end{bmatrix}$

8. $\begin{bmatrix} 3 & 1 & 0 \\ 2 & 1 & 1 \\ 1 & 0 & 1 \end{bmatrix}$

9. $\begin{bmatrix} 2 & -1 & 1 \\ 1 & 2 & -2 \\ 3 & 1 & 0 \end{bmatrix}$

10. $\begin{bmatrix} 2 & 3 & 2 \\ 3 & 1 & -2 \\ -1 & 0 & 1 \end{bmatrix}$

11. $\begin{bmatrix} 1 & 1 & -1 & 2 \\ 0 & 2 & 0 & -1 \\ -1 & 2 & 2 & -2 \\ 0 & -1 & 0 & 1 \end{bmatrix}$

12. If A is nonsingular, show that $(A^T)^{-1} = (A^{-1})^T$.

13. Let A and B be nonsingular matrices of the same order. If $C = AB$, show that $C^{-1} = B^{-1}A^{-1}$.

14. Let A be nonsingular. Show that the inverse of A^{-1} is A.

15. The matrix I is its own inverse, since $I \cdot I = I$.
    (a) Find at least two second-order matrices (other than I) that have this property.
    (b) If $A = A^{-1}$, show that det $A = \pm 1$.

16. A real nonsingular matrix A is said to be *orthogonal* if $A^{-1} = A^T$.
    (a) If A is orthogonal, show that det $A = \pm 1$.

(b) Show that A is orthogonal if and only if the row (column) vectors of A are mutually orthogonal unit vectors.

17. If $aA^2 + bA + cI = 0$, where $c \neq 0$, show that A is nonsingular (and therefore has an inverse).

18. If A is an $n \times 1$ matrix and B is a $1 \times n$ matrix, $n > 1$, show that the $n \times n$ matrix AB is singular (and therefore does not have an inverse). (Show that the row vectors of AB are all multiples of the single row vector of B.)

Additional Exercises for Chapter 2

1. Solve by reducing to row-echelon form.

    (a) $2x_1 - 3x_2 - 5x_3 = -4$
    $x_1 + x_2 + 3x_3 = 1$
    $5x_1 \qquad + 4x_3 = -1$
    $4x_1 - x_2 + \cdot x_3 = -2$

    (b) $-x_1 + x_2 + x_3 - 3x_4 = 0$
    $2x_1 - x_2 + x_3 + x_4 = 0$
    $x_1 - 2x_2 \qquad + 4x_4 = 0$

    (c) $x_1 + x_2 + x_3 = 0$
    $-x_1 + 3x_2 + x_3 = 2$
    $2x_2 - x_3 = 2$

    (d) $3x_1 - 5x_2 - x_3 = 0$
    $2x_1 - x_2 - x_3 = 0$
    $x_1 + 3x_2 - x_3 = 0$

2. Solve in two ways: by Gauss reduction and by Cramer's rule.

    (a) $2x_1 - x_2 = 5$
    $x_1 + 3x_2 = -1$

    (b) $-2x_1 + 3x_2 = 13$
    $5x_1 + 4x_2 = 2$

    (c) $x_1 - x_2 \qquad = 5$
    $2x_2 + x_3 = -3$
    $3x_1 + x_2 - 4x_3 = 0$

    (d) $2x_1 + x_2 - x_3 = 5$
    $-x_1 - x_2 + 2x_3 = -3$
    $x_1 + 2x_2 - x_3 = 0$

3. Find $A^T B A$ if
$$A = \begin{bmatrix} 2 & 1 & 0 \\ 0 & 0 & 1 \\ -1 & 2 & 0 \end{bmatrix},$$
$$B = \begin{bmatrix} -7 & 0 & 6 \\ 0 & 5 & 0 \\ 6 & 0 & 2 \end{bmatrix}$$

4. Evaluate the determinant by reducing to triangular form.

    (a) $\begin{vmatrix} -2 & 1 & 1 & -2 \\ 2 & 1 & 2 & -1 \\ 9 & -1 & 2 & 6 \\ 4 & 1 & 2 & 1 \end{vmatrix}$

    (b) $\begin{vmatrix} 1 & 5 & 0 & -2 \\ 1 & -4 & 1 & -2 \\ 3 & 3 & 2 & 2 \\ 1 & 5 & 0 & -1 \end{vmatrix}$

5. Evaluate the determinant by using Theorem 2.8.

    (a) $\begin{vmatrix} -2 & 1 & 5 \\ 3 & 0 & 0 \\ 1 & 2 & 4 \end{vmatrix}$

(b) $\begin{vmatrix} 1 & 2 & 3 \\ 4 & -1 & 1 \\ 2 & 0 & 1 \end{vmatrix}$

(c) The determinant of Exercise 4(a).
(d) The determinant of Exercise 4(b).

6. Find the value of $\delta(n-1, n-2, n-3, \ldots, 2, 1)$.

7. A *permutation matrix* has one nonzero element in each row and one nonzero element in each column, and each nonzero element is equal to one. Show that the determinant of such a matrix is either 1 or -1.

8. Let (x_1, y_1, z_1), (x_2, y_2, z_2), and (x_3, y_3, z_3) be distinct noncollinear points. Show that the equation of the plane through the points is

$$\begin{vmatrix} x & y & z & 1 \\ x_1 & y_1 & z_1 & 1 \\ x_2 & y_2 & z_2 & 1 \\ x_3 & y_3 & z_3 & 1 \end{vmatrix} = 0.$$

9. Show that $(cA)^T = cA^T$ and that $(A - B)^T = A^T - B^T$.

10. (a) If A is any square matrix, show that $A + A^T$ is symmetric and $A - A^T$ is skew-symmetric. (Use the results of Exercise 9.)
    (b) Show that any square matrix can be written as the sum of a symmetric and a skew-symmetric matrix.
    (c) Write the given matrix as the sum of a symmetric and a skew-symmetric matrix.

    (i) $\begin{bmatrix} 2 & -1 \\ 3 & 5 \end{bmatrix}$ (ii) $\begin{bmatrix} 0 & -2 & 3 \\ 0 & -4 & 2 \\ 5 & -1 & 4 \end{bmatrix}$

11. If A and B commute, show that A^2 and B^2 commute.

12. Suppose A and B are symmetric, of the same size.
    (a) Is AB necessarily symmetric?
    (b) If A and B commute, is AB necessarily symmetric?

13. If A and B are orthogonal matrices (Exercise 16, Section 2.12), show that AB is orthogonal.

14. If A is an $n \times n$ matrix with $n > 2$, and if every submatrix of A of order 2 is singular, show that A is singular.

15. If A is nonsingular and symmetric, show that A^{-1} is symmetric. If A is skew-symmetric, is A^{-1} skew-symmetric?

16. Find the inverse of each matrix in Exercise 10(c).

17. A chemical company has 3 products, A, B, and C, each of which must be processed through 2 plants. Plant 1 has 80 hours of operating time available per week, while Plant 2 has 60 hours available. The numbers of hours required for each ton of A, B, and C in each plant is shown in the accompanying table. What are the possible amounts of chemical C that can be produced in a week if both plants are used to capacity?

Hours required per ton			
Plant	A	B	C
1	4	3	2
2	1	2	4

18. A factory uses 3 machines to make 4 products. The numbers of hours of operation of each machine necessary to make one unit of each product are shown in the table below. If each machine is to be utilized 40 hours, how many units of each product can be produced? (Assume fractional units are permissible.)

Machine	Product			
	1	2	3	4
1	2	5	4	1
2	1	1	7	0
3	0	2	6	4

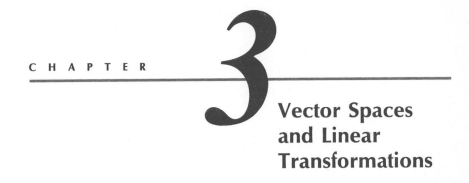

CHAPTER 3

Vector Spaces and Linear Transformations

3.1
VECTOR SPACES

In Section 2.4 we defined R^n to be the set of all ordered n-tuples of real numbers. We used the notation

$$\mathbf{u} = (u_1, u_2, \ldots, u_n), \qquad \mathbf{v} = (v_1, v_2, \ldots, v_n)$$

for elements, or vectors, in R^n. If $\mathbf{u}$ is in R^n and c is a real number, the product of the number c and the vector $\mathbf{u}$ is defined as

$$c\mathbf{u} = (cu_1, cu_2, \ldots, cu_n).$$

This product is again an element of R^n. We indicate this fact by saying that R^n is *closed under the operation of multiplication by a number*. The sum $\mathbf{u} + \mathbf{v}$ of two elements of R^n is defined as

$$\mathbf{u} + \mathbf{v} = (u_1 + v_1, u_2 + v_2, \ldots, u_n + v_n).$$

Since this sum is an element of R^n, we say that R^n is *closed under the operation of addition*. We now alter our definition of R^n slightly. We define R^n to be the set of all ordered n-tuples of real numbers, *together with two operations*, "multiplication by a number" and "addition." Thus R^n is not merely a set, but a set with which two operations are associated. The set is closed under these operations; that is, performance of either operation

129

always leads to an element of the set. Notice that R^n possesses a special element,

$$\mathbf{0} = (0, 0, \ldots, 0)$$

with the property that

$$\mathbf{u} + \mathbf{0} = \mathbf{u}$$

for every element $\mathbf{u}$ in R^n. The element $\mathbf{0}$ is called the *zero element* of R^n. Associated with every element $\mathbf{u}$ of R^n is an element

$$-\mathbf{u} = (-1)\mathbf{u} = (-u_1, -u_2, \ldots, -u_n)$$

with the property that

$$\mathbf{u} + (-\mathbf{u}) = \mathbf{0}.$$

Many other sets of objects that are frequently encountered in mathematics possess properties similar to those of R^n. Of course, two operations ("multiplication by a number" and "addition") must be associated with the set.

For example, let us consider the set of all real-valued functions defined on an interval $\mathscr{I}$. If f is such a function and c is a real number, then the product of c and f, cf, is defined to be the function whose value at each point x in $\mathscr{I}$ is $cf(x)$. The sum $f + g$ of two functions is defined to be the function with the value $f(x) + g(x)$ at each x in $\mathscr{I}$. Thus the set of functions defined on $\mathscr{I}$ is closed under the operations of addition and multiplication by a number. The zero function 0 (we use the symbol 0 for both the number zero and the zero function) whose values are all zero, has the property that

$$f + 0 = f$$

for every function f. Also, if we define

$$-f = (-1)f,$$

then

$$f + (-f) = 0$$

for every function f.

The similarities between the set of vectors in R^n and the set of functions arise not so much because of the nature of the *elements* of the sets, but because of the nature of the *operations* defined on the sets. It is therefore more efficient to consider, in an abstract way, sets of objects whose elements are not defined, but on which are defined operations of "multiplication by a number" and "addition." Sets possessing the properties described above, in addition to certain others, are called *vector spaces* or *linear spaces*. We now formulate a precise definition.

Let V be a nonempty collection or set of objects. (We denote elements of V by $\mathbf{u}$, $\mathbf{v}$, $\mathbf{w}$, and so on. We denote real numbers by a, b, c, and so on.) For every real number a and every element $\mathbf{u}$ of V, let the "product" $a\mathbf{u}$ be defined and be an element of V. For every ordered pair $(\mathbf{u}, \mathbf{v})$ of elements of V let the "sum" $\mathbf{u} + \mathbf{v}$ be defined and be an element of V. Let the operations of "multiplication by a number" and "addition" be such that the following properties hold. (Here the symbol $=$ means "is the same as.")

Property 1 For all $\mathbf{u}$, $\mathbf{v}$, and $\mathbf{w}$ in V,

$$\mathbf{v} + \mathbf{u} = \mathbf{u} + \mathbf{v}, \qquad (\mathbf{u} + \mathbf{v}) + \mathbf{w} = \mathbf{u} + (\mathbf{v} + \mathbf{w}).$$

Property 2 There exists an element $\mathbf{0}$ of V, called the *zero element*, with the property that

$$\mathbf{u} + \mathbf{0} = \mathbf{u}$$

for every element $\mathbf{u}$ of V.

Property 3 For every element $\mathbf{u}$ of V there exists an element $-\mathbf{u}$ of V, called the *additive inverse* of $\mathbf{u}$, such that

$$\mathbf{u} + (-\mathbf{u}) = \mathbf{0}.$$

Property 4 For all $\mathbf{u}$ and $\mathbf{v}$ in V and for all real numbers a, b, and c,

$$c(\mathbf{u} + \mathbf{v}) = c\mathbf{u} + c\mathbf{v}, \qquad (a + b)\mathbf{u} = a\mathbf{u} + b\mathbf{u},$$
$$a(b\mathbf{u}) = (ab)\mathbf{u}, \qquad\qquad 1\mathbf{u} = \mathbf{u}.$$

Then V, together with the operations of "multiplication by a number" and "addition," is called a *vector space*, or a *linear space, over the real numbers.* More briefly, we refer to V as a *real vector space*, or a *real linear space.*

If $\mathbf{v}_1$ and $\mathbf{v}_2$ are in V and if c_1 and c_2 are real numbers, then

$$c_1\mathbf{v}_1 + c_2\mathbf{v}_2$$

is in V. This is because $c_1\mathbf{v}_1$ and $c_2\mathbf{v}_2$ are in V, and the sum of two elements of V is in V. If $\mathbf{v}_1$, $\mathbf{v}_2$, and $\mathbf{v}_3$ are in V, then we see that

$$c_1\mathbf{v}_1 + c_2\mathbf{v}_2 + c_3\mathbf{v}_3 = (c_1\mathbf{v}_1 + c_2\mathbf{v}_2) + c_3\mathbf{v}_3$$

is in V. It can be shown, by mathematical induction, that if $\mathbf{v}_1, \mathbf{v}_2, \ldots, \mathbf{v}_m$ are elements of V (not necessarily all different) and if $c_1, c_2, \ldots, c_m$ are real numbers, then

$$c_1\mathbf{v}_1 + c_2\mathbf{v}_2 + \cdots + c_m\mathbf{v}_m$$

is in V.

The important properties

$$0\mathbf{u} = \mathbf{0}, \tag{3.1}$$

$$(-\mathbf{u}) = (-1)\mathbf{u}, \tag{3.2}$$

$$c\mathbf{0} = \mathbf{0}, \tag{3.3}$$

which hold for every element $\mathbf{u}$ of V and for every number c, follow from the definition of a vector space (Exercise 1).

It is easy to see that R^n is a real vector space. Another example of a real vector space is the set of all real-valued functions defined on an interval (together with the operations defined previously). The elements of a vector space need not be "vectors" in the sense of elements of R^n. In the last example they are functions. In the exercises are examples of vector spaces whose elements are numbers, matrices, infinite sequences, and other mathematical objects. By studying vector spaces in general we achieve considerable efficiency, because the facts that we discover apply to all the examples.

It should be emphasized that not every collection of mathematical objects with two operations is a vector space. For example, let S be the set of all unit vectors (length 1) in R^3. Then $(1, 0, 0)$ and $(0, 1, 0)$ are in S but the sum, $(1, 1, 0)$, is not. Hence S is not a vector space. Other examples appear in the exercises.

By allowing the numbers a, b, c, and so on, in our formal definition of a vector space to be complex, we can define a *vector space over the complex numbers*, or a *complex vector space*. As an important example of such a space, let us consider the set C^n of all ordered n-tuples of complex numbers. If

$$\mathbf{u} = (u_1, u_2, \ldots, u_n), \qquad \mathbf{v} = (v_1, v_2, \ldots, v_n)$$

are elements of C^n (the quantities u_i and v_i are complex numbers) we define

$$\mathbf{u} + \mathbf{v} = (u_1 + v_1, u_2 + v_2, \ldots, u_n + v_n)$$

and for every complex number c we define

$$c\mathbf{u} = (cu_1, cu_2, \ldots, cu_n).$$

Then it may be verified that C^n, together with the indicated operations, is a complex vector space.

The set of real numbers and the set of complex numbers are examples of mathematical entities known as *fields*. In more advanced treatments of linear algebra, vector spaces over arbitrary fields are considered. The elements of a field we call *scalars*. We shall restrict our attention only to real and complex numbers as scalars since, in applications, the important vector spaces are those over the fields of real and complex numbers.

———————————————— **Exercises for Section 3.1** ————————————————

1. Show that the Properties (3.1)–(3.3) hold for every vector space.

In Exercises 2–13, determine if the given set constitutes a real vector space. In each case the operations of "multiplication by a number" and "addition" are understood to be the usual operations associated with the elements of the set.

2. The set of all geometric vectors of the form $v = v_1 i + v_2 j + v_3 k$, where i, j, and k are mutually orthogonal unit vectors.

3. The set of all real numbers.

4. The set of all elements of R^3 with first component 0.

5. The set of all elements of R^3 with first component 1.

6. The set of all 2×2 matrices with real elements.

7. The set of all nonsingular 2×2 matrices with real elements.

8. The set of all singular 2×2 matrices with real elements.

9. The set of all continuous functions defined on an interval $\mathscr{I}$.

10. The set of all polynomials of degree ≤ 2.

11. The set of all solutions of the homogeneous equation $Ax = 0$. (Here A is an $m \times n$ matrix; x and 0 are column vectors.)

12. The set of all convergent infinite sequences of real numbers.

13. The set of all convergent infinite series of real numbers.

14. Is the set of all ordered n-tuples of real numbers a complex vector space?

15. Is the set of all ordered n-tuples of complex numbers a real vector space?

16. If V is a vector space over the complex numbers, show that the set of elements of V constitutes a vector space over the real numbers.

17. Show that a vector space has only one zero element. That is, show that if $v + 0 = v$ and $v + 0' = v$ for every element v, then $0' = 0$.

18. If u, v, and w are elements of a vector space and $u + v = u + w$, show that $v = w$.

3.2
SUBSPACES

In order to illustrate the main concept of this section, let us consider the set of all elements in R^3 whose first components are zero. If

$$x = (0, x_2, x_3), \qquad y = (0, y_2, y_3),$$

then

$$cx = (0, cx_2, cx_3)$$

for every real number c and

$$x + y = (0, x_2 + y_2, x_3 + y_3).$$

Also, the zero element $(0, 0, 0)$ of R^3 belongs to this set. It is now easy to verify that the set of all elements of R^3 with first component 0 forms a real vector space under the same operations associated with R^3.

More generally, let V be any vector space and suppose that some subset of V is a vector space under the same operations and the same field associated with V. Then this subset (together with the two operations) is called a *subspace* of V. In particular, V is a subspace of itself. Also, the set whose single element is the zero element of V is a subspace of V. Necessary and sufficient conditions that a subset U of V constitute a vector space are given by the following theorem.

Theorem 3.1   Let V be a vector space and let U be a subset of V. Then U is a subspace of V if and only if it is nonempty and is closed under the two operations of V: if $\mathbf{v}$ is in U so is $c\mathbf{v}$; if $\mathbf{v}_1$ and $\mathbf{v}_2$ are in U so is $\mathbf{v}_1 + \mathbf{v}_2$.

Proof   Suppose first that U is a subspace of V. It must contain a zero element and hence is nonempty. Since it is a vector space, it is closed under the two operations of V. Hence the conditions are necessary.

Now suppose that U is nonempty and closed under the two operations. Since it is nonempty, it contains an element $\mathbf{v}$. Then $c\mathbf{v}$ is in U for every scalar c. In particular $(-1)\mathbf{v}$ is in U. Hence $\mathbf{v} + (-1)\mathbf{v} = 0$ is in U, so U contains the zero element. The remaining properties of a vector space hold for the elements of U because these elements belong to V. Thus U is a vector space. The conditions are sufficient.

As another example, let $\mathscr{F}$ be the vector space of all real valued functions defined on some interval $\mathscr{I}$. Then the set of all *continuous* real valued functions defined on $\mathscr{I}$ forms a subspace of $\mathscr{F}$: if f and g are continuous, $f + g$ and cf are also continuous.

In many applications, we wish to describe a subspace of a certain larger space. For example (Exercise 6), if A is an $n \times n$ matrix, the solutions of the equation $A\mathbf{x} = \mathbf{0}$ form a subspace of R^n. As we shall see in Chapter 5, the solutions of certain differential equations constitute a subspace of a class of functions.

Now let V be an arbitrary vector space and let $\mathbf{v}_1, \mathbf{v}_2, \ldots, \mathbf{v}_m$ be elements of V. If there exist scalars $c_1, c_2, \ldots, c_m$ such that

$$\mathbf{v} = c_1\mathbf{v}_1 + c_2\mathbf{v}_2 + \cdots + c_m\mathbf{v}_m,$$

we say that $\mathbf{v}$ is a *linear combination* of the elements $\mathbf{v}_1, \mathbf{v}_2, \ldots, \mathbf{v}_m$. We also use the summation notation

$$\mathbf{v} = \sum_{i=1}^{m} c_i\mathbf{v}_i.$$

If *every* element of V is a linear combination of the elements $\mathbf{v}_1, \mathbf{v}_2, \ldots, \mathbf{v}_m$ of V, we say that V is *spanned* by these elements. We also say that the elements *span* V.

Suppose that $\mathbf{u}_1, \mathbf{u}_2, \ldots, \mathbf{u}_k$ are any k elements of a vector space V. Then the set of all linear combinations of these elements, that is, the set of all elements of V of the form

$$c_1\mathbf{u}_1 + c_2\mathbf{u}_2 + \cdots + c_k\mathbf{u}_k, \tag{3.4}$$

is a subspace of V. This follows because if

$$\mathbf{v} = a_1\mathbf{u}_1 + a_2\mathbf{u}_2 + \cdots + a_k\mathbf{u}_k, \qquad \mathbf{w} = b_1\mathbf{u}_1 + b_2\mathbf{u}_2 + \cdots + b_k\mathbf{u}_k.$$

then

$$c\mathbf{v} = (ca_1)\mathbf{u}_1 + (ca_2)\mathbf{u}_2 + \cdots + (ca_k)\mathbf{u}_k$$

and

$$\mathbf{v} + \mathbf{w} = (a_1 + b_1)\mathbf{u}_1 + (a_2 + b_2)\mathbf{u}_2 + \cdots + (a_k + b_k)\mathbf{u}_k$$

are again of the form (3.4). The elements $\mathbf{u}_1, \mathbf{u}_2, \ldots, \mathbf{u}_k$ are in the subspace, because

$$\mathbf{u}_1 = 1\mathbf{u}_1 + 0\mathbf{u}_2 + \cdots + 0\mathbf{u}_k, \qquad \mathbf{u}_2 = 0\mathbf{u}_1 + 1\mathbf{u}_2 + \cdots + 0\mathbf{u}_k,$$

and so on. The subspace is evidently spanned by these k elements. Also, any subspace that contains the k elements must contain every linear combination of them. Thus we may speak of *the* subspace spanned by the k elements.

Example 1   Let $V = R^3$, and let $\mathbf{v}_1 = (1, 1, 0)$, $\mathbf{v}_2 = (0, 1, 2)$. Then the subspace of V that is spanned by $\mathbf{v}_1$ and $\mathbf{v}_2$ is the set of all vectors of the form $c_1\mathbf{v}_1 + c_2\mathbf{v}_2$ or $(c_1, c_1 + c_2, 2c_2)$. A geometric interpretation of this subspace (obtained by regarding the components of a vector as the rectangular coordinates of a point) is the plane that passes through the origin and the points $(1, 1, 0)$, $(0, 1, 2)$. The parametric equations of the plane are

$$x_1 = c_1, \qquad x_2 = c_1 + c_2, \qquad x_3 = 2c_2.$$

Hence $2x_2 = 2c_1 + 2c_2 = 2x_1 + x_3$, or

$$2x_1 - 2x_2 + x_3 = 0$$

for each point on the plane.

Example 2   Again let $V = R^3$, and let $\mathbf{v} = (1, -1, 2)$. Then the set of all vectors of the form $c\mathbf{v}$, where c is an arbitrary real number, is a subspace of V. This subspace consists of all vectors of the form $(c, -c, 2c)$. The points that correspond to these vectors lie on the line through the origin with parametric equations

$$x_1 = c, \qquad x_2 = -c, \qquad x_3 = 2c.$$

The line is also described by the equations

$$\frac{x_1}{1} = \frac{x_2}{-1} = \frac{x_3}{2}.$$

We conclude this section with some facts about subspaces.

Theorem 3.2   Let U and V be subspaces of R^n spanned by $u_1, u_2, \ldots, u_r$ and $v_1, v_2, \ldots, v_s$ respectively. If every u is a linear combination of the v's, then U is a subspace of V. If every u is a linear combination of the v's and every v is a linear combination of the u's, then $U = V$.

Proof   Let x be any element of U. Then $x = a_1 u_1 + a_2 u_2 + \cdots + a_r u_r$. If each u is a linear combination of the v's, then x is a linear combination of the v's, so x is in V. Thus U is a subspace of V. If, in addition, every v is a linear combination of the u's, then V is a subspace of U so $U = V$.

Example 3   Let U be the subspace of R^3 spanned by $u_1 = (2, 0, -1)$ and $u_2 = (1, 1, 1)$. If

$$v_1 = u_1 - 2u_2 = (0, -2, -3)$$

then the subspace V of R^3 spanned by v_1 is a subspace of U. Let W be the subspace of R^3 spanned by

$$w_1 = u_1 + u_2 = (3, 1, 0), \qquad w_2 = 2u_2 = (2, 2, 2).$$

Then

$$u_2 = \frac{1}{2} w_2, \qquad u_1 = w_1 - u_2 = w_1 - \frac{1}{2} w_2$$

so each u is a linear combination of the w's and therefore W and U are the same.

Exercises for Section 3.2

1. Show that the set of all elements of R^2 of the form $(a, -a)$, where a is any real number, is a subspace of R^2. Give a geometric interpretation of the subspace.

2. Show that the set of all elements of R^2 of the form $(1, a)$, where a is any real number, is not a subspace of R^2.

3. Show that the set of all elements of R^3 of the form $(a + b, -a, 2b)$, where a and b are any real numbers, is a subspace of R^3. Show that the geometric interpretation of this subspace is a plane and find its equation.

4. Let U be the subspace of R^3 that is spanned by the element $(1, 3, -2)$. Show that the geometric interpretation of U is a line through the origin, and find the equations of the line.

5. Let U be the subspace of R^3 that is spanned by the vectors $(-2, 1, 1)$ and $(1, -1, 3)$. Show

that the geometric interpretation of this subspace is a plane, and find its equation.

6. Let A be a 2×2 real matrix. Show that the set of all (real) solutions of the equation $A\mathbf{x} = \mathbf{0}$ is a subspace of R^2.

7. Let $\mathscr{F}$ be the space of all functions defined on $[a, b]$.
    (a) Let V_1 be the set of all functions f in $\mathscr{F}$ for which $f(a) = 0$. Is V_1 a subspace of $\mathscr{F}$?
    (b) Let V_2 be the set of all functions g in $\mathscr{F}$ for which $g(a) = g(b)$. Is V_2 a subspace of $\mathscr{F}$?
    (c) Let V_3 be the set of all functions h in $\mathscr{F}$ for which $h(a) = 1$. Is V_3 a subspace of $\mathscr{F}$?

8. Let $\mathscr{F}$ be the space of all functions defined on an interval $\mathscr{I}$. Show that the set of all functions f in $\mathscr{F}$ that satisfy the condition $f''(x) + 2f(x) = 0$ for all x in $\mathscr{I}$ is a subspace of $\mathscr{F}$.

9. Let $\mathscr{F}$ be as in Exercise 8. Show that the set of all functions g in $\mathscr{F}$ that satisfy the condition $e^x g'(x) - (\sin x)g(x) = 0$ for all x in $\mathscr{I}$ constitutes a subspace of $\mathscr{F}$.

10. Let A be an $m \times n$ real matrix and $\mathbf{b}$ a real m-dimensional column vector. Show that the equation $A\mathbf{x} = \mathbf{b}$ has a solution if and only if $\mathbf{b}$ belongs to the subspace of R^m that is spanned by the column vectors of A.

11. Let U be a subspace of R^n. Let $U^{\perp}$ be the set of all elements of R^n that are orthogonal to every element of U; that is, $\mathbf{v}$ is in $U^{\perp}$ if $\mathbf{v} \cdot \mathbf{u} = 0$ for every element $\mathbf{u}$ of U. Show that $U^{\perp}$ is a subspace of R^n. (The subspace $U^{\perp}$ is called the *orthogonal complement* of U.)

12. Let V be the set of all convergent infinite sequences, and let U be the set of all infinite sequences that converge to zero. Verify that V is a vector space, and then show that U is a subspace of V.

13. Let U be the subspace of R^3 spanned by $\mathbf{u}_1 = (1, 2, 3)$, $\mathbf{u}_2 = (2, -1, 0)$. Let V be the subspace spanned by $\mathbf{v}_1 = (1, 2, 3)$, $\mathbf{v}_2 = (2, 4, 6)$. Is V a subspace of U? Are U and V the same?

14. Let U be the subspace of R^3 spanned by $(1, 2, 3)$ and $(2, -1, 0)$. Let V be the subspace spanned by $(5, 0, 3)$ and $(0, 5, 6)$. Is V a subspace of U? Are U and V the same?

3.3 LINEAR DEPENDENCE

Let V be a vector space. In what follows, the word "number" refers to a real or complex number, according to whether V is real or complex. A finite set $\{\mathbf{v}_1, \mathbf{v}_2, \ldots, \mathbf{v}_m\}$ of elements of V is said to be *linearly dependent* (we also say that the elements $\mathbf{v}_1, \mathbf{v}_2, \ldots, \mathbf{v}_m$ are linearly dependent) if there exist numbers $c_1, c_2, \ldots, c_m$ not all zero such that

$$c_1\mathbf{v}_1 + c_2\mathbf{v}_2 + \cdots + c_m\mathbf{v}_m = \mathbf{0}.$$

This condition is always satisfied if the numbers c_i are all zero. The restriction that these numbers are not all zero is essential.

Notice how we have used the properties of a vector space in the definition of linear dependence. Each of the products $c_i\mathbf{v}_i$ must be defined and be an element of V. The sum of the m products must also be defined and be an element of V. We also need the presence of $\mathbf{0}$, the zero element.

If the set $\{\mathbf{v}_1, \mathbf{v}_2, \ldots, \mathbf{v}_m\}$ of elements of V is not linearly dependent, it is said to be *linearly independent*.

A linearly dependent set can be characterized in another way.

Theorem 3.3   The set $\{v_1, v_2, \ldots, v_m\}$ is linearly dependent if and only if at least one element of the set is a linear combination of the others.

Proof   Suppose that v_i is a linear combination of the other elements such that

$$v_i = a_1 v_1 + \cdots + a_{i-1} v_{i-1} + a_{i+1} v_{i+1} + \cdots + a_m v_m.$$

Then

$$a_1 v_1 + \cdots + a_{i-1} v_{i-1} + (-1)v_i + a_{i+1} v_{i+1} + \cdots + a_m v_m = 0,$$

and since the coefficient of v_i is not zero, the set is linearly dependent. Next suppose that

$$c_1 v_1 + c_2 v_2 + \cdots + c_m v_m = 0$$

and that c_i, say, is not zero. Then we may write

$$v_i = -\frac{c_1}{c_i} v_1 - \cdots - \frac{c_{i-1}}{c_i} v_{i-1} - \frac{c_{i+1}}{c_i} v_{i+1} - \cdots - \frac{c_m}{c_i} v_m,$$

hence v_i is a linear combination of the other elements.
We consider some examples.

Example 1   Let $v_1 = (1, -1, 3)$ and $v_2 = (2, 1, 0)$ be elements of R^3. Then the condition

$$c_1 v_1 + c_2 v_2 = 0$$

is satisfied if and only if each component is zero:

$$c_1 + 2c_2 = 0,$$
$$-c_1 + c_2 = 0,$$
$$3c_1 + 0c_2 = 0.$$

But this system for c_1 and c_2 is satisfied if and only if $c_1 = c_2 = 0$. Hence v_1 and v_2 are linearly independent. If $v_3 = (-2, 2, -6)$, then v_1 and v_3 are linearly dependent because $v_3 = -2v_1$ or $2v_1 + v_3 = 0$.

Example 2   Let $v_1 = (1, 1, 3)$, $v_2 = (2, 1, -1)$, and $v_3 = (3, 1, -5)$. To determine if the set of vectors $\{v_1, v_2, v_3\}$ is linearly dependent, we must find out if the equation

$$c_1 v_1 + c_2 v_2 + c_3 v_3 = 0$$

is true for any constants other than $c_1 = c_2 = c_3 = 0$. This vector equation corresponds to the system

$$c_1 + 2c_2 + 3c_3 = 0,$$
$$c_1 + c_2 + c_3 = 0,$$
$$3c_1 - c_2 - 5c_3 = 0.$$

Reducing the system to row echelon form, we find at the first step that

$$c_1 + 2c_2 + 3c_3 = 0,$$
$$-c_2 - 2c_3 = 0,$$
$$-7c_2 - 14c_3 = 0.$$

This system reduces to

$$c_1 - c_3 = 0,$$
$$c_2 + 2c_3 = 0,$$
$$0 = 0.$$

Setting $c_3 = a$, where a is any number, we see that the system is satisfied if $c_1 = a$ and $c_2 = -2a$. In particular, we see by taking $a = 1$ that a solution is $c_1 = 1, c_2 = -2$, and $c_3 = 1$. Hence the set of vectors $\{v_1, v_2, v_3\}$ is linearly dependent. However, the subset $\{v_1, v_2\}$ is linearly independent, as we shall now show. The equation

$$c_1 v_1 + c_2 v_2 = 0$$

corresponds to the system

$$c_1 + 2c_2 = 0,$$
$$c_1 + c_2 = 0,$$
$$3c_1 - c_2 = 0,$$

which reduces to

$$c_1 + 2c_2 = 0,$$
$$-c_2 = 0,$$
$$-7c_2 = 0,$$

or

$$c_1 = 0,$$
$$c_2 = 0,$$
$$0 = 0.$$

Thus the equation is satisfied only if $c_1 = c_2 = 0$, so the vectors are linearly independent. It can be shown in like fashion that each of the subsets $\{v_1, v_3\}$ and $\{v_2, v_3\}$ is also linearly independent.

We now consider the general case of a set of n vectors in R^m. One possibility for determining if the vectors are linearly dependent is to proceed as in Example 1. Let $v_1, v_2, \ldots, v_n$ be elements of R^m, and let the m components of v_j be denoted by $v_{1j}, v_{2j}, \ldots, v_{mj}$. Then the condition

$$c_1 v_1 + c_2 v_2 + \cdots + c_n v_n = 0$$

can be written as

$$c_1(v_{11}, v_{21}, \ldots, v_{m1}) + c_2(v_{12}, v_{22}, \ldots, v_{m2}) + \cdots$$
$$+ c_n(v_{1n}, v_{2n}, \ldots, v_{mn}) = (0, 0, \cdots, 0),$$

or, using column vector representations, as

$$c_1 \begin{bmatrix} v_{11} \\ v_{21} \\ \vdots \\ v_{m1} \end{bmatrix} + c_2 \begin{bmatrix} v_{12} \\ v_{22} \\ \vdots \\ v_{m2} \end{bmatrix} + \cdots + c_n \begin{bmatrix} v_{1n} \\ v_{2n} \\ \vdots \\ v_{mn} \end{bmatrix} = \begin{bmatrix} 0 \\ 0 \\ \vdots \\ 0 \end{bmatrix}.$$

In either case, the vector equation corresponds to the system

$$c_1 v_{11} + c_2 v_{12} + \cdots + c_n v_{1n} = 0,$$
$$\cdots\cdots\cdots\cdots\cdots\cdots\cdots\cdots\cdots\cdots\cdots \tag{3.5}$$
$$c_1 v_{m1} + c_2 v_{m2} + \cdots + c_n v_{mn} = 0.$$

The vectors $v_1, v_2, \ldots, v_n$ are linearly dependent if and only if this system has a nontrivial solution. Because of our knowledge of systems of linear equations, certain facts stand out immediately.

Theorem 3.4 Let $v_1, v_2, \ldots, v_n$ be elements of R^m. If $n > m$ the elements are linearly dependent. If $n = m$ the elements are linearly dependent if and only if

$$\det[v_1, v_2, \ldots, v_n] = 0.$$

Proof If $n > m$ the system of equations (3.5) has more unknowns than equations, and hence has a nontrivial solution (Theorem 2.1). If $n = m$ the system has a nontrivial solution if and only if the coefficient matrix of the system is singular (Theorem 2.12).

The next fact will be needed in the sequel.

Theorem 3.5 Let each of the elements $v_1, v_2, \ldots, v_n$ of a vector space V be a linear combination of the elements $u_1, u_2, \ldots, u_m$ of V. If $m < n$, then the elements $v_1, v_2, \ldots, v_n$ are linearly dependent.

Proof   We have

$$\mathbf{v}_i = \sum_{j=1}^{m} a_{ij}\mathbf{u}_j, \qquad 1 \le i \le n.$$

The condition

$$\sum_{i=1}^{n} c_i\mathbf{v}_i = \mathbf{0}$$

is equivalent to the condition

$$\sum_{i=1}^{n} c_i \sum_{j=1}^{m} a_{ij}\mathbf{u}_j = \mathbf{0} \qquad \text{or} \qquad \sum_{j=1}^{m} \left(\sum_{i=1}^{n} a_{ij}c_i \right)\mathbf{u}_j = \mathbf{0}.$$

This is satisfied if

$$\sum_{i=1}^{n} a_{ij}c_i = 0, \qquad 1 \le j \le m.$$

But this is a system of m equations for the n quantities $c_1, c_2, \ldots, c_n$. Since $m < n$ (fewer equations than unknowns), the system has a nontrivial solution (Theorem 2.1). Hence the elements $\mathbf{v}_1, \mathbf{v}_2, \ldots, \mathbf{v}_n$ are linearly dependent.

Example 3   We shall use Theorem 3.5 to prove that any set of more than two vectors in R^2 is linearly dependent. To do this, we first show that every vector in R^2 is a linear combination of the two particular vectors $\mathbf{e}_1 = (1, 0)$ and $\mathbf{e}_2 = (0, 1)$. If $\mathbf{v} = (a, b)$ is any element of R^2, then

$$\mathbf{v} = a(1, 0) + b(0, 1) = a\mathbf{e}_1 + b\mathbf{e}_2.$$

If $\{\mathbf{v}_1, \mathbf{v}_2, \ldots, \mathbf{v}_n\}$ is a set of vectors in R^2, we know that each of these vectors is a linear combination of $\mathbf{e}_1$ and $\mathbf{e}_2$. By Theorem 3.5, if $n > 2$ the vectors are linearly dependent.

Exercises for Section 3.3

1. Determine whether the given vectors are linearly dependent or linearly independent.
   (a) $(2, -1)$, $(-4, 2)$
   (b) $(3, -1)$, $(2, 1)$
   (c) $(2, 1)$, $(3, 0)$, $(1, 4)$
   (d) $(-1, 0, 3)$, $(2, 0, -6)$
   (e) $(1, 2, 3)$, $(2, -1, 0)$
   (f) $(2, -1, 1)$, $(2, 0, 3)$, $(1, 1, -2)$
   (g) $(2, -1, 1)$, $(2, -3, -2)$, $(2, 3, 7)$
   (h) $(1, 0, 2, -2)$, $(2, 1, 0, 1)$
   (i) $(2, -1, 0, 1)$, $(4, -2, 0, 2)$
   (j) $(0, 0, 0, 1)$, $(4, 0, 0, 2)$, $(1, 1, 0, 1)$

2. If S is any finite set of elements of a vector space V that contains the zero element of V, show that S is linearly dependent.

3. Show that any (finite) set of more than n elements of R^n is linearly dependent.

4. (a) Let S be a (finite) set of linearly independent elements of a vector space. Show that every subset of S is linearly independent.

(b) If S is a (finite) set of linearly dependent elements of a vector space, is every subset of S linearly dependent?

(c) Let S be a (finite) set of linearly dependent elements of a vector space V. Show that any (finite) set of elements of V that contains S is linearly dependent.

(d) If S is a (finite) set of linearly independent elements of V, is every finite set that contains S linearly independent?

5. (a) Let $\mathbf{x}$ and $\mathbf{y}$ be linearly independent elements of a vector space. If $\mathbf{u} = a\mathbf{x} + b\mathbf{y}$ and $\mathbf{v} = c\mathbf{x} + d\mathbf{y}$, show that $\mathbf{u}$ and $\mathbf{v}$ are linearly independent if and only if $ad - bc \neq 0$.

(b) Let $\mathbf{x}_1, \mathbf{x}_2, \ldots, \mathbf{x}_n$ be linearly independent elements of a vector space and let

$$\mathbf{u}_i = \sum_{j=1}^{n} a_{ij}\mathbf{x}_j, \qquad 1 \leq i \leq n.$$

Show that the elements $\mathbf{u}_1, \mathbf{u}_2, \ldots, \mathbf{u}_n$ are linearly independent if and only if $\det A \neq 0$.

6. Let $\mathbf{x}_1, \mathbf{x}_2, \ldots, \mathbf{x}_n$ be linearly independent elements of a vector space, and suppose that $\mathbf{y}$ can be expressed as a linear combination of these elements. Show that $\mathbf{y}$ can be expressed as a linear combination in only one way; that is, if

$$\mathbf{y} = \sum_{i=1}^{n} a_i \mathbf{x}_i \qquad \text{and} \qquad \mathbf{y} = \sum_{i=1}^{n} b_i \mathbf{x}_i,$$

then $a_i = b_i$, $1 \leq i \leq n$.

7. Let $\mathbf{x}$, $\mathbf{y}$, and $\mathbf{z}$ be elements of a vector space. If the sets $\{\mathbf{x}, \mathbf{y}\}$ and $\{\mathbf{y}, \mathbf{z}\}$ are linearly independent, is the set $\{\mathbf{x}, \mathbf{y}, \mathbf{z}\}$ necessarily linearly independent?

8. Let $\mathbf{v}_1, \mathbf{v}_2, \ldots, \mathbf{v}_k$ be mutually orthogonal vectors in R^n, so that $\mathbf{v}_i \cdot \mathbf{v}_j = 0$ if $i \neq j$. If none of the vectors is the zero vector, show that they are linearly independent.

9. Sequences $(1, 0, 0, 0, \ldots)$ and $(0, 1, 0, 0, \ldots)$ are elements of the vector space of convergent sequences. Are they linearly independent?

10. Let $\mathbf{v}_1, \mathbf{v}_2, \ldots, \mathbf{v}_n$ be elements of R^m. Theorem 3.4 discusses the cases $n > m$ and $n = m$. If $n < m$, what can be said about the linear dependence of the elements?

11. The set of all polynomials of degree ≤ 2 is a vector space (Exercise 10, Section 3.1). Use the definition of linear independence to determine whether the given set of elements of this space is linearly independent.

(a) $f_0(x) = 1$, $f_1(x) = x$, $f_2(x) = x^2$

(b) $g_0(x) = 2 - x^2$, $g_1(x) = 3x$, $g_2(x) = x^2 + x - 2$

12. The set of all 2×2 real matrices constitutes a real vector space (Exercise 6, Section 3.1). Determine whether the given set of elements is linearly independent.

(a) $\begin{bmatrix} 1 & 2 \\ 3 & 4 \end{bmatrix}$, $\begin{bmatrix} 2 & 1 \\ 0 & -1 \end{bmatrix}$, $\begin{bmatrix} 4 & 5 \\ 6 & 7 \end{bmatrix}$

(b) $\begin{bmatrix} 2 & 3 \\ 1 & 1 \end{bmatrix}$, $\begin{bmatrix} -1 & 2 \\ 0 & 0 \end{bmatrix}$, $\begin{bmatrix} 1 & 0 \\ 0 & 1 \end{bmatrix}$

3.4 WRONSKIANS

In this section we shall concern ourselves with a method for establishing the linear independence of a set of *functions*. In order to illustrate the basic idea, let us consider the pair of functions f and g, where

$$f(x) = 3x^2, \qquad g(x) = 2x,$$

for all x. If

$$c_1 f + c_2 g = 0,$$

then we must have

$$c_1 f(x) + c_2 g(x) = 0$$

and

$$c_1 f'(x) + c_2 g'(x) = 0$$

for all x. That is,

$$3c_1 x^2 + 2c_2 x = 0,$$

$$6c_1 x + 2c_2 = 0$$

for all x. For any fixed x, this is a system of equations for c_1 and c_2. The determinant of the system is

$$6x^2 - 12x^2 = -6x^2.$$

Since this determinant is not zero when $x = 1$ (or for any other value of x except $x = 0$), c_1 and c_2 must both be zero. Consequently the functions f and g must be linearly independent.

We now generalize the above procedure. Let $\mathscr{I}$ be a fixed interval. Let $\mathscr{F}^n$, where n is a nonnegative integer, denote the vector space of all functions that are defined and possess at least n derivatives on $\mathscr{I}$. (Here $\mathscr{F}^0$ is simply the set of all functions defined on $\mathscr{I}$.)

Let $f_1, f_2, \ldots, f_m$ be functions that belong to the space $\mathscr{F}^{m-1}$. Associated with this set of functions at each point x of the interval $\mathscr{I}$ is the determinant

$$\begin{vmatrix} f_1(x) & f_2(x) & \cdots & f_m(x) \\ f_1'(x) & f_2'(x) & \cdots & f_m'(x) \\ \vdots & \vdots & & \vdots \\ f_1^{(m-1)}(x) & f_2^{(m-1)}(x) & \cdots & f_m^{(m-1)}(x) \end{vmatrix}.$$

This determinant is called the *Wronskian* of the functions $f_1, f_2, \ldots, f_m$ at the point x. We shall denote it by the symbol

$$W(x; f_1, f_2, \ldots, f_m)$$

or sometimes simply by $W(x)$ when it is clear what functions are involved. The Wronskian of a set of functions provides a test for the linear independence of the functions, as we shall now see.

Theorem 3.6   Let the functions $f_1, f_2, \ldots, f_m$ belong to the space $\mathscr{F}^{m-1}$, relative to an interval $\mathscr{I}$. If the functions are linearly dependent on $\mathscr{I}$, then the Wronskian $W(x; f_1, f_2, \ldots, f_m)$ is zero at each point x of $\mathscr{I}$. Hence if the Wronskian is not zero at even one point of $\mathscr{I}$, the functions are linearly independent.

Proof   If the functions are linearly dependent there exist numbers $c_1, c_2, \ldots, c_m$, not all zero, such that

$$c_1 f_1(x) + c_2 f_2(x) + \cdots + c_m f_m(x) = 0$$

for all x in $\mathscr{I}$. Since the function $c_1 f_1 + c_2 f_2 + \cdots + c_m f_m$ is the zero function, its derivatives must also be the zero function. Hence we have the m relations

$$c_1 f_1(x) + c_2 f_2(x) + \cdots + c_m f_m(x) = 0,$$

$$c_1 f_1'(x) + c_2 f_2'(x) + \cdots + c_m f_m'(x) = 0,$$

$$\cdots\cdots\cdots\cdots\cdots\cdots\cdots\cdots\cdots\cdots\cdots\cdots\cdots\cdots$$

$$c_1 f_1^{(m-1)}(x) + c_2 f_2^{(m-1)}(x) + \cdots + c_m f_m^{(m-1)}(x) = 0$$

for all x in $\mathscr{I}$. For each fixed x this system is a linear homogeneous system of equations that is satisfied by $c_1, c_2, \ldots, c_m$. These numbers are not all zero, so by Theorem 2.12 the determinant of the system must be zero for every x in $\mathscr{I}$. But this determinant is the Wronskian of the functions. If the Wronskian does not vanish at some point, the functions cannot be linearly dependent. Hence, they must be linearly independent.

As an example, we consider the three functions f_1, f_2, and f_3, where

$$f_1(x) = \cos x, \qquad f_2(x) = \sin x, \qquad f_3(x) = x$$

for all x. The Wronskian is

$$W(x) = \begin{vmatrix} \cos x & \sin x & x \\ -\sin x & \cos x & 1 \\ -\cos x & -\sin x & 0 \end{vmatrix} = x.$$

Since $W(x)$ is not zero for *all* x, the functions are linearly independent.

Theorem 3.6 states that if a set of functions is linearly dependent, then the Wronskian of the functions is identically zero. The converse is not true. If the Wronskian of a set of functions is identically zero the functions need not be linearly dependent. This can be seen from the following example. Let

$$g_1(x) = x^2,$$

$$g_2(x) = x|x| = \begin{cases} x^2, & x \geq 0, \\ -x^2, & x < 0. \end{cases}$$

We note that $g_2'(0)$ exists and is equal to zero. When $x \geq 0$, we have

$$W(x) = \begin{vmatrix} x^2 & x^2 \\ 2x & 2x \end{vmatrix} = 0$$

and when $x < 0$ we have

$$W(x) = \begin{vmatrix} x^2 & -x^2 \\ 2x & -2x \end{vmatrix} = 0.$$

Hence $W(x) = 0$ for all x. But the functions g_1 and g_2 are linearly independent because if

$$c_1 g_1(x) + c_2 g_2(x) = 0$$

for all x, we see by setting $x = 1$ and $x = -1$ that $c_1 + c_2 = 0$ and $c_1 - c_2 = 0$. Hence $c_1 = c_2 = 0$. Thus the functions g_1 and g_2 are linearly independent even though their Wronskian vanishes identically.

We now use Theorem 3.6 to establish the linear independence, on any interval, of a set of exponential functions of the form

$$f_1(x) = e^{r_1 x}, \qquad f_2(x) = e^{r_2 x}, \ldots, \qquad f_n(x) = e^{r_n x}, \tag{3.6}$$

where $r_i \neq r_j$ when $i \neq j$. The fact that these functions are linearly independent will be of some interest to us later in our study of linear differential equations.

The Wronskian of the functions (3.6) is

$$W(x) = e^{(r_1 + r_2 + \cdots + r_n)x} D_n,$$

where

$$D_n = \begin{vmatrix} 1 & 1 & \cdots & 1 \\ r_1 & r_2 & \cdots & r_n \\ r_1^2 & r_2^2 & \cdots & r_n^2 \\ \vdots & \vdots & & \vdots \\ r_1^{n-1} & r_2^{n-1} & \cdots & r_n^{n-1} \end{vmatrix}. \tag{3.7}$$

The determinant (3.7) is known as *Vandermonde's determinant*.

When $n = 2$, we have

$$D_2 = \begin{vmatrix} 1 & 1 \\ r_1 & r_2 \end{vmatrix} = r_2 - r_1.$$

When $n = 3$, we leave it to the reader to show that

$$D_3 = \begin{vmatrix} 1 & 1 & 1 \\ r_1 & r_2 & r_3 \\ r_1^2 & r_2^2 & r_3^2 \end{vmatrix} = (r_2 - r_1)(r_3 - r_1)(r_3 - r_2). \tag{3.8}$$

In general it can be shown (Exercise 5) that

$$D_n = (r_2 - r_1)[(r_3 - r_1)(r_3 - r_2)][(r_4 - r_1)(r_4 - r_2)(r_4 - r_3)] \cdots$$
$$\times [(r_n - r_1)(r_n - r_2) \cdots (r_n - r_{n-1})] \tag{3.9}$$

and hence that $D_n \neq 0$. This means that the exponential functions (3.6) are linearly independent.

--- **Exercises for Section 3.4** ---

1. Compute the Wronskian of the given set of functions. Then determine whether the functions are linearly dependent or linearly independent.

   (a) e^{ax}, e^{bx}, $a \neq b$, x in any interval
   (b) $\cos ax$, $\sin ax$, $a \neq 0$, x in any interval
   (c) 1, x, x^2, x in any interval
   (d) $x + 1$, $x + 2$, $x + 3$, all x
   (e) 1, x^{-1}, x^{-2}, $x > 0$
   (f) x, e^x, xe^x, x in any interval
   (g) x^m, $|x|^m$, m a positive integer, all x
   (h) $x^2 - x$, $x^2 + x$, x^2, all x

2. (a) If the Wronskian of a set of functions is identically zero, the functions must be linearly dependent. True or false?

   (b) If a set of functions is linearly dependent, the Wronskian must be identically zero. True or false?

   (c) If the Wronskian of a set of functions is zero at some points and not zero at some points, the functions are linearly independent. True or false?

3. Derive formula (3.8).

4. Use formula (3.9) to find the Wronskian of the functions e^x, e^{2x}, e^{3x}, e^{-x}.

5. Prove formula (3.9). Use mathematical induction. Observe that if

$$P(r) = \begin{vmatrix} 1 & 1 & \cdots & 1 & 1 \\ r_1 & r_2 & \cdots & r_k & r \\ \vdots & \vdots & & \vdots & \vdots \\ r_1^k & r_2^k & \cdots & r_k^k & r^k \end{vmatrix},$$

then P is a polynomial of degree k with zeros $r_1, r_2, \ldots, r_k$ and $P(r_{k+1}) = D_{k+1}$.

6. Show that

   (a) $W(x; gf_1, gf_2) = [g(x)]^2 W(x; f_1, f_2)$

   (b) $W(x; gf_1, gf_2, gf_3)$
       $$= [g(x)]^3 W(x; f_1, f_2, f_3)$$

   (c) $W(x; gf_1, gf_2, \ldots, gf_n)$
       $$= [g(x)]^n W(x; f_1, f_2, \ldots, f_n)$$

7. Let $f_m(x) = x^m$ for all x and for m a non-negative integer. For every fixed n show that the functions $f_0, f_1, \ldots, f_n$ are linearly independent.

8. Use the result of Exercise 7 to show that a polynomial is the zero function if and only if all its coefficients are zero.

9. Use Theorem 3.5 to show that any set of $n + 2$ polynomials whose degrees are all at most n is linearly dependent.

10. (a) Let f_1 and f_2 be functions such that $W(x; f_1, f_2) \neq 0$ for x in an interval (a, b). Show that the equation $W(x; f_1, f_2, y) = 0$ is a differential equation of order two with f_1 and f_2 as solutions.

    (b) Find a second order differential equation that has $f_1(x) = e^x$ and $f_2(x) = xe^x$ as solutions for all x.

3.5
DIMENSION

Let V be a vector space. If for some positive integer n there exists a set of n linearly independent elements of V and if every set of more than n elements is linearly dependent, then V is said to be a *finite-dimensional vector space* and to have *dimension n*. (In addition, the vector space whose only element is a zero element is said to be finite-dimensional and to have dimension zero.)

If V is a vector space of dimension n, any set of n linearly independent elements of V is called a *basis* for V. The significance and importance of the notion of a basis is explained in the next three theorems.

Theorem 3.7   If V is a vector space of dimension n, then every basis for V spans V.

Proof   Let $\{v_1, v_2, \ldots, v_n\}$ be a basis for V. Let v be any element of V. We must show that v is a linear combination of the n linearly independent elements. We know that the set $\{v, v_1, v_2, \ldots, v_n\}$ is linearly dependent, since it has $n + 1$ elements and V has dimension n. Hence there exist scalars c_0, $c_1, \ldots, c_n$, not all zero, such that

$$c_0 v + c_1 v_1 + \cdots + c_n v_n = 0.$$

We claim that $c_0 \neq 0$. Otherwise, we would have

$$c_1 v_1 + \cdots + c_n v_n = 0,$$

and since $v_1, v_2, \ldots, v_n$ are linearly independent, this means that $c_1 = c_2 = \cdots = c_n = 0$ also. Hence we may write

$$v = \sum_{i=1}^{n} \left(-\frac{c_i}{c_0} \right) v_i;$$

therefore, v is a linear combination of the elements of the linearly independent set of n elements.

Theorem 3.8   If there is a set of n linearly independent elements of V that spans V, then V has dimension n.

Proof   Suppose that there exists a set of n linearly independent elements of V that spans V. Then every element of V is a linear combination of these n elements. By Theorem 3.5, every finite set with more than n elements is linearly dependent. Hence V has dimension n.

Theorem 3.9   If V has dimension n, then every set of linearly independent elements that spans V has exactly n elements.

Proof   Let $\{v_1, v_2, \ldots, v_k\}$ be a linearly independent set of k elements that spans V. We must prove that $k = n$. It is impossible that $k > n$, by the definition of a vector space of dimension n. On the other hand, if $k < n$, then V has dimension k by Theorem 3.8, and this contradicts the hypothesis that V has dimension n. Hence we are left with the only alternative, $k = n$.

We shall use Theorem 3.8 to prove that R^n is an n-dimensional vector space. To accomplish this, we have to exhibit a set of n linearly independent elements of R^n and then show that every element of R^n is a linear combination of these n elements. Let

$$e_1 = (1, 0, 0, 0, \ldots, 0),\ e_2 = (0, 1, 0, 0, \ldots, 0), \ldots, e_n = (0, 0, 0, 0, \ldots, 1).$$

The $n \times n$ matrix with these vectors as its column vectors is the identity matrix I, which is nonsingular. By Theorem 3.4, the vectors are linearly independent. If $\mathbf{x} = (x_1, x_2, \ldots, x_n)$ is any element of R^n, we can write

$$\mathbf{x} = x_1 \mathbf{e}_1 + x_2 \mathbf{e}_2 + \cdots + x_n \mathbf{e}_n.$$

Hence R^n is n-dimensional and $\{\mathbf{e}_1, \mathbf{e}_2, \ldots, \mathbf{e}_n\}$ is a basis for R^n. We call this particular basis the *standard basis* for R^n. In similar fashion it can be shown (Exercise 3) that the complex vector space C^n is n-dimensional.

Having shown that R^n is n-dimensional, we know that any set of n linearly independent elements constitutes a basis for R^n. For R^2, the vectors $\mathbf{e}_1 = (1, 0)$ and $\mathbf{e}_2 = (0, 1)$ constitute a basis. But the vectors $(2, -1)$ and $(-2, 2)$ also form a basis, since they are linearly independent.

Geometrically, the vectors in a one-dimensional subspace of R^2 or R^3 correspond to the points on a line through the origin of a rectangular coordinate-system. A two-dimensional subspace of R^3 corresponds to a plane through the origin.

The set of all real valued functions defined on an interval $\mathscr{I}$ is a real vector space, but not a finite-dimensional space. In fact, we can show that for every positive integer n there exists a set of n linearly independent elements of the space. Thus there can be no largest set of linearly independent elements. Let $f_m(x) = e^{mx}$ for x in $\mathscr{I}$ and for every positive integer m. Then for each positive integer n the functions $f_1, f_2, \ldots, f_n$ are linearly independent, as Section 3.4 indicated. The set of all functions of the form $af_1 + bf_2$, where a and b are any real numbers, is a two-dimensional subspace of the space considered here. For f_1 and f_2 are linearly independent elements of the subspace, and every element of the subspace is a linear combination of f_1 and f_2.

Let us consider the subspace of R^n that is spanned by the vectors $\mathbf{v}_1, \mathbf{v}_2, \ldots, \mathbf{v}_r$. Suppose we wish to determine the dimension of this subspace, and to find a basis for it. A tool for doing both is provided by the next theorem.

Theorem 3.10   Let A be an $m \times n$ matrix and let B be a matrix obtained from A by performing a finite sequence of elementary row (column) operations (Section 2.1). Then the subspace of R^n that is spanned by the row (column) vectors of A is the same as the subspace spanned by the row (column) vectors of B.

Proof   We give the proof for row vectors only; the one for column vectors is practically the same. Suppose that B is obtained from A by a sequence of elementary row operations:

$$A \rightarrow A_1 \rightarrow A_2 \rightarrow \cdots \rightarrow A_{k-1} \rightarrow B$$

Here a single row operation is performed on A to obtain A_1, a single row operation is performed on A_1 to obtain A_2, and so on. Since each row vector of A_{j+1} is a linear combination of the row vectors of A_j, the subspace spanned by the row vectors of A_{j+1} is a subspace of the space spanned by the row vectors of A_j. Let V_A and V_B be the subspaces of R^n spanned by the row vectors of A and B, respectively. Then by Theorem 3.2, V_B is a subspace of V_A. However, each elementary row operation is reversible; that is, A_j can be obtained from A_{j+1} by means of an elementary row operation. Thus V_A is also a subspace of V_B. We conclude that $V_A = V_B$.

Example   Let us determine the dimension of the subspace of R^4 that is spanned by the vectors

$$\mathbf{v}_1 = (1, 0, -1, 2), \qquad \mathbf{v}_2 = (1, 1, 3, 0), \qquad \mathbf{v}_3 = (-1, 1, 5, -4)$$

and find a basis for this subspace. First we form the matrix

$$A = \begin{bmatrix} 1 & 0 & -1 & 2 \\ 1 & 1 & 3 & 0 \\ -1 & 1 & 5 & -4 \end{bmatrix}$$

whose row vectors are the given vectors. Now we reduce the matrix to row-echelon form by means of elementary row operations. Subtracting the first row from the second and adding the first row to the third, we obtain the matrix

$$\begin{bmatrix} 1 & 0 & -1 & 2 \\ 0 & 1 & 4 & -2 \\ 0 & 1 & 4 & -2 \end{bmatrix}$$

Next, we subtract the second row from the third, obtaining the matrix

$$B = \begin{bmatrix} 1 & 0 & -1 & 2 \\ 0 & 1 & 4 & -2 \\ 0 & 0 & 0 & 0 \end{bmatrix}$$

The row vectors $\mathbf{w}_1 = (1, 0, -1, 2)$, $\mathbf{w}_2 = (0, 1, 4, -2)$ are linearly independent, because the condition

$$c_1 \mathbf{w}_1 + c_2 \mathbf{w}_2 = \mathbf{0}$$

requires that $c_1 = c_2 = 0$. Hence the subspace of R^4 that is spanned by the row vectors of B (and also by the row vectors of A) has dimension 2, and the vectors $\mathbf{w}_1$ and $\mathbf{w}_2$ constitute a basis for this subspace.

Any two independent vectors from the original set of four also constitute a basis, but it is not readily apparent which pairs are independent. An alternative approach to finding the dimension of and a basis for the subspace spanned by a set of vectors is explored in the set of additional exercises at the end of this chapter. This approach determines a basis consisting of vectors of the original set.

———————————————————— Exercises for Section 3.5 ————————————

1. If V is a finite-dimensional vector space with dimension n, show that any subspace U of V is also finite-dimensional and that its dimension cannot exceed n. If U has dimension n, show that $U = V$.

2. If V is an n-dimensional vector space, can a basis for V have
   (a) more than n elements?
   (b) fewer than n elements?

3. Show that the complex vector space C^n is n-dimensional.

4. Determine if the set of vectors $\{\mathbf{u}, \mathbf{v}, \mathbf{w}\}$ constitutes a basis for R^3.
   (a) $\mathbf{u} = (1, 1, 0)$, $\mathbf{v} = (2, 0, 1)$,
       $\mathbf{w} = (1, -1, 1)$
   (b) $\mathbf{u} = (0, 1, -1)$, $\mathbf{v} = (1, 0, 1)$,
       $\mathbf{w} = (2, 2, 2)$

5. Show that the set of all 2×2 matrices forms a finite-dimensional vector space. Determine the dimension and find a basis for the space.

6. Show that the vector space of all convergent infinite sequences is not finite-dimensional.

7. Show that the set of all real polynomials of degree <2 is a finite-dimensional vector space. Determine the dimension and find at least two bases.

8. Is the space of all real valued continuous functions defined on the interval $[0, 1]$ finite-dimensional?

9. Let $\mathbf{v} = (v_1, v_2)$ be an arbitrary element of R^2. Find numbers a and b such that $\mathbf{v} = a\mathbf{x} + b\mathbf{y}$, where $\{\mathbf{x}, \mathbf{y}\}$ is the indicated basis for R^2.
   (a) $\mathbf{x} = (2, -1)$, $\mathbf{y} = (1, 1)$
   (b) $\mathbf{x} = (1, 3)$, $\mathbf{y} = (2, 0)$

10. Let $\mathscr{P}^n$ be the space of all (real) polynomials of degree $<n$. Show that $\mathscr{P}^n$ has dimension n.

11. Show that the set of all polynomials constitutes a vector space that is not finite-dimensional.

12. If a matrix is in row-echelon form, show that the dimension of the subspace spanned by its row vectors is equal to the number of nonzero row vectors.

13. Find the dimension of and a basis for the subspace spanned by the given set of vectors.
    (a) $(2, -1, 1)$, $(1, 1, 1)$, $(4, -5, 1)$
    (b) $(2, 3, -1)$, $(0, 1, 1)$, $(-2, -2, 2)$,
        $(-1, 0, 2)$
    (c) $(1, 3, 0, -2)$, $(3, 1, 2, 0)$, $(1, -1, 1, 1)$,
        $(1, 0, 1, 0)$
    (d) $(1, 0, 1, 1)$, $(0, 4, 0, 2)$, $(1, 2, 1, 2)$,
        $(3, 2, 3, 4)$

3.6
ORTHOGONAL
BASES

In R^n (or C^n), any set of nonzero mutually orthogonal elements is linearly independent. To see this, suppose that $\mathbf{v}_1, \mathbf{v}_2, \ldots, \mathbf{v}_k$ are such elements, and that

$$c_1 \mathbf{v}_1 + c_2 \mathbf{v}_2 + \cdots + c_k \mathbf{v}_k = \mathbf{0}.$$

Taking the scalar product of both members of this equation with $\mathbf{v}_i$, where i is any integer such that $1 \leq i \leq k$, we have

$$c_1(\mathbf{v}_1 \cdot \mathbf{v}_i) + c_2(\mathbf{v}_2 \cdot \mathbf{v}_i) + \cdots + c_k(\mathbf{v}_k \cdot \mathbf{v}_i) = 0.$$

Because of the orthogonality of the elements, this equation reduces to

$$c_i(\mathbf{v}_i \cdot \mathbf{v}_i) = 0.$$

But since $v_i \neq 0$ then $v_i \cdot v_i > 0$ and hence $c_i = 0$ for $1 \leq i \leq k$. Thus the elements must be linearly independent.

If $k = n$, the set of mutually orthogonal vectors spans R^n (or C^n) and forms a basis for the space. Such bases are convenient. For suppose that w is any vector and we wish to calculate the constants c_i such that

$$w = c_1 v_1 + c_2 v_2 + \cdots + c_n v_n.$$

Taking the scalar product of both sides with v_i, we see that

$$w \cdot v_i = c_i v_i \cdot v_i$$

because of the orthogonality, and hence

$$c_i = \frac{v_i \cdot w}{\|v_i\|^2}.$$

If the vectors v_i are unit vectors, this formula simplifies even further to

$$c_i = v_i \cdot w.$$

A basis $\mathcal{B} = \{v_1, v_2, \ldots, v_n\}$ for R^n (or C^n) is called an *orthogonal basis* if $v_i \cdot v_j = 0$ for $i \neq j$. The basis $\mathcal{B}$ is called an *orthonormal basis* if it is orthogonal and if in addition its elements are unit vectors. Thus $\mathcal{B}$ is an orthonormal basis if $v_i \cdot v_j = \delta_{ij}$. The standard basis is an example of an orthonormal basis for R^n. Another example of an orthonormal basis for R^3 is $\{v_1, v_2, v_3\}$ where

$$v_1 = \frac{1}{3}(2, 2, 1), \qquad v_2 = \frac{1}{3}(2, -1, -2), \qquad v_3 = \frac{1}{3}(-1, 2, -2).$$

If a basis for a subspace is known, it is always possible to find an orthogonal basis, as the next theorem shows. In order to gain some insight into the method explained in the theorem, let us examine an example from R^3. Suppose that v_1 and v_2 are any two linearly independent vectors in R^3, and let $\mathcal{V}$ be the subspace spanned by them. The situation is illustrated in Figure 3.1. Geometrically, the subspace consists of the plane through the origin containing v_1 and v_2. This plane is shown in Figure 3.2. Any two mutually perpendicular nonzero vectors u_1 and u_2 in this plane emanating from the origin will form an orthogonal basis for $\mathcal{V}$. One possibility is to choose $u_1 = v_1$ and then choose u_2 to be the vector projection of v_2 on a line perpendicular to u_1. This choice is shown in Figure 3.2. From the figure, we see that

$$u_2 = v_2 + cu_1,$$

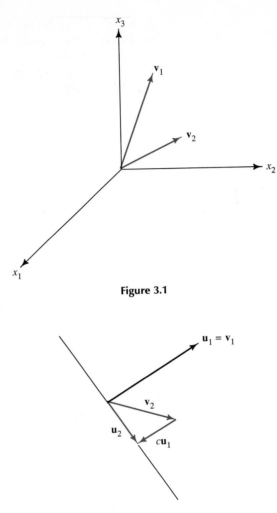

Figure 3.1

Figure 3.2

where c is a constant. This constant can be determined from the condition $\mathbf{u}_1 \cdot \mathbf{u}_2 = 0$. Thus

$$\mathbf{u}_1 \cdot \mathbf{u}_2 = \mathbf{u}_1 \cdot (\mathbf{v}_2 + c\mathbf{u}_1) = \mathbf{u}_1 \cdot \mathbf{v}_2 + c\|\mathbf{u}_1\|^2 = 0$$

so

$$c = -\frac{\mathbf{u}_1 \cdot \mathbf{v}_2}{\|\mathbf{u}_1\|^2}.$$

Hence

$$\mathbf{u}_2 = \mathbf{v}_2 - \frac{\mathbf{u}_1 \cdot \mathbf{v}_2}{\|\mathbf{u}_1\|^2}\mathbf{u}_1.$$

The following theorem shows that a subspace of R^n (or C^n) always has an orthogonal basis. It also shows how such a basis can be constructed.

Theorem 3.11 Let $\mathbf{v}_1, \mathbf{v}_2, \ldots, \mathbf{v}_m$, $m > 1$, be linearly independent elements of R^n (or C^n). Then the subspace of R^n (or C^n) that is spanned by these elements has an orthogonal basis $(\mathbf{u}_1, \mathbf{u}_2, \ldots, \mathbf{u}_m)$, where

$$\mathbf{u}_1 = c_1 \mathbf{v}_1,$$

$$\mathbf{u}_2 = c_2 \left\{ \mathbf{v}_2 - \frac{\mathbf{u}_1 \cdot \mathbf{v}_2}{\|\mathbf{u}_1\|^2} \mathbf{u}_1 \right\},$$

$$\mathbf{u}_3 = c_3 \left\{ \mathbf{v}_3 - \frac{\mathbf{u}_1 \cdot \mathbf{v}_3}{\|\mathbf{u}_1\|^2} \mathbf{u}_1 - \frac{\mathbf{u}_2 \cdot \mathbf{v}_3}{\|\mathbf{u}_2\|^2} \mathbf{u}_2 \right\}, \qquad (3.10)$$

$$\cdots\cdots\cdots\cdots\cdots\cdots\cdots\cdots$$

$$\mathbf{u}_m = c_m \left\{ \mathbf{v}_m - \frac{\mathbf{u}_1 \cdot \mathbf{v}_m}{\|\mathbf{u}_1\|^2} \mathbf{u}_1 - \cdots - \frac{\mathbf{u}_{m-1} \cdot \mathbf{v}_m}{\|\mathbf{u}_{m-1}\|^2} \mathbf{u}_{m-1} \right\},$$

and $c_1, c_2, \ldots, c_m$ are any nonzero constants.

Proof Using the formulas (3.10) we first define $\mathbf{u}_1$. If $\mathbf{u}_1$ is not the zero vector then $\|\mathbf{u}_1\| \neq 0$ and we can define $\mathbf{u}_2$. If $\mathbf{u}_2$ is not the zero vector we can define $\mathbf{u}_3$, and so on. We proceed by induction. Certainly $\mathbf{u}_1$ is not the zero vector because $\mathbf{v}_1$ is not the zero vector. (See Exercise 2, Section 3.3.) Suppose that none of $\mathbf{u}_1, \mathbf{u}_2, \ldots, \mathbf{u}_k$ is the zero vector, where k is any positive integer less than m. If $\mathbf{u}_{k+1} = 0$ then $\mathbf{v}_{k+1}$ is a linear combination of $\mathbf{u}_1, \mathbf{u}_2, \ldots, \mathbf{u}_k$ and hence is a linear combination of $\mathbf{v}_1, \mathbf{v}_2, \ldots, \mathbf{v}_k$. By Theorem 3.5, the set $\{\mathbf{v}_1, \mathbf{v}_2, \ldots, \mathbf{v}_{k+1}\}$ is linearly dependent. This is impossible. Hence each of the vectors $\mathbf{u}_1, \mathbf{u}_2, \ldots, \mathbf{u}_m$ is defined and is not the zero vector. It is not hard to show by induction that $\mathbf{u}_i \cdot \mathbf{u}_j = 0$ for $j < i$ and $i = 2, 3, \ldots, m$. As explained earlier, the mutually orthogonal nonzero vectors are linearly independent.

The method that was used in the proof of the theorem to find an orthogonal basis is known as the *Gram-Schmidt orthogonalization procedure*. In finding a basis in a particular situation, we may choose the constants c_i in such a way as to simplify the calculations. The following example provides an illustration.

Example We determine an orthonormal basis for the subspace of R^3 that is spanned by the vectors $(2, 2, 1)$ and $(1, -1, 3)$. Setting $\mathbf{v}_1 = (2, 2, 1)$, $\mathbf{v}_2 = (1, -1, 3)$, and using the Gram-Schmidt procedure, we find that $\mathbf{u}_1 = c_1 \mathbf{v}_1 = c_1(2, 2, 1)$, where c_1 may be any nonzero constant. Choosing $c_1 = 1$, we have

$$\mathbf{u}_1 = (2, 2, 1).$$

Next, we find that

$$\mathbf{u}_2 = \left\{ c_2(1, -1, 3) - \frac{3}{9}(2, 2, 1) \right\} = c_2 \frac{1}{3}(1, -5, 8).$$

Choosing $c_2 = 3$, we have

$$\mathbf{u}_2 = (1, -5, 8).$$

The vectors $\mathbf{u}_1$ and $\mathbf{u}_2$ form an orthogonal basis. If each vector is multiplied by the reciprocal of its length, the resulting vectors

$$\frac{1}{3}(2, 2, 1), \qquad \frac{1}{3\sqrt{10}}(1, -5, 8)$$

form an orthonormal basis. It should be pointed out that this is not the only orthonormal basis for the subspace. If we apply the orthogonalization process to $\mathbf{v}_1 = (1, -1, 3)$, $\mathbf{v}_2 = (2, 2, 1)$ (thereby reversing the order of the given vectors) we find that $\mathbf{u}_1 = (1, -1, 3)$ and

$$\mathbf{u}_2 = c_2 \left\{ (2, 2, 1) - \frac{3}{11}(1, -1, 3) \right\} = c_2 \frac{1}{11}(19, 25, 2).$$

Choosing $c_2 = 11$ yields

$$\mathbf{u}_2 = (19, 25, 2).$$

Then the vectors

$$\frac{1}{\sqrt{11}}(1, -1, 3), \qquad \frac{1}{3\sqrt{110}}(19, 25, 2)$$

form an orthonormal basis.

Exercises for Section 3.6

In Exercises 1–8, use the Gram-Schmidt procedure to find an orthonormal basis for the given subspace.

1. The subspace of R^2 spanned by $(4, -3)$

2. The subspace of R^3 spanned by $(2, 2, -1)$

3. The subspace of R^3 spanned by $(1, 2, -2)$ and $(1, 1, -1)$

4. The subspace of R^3 spanned by $(1, 1, 1)$ and $(6, 3, 2)$

5. The subspace of R^3 spanned by $(1, 0, 3)$ and $(2, 2, 1)$

6. The subspace of R^4 spanned by $(1, 0, 1, 0)$, $(0, 1, 1, -1)$, and $(1, 0, 0, 1)$

7. The subspace of R^4 spanned by $(0, 1, 2, 2)$, $(1, 1, 1, 0)$, and $(2, 0, 1, 2)$

8. The subspace of R^5 spanned by $(0, 1, 1, 0, 0)$, $(1, 0, 1, 0, 0)$, and $(0, 0, 1, 1, 1)$

9. Find an orthogonal basis for the subspace spanned by the given vectors.
   (a) $(2, -1, 5)$, $(0, -1, 2)$, $(2, 1, 1)$
   (b) $(4, 3, -1)$, $(1, 0, -1)$, $(2, 3, 1)$
   (c) $(1, -1, 0, 1)$, $(2, -1, 1, 1)$, $(1, 0, 1, 0)$, $(0, 1, 1, -1)$
   (d) $(1, 1, 0, 0)$, $(3, 5, 2, 1)$, $(-1, 0, 1, 1)$, $(2, 2, 0, -1)$

10. Let V be the space of continuous functions defined on an interval $[a, b]$. We define the scalar product $\langle f, g \rangle$ of two elements f and g of this space by

$$\langle f, g \rangle = \int_a^b f(x)g(x)\,dx$$

and the norm of an element f by $\|f\| = \sqrt{\langle f, f \rangle}$. Show that the following properties hold:

(a) $\langle f, f \rangle \geq 0$
(b) $\langle g, f \rangle = \langle f, g \rangle$
(c) $\langle cf, g \rangle = c\langle f, g \rangle = \langle f, cg \rangle$
(d) $\langle f + g, h \rangle = \langle f, h \rangle + \langle g, h \rangle$

11. Two continuous functions f and g on $[a, b]$ are said to be *orthogonal* if $\langle f, g \rangle = 0$ (see the previous exercise). If the continuous functions $f_1, f_2, \ldots, f_n$ are mutually orthogonal (that is, $\langle f_i, f_j \rangle = 0$ when $i \neq j$), show that they are linearly independent.

12. Verify that the set of all polynomials of degree <3, with domain restricted to $[0, 1]$, is a vector space of dimension 3. Defining a scalar product on this space as

$$\langle f, g \rangle = \int_0^1 f(x)g(x)\,dx$$

(see Exercise 11), find an orthogonal basis for this space. Suggestion: one basis consists of the functions $f_0(x) = 1$, $f_1 = x$, $f_2 = x^2$.

3.7 LINEAR TRANSFORMATIONS

Let A and B be arbitrary sets. A rule of correspondence that assigns to each element of A exactly one element of B is called a function, or mapping, from A into B. Alternatively, a function from A into B may be defined as a collection of ordered pairs (a, b), where a is in A, b is in B, and each element of A occurs in exactly one ordered pair. The set A is called the *domain* of the mapping. We say that the mapping maps a into b. We denote mappings by capital letters S, T, and so on. If T is a mapping from A into B, we write

$$T: A \to B.$$

To indicate that T maps a into b, we write $b = Ta$ or $b = T(a)$. We call b the *image* of a under T.

We shall be concerned with mappings from a vector space into a vector space. A mapping from R^1 into R^1 is usually called a function of a real variable. A mapping from R^2 into R^1 is usually called a function of two real variables. A mapping from R^2 into R^2 can be described by an ordered pair of functions of two real variables. The equations

$$y_1 = f_1(x_1, x_2), \qquad y_2 = f_2(x_1, x_2)$$

describe such a mapping, in which (x_1, x_2) is mapped into (y_1, y_2).

Let V and W be vector spaces, both real or both complex, and let T be a mapping from V into W. The mapping T is said to be a *linear transformation*, or a *linear mapping*, or a *linear operator* if it has the following properties.

Property 1   For every number c and every element $\mathbf{v}$ of V,

$$T(c\mathbf{v}) = cT\mathbf{v}.$$

Property 2   For every pair $\{\mathbf{u}, \mathbf{v}\}$ of elements of V,

$$T(\mathbf{u} + \mathbf{v}) = T\mathbf{u} + T\mathbf{v}.$$

Notice the essential uses of the properties of a vector space in the definition of a linear transformation. If $\mathbf{u}$ and $\mathbf{v}$ are in V, the product $c\mathbf{v}$ and the sum $\mathbf{u} + \mathbf{v}$ are defined and are elements of V. Since $T\mathbf{u}$ and $T\mathbf{v}$ are in W, the product $cT\mathbf{v}$ and the sum $T\mathbf{u} + T\mathbf{v}$ are defined and are elements of W.

If $\mathbf{v}$ is any element of V, we see by setting $c = 0$ in Property 1 that

$$T\mathbf{0} = T(0\mathbf{v}) = 0T\mathbf{v} = \mathbf{0}.$$

Thus the zero element of V is mapped into the zero element of W by the linear transformation T.

The set of all elements of V that are mappeed into the zero element of W by T form a subspace of V. To see this, suppose that $T\mathbf{u} = \mathbf{0}$ and $T\mathbf{v} = \mathbf{0}$. Then

$$T(c\mathbf{u}) = cT\mathbf{u} = c\mathbf{0} = \mathbf{0}$$

and

$$T(\mathbf{u} + \mathbf{v}) = T\mathbf{u} + T\mathbf{v} = \mathbf{0} + \mathbf{0} = \mathbf{0}.$$

This subspace of V is called the *kernel*, or *null space*, of the linear transformation T.

The set of all elements $\mathbf{w}$ of W such that $T\mathbf{v} = \mathbf{w}$ for at least one element $\mathbf{v}$ of V is a subspace of W. For if $\mathbf{x} = T\mathbf{u}$ and $\mathbf{y} = T\mathbf{v}$, then $c\mathbf{x} = T(c\mathbf{u})$ and $\mathbf{x} + \mathbf{y} = T(\mathbf{u} + \mathbf{v})$. This subspace of W is called the *range* of the linear transformation T.

In order to illustrate these concepts, we consider some examples.

Example 1   First we describe a linear transformation from R^n into R^m. Let us represent the elements of R^n and R^m by their associated column vectors, and let A be an $m \times n$ real matrix. If $\mathbf{x}$ is in R^n, then $A\mathbf{x}$ is in R^m and we write

$$T\mathbf{x} = A\mathbf{x}.$$

It is not hard to see that T is a linear transformation. For if $\mathbf{u}$ and $\mathbf{v}$ are in R^n, and c is a real number, then

$$T(c\mathbf{u}) = A(c\mathbf{u}) = cA\mathbf{u} = cT(\mathbf{u})$$

and

$$T(\mathbf{u} + \mathbf{v}) = A(\mathbf{u} + \mathbf{v}) = A\mathbf{u} + A\mathbf{v} = T(\mathbf{u}) + T(\mathbf{v}).$$

The kernel of T is the set of all solutions of the homogeneous equation

$$A\mathbf{x} = \mathbf{0}.$$

The range of T is the set of all elements $\mathbf{y}$ in R^m for which the equation

$$A\mathbf{x} = \mathbf{y}$$

has a solution.

This example shows that if A is an $m \times n$ matrix and if $\mathbf{x}$ is the column vector representation of a vector in R^n, then the formula $T\mathbf{x} = A\mathbf{x}$ defines a linear transformation from R^n into R^m. Actually, *every* linear transformation from R^n into R^m can be described in this way. To see this, suppose the elements

$$\mathbf{e}_1 = (1, 0, 0, \ldots, 0), \qquad \mathbf{e}_2 = (0, 1, 0, \ldots, 0), \ldots, \qquad \mathbf{e}_n = (0, 0, 0, \ldots, 1)$$

of the standard basis for R^n have the images

$$T\mathbf{e}_1 = \mathbf{a}_1, \qquad T\mathbf{e}_2 = \mathbf{a}_2, \ldots, \qquad T\mathbf{e}_n = \mathbf{a}_n$$

under T. Then if $\mathbf{x} = (x_1, x_2, \ldots, x_n)$ is any element of R^n, we may write

$$\mathbf{x} = x_1\mathbf{x}_1 + x_2\mathbf{e}_2 + \cdots + x_n\mathbf{e}_n$$

and the image of $\mathbf{x}$ is

$$\begin{aligned} T\mathbf{x} &= T(x_1\mathbf{e}_1 + x_2\mathbf{e}_2 + \cdots + x_n\mathbf{e}_n) \\ &= x_1 T\mathbf{e}_1 + x_2 T\mathbf{e}_2 + \cdots + x_n T\mathbf{e}_n \\ &= x_1\mathbf{a}_1 + x_2\mathbf{a}_2 + \cdots + x_n\mathbf{a}_n \end{aligned}$$

According to Theorem 2.4, $T\mathbf{x} = A\mathbf{x}$ where $A = [\mathbf{a}_1, \mathbf{a}_2, \ldots, \mathbf{a}_n]$

Example 2   Let $T: R^2 \to R^3$ be such that $T\mathbf{e}_1 = T(1, 0) = (1, 2, 3)$ and $T\mathbf{e}_2 = T(0, 1) = (5, 0, -2)$. Then T may be written in the form $T\mathbf{x} = A\mathbf{x}$, where A is the matrix

$$A = \begin{bmatrix} 1 & 5 \\ 2 & 0 \\ 3 & -2 \end{bmatrix}.$$

Example 3   Let $T: R^2 \to R^3$ be such that $T(2, 5) = (2, -3, 1)$ and $T(1, 4) = (1, -1, 0)$. Since Te_1 and Te_2 are not specified, we cannot write down the matrix A of the transformation immediately. However, since

$$A \begin{bmatrix} 2 & 1 \\ 5 & 4 \end{bmatrix} = \begin{bmatrix} 2 & 1 \\ -3 & -1 \\ 1 & 0 \end{bmatrix}$$

we have

$$A = \begin{bmatrix} 2 & 1 \\ -3 & -1 \\ 1 & 0 \end{bmatrix} \begin{bmatrix} 2 & 1 \\ 5 & 4 \end{bmatrix}^{-1}$$

$$= \begin{bmatrix} 2 & 1 \\ -3 & -1 \\ 1 & 0 \end{bmatrix} \frac{1}{3} \begin{bmatrix} 4 & -1 \\ -5 & 2 \end{bmatrix} = \frac{1}{3} \begin{bmatrix} 3 & 0 \\ -7 & 1 \\ 4 & -1 \end{bmatrix}.$$

We see from this formula that $Te_1 = \frac{1}{3}(3, -7, 4)$ and $Te_2 = \frac{1}{3}(0, 1, -1)$.

Example 4   Let $\mathscr{F}$ be the set of all functions defined on an interval $\mathscr{I}$ and let $\mathscr{F}'$ be the set of all differentiable functions on $\mathscr{I}$. We define a transformation $T: \mathscr{F}' \to \mathscr{F}$ by the equation $Tf = f'$, where f' is the derivative of the function f. This transformation is linear, because

$$T(cf) = (cf)' = cf' = cTf$$

and

$$T(f + g) = (f + g)' = f' + g' = Tf + Tg$$

for all differentiable functions f and g. The kernel of T is the set of all functions whose derivative is the zero function, that is, the set of constant functions. The range of T consists of those functions that are the derivative of some function. Thus g is in the range of T if there is a function f such that $f' = g$.

More examples of linear transformations are presented in the exercises.

Example 5   As an example of a *nonlinear* transformation, let us consider the mapping T from R^2 into R^2 that is defined by the formula

$$T(x_1, x_2) = (x_1 + x_2, x_1 x_2).$$

Then $T(1, 1) = (2, 1)$ and $T[2(1, 1)] = T(2, 2) = (4, 4)$ so

$$2T(1, 1) \neq T[2(1, 1)].$$

Hence T is nonlinear.

Exercises for Section 3.7

1. Which of the following transformations from R^2 into R^2 are linear?
   (a) $T(x_1, x_2) = (2x_1 - x_2, x_1 + x_2)$
   (b) $T(x_1, x_2) = (x_2, -x_1)$
   (c) $T(x_1, x_2) = (x_2 + 1, x_1 - 1)$
   (d) $T(x_1, x_2) = (x_2^2, x_1 + x_2)$

2. Which of the following transformations from R^3 into R^2 are linear?
   (a) $T(x_1, x_2, x_3) = (x_2 - x_3, x_1 + x_2)$
   (b) $T(x_1, x_2, x_3) = (x_3, 1)$

3. Let T be the linear transformation from R^2 into R^2 that is defined by the relations

   $$T\mathbf{x} = A\mathbf{x}, \qquad A = \begin{bmatrix} 2 & -3 \\ -1 & 2 \end{bmatrix}.$$

   Find $T\mathbf{x}$ if $\mathbf{x}$ is the given point and determine if the point belongs to the kernel of T.
   (a) $(1, 2)$ (b) $(0, 0)$ (c) $(-3, 1)$

4. Let T be the transformation of Exercise 3. Find all points, if any, that are mapped into the point $(2, 3)$ by T.

5. Let T be the linear transformation from R^2 into R^2 that is defined by the relations

   $$T\mathbf{x} = A\mathbf{x}, \qquad A = \begin{bmatrix} 3 & -2 \\ -6 & 4 \end{bmatrix}.$$

   Which, if any, of the given points belong to the kernel of T?
   (a) $(1, 1)$ (b) $(2, 3)$ (c) $(4, 6)$

6. If T is the transformation of Exercise 5, which, if any, of the given points belong to the range of T?

7. (a) Let $T: R^2 \to R^3$ be such that $T\mathbf{e}_1 = (-1, 0, 2)$, and $T\mathbf{e}_2 = (2, 3, 0)$. Find a matrix A such that $T\mathbf{x} = A\mathbf{x}$, and find $T(4, 5)$.
   (b) Let $T: R^2 \to R^2$ be such that $T\mathbf{e}_1 = (2, 3)$ and $T\mathbf{e}_2 = (-2, 1)$. Find a matrix A such that $T\mathbf{x} = A\mathbf{x}$, and find $T(7, 3)$.
   (c) Let $T: R^3 \to R^2$ be such that $T\mathbf{e}_1 = (1, -2)$, $T\mathbf{e}_2 = (0, 3)$, and $T\mathbf{e}_3 = (1, 1)$. Find a matrix A such that $T\mathbf{x} = A\mathbf{x}$, and find $T(1, 2, 3)$.
   (d) Let $T: R^3 \to R^3$ be such that $T\mathbf{e}_1 = (1, 1, 0)$, $T\mathbf{e}_2 = (0, 1, -1)$, and $T\mathbf{e}_3 = (2, 0, 1)$. Find a matrix A such that $T\mathbf{x} = A\mathbf{x}$, and find $T(2, 2, -1)$.

8. Let f be a function of a single real variable that maps R^1 into R^1. Show that f is a linear transformation if and only if $f(x) = mx$ for some number m. Suggestion: let $m = f(1)$.

9. If T is a mapping from U into V such that $T\mathbf{u} = 0$ for every element $\mathbf{u}$ of U, show that T is a linear transformation. (T is called the *zero transformation* from U into V.)

10. Let T be a mapping from a vector space V into V such that $T\mathbf{v} = \mathbf{v}$ for every element of V. Show that T is a linear transformation (called the *identity transformation on V*).

11. Let $T: R^2 \to R^2$ be defined by the relation

    $$T(x_1, x_2) = (2x_1 - x_2, -4x_1 + 2x_2).$$

    Show that T is linear. Find the kernel and range of T and give a geometric description of each.

12. (a) Let $T: R^2 \to R^2$ be such that $T(2, -1) = (3, 4)$ and $T(1, 1) = (3, 2)$. Find $T(6, 3)$. Find a matrix A such that $T\mathbf{x} = A\mathbf{x}$.
    (b) Let $T: R^2 \to R^2$ be such that $T(3, -1) = (2, 5)$ and $T(1, 2) = (-1, 1)$. Find $T(2, 3)$. Find a matrix A such that $T\mathbf{x} = A\mathbf{x}$.
    (c) Let $T: R^2 \to R^3$ be such that $T(1, 2) = (1, -1, 0)$ and $T(1, 1) = (1, 2, -1)$. Find $T(1, 4)$. Find a matrix A such that $T\mathbf{x} = A\mathbf{x}$.
    (d) Let $T: R^3 \to R^2$ be such that $T(1, 2, 0) = (-1, 2)$, $T(0, 1, -1) = (3, 1)$ and $T(1, 0, -3) = (-4, 3)$. Find $T(1, 2, 3)$. Find a matrix A such that $T\mathbf{x} = A\mathbf{x}$.

13. Let A be an $m \times n$ matrix and let $T\mathbf{x} = A\mathbf{x}$. Show that the range of T is spanned by the column vectors of A. (Use Theorem 2.4.)

14. Let $T\mathbf{x} = A\mathbf{x}$, where A is the given matrix. Find the dimension of, and a basis for, the range of T. (Use Theorem 3.10 and the result of the previous exercise.)

(a)
$$\begin{bmatrix} 1 & 0 & 2 & 3 \\ -2 & 5 & 1 & 4 \\ 0 & 2 & 2 & 4 \end{bmatrix}$$

(b)
$$\begin{bmatrix} 2 & 6 & -2 \\ 1 & 3 & -1 \\ -1 & -3 & 2 \end{bmatrix}$$

(c)
$$\begin{bmatrix} 1 & 1 & -3 \\ -2 & 2 & -4 \\ -2 & 0 & 1 \\ -3 & -1 & 4 \end{bmatrix}$$

(d)
$$\begin{bmatrix} 1 & 1 & -1 & 2 \\ 0 & 1 & 3 & -1 \\ -2 & 2 & -4 & 1 \\ 1 & 0 & -2 & 2 \end{bmatrix}$$

15. If V is the space of all continuous functions on $[0, 1]$ and if

$$Tf = \int_0^1 f(x)\, dx$$

for f in V, show that T is a linear transformation from V into R^1.

16. Let V be the space of infinite sequences. For each sequence $(x_1, x_2, x_3, \ldots)$ in V let

$$T(x_1, x_2, x_3, \ldots) = (x_2, x_3, x_4, \ldots).$$

Show that T is a linear transformation. Describe the kernel and the range of T.

3.8 PROPERTIES OF LINEAR TRANS-FORMATIONS

If T_1 and T_2 are both linear transformations from a vector space V into a vector space W, we say that T_1 is *equal* to T_2, written $T_1 = T_2$, if

$$T_1\mathbf{v} = T_2\mathbf{v}$$

for every element $\mathbf{v}$ of V.

If T_1 and T_2 are both linear transformations from V into W, we define the *sum*, $T_1 + T_2$, to be the transformation S such that

$$S\mathbf{v} = T_1\mathbf{v} + T_2\mathbf{v}$$

for every element $\mathbf{v}$ of V. It is easily verified (Exercise 8) that $T_1 + T_2$ is a linear transformation from V into W, and that $T_2 + T_1 = T_1 + T_2$.

Let V and W be vector spaces over the real (complex) numbers, and let T be a linear transformation from V into W. If c is a real (complex) number, we define the *product* cT to be the transformation P such that

$$P\mathbf{v} = cT\mathbf{v}$$

for every element $\mathbf{v}$ of V. It can be verified that cT is a linear transformation.

Now let U, V, and W be vector spaces, all real or all complex. Let T_1 and T_2 be linear transformations such that $T_1 : U \to V$ and $T_2 : V \to W$. The *product* $T_2 T_1$ is defined to be the transformation S from U into W such that

$$S\mathbf{u} = T_2(T_1\mathbf{u})$$

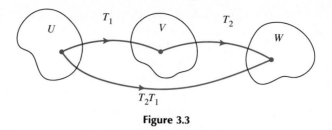

Figure 3.3

for every element **u** of U. The situation is illustrated in Figure 3.3. Notice that $T_1\mathbf{u}$ is in V, so $T_2(T_1\mathbf{u})$ is defined and is an element of W. Verification that $T_2 T_1$ is a *linear* transformation from U into W is left as an exercise.

In case $T_1: V \rightarrow V$ and $T_2: V \rightarrow V$, then $T_2 T_1$ and $T_1 T_2$ are both defined and map V into V. In general, $T_1 T_2 \neq T_2 T_1$, as we shall see in the following examples, and in the examples of the next section.

Example 1 Let A and B be $n \times n$ matrices. If we represent elements in R^n by column vectors then the transformations T_1 and T_2 that are defined by the relations

$$T_1\mathbf{x} = A\mathbf{x}, \qquad T_2\mathbf{x} = B\mathbf{x}$$

and linear transformations from R^n into R^n. Then

$$T_2 T_1 \mathbf{x} = BA\mathbf{x}$$

and

$$T_1 T_2 \mathbf{x} = AB\mathbf{x}$$

so that $T_2 T_1 \neq T_1 T_2$ unless A and B commute.

Example 2 Let V be the space of all polynomials. Define

$$T_1 f = f', \qquad T_2 f(x) = \int_0^x f(t)\, dt$$

for all f in V. Then T_1 and T_2 are linear transformations from V into V, as is readily verified. We find that

$$T_1 T_2 f = f$$

and

$$T_2 T_1 f(x) = \int_0^x f'(t)\, dt = f(x) - f(0).$$

Thus $T_1 T_2 f$ is not equal to $T_2 T_1 f$ unless $f(0) = 0$, and so $T_1 T_2 \neq T_2 T_1$.

Let T be a linear transformation from V into W. The transformation T is said to be a transformation from V *onto* W if the range of T is W. Thus T is onto if for every element $\mathbf{w}$ in W there is at least one element $\mathbf{v}$ of V such that $T\mathbf{v} = \mathbf{w}$.

The transformation T is said to be *one-to-one* if $T\mathbf{x} \neq T\mathbf{y}$ whenever $\mathbf{x} \neq \mathbf{y}$. Thus if T is one-to-one, distinct elements in V cannot map into the same element of W.

If T is both one-to-one and onto then there is associated with each element $\mathbf{w}$ of W exactly one element $\mathbf{v}$ of V such that $T\mathbf{v} = \mathbf{w}$. This association defines a transformation from W into V that is called the *inverse* of T. If $\mathbf{w} = T\mathbf{v}$ we write $\mathbf{v} = T^{-1}\mathbf{w}$. We see that

$$T^{-1}T\mathbf{v} = \mathbf{v}$$

for every $\mathbf{v}$ in V and

$$TT^{-1}\mathbf{w} = \mathbf{w}$$

for every $\mathbf{w}$ in W. It can be shown (Exercise 11) that the mapping T^{-1} is a *linear* mapping if T is linear.

Example 3   Let $T:R^n \rightarrow R^n$ be defined by the equation $T\mathbf{x} = A\mathbf{x}$, where A is an $n \times n$ nonsingular matrix. For each $\mathbf{y}$ in R^n the equation $A\mathbf{x} = \mathbf{y}$ has one and only one solution, namely $\mathbf{x} = A^{-1}\mathbf{y}$. Thus T has an inverse and the inverse transformation is described by the relation $T^{-1}\mathbf{y} = A^{-1}\mathbf{y}$.

Exercises for Section 3.8

1. Let $T_1:R^2 \rightarrow R^2$ and $T_2:R^2 \rightarrow R^3$ be defined by $T_1\mathbf{x} = A\mathbf{x}$ and $T_2\mathbf{x} = B\mathbf{x}$, where

$$A = \begin{bmatrix} 2 & 1 \\ -1 & 1 \end{bmatrix}, \qquad B = \begin{bmatrix} 3 & 1 \\ 2 & -1 \\ 1 & 1 \end{bmatrix}.$$

(a) Find $T_2 T_1 \mathbf{x}$, for arbitrary $\mathbf{x} = (x_1, x_2)$.
(b) Is $T_1 T_2$ defined? Explain.
(c) Is $T_1 + T_2$ defined? Explain.

2. Let $T_1:R^2 \rightarrow R^2$ and $T_2:R^2 \rightarrow R^2$ be defined by $T_1\mathbf{x} = A\mathbf{x}$ and $T_2\mathbf{x} = B\mathbf{x}$, where

$$A = \begin{bmatrix} 0 & 2 \\ -1 & 1 \end{bmatrix}, \qquad B = \begin{bmatrix} 1 & 1 \\ 2 & 3 \end{bmatrix}.$$

(a) Find $T_1 T_2 \mathbf{x}$ and $T_2 T_1 \mathbf{x}$ for arbitrary $\mathbf{x} = (x_1, x_2)$. Does $T_1 T_2 = T_2 T_1$?
(b) Find $(T_1 + T_2)\mathbf{x}$ for arbitrary $\mathbf{x}$.

3. Let V be the space of all polynomials. Define

$$T_1 f(x) = \int_0^x f(t)\, dt, \qquad T_2 f(x) = \int_x^1 f(t)\, dt.$$

(a) Find $T_1 T_2 f$ and $T_2 T_1 f$ if $f(x) = x^2$. Does $T_1 T_2 = T_2 T_1$?
(b) Find $(T_1 + T_2)f$ if $f(x) = x^2$. Describe the range of $T_1 + T_2$.

4. Let $T:R^2 \rightarrow R^2$ be defined by $T\mathbf{x} = A\mathbf{x}$, where

$$A = \begin{bmatrix} 3 & 1 \\ 2 & 1 \end{bmatrix}.$$

Verify that T has an inverse and describe T^{-1}.

5. Let $T_1:R^2 \rightarrow R^2$ and $T_2:R^2 \rightarrow R^2$ have the indicated properties. Find matrices A, B, and

C such that

$$T_2 T_1 \mathbf{x} = A\mathbf{x}, \qquad T_1 T_2 \mathbf{x} = B\mathbf{x},$$

$$(T_1 + T_2)\mathbf{x} = C\mathbf{x}$$

(a) $T_1 \mathbf{e}_1 = (1, 3), \quad T_1 \mathbf{e}_2 = (2, 2),$
$T_2 \mathbf{e}_1 = (-1, 1), \quad T_2 \mathbf{e}_2 = (2, -1),$
(b) $T_1 \mathbf{e}_1 = (-2, 1), \quad T_1 \mathbf{e}_2 = (1, 3),$
$T_2(-2, 1) = (3, 1), \quad T_2(1, 3) = (-1, 2)$

6. Let $T: R^n \to R^m$ and let $\{\mathbf{v}_1, \mathbf{v}_2, \dots, \mathbf{v}_n\}$ be a basis for R^n. Show that $\{T\mathbf{v}_1, T\mathbf{v}_2, \dots, T\mathbf{v}_n\}$ spans the range of T.

7. Let $T: R^3 \to R^3$ have the indicated properties. Find a basis for the range of T. (See the previous exercise.)
   (a) $T\mathbf{e}_1 = (2, 0, -1), \quad T\mathbf{e}_2 = (1, 3, 0),$
   $T\mathbf{e}_3 = (-1, 9, 2)$
   (b) $T(1, 1, 0) = (2, 0, 1),$
   $T(0, 2, -1) = (0, 2, 1),$
   $T(1, 0, 1) = (-4, 6, 1)$

8. If T_1 and T_2 are linear transformations from V into W, verify that $T_1 + T_2$ and cT_1 are linear transformations.

9. Let T_1 and T_2 be linear transformations, with $T_1: U \to V$ and $T_2: V \to W$. Verify that $T_2 T_1$ is a linear transformation.

10. Show that the set of all linear transformations from a vector space V into a vector space W forms a vector space. The operations of addition and multiplication by a number for linear transformations are as defined in the text.

11. If the linear transformation T has an inverse, show that T^{-1} is a linear transformation. Suggestion: to show that $T^{-1}(c\mathbf{w}) = cT^{-1}\mathbf{w}$, it suffices to show that $T(T^{-1}(c\mathbf{w})) = T(cT^{-1}\mathbf{w})$ because T is one-to-one.

12. Show that a linear transformation T is one-to-one if and only if the kernel of T consists of the zero vector.

13. Let T be a linear transformation from an n-dimensional space V into an m-dimensional space W.
    (a) If $m > n$, show that T cannot be a mapping from V onto W.
    (b) If $m < n$, show that T cannot be one-to-one.

3.9 TRANSFORMATIONS OF THE PLANE

We consider here only linear transformations of the type $T: R^2 \to R^2$, which map the plane into a plane. Associated with each point (x_1, x_2) is a vector with components x_1 and x_2. Using column vectors to represent points and a 2×2 matrix A to describe T, we have $\mathbf{y} = A\mathbf{x}$ or

$$\begin{bmatrix} y_1 \\ y_2 \end{bmatrix} = \begin{bmatrix} a_{11} & a_{12} \\ a_{21} & a_{22} \end{bmatrix} \begin{bmatrix} x_1 \\ x_2 \end{bmatrix}.$$

If A is the zero matrix, every point (x_1, x_2) maps into the point $(0, 0)$. If A is nonzero but singular, then the row vectors of A are linearly dependent. Hence there are constants c_1 and c_2, not both zero, such that

$$c_1 a_{11} + c_2 a_{21} = 0, \qquad c_1 a_{12} + c_2 a_{22} = 0.$$

But then $c_1 y_1 + c_2 y_2 = 0$ for all (x_1, x_2), so all points of the plane map into a straight line. In what follows we shall assume that A is nonsingular. Then T is one-to-one and onto, and has an inverse whose matrix is A^{-1}; that is, $\mathbf{x} = A^{-1}\mathbf{y}$ for all $\mathbf{y}$.

Example 1   Given the transformation $\mathbf{y} = A\mathbf{x}$, where

$$A = \begin{bmatrix} 3 & 5 \\ 1 & 2 \end{bmatrix},$$

we have

$$y_1 = 3x_1 + 5x_2, \qquad y_2 = x_1 + 2x_2.$$

Let us find the image of the circle $x_1^2 + x_2^2 = 9$. Solving for x_1 and x_2 in the equations above, we find that

$$x_1 = 2y_1 - 5y_2, \qquad x_2 = -y_1 + 3y_2.$$

Then the image of the circle is

$$x_1^2 + x_1^2 = (2y_1 - 5y_2)^2 + (-y_1 + 3y_2)^2 = 9$$

or

$$5y_1^2 - 26y_1 y_2 + 34y_2^2 = 0.$$

Since $(26)^2 - 4(5)(34) = -4 < 0$, the curve is an ellipse.

We shall be particularly interested in the five linear transformations whose matrices are

$$E_1(k) = \begin{bmatrix} k & 0 \\ 0 & 1 \end{bmatrix}, \qquad E_2(k) = \begin{bmatrix} 1 & 0 \\ 0 & k \end{bmatrix}, \qquad F_{12} = \begin{bmatrix} 0 & 1 \\ 1 & 0 \end{bmatrix},$$

$$G_{12}(c) = \begin{bmatrix} 1 & 0 \\ c & 1 \end{bmatrix}, \qquad G_{21}(c) = \begin{bmatrix} c & 1 \\ 1 & 0 \end{bmatrix}$$

where $k \neq 0$. Each matrix is obtained from the 2×2 identity matrix by performing an elementary row operation on it. The matrix $E_1(k)$ is obtained by multiplying through in the first row of I_2 by k, $E_2(k)$ by multiplying through in the second row by k, F_{12} by interchanging the two rows, $G_{12}(c)$ by adding c times the first row to the second, and $G_{21}(c)$ by adding c times the second row to the first. If $k \neq 0$, all the matrices, called *elementary matrices*, are nonsingular. The linear transformations of the plane that are associated with these matrices are called *elementary transformations*. They are fundamental in understanding the geometric properties of all linear transformations, as we shall see.

Let us consider the effect of each elementary transformation on a square in the $x_1 x_2$-plane with vertices $A(0, 0)$, $B(1, 0)$, $C(1, 1)$, and $D(0, 1)$. This square is shown in Fig. 3.4(a). Consider first the transformation $\mathbf{y} = E_1(k)\mathbf{x}$. We have

$$y_1 = kx_1, \qquad y_2 = x_2.$$

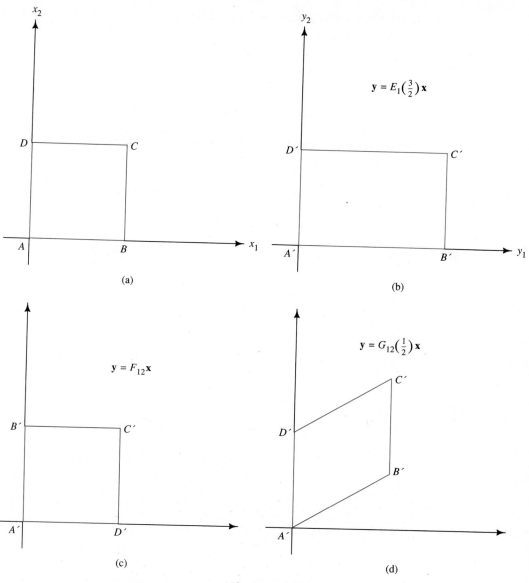

Figure 3.4

All figures are simply "stretched" by a factor k in the x_1-direction; see Fig. 3.4(b). If $k > 1$ the figure is enlarged. If $0 < k < 1$ it is shrunk. If $k < 0$ the figure is reflected in the x_2-axis and then enlarged or shrunk according as $|k| > 1$ or $|k| < 1$. The analysis of the transformation $\mathbf{y} = E_2(k)\mathbf{x}$ is similar, except that figures are stretched in the x_2-direction.

The transformation $\mathbf{y} = F_{12}\mathbf{x}$ becomes

$$y_1 = x_2, \qquad y_2 = x_1.$$

The coordinates of the point (x_1, x_2) are interchanged, so the point is reflected in the line $x_2 = x_1$. The effect on the square is shown in Fig. 3.4(c). The transformation $\mathbf{y} = G_{12}(c)\mathbf{x}$ becomes

$$y_1 = x_1, \qquad y_2 = kx_1 + x_2.$$

The first coordinate remains the same, but the second is changed by the amount kx_1. The result is a shear in the x_2-direction. If $c > 0$, points to the right of the x_2-axis are sheared upward and points to the left downward. If $c < 0$, the directions are reversed. The effect on a square, for $c > 0$, is shown in Fig. 3.4(d).
An analysis of the transformation $\mathbf{y} = G_{21}(c)\mathbf{x}$ indicated that it shears figures in the x_1-direction.

Let us now see why elementary matrices are important. If A is any 2×2 matrix, we have

$$E_1(k)A = \begin{bmatrix} k & 0 \\ 0 & 1 \end{bmatrix} \begin{bmatrix} a_{11} & a_{12} \\ a_{21} & a_{22} \end{bmatrix} = \begin{bmatrix} ka_{11} & ka_{12} \\ a_{21} & a_{22} \end{bmatrix}.$$

Thus the effect of premultiplying A by $E_1(k)$ is to perform the elementary row operation of multiplying through in the first row of A by k. Next, consider

$$G_{12}(c)A = \begin{bmatrix} 1 & 0 \\ c & 1 \end{bmatrix} \begin{bmatrix} a_{11} & a_{12} \\ a_{21} & a_{22} \end{bmatrix} = \begin{bmatrix} a_{11} & a_{12} \\ ca_{11} + a_{21} & ca_{12} + a_{22} \end{bmatrix}.$$

The effect of premultiplying A by $G_{12}(c)$ is to perform the elementary row operation of adding c times the first row to the second. We leave it to the reader to examine the products $E_2(k)A$, $F_{12}A$, and $G_{21}(c)A$. The first multiplies through in the second row of A by k, the second interchanges the rows of A, and the third adds c times the second row to the first.

The inverse of an elementary matrix is also an elementary matrix. For example, let us consider $G_{12}(c)$. We have

$$\begin{bmatrix} 1 & 0 \\ c & 1 \end{bmatrix} \begin{bmatrix} 1 & 0 \\ -c & 1 \end{bmatrix} = \begin{bmatrix} 1 & 0 \\ 0 & 1 \end{bmatrix}$$

or $G_{12}(c)G_{12}(-c) = I$ so $[G_{12}(c)]^{-1} = G_{12}(-c)$. We leave it to the reader to verify the remainder of the five relations

$$[E_1(k)]^{-1} = E_1\left(\frac{1}{k}\right), \qquad [E_2(k)]^{-1} = E_2\left(\frac{1}{k}\right), \qquad F_{12}^{-1} = F_{12}$$

$$[G_{12}(c)]^{-1} = G_{12}(-c), \qquad [G_{21}(c)]^{-1} = G_{21}(-c).$$

Our basic result about elementary matrices is as follows:

Theorem 3.12   A nonsingular matrix is either elementary or is equal to the product of (two or more) elementary matrices.

Proof   If a nonsingular matrix A is reduced to row-echelon form by a sequence of elementary row operations, the result must be the identity matrix. To see this, observe that a nonsingular square matrix in row-echelon form is upper triangular, and its diagonal elements must be nonzero. Since the first nonzero elements in the rows of a matrix in row-echelon form are ones, all diagonal elements are equal to one, and all elements above and below these ones are zero. Hence there exist elementary matrices $A_1, A_2, \ldots, A_k$ such that

$$A_k A_{k-1} \cdots A_2 A_1 A = I$$

and so

$$A = A_1^{-1} A_2^{-1} \cdots A_{k-1}^{-1} A_k^{-1}.$$

Since the inverse of an elementary matrix is an elementary matrix, A is the product of elementary matrices.

Example 2   Let us consider the matrix

$$A = \begin{bmatrix} 3 & 5 \\ 1 & 2 \end{bmatrix}$$

of Example 1. We shall reduce A to row-echelon form by performing elementary row operations. First we interchange the two rows, in effect premultiplying by F_{12}, to obtain the matrix

$$\begin{bmatrix} 1 & 2 \\ 3 & 5 \end{bmatrix}.$$

Adding -3 times the first row to the second (thus premultiplying by $G_{12}(-3)$), we have

$$\begin{bmatrix} 1 & 2 \\ 0 & -1 \end{bmatrix}.$$

Premultiplying by $E_2(-1)$, we have

$$\begin{bmatrix} 1 & 2 \\ 0 & 1 \end{bmatrix}.$$

Premultiplying by $G_{21}(-2)$, we obtain the identity matrix I_2. Thus

$$G_{21}(-2)E_2(-1)G_{12}(-3)F_{12}A = I$$

so

$$A = F_{12}G_{12}(3)E_2(-1)G_{12}(2).$$

In Example 1 we saw that the image of the circle $x_1^2 + x_2^2 = 9$ had a second degree polynomial equation. From the analysis just shown, we see that the circle was subjected to a sequence of shears, reflections, and stretches, all of which take close curves into closed curves. Thus the image of the circle (or of any circle or ellipse) must be a closed curve, and is therefore an ellipse (or a circle as a special case).

What we have done here can be generalized to n-dimensions. The number of possible elementary row operations depends on n, but there are still only three types. Also, elementary matrices corresponding to elementary column operations can be defined. We refer the reader to the books on linear algebra listed among the references in the back of this textbook.

─────────────── **Exercises for Section 3.9** ───────────────

1. Given the linear transformation $y_1 = 2x_1 + x_2$, $y_2 = 4x_1 + 3x_2$:
    (a) Find the matrix of the transformation and express it as the product of elementary matrices.
    (b) Find the image of the line $2x_1 - 3x_2 = 6$.
    (c) Find the image of the hyperbola $-x_1^2 + 4x_2^2 = 4$, and determine what kind of curve it is.

2. Given the linear transformation whose matrix is

    $$\begin{bmatrix} -2 & -1 \\ 1 & 2 \end{bmatrix}:$$

    (a) Find the image of the parabola $x_1 - 4 = 2(x_2 + 3)^2$, and determine what kind of curve it is.
    (b) Find the image of the curve $5x_1^2 + 8x_1x_2 + 5x_2^2 = 1$.
    (c) Express the matrix as the product of elementary matrices.

3. Draw a figure showing the effect of the given elementary transformation on the rectangle of Fig. 3.4(a)
    (a) $E\left(\dfrac{1}{3}\right)$ (b) $E_1(-2)$
    (c) $G_{21}\left(\dfrac{1}{2}\right)$ (d) $G_{12}(-3)$

4. Verify that premultiplication of a 2×2 matrix by the given elementary matrix has the effect indicated in the text.
    (a) $E_2(k)$ (b) F_{12} (c) $G_{21}(c)$

5. For each of the five elementary matrices, find the inverse and show that it is an elementary matrix.

6. (a) Show that a rotation of each point in the plane counterclockwise through an angle θ about the origin can be accomplished by a linear transformation with matrix

    $$\begin{bmatrix} \cos\theta & -\sin\theta \\ \sin\theta & \cos\theta \end{bmatrix}.$$

    (b) Express the matrix of Part (a) as the product of elementary matrices.

7. Show that the reflection of each point in the plane across the line through the origin with angle of inclination α can be accomplished by the linear transformation whose matrix is

    $$\begin{bmatrix} \cos 2\alpha & \sin 2\alpha \\ \sin 2\alpha & -\cos 2\alpha \end{bmatrix}.$$

8. Show that each of the five elementary matrices can be obtained by performing an elementary

column operation on the identity matrix. (For example, $E_1(k)$ can be obtained by multiplying through in the first column by k.)

9. Show that the effect of postmultiplying a matrix A by an elementary matrix has the

effect of performing an elementary column operation on A. (See the previous exercise.)

10. For 3×3 matrices, define elementary matrices that correspond to elementary row operations.

3.10
DIFFERENTIAL
OPERATORS

We now consider a class of linear transformations that are defined on spaces of functions. These transformations, or operators, are important in the study of linear differential equations, as will be seen in Chapter 5.

Let $\mathscr{I}$ be a fixed interval and let $\mathscr{F}^n$ denote the set of all functions that are defined and possess at least n derivatives on $\mathscr{I}$. By $\mathscr{F}^0$ or $\mathscr{F}$, we mean the set of all functions defined on $\mathscr{I}$. For each n, the set $\mathscr{F}^n$ forms a vector space if the usual operations for functions are used. We shall therefore speak of the space $\mathscr{F}^n$.

First we introduce the derivative operator D, writing

$$Df = f', \qquad Df(x) = f'(x)$$

for every function f in $\mathscr{F}^1$ and all x in $\mathscr{I}$. Thus D maps $\mathscr{F}^1$ into $\mathscr{F}$. This operator, already encountered in Example 4, Section 3.7, is a linear operator because

$$D(cf) = (cf)' = cf' = cDf$$

and

$$D(f + g) = (f + g)' = f' + g' = Df + Dg$$

for f and g in $\mathscr{F}^1$ and for every real number c.

The product DD is defined on $\mathscr{F}^2$ and maps $\mathscr{F}^2$ into $\mathscr{F}$. If h is in $\mathscr{F}^2$, then

$$DDh = D(Dh) = Dh' = h''.$$

We write

$$D^2 = DD,$$
$$D^3 = DD^2,$$
$$\cdots\cdots\cdots$$
$$D^n = DD^{n-1}.$$

It is easily seen that $D^n f = f^{(n)}$ for every function f in $\mathscr{F}^n$. We also define the operator D^0 by means of the formula

$$D^0 f = f.$$

Thus D^0 is the *identity operator* that maps each element of $\mathscr{F}$ into itself.

We now define a more general class of linear operators. To begin, let a_0, a_1, and a_2 be specific functions that are defined on an interval $\mathscr{I}$. We introduce an operator L,

$$L = a_0 D^2 + a_1 D + a_2,$$

where

$$Lf = a_0 D^2 f + a_1 Df + a_2 f = a_0 f'' + a_1 f' + a_2 f$$

for every function f in $\mathscr{F}^2$. Thus L maps $\mathscr{F}^2$ into $\mathscr{F}$. To see that L is a linear operator, we observe that

$$\begin{aligned} L(cf) &= a_0(cf)'' + a_1(cf') + a_2(cf) \\ &= c(a_0 f'' + a_1 f' + a_2 f) \\ &= cLf \end{aligned}$$

and that

$$\begin{aligned} L(f + g) &= a_0(f + g)'' + a_1(f + g)' + a_2(f + g) \\ &= (a_0 f'' + a_1 f' + a_2 f) + (a_0 g'' + a_1 g' + a_2 g) \\ &= Lf + Lg \end{aligned}$$

for f and g in $\mathscr{F}^2$ and for every real number c.

More generally, let $a_0, a_1, \ldots, a_n$ be specific functions defined on an interval $\mathscr{I}$. (It is assumed that a_0 is not the zero function.) We define an operator L,

$$L = a_0 D^n + a_1 D^{n-1} + a_2 D^{n-2} + \cdots + a_{n-1}D + a_n,$$

on the space $\mathscr{F}^n$ by means of the formula

$$Lf = a_0 f^{(n)} + a_1 f^{(n-1)} + a_2 f^{(n-2)} + \cdots + a_{n-1}f' + a_n f,$$

where f is in $\mathscr{F}^n$. It can be shown that L is a linear operator, and we say that L is a *linear differential operator of order n*. This operator maps $\mathscr{F}^n$ into $\mathscr{F}$.

To consider some specific examples, let L_1 and L_2 be defined by the formulas

$$L_1 f(x) = xf'(x) - f(x),$$
$$L_2 f(x) = f'(x) + 2xf(x).$$

Here L_1 and L_2 are both first-order linear differential operators defined on the space $\mathscr{F}^1$. (The interval $\mathscr{I}$ may be taken as the set of all real numbers.)

We may write

$$L_1 = xD - 1, \qquad L_2 = D + 2x.$$

Notice that

$$(L_1 + L_2)f(x) = xf'(x) - f(x) + f'(x) + 2xf(x)$$
$$= (x + 1)f'(x) + (2x - 1)f(x)$$

so that

$$L_1 + L_2 = (x + 1)D + (2x - 1).$$

Both $L_1 L_2$ and $L_2 L_1$ are defined on the space $\mathscr{F}^2$. If g is in $\mathscr{F}^2$ we have

$$L_1 L_2 g = L_1(g' + 2xg)$$
$$= x(g' + 2xg)' - (g' + 2xg)$$
$$= xg'' + (2x^2 - 1)g'$$

and

$$L_2 L_1 g = L_2(xg' - g)$$
$$= (xg' - g)' + 2x(xg' - g)$$
$$= xg'' + 2x^2 g' - 2xg.$$

We see that $L_1 L_2 \neq L_2 L_1$ because if $g(x) = 1$ for all x, then

$$L_1 L_2 g(x) = 0,$$

but

$$L_2 L_1 g(x) = -2x.$$

We recall that two operators T_1 and T_2, defined on the same space, are equal if and only if $T_1 \mathbf{u} = T_2 \mathbf{u}$ for every element $\mathbf{u}$ in the space. Let $\mathscr{I}$ be a fixed interval and let

$$L_1 = a_0 D^m + a_1 D^{m-1} + \cdots + a_{m-1} D + a_m,$$
$$L_2 = b_0 D^n + b_1 D^{n-1} + \cdots + b_{n-1} D + b_n,$$

where the a_i and b_i are functions defined on $\mathscr{I}$, and neither a_0 nor b_0 is the zero function. If k is the larger of m and n, that is, if $k = \max(m, n)$, then L_1 and L_2 are linear operators on $\mathscr{F}^k$. The following theorem characterizes equality for linear differential operators.

Theorem 3.13   The operators L_1 and L_2 are equal if and only if $m = n$ and $a_i = b_i$ for $1 \leq i \leq m$.

Proof If $m = n$ and $a_i = b_i$ for all i, it is evident that $L_1 f = L_2 f$ for every function f in $\mathscr{F}^n$ and hence $L_1 = L_2$.

Now suppose that $L_1 = L_2$. Let $f_0(x) = 1$ and $f_k(x) = x^k$ for x in $\mathscr{I}$ and for every positive integer k. Since $L_1 f_0 = a_m$ and $L_2 f_0 = b_n$, the condition $L_1 f_0 = L_2 f_0$ requires that $a_m = b_n$. Next,

$$L_1 f_1(x) = a_{m-1}(x) + x a_m(x), \qquad L_2 f_1(x) = b_{n-1}(x) + x b_n(x),$$

so that the condition $L_1 f_1 = L_2 f_1$ says that $a_{m-1} = b_{n-1}$. Since

$$L_1 f_2(x) = 2a_{m-2}(x) + 2x a_{m-1}(x) + x^2 a_m(x),$$
$$L_2 f_2(x) = 2b_{n-2}(x) + 2x b_{n-1}(x) + x^2 b_n(x),$$

we must have $a_{m-2} = b_{n-2}$.

Let $a_j = 0$ if $j < 0$ and $b_j = 0$ if $j < 0$. Then, by mathematical induction, it can be shown that $a_{m-k} = b_{n-k}$ for $k = 0, 1, 2, \ldots$, and hence that $m = n$.

The equation

$$Ly = 0,$$

where y is an unknown function, is called a *linear homogeneous differential equation* of order n. A function of the space $\mathscr{F}^n$ that satisfies the equation is called a *solution* of the differential equation. A function is a solution if and only if it belongs to the kernel of the operator L. Differential equations of this type will be discussed in Chapter 5.

Exercises for Section 3.10

1. Let $a_0(x) = 2$, $a_1(x) = x^2$, $a_2(x) = -x$ for all x, and let

    $$L = a_0 D^2 + a_1 D + a_2.$$

    Find Lf, where f is as given.
    (a) $f(x) = 4x$ (b) $f(x) = e^{3x}$
    (c) $f(x) = \sin x$

2. Let a_0 and a_1 be functions that are defined on an interval $\mathscr{I}$, and let

    $$L = a_0 D + a_1.$$

    Verify that L is a linear operator from $\mathscr{F}^1$ into $\mathscr{F}$.

3. Let $L = D^2 - 3D + 2$. Which, if any, of the following functions belongs to the kernel of L?

    (a) $f(x) = e^{2x}$ (b) $g(x) = e^{3x}$
    (c) $h(x) = 2e^x - e^{2x}$

4. Find $L_1 L_2$ and $L_2 L_1$ if
    (a) $L_1 = D + x$, $L_2 = D + 2$.
    (b) $L_1 = xD$, $L_2 = xD + 1$.
    (c) $L_1 = D - 1$, $L_2 = D^2 - xD + 1$.
    (d) $L_1 = D - \tan x$, $L_2 = D + \tan x$.

5. Let $L_1 = a_1 D + b_1$ and $L_2 = a_2 D + b_2$, where a_1, b_1, a_2, and b_2 are constants. Show that L_1 and L_2 commute; that is, that $L_1 L_2 = L_2 L_1$.

6. Let a_1, b_1, a_2, b_2 be differentiable functions on an interval $\mathscr{I}$, such that a_1 and a_2 are never zero. If

    $$L_1 = a_1 D + b_1, \qquad L_2 = a_2 D + b_2,$$

show that $L_1 L_2 = L_2 L_1$ if and only if $a_2 = ka_1$ and $b_2 = kb_1 + K$, where k and K are constants.

7. Let L_1 and L_2 be linear differential operators.
    (a) If the function f belongs to the kernel of L_2, show that f belongs to the kernel of $L_1 L_2$.
    (b) If the function g belongs to the kernel of L_1, does g necessarily belong to the kernel of $L_1 L_2$?

8. If L_1, L_2, and L_3 are linear differential operators such that $L_1 L_2$ and $L_1 L_3$ are defined, show that $L_1(L_2 + L_3) = L_1 L_2 + L_1 L_3$.

9. (a) Describe the kernel of the derivative operator D that maps $\mathscr{F}^1$ into $\mathscr{F}$.
    (b) Describe the kernel of the transformation D^n that maps $\mathscr{F}^n$ into $\mathscr{F}$.

10. Show that the kernel of the operator $L = D - 1$, which maps $\mathscr{F}^1$ into $\mathscr{F}$, is the set of all functions y of the form $y(x) = ce^x$, where c is a constant. Suggestion: multiply through in the equation $y'(x) - y(x) = 0$ by e^{-x}.

11. (a) Let $L_1 = f(x)D$, $L_2 = g(x)D$, and $L_3 = h(x)D$. Prove that $L_1(L_2 L_3) = (L_1 L_2)L_3$. What is an appropriate domain?
    (b) Consider the general case $T_1: V_1 \to V_2$, $T_2: V_2 \to V_3$, $T_3: V_3 \to V_4$. Prove that $T_1(T_2 T_3) = (T_1 T_2)T_3$.

12. (a) Let $L = a_0 D^n + a_1 D^{n-1} + \cdots + a_n$, where the a's are constants. Show that $Le^{ax} = P(a)e^{ax}$, where P is the polynomial $P(x) = a_0 x^n + a_1 x^{n-1} + \cdots a_n$.
    (b) Use the result of Part (a) to give a different proof of Theorem 3.13.
    (c) If L_1 and L_2 are linear differential operators with constant coefficients, prove that $L_1 L_2 = L_2 L_1$.

Additional Exercises for Chapter 3

1. Determine if the set of vectors $\{x_1, x_2, x_3, x_4\}$ constitutes a basis for R^4.
    (a) $x_1 = (1, 1, 0, 0)$, $x_2 = (1, -1, 1, 3)$,
         $x_3 = (0, 2, 1, -1)$, $x_4 = (0, 0, 1, 1)$
    (b) $x_1 = (0, 1, 1, 1)$, $x_2 = (1, 0, 1, 1)$,
         $x_3 = (1, 1, 0, 1)$, $x_4 = (1, 1, 1, 0)$

In Exercises 2, 3, and 4, $\mathscr{P}^3$ refers to the vector space of all polynomials of degree less than 3.

2. Which of the following sets of polynomials are bases for $\mathscr{P}^3$?
    (a) $\{x + 1, x^2, x^2 - 1\}$
    (b) $\{x^2, x + 1, x - 1\}$
    (c) $\{x^2, x^2 - 1, x^2 + 1\}$
    (d) $\{1 + x + x^2, 1 - x + x^2, 1\}$

3. Let T be the transformation from $\mathscr{P}^3$ into $\mathscr{P}^3$ that is defined by the formula $Tf(x) = xf'(x)$ for f in $\mathscr{P}^3$.
    (a) Show that T is a linear transformation.
    (b) Find the dimension of the range of T.
    (c) Find a basis for the range of T.

4. Let T be the linear transformation from $\mathscr{P}^3$ into $\mathscr{P}^3$ such that
    $$T(1 + x) = x, \qquad T(x^2) = 3 - x,$$
    $$T(x - x^2) = x^2.$$
    Find $T(x)$ and $T(1)$.

5. If the set $\{v_1, v_2, \ldots, v_n\}$ is linearly independent and if none of the scalars $c_1, c_2, \ldots, c_n$ is zero, show that the set $\{c_1 v_1, c_2 v_2, \ldots, c_n v_n\}$ is linearly independent.

6. Let T be a linear transformation from V into W and let $T(v_i) = w_i$ for $i = 1, 2, \ldots, n$. If $\{w_1, w_2, \ldots, w_n\}$ is linearly independent, show that $\{v_1, v_2, \ldots, v_n\}$ is linearly independent.

7. Let $T: R^5 \to R^5$ be defined by $Tx = Ax$, where A is the matrix

$$A = \begin{bmatrix} 1 & 2 & 0 & 4 & 1 \\ 1 & 3 & 0 & 7 & 1 \\ -1 & -1 & 1 & -2 & -2 \\ -1 & -1 & 0 & -1 & -1 \\ 0 & 2 & 1 & 5 & -1 \end{bmatrix}.$$

Determine the dimension of the kernel of T, and find a basis for it.

8. (a) Let T be a linear transformation from R^2 into R^2 such that $Tu_1 = v_1$ and $Tu_2 = v_2$. If u_1 and u_2 are linearly independent, show that $Tx = Ax$, where

$$A = [v_1, v_2][u_1, u_2]^{-1}.$$

   (b) If $T(2, 1) = (1, 1)$ and $T(1, 4) = (-1, 0)$, use the result of Part (a) to find a formula for $T(x_1, x_2)$.

9. Let $T:R^n \to R^n$ be defined by $Tx = Ax$, where A is an $n \times n$ matrix. Prove that T is one-to-one if and only if it is onto.

10. A linear transformation T from R^n into R^n is called an *orthogonal transformation* if $\|Tu\| = \|u\|$ for every u in R^n. (Thus an orthogonal transformation preserves lengths.) Prove that a linear transformation T is orthogonal if and only if $Tu \cdot Tv = u \cdot v$ for every u and v in R^n. (Thus angles are also preserved.) Suggestion: note that

$$\|u + v\|^2 = \|u\|^2 + 2u \cdot v + \|v\|^2$$

and find a similar formula for $\|Tu + Tv\|^2$.

11. Let $Tx = Ax$, where A is a real $n \times n$ matrix and x is a column vector. Referring to the previous exercise, show that if T is orthogonal, then $A^T A = I$ or $A^T = A^{-1}$. A real matrix A with this last property is called an *orthogonal matrix*. Suggestion: recall that $x \cdot y = x^T y$.

12. (a) If $T:R^2 \to R^2$ is such that $T(1, 2) = (-1, 1)$ and $T(3, 1) = (2, 5)$, find a matrix A such that $Tx = Ax$.
   (b) Let $T:R^3 \to R^3$ be such that $T(-1, 2, 0) = (1, 2, 0)$, $T(1, 0, 1) = (2, -1, 3)$ and $T(0, 1, 1) = (7, 4, 6)$. Find bases for the kernel and the range of T.
   (c) Find a basis for the range of the transformation of Exercise 7.

13. If a is a fixed element of R^n and if $Tx = a \cdot x$ for every x in R^n, show that T is a linear trans-

formation from R^n into R^1. Show that every linear transformation from R^n into R^1 can be written in this form.

14. Let V be an n-dimensional vector space with basis $\{v_1, v_2, \ldots, v_n\}$. Let

$$x = a_1 v_1 + a_2 v_2 \cdots + a_n v_n,$$

$$y = b_1 v_1 + b_2 v_2 + \cdots + b_n v_n$$

Show that x and y are linearly independent if and only if the elements

$$(a_1, a_2, \ldots, a_n), \quad (b_1, b_2, \ldots, b_n)$$

of R^n are linearly independent.

15. Let X be an $m \times n$ matrix and Y be a $k \times n$ matrix. Show that the row vectors of Y are linear combinations of the row vectors of X if and only if there is a $k \times m$ matrix B such that $Y = BX$. State and prove a similar result about column vectors.

16. Let X, Y, and B be matrices such that $Y = BX$. If the column vectors of X are linearly dependent, show that the column vectors of Y are linearly dependent.

17. The *row rank* of an $m \times n$ matrix A is defined to be the dimension of the subspace of R^n spanned by the row vectors of A, and the *column rank* is the dimension of the subspace of R^m spanned by the column vectors. Let B be the matrix obtained from A by reducing it to row-echelon form. Each nonzero row of B has a first nonzero element that is a 1. Let $n_1, n_2, \ldots, n_k$ be the numbers of the columns of B that contain these 1s. (Note that k is the number of these columns.)
   (a) Prove that the row rank of A is k.
   (b) Prove that the column rank of A is also k. (The common value is called the *rank* of the matrix A.)
   (c) Prove that the column vectors of A numbered $n_1, n_2, \ldots, n_k$ constitute a basis for the subspace spanned by the column vectors of A.
   (d) Show that elementary row and column operations do not change the rank of a matrix.

Suggestion: in Parts (b) and (c), consider the equation

$$c_1 \mathbf{v}_1 + c_2 \mathbf{v}_2 + \cdots + c_n \mathbf{v}_n = \mathbf{0},$$

where the vectors $\mathbf{v}_i$ are the column vectors of A. This vector equation corresponds to a system of m equations for the n constants c_1, $c_2, \ldots, c_n$. The matrix B is the row-echelon form of the coefficient matrix of this system. Show that $c_{n_1} \mathbf{v}_{n_1} + \cdots + c_{n_k} \mathbf{v}_{n_k} = \mathbf{0}$ if and only if $c_{n_1} = \cdots = c_{n_k} = 0$, and hence that the vectors $\mathbf{v}_{n_1}, \ldots, \mathbf{v}_{n_k}$ are linearly independent. Then show that every other column vector is a linear combination of these column vectors.

18. For each given matrix, determine its rank and find a set of column vectors that constitutes a basis for the subspace spanned by the column vectors. (Use the technique outlined in Exercise 17.)

(a) $\begin{bmatrix} 2 & -1 & 3 \\ 1 & -1 & 2 \\ 0 & 1 & -1 \end{bmatrix}$

(b) $\begin{bmatrix} 1 & -2 & 3 & 2 \\ 2 & -4 & 6 & 4 \\ -1 & 2 & -3 & -2 \end{bmatrix}$

(c) $\begin{bmatrix} 1 & -1 & 0 & 2 \\ 1 & 3 & 2 & -4 \\ 3 & 1 & 2 & 0 \\ 4 & 0 & 2 & 2 \end{bmatrix}$

(d) $\begin{bmatrix} 0 & 3 & 4 & 1 & 1 \\ -2 & 1 & -1 & -2 & 0 \\ 1 & 1 & 1 & 0 & -1 \\ 0 & 2 & 3 & 1 & 1 \end{bmatrix}$

19. Let T be a linear transformation from R^n into R^m. The *rank* of T is defined to be the dimension of the range of T. Let A be the $m \times n$ matrix such that $T\mathbf{x} = A\mathbf{x}$. Show that the rank of T is equal to the rank of A. (See Exercise 17.)

20. Show that a matrix has rank r (See Exercise 17) if and only if it has a square nonsingular submatrix of order r, and every square submatrix of order greater than r is singular. Suggestion: consider submatrices with r rows whose row vectors are row vectors of the original matrix.

4

Characteristic
Values

4.1
CHARACTERISTIC
VALUES

Let T be a linear transformation from a vector space V into itself. The problem we wish to consider is that of finding all nonzero vectors $\mathbf{x}$ and all scalars λ such that $T\mathbf{x} = \lambda\mathbf{x}$. We are particularly interested in the case where V is C^n or R^n. Applications that give rise to this problem are presented in Section 4.2 and in Chapters 6 and 10.

Let T be a linear transformation from C^n into C^n defined by the equation

$$T\mathbf{x} = A\mathbf{x},$$

where A is an $n \times n$ matrix and $\mathbf{x}$ is a column vector. If there is a complex number λ and a nonzero vector $\mathbf{x}$ such that $T\mathbf{x} = \lambda\mathbf{x}$, then λ and $\mathbf{x}$ must satisfy the condition $A\mathbf{x} = \lambda\mathbf{x}$ or

$$(\lambda I - A)\mathbf{x} = \mathbf{0}, \tag{4.1}$$

where I is the $n \times n$ identity matrix. According to Theorem 2.12, this equation has a nontrivial solution if and only if λ is such that

$$\det(\lambda I - A) = 0. \tag{4.2}$$

Thus we must find all the roots of Eq. (4.2), and for each root λ we must find the nonzero vector solutions of Eq. (4.1).

These considerations lead us to formalize certain ideas. For any $n \times n$ matrix A, a complex number λ for which the condition (4.2) holds is called a *characteristic value* (*eigenvalue*) of A. Although in some applications we are interested in situations where λ must be real, there are other problems where complex characteristic values are important. (See, for example, Section 6.6.) If λ is a characteristic value of A, a nonzero vector $\mathbf{x}$, which may be complex, such that

$$(\lambda I - A)\mathbf{x} = \mathbf{0}$$

is called a *characteristic vector* (*eigenvector*) of A corresponding to the characteristic value λ.

Example 1   Let

$$A = \begin{bmatrix} 1 & -3 \\ -2 & 2 \end{bmatrix}.$$

Then

$$\det(\lambda I - A) = \begin{vmatrix} \lambda - 1 & 3 \\ 2 & \lambda - 2 \end{vmatrix} = (\lambda + 1)(\lambda - 4).$$

The characteristic values of A are therefore $\lambda_1 = -1$ and $\lambda_2 = 4$. When $\lambda = -1$ the equation $(\lambda I - A)\mathbf{x} = \mathbf{0}$ corresponds to the system

$$-2x_1 + 3x_2 = 0,$$
$$2x_1 - 3x_2 = 0.$$

The solutions of the system are the ordered pairs $(3a, 2a)$ where a is an arbitrary constant. The characteristic vectors of A corresponding to $\lambda = -1$ are the nonzero vectors of the form $(3a, 2a)$.

When $\lambda = 4$ the equation $(\lambda I - A)\mathbf{x} = \mathbf{0}$ corresponds to the system

$$3x_1 + 3x_2 = 0,$$
$$2x_1 + 2x_2 = 0.$$

This time the characteristic vectors are of the form $(a, -a)$, with $a \neq 0$.

Example 2   Let

$$A = \begin{bmatrix} 1 & 2 \\ 0 & 1 \end{bmatrix}.$$

Then

$$\det(\lambda I - A) = \begin{vmatrix} \lambda - 1 & -2 \\ 0 & \lambda - 1 \end{vmatrix} = (\lambda - 1)^2,$$

so $\lambda_1 = 1$ is the only characteristic value. The equation $(\lambda I - A)\mathbf{x} = \mathbf{0}$ becomes

$$0x_1 - 2x_2 = 0,$$
$$0x_1 + 0x_2 = 0,$$

and we see that the characteristic vectors are of the form $(a, 0)$, with $a \neq 0$.

Example 3 Let

$$A = \begin{bmatrix} 1 & 1 & -1 \\ -1 & 3 & -1 \\ -1 & 1 & 1 \end{bmatrix}.$$

Calculation shows that

$$\det(\lambda I - A) = \begin{vmatrix} \lambda - 1 & -1 & 1 \\ 1 & \lambda - 3 & 1 \\ 1 & -1 & \lambda - 1 \end{vmatrix} = (\lambda - 1)(\lambda - 2)^2.$$

The characteristic values are therefore $\lambda_1 = 1$ and $\lambda_2 = 2$. The equation $(\lambda_1 I - A)\mathbf{x} = \mathbf{0}$ corresponds to the system

$$-x_2 + x_3 = 0,$$
$$x_1 - 2x_2 + x_3 = 0,$$
$$x_1 - x_2 \qquad = 0,$$

which reduces to

$$x_1 - x_3 = 0,$$
$$x_2 - x_3 = 0,$$
$$0 = 0.$$

Thus the characteristic vectors are (a, a, a), with $a \neq 0$.
The equation $(\lambda_2 I - A)\mathbf{x} = \mathbf{0}$ corresponds to the system

$$x_1 - x_2 + x_3 = 0,$$
$$x_1 - x_2 + x_3 = 0,$$
$$x_1 - x_2 + x_3 = 0,$$

or

$$x_1 - x_2 + x_3 = 0,$$
$$0 = 0,$$
$$0 = 0.$$

$$\begin{bmatrix} a-b \\ a \\ b \end{bmatrix}$$

The characteristic vectors are $(a - b, a, b)$, with a and b not both zero. In particular, $(a, a, 0)$ and $(-b, 0, b)$ are linearly independent characteristic vectors.

Let us now consider the general case of an $n \times n$ matrix A. Associated with A is the function p, where

$$p(\lambda) = \det(\lambda I - A).$$

This function p is a polynomial of degree n, called the *characteristic polynomial* of A. To see this, let us write

$$p(\lambda) = \begin{vmatrix} \lambda - a_{11} & -a_{12} & \cdots & -a_{1n} \\ -a_{21} & \lambda - a_{22} & \cdots & -a_{2n} \\ \vdots & \vdots & & \vdots \\ -a_{n1} & -a_{n2} & \cdots & \lambda - a_{nn} \end{vmatrix}.$$

Recalling the definition of a determinant as the sum of products involving one element from each row and column, we observe that

$$(\lambda - a_{11})(\lambda - a_{22}) \cdots (\lambda - a_{nn})$$

is a product in the sum. Also, this is the only product that introduces the nth power of λ. Thus p has the form

$$p(\lambda) = \lambda^n + a_1 \lambda^{n-1} + a_2 \lambda^{n-2} + \cdots + a_n.$$

The zeros of p are the characteristic values of A. The equation $p(\lambda) = 0$ is called the *characteristic equation* of A.

Some of the zeros of p may be complex. For this case we have the following theorem.

Theorem 4.1   If $\lambda = \alpha + i\beta$, $\beta \neq 0$, is a complex characteristic value of the real matrix A, with corresponding characteristic vector $\mathbf{x}$, then $\bar{\lambda} = \alpha - i\beta$ is also a characteristic value, with corresponding characteristic vector $\bar{\mathbf{x}}$.

Proof   If A is real then p has real coefficients. Hence if $\lambda = \alpha + i\beta$, $\beta \neq 0$, is a complex zero of p, then $\bar{\lambda} = \alpha - i\beta$ is also a zero. Since $A\mathbf{x} = \lambda\mathbf{x}$, we see upon taking complex conjugates that $A\bar{\mathbf{x}} = \bar{\lambda}\bar{\mathbf{x}}$. Hence $\bar{\mathbf{x}}$ is a characteristic vector corresponding to $\bar{\lambda}$.

Example 4   Let

$$A = \begin{bmatrix} 3 & -2 & 4 \\ 4 & -3 & 4 \\ -2 & 0 & -3 \end{bmatrix}.$$

Then

$$\det(\lambda I - A) = \begin{vmatrix} \lambda - 3 & 2 & -4 \\ -4 & \lambda + 3 & -4 \\ 2 & 0 & \lambda + 3 \end{vmatrix} = (\lambda + 1)(\lambda^2 + 2\lambda + 5).$$

The characteristic values are $\lambda_1 = -1$, $\lambda_2 = -1 + 2i$, and $\lambda_3 = \bar{\lambda}_2 = -1 - 2i$. The equation $(\lambda_1 I - A)\mathbf{x} = \mathbf{0}$, in component form, becomes

$$-4x_1 + 2x_2 - 4x_3 = 0$$
$$-4x_1 + 2x_2 - 4x_3 = 0$$
$$2x_1 \qquad + 2x_3 = 0.$$

This system simplifies to

$$x_2 = 0$$
$$x_1 + x_3 = 0.$$

The characteristic vectors are $(a, 0, -a) = a(1, 0, -1)$, $a \neq 0$.
The equation $(\lambda_2 I - A)\mathbf{x} = \mathbf{0}$ becomes

$$(-4 + 2i)x_1 + \qquad 2x_2 - \qquad 4x_3 = 0$$
$$-4x_1 + (2 + 2i)x_2 - \qquad 4x_3 = 0$$
$$2x_1 \qquad + (2 + 2i)x_3 = 0.$$

This system reduces to

$$x_1 + (1 + i)x_3 = 0$$
$$x_2 + (1 + i)x_3 = 0.$$

Setting $x_3 = a$, we have $x_1 = x_2 = -a(1 + i)$. Thus the characteristic vectors corresponding to λ_2 are the nonzero complex multiples of $(-1 - i, -1 - i, 1)$. Because of Theorem 4.1 we can assert without further ado that the characteristic vectors corresponding to λ_3 are the nonzero complex multiples of $(-1 + i, -1 + i, 1)$.

Exercises for Section 4.1

In Exercises 1–14, determine the characteristic values of the given matrix, and find the corresponding characteristic vectors.

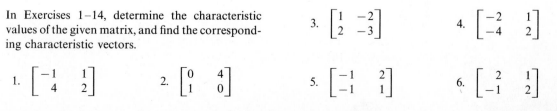

1. $\begin{bmatrix} -1 & 1 \\ 4 & 2 \end{bmatrix}$

2. $\begin{bmatrix} 0 & 4 \\ 1 & 0 \end{bmatrix}$

3. $\begin{bmatrix} 1 & -2 \\ 2 & -3 \end{bmatrix}$

4. $\begin{bmatrix} -2 & 1 \\ -4 & 2 \end{bmatrix}$

5. $\begin{bmatrix} -1 & 2 \\ -1 & 1 \end{bmatrix}$

6. $\begin{bmatrix} 2 & 1 \\ -1 & 2 \end{bmatrix}$

7. $\begin{bmatrix} 1 & 2 & -1 \\ 0 & -2 & 0 \\ 0 & -5 & 2 \end{bmatrix}$ 8. $\begin{bmatrix} 3 & 4 & 2 \\ 2 & 1 & 2 \\ -2 & -2 & -1 \end{bmatrix}$

9. $\begin{bmatrix} 2 & -2 & 1 \\ 1 & -1 & 1 \\ -3 & 2 & -2 \end{bmatrix}$ 10. $\begin{bmatrix} 1 & 2 & 3 \\ -1 & 4 & 3 \\ 1 & -2 & -1 \end{bmatrix}$

11. $\begin{bmatrix} 2 & 1 & 1 \\ 1 & 2 & 1 \\ -2 & -2 & -1 \end{bmatrix}$ 12. $\begin{bmatrix} -1 & 1 & 2 \\ -1 & 1 & 1 \\ -2 & 1 & 3 \end{bmatrix}$

13. $\begin{bmatrix} 0 & 1 & 1 \\ -1 & 1 & 1 \\ -1 & 1 & 1 \end{bmatrix}$ 14. $\begin{bmatrix} 4 & 2 & -5 \\ 5 & 1 & -5 \\ 5 & 3 & -6 \end{bmatrix}$

15. Show that a square matrix is singular if and only if zero is a characteristic value.

16. Show that the characteristic values of a triangular matrix are the diagonal elements. In particular, show that the characteristic values of a diagonal matrix are the diagonal elements. Find a 3×3 diagonal matrix whose characteristic values are $3, -1, 5$.

17. If A is nonsingular, show that the characteristic values of A^{-1} are the reciprocals of A, and that A and A^{-1} have the same characteristic vectors.

18. Let A be a square matrix. Show that A and A^T have the same characteristic values. Do they have the same characteristic vectors?

19. (a) Given a polynomial f of the form

$$f(x) = x^n + b_1 x^{n-1} + b_2 x^{n-2} + \cdots + b_{n-1} x + b_n$$

show that the matrix below (called the *companion matrix* of f) has f as its characteristic polynomial.

$$\begin{bmatrix} 0 & 1 & 0 & \cdots & 0 \\ 0 & 0 & 1 & \cdots & 0 \\ \cdots & \cdots & \cdots & \cdots & \cdots \\ 0 & 0 & 0 & \cdots & 1 \\ -b_n & -b_{n-1} & -b_{n-2} & \cdots & -b_1 \end{bmatrix}$$

(b) Find the companion matrix of the polynomial $x^4 - 2x^3 - 5x^2 + 6x + 7$.

4.2
AN
APPLICATION

A company uses a certain kind of machine. Suppose that 80% of the new machines survive one year, that 60% of the one-year-old machines survive another year, and that no machines last longer than three years. Let x_{0t}, x_{1t}, and x_{2t} be the numbers of new machines, one-year-old machines, and two-year-old machines at the beginning of year number t, and let

$$\mathbf{x}_t = \begin{bmatrix} x_{0t} \\ x_{1t} \\ x_{2t} \end{bmatrix}.$$

Then the vectors

$$\mathbf{x}_0 = \begin{bmatrix} x_{00} \\ x_{10} \\ x_{20} \end{bmatrix}, \qquad \mathbf{x}_1 = \begin{bmatrix} x_{01} \\ x_{11} \\ x_{21} \end{bmatrix}$$

describe the numbers of the machines in the three age categories at the beginnings of years zero and one, respectively. If the total number of machines stays the same, then

$$x_{00} + x_{10} + x_{20} = x_{01} + x_{11} + x_{21}.$$

From the information provided at the beginning of this section, we see that

$$x_{11} = 0.80x_{00},$$

$$x_{21} = 0.60x_{10}.$$

We must determine x_{01}, the number of new machines to be purchased at the beginning of year one. Since 20% of the one-year-old machines, 40% of the two-year-old ones, and all of the three-year-old ones must be replaced, we have

$$x_{01} = 0.20x_{00} + 0.40x_{10} + 1.00x_{20}.$$

Thus

$$\mathbf{x}_1 = A\mathbf{x}_0$$

where

$$A = \begin{bmatrix} 0.20 & 0.40 & 1.00 \\ 0.80 & 0 & 0 \\ 0 & 0.60 & 0 \end{bmatrix}. \tag{4.3}$$

The question arises of how many machines are in the three age categories if the sizes remain the same from year to year. For such a configuration, $\mathbf{x}_1 = \mathbf{x}_0$, so that $A\mathbf{x}_0 = \mathbf{x}_0$. Thus $\lambda = 1$ is a characteristic value of A. We leave it to the reader to show that

$$\det(\lambda I - A) = \lambda^3 - 0.20\lambda^2 - 0.32\lambda - 0.48$$

and that $\lambda = 1$ is the only real characteristic value. We also ask the reader to show that the characteristic vectors are proportional to

$$(0.48, 0.80, 1.00).$$

The percentage of machines that must be replaced each year is therefore

$$\frac{0.48}{0.48 + 0.80 + 1.00} \times 100 = 21.05\%.$$

Now suppose that it is desired that the total number of machines *increase* each year, by a factor k. Then

$$x_{01} = 0.20x_{00} + 0.40x_{10} + x_{20} + k(x_{00} + x_{10} + x_{20})$$

$$= (0.20 + k)x_{00} + (0.40 + k)x_{10} + (1 + k)x_{20}$$

so the transition matrix A becomes

$$A = \begin{bmatrix} 0.20 + k & 0.40 + k & 1.00 + k \\ 0.80 & 0 & 0 \\ 0 & 0.60 & 0 \end{bmatrix}. \tag{4.4}$$

The three age groups will not, of course, remain constant in size, but we may ask if the percentages of the sizes of the groups can remain the same. Thus, we ask if there is a vector $\mathbf{x}_0$ such that $A\mathbf{x}_0 = \lambda\mathbf{x}_0$. We leave it to the reader to show that $\lambda = 1 + k$ is a characteristic value of A and that the corresponding characteristic vectors are multiples of

$$(0.48, 0.80(1 + k), (1 + k)^2).$$

Exercises for Section 4.2

1. Find the characteristic polynomial of the matrix (4.3) and show that $\lambda = 1$ is the only real characteristic value.

2. Find the characteristic vectors of the matrix (4.3) corresponding to $\lambda = 1$.

3. In the "no growth" model (4.3), find the percentage of one-year-old machines.

4. Find the characteristic polynomial of the matrix (4.4) and show that $\lambda = 1 + k$ is the only real characteristic value.

5. Find the characteristic vectors of the matrix (4.4) corresponding to $\lambda = k + 1$.

6. Assuming a 10% growth rate ($k = 0.10$), what percentage of the machines are new each year?

7. (a) Suppose that all machines survive one year, but that 30% of two-year-old machines must be replaced. Show that the transition matrix has the form

$$\begin{bmatrix} 0 & 0.30 & 1 \\ 1 & 0 & 0 \\ 0 & 0.70 & 0 \end{bmatrix}.$$

   (b) Show that $\lambda = 1$ is a characteristic value. Find the corresponding characteristic vectors. Find the percentage of new machines if the proportions of the three age groups remains constant.

8. Assuming a 20% annual increase in the total number of machines under the assumptions of Exercise 7, find the transition matrix and find the percentage of machines that are new, assuming that the three age groups remain in the same proportions each year.

4.3 DIAGONALIZATION

If A and B are square matrices of the same size, we say that A is *similar* to B if there exists a nonsingular matrix K such that

$$B = K^{-1}AK.$$

We now investigate the possibility that a given $n \times n$ matrix A is similar to a diagonal matrix.

Let $D = \mathrm{diag}(\lambda_1, \lambda_2, \ldots, \lambda_n)$. Then A is similar to D if and only if there is a nonsingular matrix K such that

$$D = K^{-1}AK$$

or

$$KD = AK.$$

Let $K = [\mathbf{k}_1, \mathbf{k}_2, \ldots, \mathbf{k}_n]$. Then the last condition can be written as

$$[\lambda_1 \mathbf{k}_1, \lambda_2 \mathbf{k}_2, \ldots, \lambda_n \mathbf{k}_n] = [A\mathbf{k}_1, A\mathbf{k}_2, \ldots, A\mathbf{k}_n].$$

This matrix equation corresponds to the system of vector equations

$$A\mathbf{k}_1 = \lambda_1 \mathbf{k}_1, \qquad A\mathbf{k}_2 = \lambda_2 \mathbf{k}_2, \ldots, \qquad A\mathbf{k}_n = \lambda_n \mathbf{k}_n.$$

Thus the numbers $\lambda_1, \lambda_2, \ldots, \lambda_n$ are characteristic values of A and the vectors $\mathbf{k}_1, \mathbf{k}_2, \ldots, \mathbf{k}_n$ are corresponding characteristic vectors. Furthermore, since K is nonsingular, the characteristic vectors must be linearly independent according to Theorem 3.3. Our results may be summarized as follows.

Theorem 4.2   The $n \times n$ matrix A is similar to a diagonal matrix if and only if it possesses a set of n linearly independent characteristic vectors $\mathbf{k}_1, \mathbf{k}_2, \ldots, \mathbf{k}_n$. If $K = [\mathbf{k}_1, \mathbf{k}_2, \ldots, \mathbf{k}_n]$, then

$$K^{-1}AK = \mathrm{diag}(\lambda_1, \lambda_2, \ldots, \lambda_n).$$

An $n \times n$ matrix that is similar to a diagonal matrix is said to be *diagonalizable*. Such a matrix possesses a *complete set of characteristic vectors*; that is, it has a set of n linearly independent characteristic vectors.

Example 1   The matrix

$$A = \begin{bmatrix} 1 & -3 \\ -2 & 2 \end{bmatrix}$$

of Section 4.1 had $\lambda_1 = -1$ and $\lambda_2 = 4$ as characteristic values. The characteristic vectors corresponding to λ_1 were $(3a, 2a)$, $a \neq 0$, and those corresponding to λ_2 were $(a, -a)$, $a \neq 0$. In particular, by taking $a = 1$, we see that $\mathbf{k}_1 = (3, 2)$ and $\mathbf{k}_2 = (1, -1)$ are characteristic vectors. These vectors are linearly independent, for if

$$K = \begin{bmatrix} 3 & 1 \\ 2 & -1 \end{bmatrix}$$

then $\det(K) \neq 0$. By Theorem 4.2

$$K^{-1}AK = \begin{bmatrix} -1 & 0 \\ 0 & 4 \end{bmatrix}.$$

Not every square matrix is similar to a diagonal matrix, since not every square matrix has a complete set of characteristic vectors. Two kinds of matrices that are diagonalizable will be examined in this section and in the next. We shall need the following result.

Theorem 4.3   Let $\lambda_1, \lambda_2, \ldots, \lambda_r$ be *distinct* characteristic values of an $n \times n$ matrix A, and let $\mathbf{k}_1, \mathbf{k}_2, \ldots, \mathbf{k}_r$ be corresponding characteristic vectors. Then these vectors are linearly independent.

Proof   We shall use induction. If $r = 1$ the set $\{\mathbf{k}_1\}$ is linearly independent, since $\mathbf{k}_1$ is not the zero vector. If the theorem is true for $r = m - 1$ then the set $\{\mathbf{k}_1, \mathbf{k}_2, \ldots, \mathbf{k}_{m-1}\}$ is linearly independent. This means that

$$c_1\mathbf{k}_1 + c_2\mathbf{k}_2 + \cdots + c_{m-1}\mathbf{k}_{m-1} = \mathbf{0} \tag{4.5}$$

if and only if $c_i = 0$ for each i. We want to show that the set $\{\mathbf{k}_1, \ldots, \mathbf{k}_{m-1}, \mathbf{k}_m\}$ is linearly independent. Suppose that

$$a_1\mathbf{k}_1 + a_2\mathbf{k}_2 + \cdots + a_{m-1}\mathbf{k}_{m-1} + a_m\mathbf{k}_m = \mathbf{0}. \tag{4.6}$$

We shall first show that $a_m = 0$. If $a_m \neq 0$ then

$$\mathbf{k}_m = b_1\mathbf{k}_1 + b_2\mathbf{k}_2 + \cdots + b_{m-1}\mathbf{k}_{m-1} \tag{4.7}$$

where $b_i = -a_i/a_m$ for each i. Premultiplying both sides of this last equation by $\lambda_m I - A$, and remembering that $A\mathbf{k}_i = \lambda_i\mathbf{k}_i$, we have

$$\mathbf{0} = (\lambda_1 - \lambda_m)b_1\mathbf{k}_1 + (\lambda_2 - \lambda_m)b_2\mathbf{k}_2 + \cdots + (\lambda_{m-1} - \lambda_m)b_{m-1}\mathbf{k}_{m-1}.$$

But by the induction hypothesis (4.5) we must have $(\lambda_i - \lambda_m)b_i = 0$ for each i. Since the characteristic values are distinct, $b_i = 0$ for each i. Then, according to formula (4.7), $\mathbf{k}_m = \mathbf{0}$. But this is impossible, since $\mathbf{k}_m$ is a characteristic vector. We must conclude that $a_m = 0$. But if $a_m = 0$ in Eq. (4.6) we must have $a_i = 0$ for $i = 1, 2, \ldots, m - 1$ also, because of the induction hypothesis. We conclude that the set $\{\mathbf{k}_1, \mathbf{k}_2, \ldots, \mathbf{k}_m\}$ is linearly independent. Thus if the theorem is true for $r = m - 1$ it is true for $r = m$. Since it is true for $r = 1$ it is true for every r, $1 \leq r \leq n$.

Corollary   If the $n \times n$ matrix A possesses n distinct characteristic values, then it possesses a complete set of characteristic vectors and is diagonalizable.

Proof   According to Theorem 4.3, with $r = n$, the matrix A possesses a complete set of characteristic vectors. According to Theorem 4.2, it is diagonalizable.

An $n \times n$ matrix need not have n distinct characteristic values. Examples 2 and 3 of Section 4.1 illustrate this fact. In general, if $\lambda_1, \lambda_2, \ldots, \lambda_r$ are the *distinct* characteristic values of an $n \times n$ matrix A (so that $r \leq n$) then the characteristic polynomial of A has the form

$$p(\lambda) = (\lambda - \lambda_1)^{m_1}(\lambda - \lambda_2)^{m_2} \cdots (\lambda - \lambda_r)^{m_r}$$

where $m_1 + m_2 + \cdots + m_r = n$. The integer m_i is called the *multiplicity* of the characteristic value λ_i.

Let us denote by V_i the set of all characteristic vectors that correspond to λ_i, together with the zero vector. Then V_i is a vector space, since it is the kernel of the matrix $\lambda_i I - A$ (see Section 3.8). Notice that the zero vector must be specifically included in the description of V_i because a characteristic vector is by definition not the zero vector. Let us denote the dimension of the space V_i by n_i. It can be shown, although we shall not prove it here, that

$$1 \le n_i \le m_i$$

for each i. That is, the dimension of V_i cannot exceed the multiplicity of λ_i. In Example 2 of Section 4.1 the multiplicity of λ_1 was 2 but the dimension n_1 of V_1 was 1. However, in Example 3 of Section 4.1 the multiplicity of λ_2 and the dimension of V_2 were both 2. It can be shown that a square matrix possesses a complete set of characteristic vectors if and only if $m_i = n_i$ for each i. In the next section we shall discuss a type of matrix that always possesses a complete set of characteristic vectors even though its characteristic values may not all be distinct.

We are interested in similar matrices because some problems involving a matrix can be made more tractable by finding a "simpler" matrix similar to the original one. An example is presented in Chapter 10. A diagonal matrix is regarded as simple. However, if the matrix does not have a complete set of characteristic vectors, it is not similar to a diagonal matrix.

Even if a matrix A is similar to a diagonal matrix, so that $K^{-1}AK = D$, D and K will not be real if A has nonreal characteristic values. However, A is similar to a real matrix which, although not diagonal, is usually simpler than A. We shall discuss matrices of order two. The method can be extended to larger matrices.

Suppose that A is a 2×2 real matrix with complex characteristic values $\lambda_1 = \alpha + i\beta$ and $\lambda_2 = \alpha - i\beta$, where $\beta \ne 0$. If $\mathbf{k}_1 = \mathbf{u} + i\mathbf{v}$ is a characteristic vector corresponding to λ_1, then $\mathbf{k}_2 = \mathbf{u} - i\mathbf{v}$ is a characteristic vector corresponding to λ_2. (Here $\mathbf{u}$ and $\mathbf{v}$ are real, and linearly independent). We shall show that the real matrix $K = [\mathbf{u}, \mathbf{v}]$, with $\mathbf{u}$ and $\mathbf{v}$ as column vectors, has the property that

$$K^{-1}AK = \begin{bmatrix} \alpha & \beta \\ -\beta & \alpha \end{bmatrix}. \tag{4.8}$$

This relationship is equivalent to

$$AK = K\begin{bmatrix} \alpha & \beta \\ -\beta & \alpha \end{bmatrix}.$$

Now

$$\mathbf{u} = \frac{1}{2}(\mathbf{k}_1 + \mathbf{k}_2), \qquad \mathbf{v} = \frac{1}{2i}(\mathbf{k}_1 - \mathbf{k}_2)$$

so

$$Au = \frac{1}{2}(Ak_1 + Ak_2) = \frac{1}{2}(\lambda_1 k_1 + \lambda_2 k_2) = \alpha u - \beta v,$$

$$Av = \frac{1}{2i}(Ak_1 - Ak_2) = \frac{1}{2i}(\lambda_1 k_1 - \lambda_2 k_2) = \beta u + \alpha v.$$

Then

$$AK = A[u, v] = [Au, Av] = [\alpha u - \beta v, \beta u + \alpha v].$$

However, using Theorem 2.4, we see that

$$K\begin{bmatrix} \alpha & \beta \\ -\beta & \alpha \end{bmatrix} = [u, v]\begin{bmatrix} \alpha & \beta \\ -\beta & \alpha \end{bmatrix} = [\alpha u - \beta v, \beta u + \alpha v].$$

Thus

$$AK = K\begin{bmatrix} \alpha & \beta \\ -\beta & \alpha \end{bmatrix}$$

as we wished to show.

Example 2 Let

$$A = \begin{bmatrix} -1 & 3 \\ -6 & 5 \end{bmatrix}.$$

We have

$$\det(\lambda I - A) = \begin{vmatrix} \lambda + 1 & -3 \\ 6 & \lambda - 5 \end{vmatrix} = \lambda^2 - 4\lambda + 13$$

so the characteristic values are $\lambda_1 = 2 + 3i$ and $\lambda_2 = 2 - 3i$. The system $(\lambda_1 I - A)k = 0$ becomes

$$(3 + 3i)k_1 - 3k_2 = 0$$
$$6k_1 + (-3 + 3i)k_2 = 0.$$

A characteristic vector is $k = (1, 1 + i)$. Its real and imaginary parts are $u = (1, 1)$, $v = (0, 1)$. Thus the real matrix

$$K = \begin{bmatrix} 1 & 0 \\ 1 & 1 \end{bmatrix}$$

has the property that

$$K^{-1}AK = \begin{bmatrix} 2 & 3 \\ -3 & 2 \end{bmatrix}.$$

Finally, let us consider a 2×2 matrix A that has a single characteristic value λ_1 of multiplicity two. If $\lambda_1 I - A$ is the zero matrix, there exist two linearly independent characteristic values of A (indeed, every nonzero vector is a characteristic vector), and A is similar to the diagonal matrix $D = \text{diag}(\lambda_1, \lambda_1)$. But if $\lambda_1 I - A$ is not the zero matrix, the subspace of characteristic vectors (the kernel of $\lambda_1 I - A$) has dimension one, and all characteristic vectors are multiples of a nonzero vector $\mathbf{k}_1$. We claim that there is a nonsingular matrix K such that

$$K^{-1}AK = \begin{bmatrix} \lambda_1 & 1 \\ 0 & \lambda_2 \end{bmatrix}. \qquad (4.9)$$

Let $\mathbf{v}$ be any vector such that $\mathbf{k}_1$ and $\mathbf{v}$ are linearly independent, and let $K = [\mathbf{k}_1 \ \mathbf{v}]$. Then Equation (4.9) is equivalent to

$$AK = K \begin{bmatrix} \lambda_1 & 1 \\ 0 & \lambda_1 \end{bmatrix}$$

or

$$A[\mathbf{k}_1, \mathbf{v}] = [\mathbf{k}_1, \mathbf{v}] \begin{bmatrix} \lambda_1 & 1 \\ 0 & \lambda_1 \end{bmatrix}$$

or

$$[A\mathbf{k}_1, A\mathbf{v}] = [\lambda_1 \mathbf{k}_1, \mathbf{k}_1 + \lambda_1 \mathbf{v}]$$

or, since $\mathbf{k}_1$ is a characteristic vector,

$$[\lambda_1 \mathbf{k}_1, A\mathbf{v}] = [\lambda_1 \mathbf{k}_1, \mathbf{k}_1 + \lambda_1 \mathbf{v}].$$

This equation requires that

$$A\mathbf{v} = \mathbf{k}_1 + \lambda_1 \mathbf{v}$$

or

$$(\lambda_1 I - A)\mathbf{v} = -\mathbf{k}_1.$$

If $\mathbf{k}_1$ is in the one-dimensional subspace spanned by the column vectors of $\lambda_1 I - A$, then this equation has a solution. (See Exercise 20, Section 2.5.) Also, $\mathbf{v}$ and $\mathbf{k}_1$ are linearly independent, since $(\lambda_1 - A)\mathbf{v}$ is not $\mathbf{0}$, and so $\mathbf{v}$ is not a characteristic vector and hence not a multiple of $\mathbf{k}_1$. Thus K is a nonsingular matrix which has the property (4.9).

We shall show that $\mathbf{k}_1$ must be in the one-dimensional subspace spanned by the column vectors of $\lambda_1 I - A$. If it is not, there exists a nonzero vector $\mathbf{w}$ of this subspace, and by Exercise 20, Section 2.5, there is a vector $\mathbf{u}$ such that

$$(\lambda_1 I - A)\mathbf{u} = \mathbf{w}.$$

Now $\mathbf{u} = c_1\mathbf{k}_1 + c_2\mathbf{w}$, where $c_2 \neq 0$, since $\mathbf{k}_1$ and $\mathbf{w}$ are linearly independent. Hence

$$(\lambda_1 I - A)(c_1\mathbf{k}_1 + c_2\mathbf{w}) = \mathbf{w}$$

or, since $\mathbf{k}_1$ is a characteristic vector,

$$(\lambda_1 I - A)(c_2\mathbf{w}) = \mathbf{w}.$$

Then

$$A\mathbf{w} = \left(\lambda_1 - \frac{1}{c_2}\right)\mathbf{w}.$$

But this means that $\lambda_1 - \dfrac{1}{c_2}$ is a characteristic value of A, which is false since λ_1 is the only characteristic value.

Example 3   Let

$$A = \begin{bmatrix} 5 & 3 \\ -3 & -1 \end{bmatrix}.$$

The characteristic values of A are $\lambda_1 = \lambda_2 = 2$. The equation $(\lambda_1 I - A)\mathbf{k} = \mathbf{0}$ corresponds to the system

$$-3k_1 - 3k_2 = 0, \qquad 3k_1 + 3k_2 = 0.$$

One solution is $\mathbf{k} = (1, -1)$. The second column vector $\mathbf{v}$ of the matrix K must satisfy $(\lambda_1 I - A)\mathbf{v} = -\mathbf{k}$, so that

$$-3v_1 - 3v_2 = -1, \qquad 3v_1 + 3v_2 = 1.$$

One solution is $\mathbf{v} = (\frac{1}{3}, 0)$. The matrix

$$K = [\mathbf{k}, \mathbf{v}] = \begin{bmatrix} 1 & \frac{1}{3} \\ -1 & 0 \end{bmatrix}$$

has the property that

$$K^{-1}AK = \begin{bmatrix} 2 & 1 \\ 0 & 2 \end{bmatrix}.$$

If a matrix does not have a complete set of characteristic vectors, it is not similar to a diagonal matrix. It is, however, always similar to a slightly more complicated matrix, called the *Jordan canonical form* of the matrix. Discussions of the Jordan canonical form are presented in most books on linear algebra. If A has r distinct characteristic values, the Jordan matrix to which it is similar has the form

$$\begin{bmatrix} J_1 & 0 & \cdots & 0 \\ 0 & J_2 & \cdots & 0 \\ \multicolumn{4}{c}{\dotfill} \\ 0 & 0 & \cdots & J_r \end{bmatrix}$$

where J_i is a square submatrix associated with the characteristic value λ_i. It is of order m_i, the multiplicity of λ_i, and has the form

$$J_i = \begin{bmatrix} \lambda_i & x & & & \\ & \lambda_i & x & & \\ & & \lambda_i & x & \\ & & & \ddots & \\ & & & & \lambda_i \end{bmatrix}.$$

All the diagonal elements are equal to λ_i, the elements just above the main diagonal (marked by x's) are either 0 or 1 (not necessarily all 0 or all 1), and all other elements are zero. The number of ones above the main diagonal is equal to $m_i - n_i$, where m_i is the multiplicity of the characteristic value and n_i is the dimension of the subspace of characteristic vectors associated with λ_i.

---------------------------------- **Exercises for Section 4.3** ----------------------------------

In Exercises 1–10, find the characteristic values of the given matrix. Determine the multiplicity of each characteristic value and the dimension of the space of characteristic vectors associated with it. Is the matrix similar to a diagonal matrix? If it is, find a matrix K and diagonal matrix D such that $K^{-1}AK = D$, where A is the given matrix.

1. $\begin{bmatrix} 2 & 3 \\ 0 & 2 \end{bmatrix}$ 2. $\begin{bmatrix} 1 & -1 \\ 1 & 3 \end{bmatrix}$

3. $\begin{bmatrix} 0 & 1 \\ 2 & 1 \end{bmatrix}$ 4. $\begin{bmatrix} 5 & 2 \\ -4 & -1 \end{bmatrix}$

5. $\begin{bmatrix} 2 & -2 & 1 \\ 1 & -1 & 1 \\ -3 & 2 & -2 \end{bmatrix}$ 6. $\begin{bmatrix} 2 & 1 & 1 \\ 1 & 2 & 1 \\ -2 & -2 & -1 \end{bmatrix}$

7. $\begin{bmatrix} 1 & 2 & 3 \\ -1 & 4 & 3 \\ 1 & -2 & -1 \end{bmatrix}$ 8. $\begin{bmatrix} 3 & 0 & 0 \\ 0 & 3 & 0 \\ 0 & 0 & 3 \end{bmatrix}$

9. $\begin{bmatrix} -1 & 1 & 2 \\ -1 & 1 & 1 \\ -2 & 1 & 3 \end{bmatrix}$ 10. $\begin{bmatrix} 0 & 1 & 1 \\ -1 & 1 & 1 \\ -1 & 1 & 1 \end{bmatrix}$

11. Find a real matrix K such that $K^{-1}AK$ has the form (4.8) if A is the given matrix.

(a) $\begin{bmatrix} 1 & 2 \\ -4 & -3 \end{bmatrix}$ (b) $\begin{bmatrix} 5 & -1 \\ 5 & 1 \end{bmatrix}$

(c) $\begin{bmatrix} 0 & 1 \\ -2 & -2 \end{bmatrix}$ (d) $\begin{bmatrix} 5 & -4 \\ 2 & 1 \end{bmatrix}$

12. Find a real matrix K such that $K^{-1}AK$ has the form (4.9) if A is the matrix of
    (a) Exercise 1.   (b) Exercise 2.

13. If A is similar to a scalar matrix show that A is a scalar matrix.

14. (a) If A is similar to B, show that B is similar to A.
    (b) If A is similar to B and B is similar to C, show that A is similar to C.

15. If A is similar to B, show that A and B have the same characteristic polynomial and hence the same characteristic values.

16. If A and B are similar, so that $B = K^{-1}AK$, show that $\mathbf{x}$ is a characteristic vector of A if and only if $K^{-1}\mathbf{x}$ is a characteristic vector of B.

17. It can be shown that every square matrix is similar to an upper triangular matrix. Prove this for matrices of order 2. Suggestion: A has at least one characteristic value, λ_1 (which may be complex). Let $\mathbf{k}_1$ be a corresponding characteristic vector. Let $\mathbf{k}_2$ be any vector such that $\mathbf{k}_1$ and $\mathbf{k}_2$ are linearly independent. Let $K = [\mathbf{k}_1, \mathbf{k}_2]$. Show that $K^{-1}AK$ is upper triangular.

18. Let A be a 3×3 real matrix with characteristic values $\lambda_1 = \alpha + i\beta$, $\lambda_2 = \alpha - i\beta$, λ_3, where $\beta \neq 0$ and λ_3 is real. Show that there

is a real matrix K such that $K^{-1}AK$ is equal to the real matrix

$$\begin{bmatrix} \alpha & \beta & 0 \\ -\beta & \alpha & 0 \\ 0 & 0 & \lambda_3 \end{bmatrix}.$$

19. Let A be a real 4×4 matrix with distinct characteristic values $\alpha + i\beta$, $\alpha - i\beta$, $\gamma + i\delta$, and $\gamma - i\delta$, where $\beta \neq 0$ and $\delta \neq 0$. Show that there is a real matrix K such that $K^{-1}AK$ is equal to the real matrix

$$\begin{bmatrix} \alpha & \beta & 0 & 0 \\ -\beta & \alpha & 0 & 0 \\ 0 & 0 & \gamma & \delta \\ 0 & 0 & -\gamma & \delta \end{bmatrix}.$$

20. Each of the 3×3 matrices below is in Jordan canonical form and each has a single characteristic value with multiplicity three. Verify that the number of independent characteristic vectors is as stated in the discussion in the text.

(a) $\begin{bmatrix} 2 & 0 & 0 \\ 0 & 2 & 0 \\ 0 & 0 & 2 \end{bmatrix}$ (b) $\begin{bmatrix} 2 & 1 & 0 \\ 0 & 2 & 0 \\ 0 & 0 & 2 \end{bmatrix}$

(c) $\begin{bmatrix} 2 & 1 & 0 \\ 0 & 2 & 1 \\ 0 & 0 & 2 \end{bmatrix}$

4.4
REAL
SYMMETRIC
MATRICES

Real symmetric matrices arise in a number of applications. (We recall that A is symmetric if $A^T = A$; that is, if $a_{ji} = a_{ij}$ for all i and j.) In this section we shall describe some of the special properties of the characteristic values and vectors of such matrices. We begin with a preliminary result.

Lemma Let $\mathbf{x}$ and $\mathbf{y}$ be n-dimensional (real or complex) column vectors and let A be an $n \times n$ symmetric matrix. Then

$$\mathbf{x}^T A\mathbf{y} = \mathbf{y}^T A\mathbf{x}.$$

Proof Both members of the above equation are 1×1 matrices, and hence are symmetric. We have

$$\mathbf{x}^T A\mathbf{y} = (\mathbf{x}^T A\mathbf{y})^T = \mathbf{y}^T A^T\mathbf{x} = \mathbf{y}^T A\mathbf{x},$$

where the last equality follows from the symmetry of A.

Theorem 4.4   Let A be a real symmetric matrix. Then all the characteristic values of A are real, and (real) characteristic vectors of A that correspond to different characteristic values are orthogonal.

Proof   Suppose that $\lambda = \alpha + i\beta$ is a characteristic value of A and let $\mathbf{x}$ be a corresponding characteristic vector. Then

$$A\mathbf{x} = \lambda\mathbf{x}, \qquad A\bar{\mathbf{x}} = \bar{\lambda}\bar{\mathbf{x}}.$$

Premultiplying in the first equation by $\bar{\mathbf{x}}^T$ and in the second by $\mathbf{x}^T$, we have

$$\bar{\mathbf{x}}^T A\mathbf{x} = \lambda\bar{\mathbf{x}}^T\mathbf{x}, \qquad \mathbf{x}^T A\bar{\mathbf{x}} = \bar{\lambda}\mathbf{x}^T\bar{\mathbf{x}}.$$

Making use of the lemma, we see that

$$\lambda\bar{\mathbf{x}}^T\mathbf{x} = \bar{\lambda}\mathbf{x}^T\bar{\mathbf{x}}.$$

But $\bar{\mathbf{x}}^T\mathbf{x} = \mathbf{x}^T\bar{\mathbf{x}} = [\mathbf{x} \cdot \mathbf{x}]$; therefore

$$(\lambda - \bar{\lambda})\bar{\mathbf{x}}^T\mathbf{x} = 2i\beta\bar{\mathbf{x}}^T\mathbf{x} = [0].$$

Since $\mathbf{x}$ is a characteristic vector, $\|\mathbf{x}\| > 0$, so that we must have $\beta = 0$. Hence λ is real.

Next let λ_1 and λ_2 be distinct characteristic values of A, and let $\mathbf{x}$ and $\mathbf{y}$ be corresponding real characteristic vectors. Then

$$A\mathbf{x} = \lambda_1\mathbf{x}, \qquad A\mathbf{y} = \lambda_2\mathbf{y}.$$

Premultiplying in the first of these equations by $\mathbf{y}^T$ and in the second by $\mathbf{x}^T$, we have

$$\mathbf{y}^T A\mathbf{x} = \lambda_1\mathbf{y}^T\mathbf{x}, \qquad \mathbf{x}^T A\mathbf{y} = \lambda_2\mathbf{x}^T\mathbf{y}.$$

Again making use of the lemma, we have

$$(\lambda_1 - \lambda_2)\mathbf{x}^T\mathbf{y} = [0].$$

Since $\lambda_1 \neq \lambda_2$, we must have $\mathbf{x}^T\mathbf{y} = [0]$ or $\mathbf{x} \cdot \mathbf{y} = 0$, as we wished to prove. Our next result is more difficult to prove and we only state it.[1]

Theorem 4.5   A real symmetric matrix possesses a complete set of characteristic vectors. In fact, it possesses a complete orthonormal set of characteristic vectors. We recall that a set of vectors $\{\mathbf{k}_1, \mathbf{k}_2, \ldots, \mathbf{k}_n\}$ is orthonormal if $\|\mathbf{k}_i\| = 1$ for all i and $\mathbf{k}_i \cdot \mathbf{k}_j = 0$ if $i \neq j$. If A is real and symmetric, the multiplicity

---

[1] See, for example, Fraleigh and Beauregard (1990).

m_i of each characteristic value λ_i must be equal to n_i, the dimension of the corresponding space V_i of characteristic vectors. Let $\lambda_1, \lambda_2, \ldots, \lambda_r$ be the distinct characteristic values of A. If a basis for each of the corresponding spaces $V_1, V_2, \ldots, V_r$ can be found, then the orthogonalization procedure of Section 3.6 can be used to produce an orthonormal basis for each of the spaces. The set of all n vectors in the r orthonormal bases constitutes a complete orthonormal set of characteristic vectors for A.

Example 1   The matrix

$$A = \begin{bmatrix} -7 & 0 & 6 \\ 0 & 5 & 0 \\ 6 & 0 & 2 \end{bmatrix}$$

is real and symmetric. Its characteristic polynomial is

$$\begin{vmatrix} \lambda + 7 & 0 & -6 \\ 0 & \lambda - 5 & 0 \\ -6 & 0 & \lambda - 2 \end{vmatrix} = (\lambda - 5)^2(\lambda + 10).$$

The characteristic vectors corresponding to $\lambda_1 = -10$ are

$$(2a, 0, -a) = a(2, 0, -1)$$

with $a \neq 0$. The space V_1 is one-dimensional, and the single vector $\mathbf{k}_1 = (2, 0, -1)/\sqrt{5}$ constitutes an orthonormal basis for V_1. The characteristic vectors corresponding to $\lambda_2 = 5$ are the nonzero vectors of the form

$$(a, b, 2a) = a(1, 0, 2) + b(0, 1, 0).$$

The space V_2 is two-dimensional. The vectors $(1, 0, 2)$ and $(0, 1, 0)$ constitute a basis for V_2. The unit vectors $\mathbf{k}_2 = (1, 0, 2)/\sqrt{5}$, $\mathbf{k}_3 = (0, 1, 0)$ form an orthonormal basis for V_2. The matrix

$$K = [\mathbf{k}_1, \mathbf{k}_2, \mathbf{k}_3] = \begin{bmatrix} 2/\sqrt{5} & 1/\sqrt{5} & 0 \\ 0 & 0 & 1 \\ -1/\sqrt{5} & 2/\sqrt{5} & 0 \end{bmatrix}$$

has the property that

$$K^{-1}AK = \text{diag}(-10, 5, 5).$$

A matrix K with the property that $K^T = K^{-1}$ is called an *orthogonal matrix*. There is good reason for this name. If

$$K = [\mathbf{k}_1, \mathbf{k}_2, \ldots, \mathbf{k}_n]$$

then the element in the *ith* row and *j*th column of $K^T K$ is $\mathbf{k}_i \cdot \mathbf{k}_j$. Thus $K^T = K^{-1}$ if and only if the column vectors of K form an orthonormal set. (Similarly, by considering KK^T, we see that K is orthogonal if and only if its row vectors form an orthonormal set.) The matrix K in the preceding example is an orthogonal matrix, so

$$K^{-1} = K^T = \begin{bmatrix} 2/\sqrt{5} & 0 & -1/\sqrt{5} \\ 1/\sqrt{5} & 0 & 2/\sqrt{5} \\ 0 & 1 & 0 \end{bmatrix}$$

and $K^T AK = \operatorname{diag}(-10, 5, 5)$.

In Section 2.7 it was pointed out that associated with a real symmetric matrix A is a quadratic form f, where

$$f(\mathbf{x}) = \mathbf{x}^T A \mathbf{x} = \sum_{i=1}^{n} \sum_{j=1}^{n} a_{ij} x_i x_j.$$

A quadratic form f is said to be *positive definite* if $f(\mathbf{x}) > 0$ whenever $\mathbf{x} \neq \mathbf{0}$. A criterion for determining whether a quadratic form is positive definite is provided by the following theorem.

Theorem 4.6   The quadratic form $\mathbf{x}^T A \mathbf{x}$ is positive definite if and only if all the characteristic values of A are positive.

Proof   Let K be an orthogonal matrix such that $K^T AK = D$, where $D = \operatorname{diag}(\lambda_1, \lambda_2, \dots, \lambda_n)$. Let $\mathbf{y} = K^T \mathbf{x}$. Then $\mathbf{x} = K\mathbf{y}$ and we have

$$\mathbf{x}^T A \mathbf{x} = \mathbf{y}^T K^T AK\mathbf{y} = \mathbf{y}^T D\mathbf{y} = \lambda_1 y_1^2 + \lambda_2 y_2^2 + \cdots + \lambda_n y_n^2.$$

But the quadratic form in $\mathbf{y}$ is positive definite if and only if the characteristic values are all positive. To see this, suppose λ_j is not positive. Let $y_j = 1$ and $y_i = 0$ for $i \neq j$. Then $\mathbf{y}^T D\mathbf{y} = \lambda_j \leq 0$ but $\mathbf{y} \neq \mathbf{0}$. But $\mathbf{y} = K^T \mathbf{x}$ is $\mathbf{0}$ if and only if $\mathbf{x} = \mathbf{0}$, so $\mathbf{x}^T A\mathbf{x}$ is positive definite if and only if $\mathbf{y}^T D\mathbf{y}$ is positive definite.

The next theorem, whose proof we omit (see Strang (1988)), provides an alternative criterion for determining whether a quadratic form is positive definite that does not require finding the characteristic values.

Theorem 4.7   The quadratic form $\mathbf{x}^T A \mathbf{x}$ is positive definite if and only if the n determinants

$$|a_{11}|, \quad \begin{vmatrix} a_{11} & a_{12} \\ a_{21} & a_{22} \end{vmatrix}, \quad \begin{vmatrix} a_{11} & a_{12} & a_{13} \\ a_{21} & a_{22} & a_{23} \\ a_{31} & a_{32} & a_{33} \end{vmatrix}, \dots, \quad \det(A)$$

are all positive. (These determinants are called the *principal leading minors* of A.)

Example 2   Consider the quadratic forms

$$f(\mathbf{x}) = 2x_1^2 + 3x_2^2 + 4x_3^2 + 4x_2x_3 + 2x_3x_1 - 2x_1x_2$$

$$g(\mathbf{x}) = 2x_1^2 + x_2^2 + 5x_3^3 + 4x_2x_3 + 2x_3x_1 + 6x_1x_2.$$

Associated with f and g are the real symmetric matrices

$$A = \begin{bmatrix} 2 & -1 & 1 \\ -1 & 3 & 2 \\ 1 & 2 & 4 \end{bmatrix}, \qquad B = \begin{bmatrix} 2 & 3 & 1 \\ 3 & 1 & 2 \\ 1 & 2 & 5 \end{bmatrix}.$$

For the matrix A we find that

$$|a_{11}| = |2| > 0, \qquad \begin{vmatrix} 2 & -1 \\ -1 & 3 \end{vmatrix} = 5 > 0, \qquad \det(A) = 5 > 0$$

so the quadratic form f is positive definite, but since

$$\begin{vmatrix} 2 & 3 \\ 3 & 1 \end{vmatrix} = -4$$

the quadratic form g is not positive definite.

Exercises for Section 4.4

In Exercises 1–8, verify that the given matrix A is real and symmetric. Then find an orthogonal matrix K and a diagonal matrix D such that $K^T AK = D$.

1. $\begin{bmatrix} 7 & 6 \\ 6 & -2 \end{bmatrix}$ 2. $\begin{bmatrix} 52 & 14 \\ 14 & 148 \end{bmatrix}$

3. $\begin{bmatrix} 3 & 2 \\ 2 & -1 \end{bmatrix}$ 4. $\begin{bmatrix} 0 & -2 \\ -2 & 2 \end{bmatrix}$

5. $\begin{bmatrix} 4 & -12 & 6 \\ -12 & 36 & -18 \\ 6 & -18 & 9 \end{bmatrix}$

6. $\begin{bmatrix} 1 & 1 & -2 \\ 1 & 1 & 2 \\ -2 & 2 & 8 \end{bmatrix}$

7. $\begin{bmatrix} 1 & 1 & -2 \\ 1 & 1 & -2 \\ -2 & -2 & 4 \end{bmatrix}$

8. $\begin{bmatrix} 4 & -2 & -2 \\ -2 & 4 & -2 \\ -2 & -2 & 4 \end{bmatrix}$

9. If A is an $n \times n$ real matrix that possesses an orthonormal set of n real characteristic vectors, show that A is symmetric.

10. Let λ_1 be a real characteristic value of A with $\mathbf{x}$ a corresponding characteristic vector. Let λ_2 be a real characteristic value of A^T, with $\mathbf{y}$ a corresponding characteristic vector. If $\lambda_1 \neq \lambda_2$, show that $\mathbf{x}$ and $\mathbf{y}$ are orthogonal.

11. (a) If D is a real diagonal matrix with non-negative elements, show that D can be expressed as $D = D_1^2$, where D_1 is real and diagonal.
    (b) If A is a real symmetric matrix with nonnegative characteristic values, show that there is a real symmetric matrix B such that $A = B^2$.

12.   Let A be an $n \times n$ orthogonal matrix.
   (a)   If λ is a characteristic value of A, show that $|\lambda| = 1$.
   (b)   If n is odd and det $A = 1$, show that $\lambda = 1$ is a characteristic value of A.

13.   For each of the matrices in Exercises 1–8, write out the quadratic form associated with the matrix, and determine whether it is positive definite. Use both Theorem 4.6 and Theorem 4.7.

**4.5
FUNCTIONS OF
MATRICES**

If A is a square matrix of order n, then the product AA is defined and is again a square matrix of order n. We define positive integral powers of A in a natural way:

$$A^1 = A, \qquad A^2 = AA, \qquad A^3 = AA^2, \dots, \qquad A^m = AA^{m-1}, \dots .$$

We also define A to the zero power as

$$A^0 = I,$$

where I is the identity matrix of order n. It can be shown that the law of exponents

$$A^m A^k = A^{m+k}$$

holds for all nonnegative integers m and k.

If q is a polynomial function, defined by the formula

$$q(x) = c_0 + c_1 x + c_2 x^2 + \cdots + c_m x^m$$

for all real numbers x, we define $q(A)$, where A is a square matrix, by the relation

$$q(A) = c_0 I + c_1 A + c_2 A^2 + \cdots + c_m A^m.$$

Note that $q(A)$ is a square matrix of the same size as A. If f, g, q, and r are polynomials such that

$$f(x) = q(x) + r(x), \qquad g(x) = q(x)r(x)$$

for all x, then it can be shown

$$f(A) = q(A) + r(A)$$

$$g(A) = g(A)r(A)$$

for every square matrix A. Since $q(x)\, r(x) = r(x)\, q(x)$ for all x, it follows that

$$q(A)\, r(A) = r(A)\, q(A).$$

In particular, if p is the characteristic polynomial of the square matrix A,

$$p(\lambda) = \det(\lambda I - A),$$

then we can consider $p(A)$. A famous theorem is as follows. (Exercise 10 outlines a proof.)

Theorem 4.8 *(Cayley-Hamilton theorem)* If A is a square matrix with characteristic polynomial p, then $p(A) = 0$.

This result is often described by saying that a square matrix satisfies its characteristic polynomial. We shall verify the theorem for the matrix

$$A = \begin{bmatrix} 1 & -3 \\ -2 & 4 \end{bmatrix}.$$

Here

$$p(\lambda) = \begin{vmatrix} \lambda - 1 & 3 \\ 2 & \lambda - 4 \end{vmatrix} = \lambda^2 - 5\lambda - 2.$$

Since

$$A^2 = \begin{bmatrix} 7 & -15 \\ -10 & 22 \end{bmatrix},$$

we have

$$A^2 - 5A - 2I = \begin{bmatrix} 7 & -15 \\ -10 & 22 \end{bmatrix} - 5\begin{bmatrix} 1 & -3 \\ -2 & 4 \end{bmatrix} - 2\begin{bmatrix} 1 & 0 \\ 0 & 1 \end{bmatrix}$$

$$= \begin{bmatrix} 0 & 0 \\ 0 & 0 \end{bmatrix}$$

as the theorem asserts. The Cayley-Hamilton theorem can be applied to calculate higher powers of A. From the relation

$$A^2 - 5A - 2I = 0,$$

we see that

$$A^2 = 5A + 2I.$$

Then, multiplying both sides by A,

$$A^3 = 5A^2 + 2A = 5(5A + 2I) + 2A = 27A + 10I,$$

$$A^4 = 27A^2 + 10A = 27(5A + 2I) + 10A = 145A + 54I,$$

and so on. In particular,

$$A^4 = 145\begin{bmatrix} 1 & -3 \\ -2 & 4 \end{bmatrix} + 54\begin{bmatrix} 1 & 0 \\ 0 & 1 \end{bmatrix} = \begin{bmatrix} 199 & -435 \\ -290 & 634 \end{bmatrix}.$$

We can also find the inverse of A from the equation satisfied by A. We have

$$I = \frac{1}{2}(A^2 - 5A)$$

and multiplication of both sides of this equation by A^{-1} yields

$$A^{-1} = \frac{1}{2}(A - 5I).$$

Calculation shows that

$$A^{-1} = \frac{1}{2}\begin{bmatrix} -4 & -3 \\ -2 & -1 \end{bmatrix}.$$

Exercises for Section 4.5

1. If $f(r) = 2r^2 - 3r + 4$, find $f(A)$ when

   (a) $A = \begin{bmatrix} 3 & 1 \\ -2 & -1 \end{bmatrix}.$

   (b) $A = \begin{bmatrix} 0 & -1 & 2 \\ 1 & 0 & 0 \\ 2 & 1 & -1 \end{bmatrix}.$

2. With A as given, find the indicated power of A by using the Cayley-Hamilton theorem.

   (a) $A = \begin{bmatrix} 1 & 1 \\ 1 & -2 \end{bmatrix};\ A^3$

   (b) $A = \begin{bmatrix} -4 & 2 \\ -1 & 1 \end{bmatrix};\ A^4$

   (c) $A = \begin{bmatrix} 2 & 4 \\ 1 & 2 \end{bmatrix};\ A^4$

   (d) $A = \begin{bmatrix} 1 & 0 & -2 \\ 0 & 0 & 3 \\ -2 & 1 & 0 \end{bmatrix};\ A^4$

3. Use the Cayley-Hamilton theorem to find the inverse of each nonsingular matrix in Exercise 2.

4. Verify the Cayley-Hamilton theorem for the matrices of Exercise 2.

5. Let λ be a characteristic value of A, with $\mathbf{x}$ as a corresponding characteristic vector. If f is a polynomial, show that $f(\lambda)$ is a characteristic value of $f(A)$ with $\mathbf{x}$ as a corresponding characteristic vector. Suggestion: if $A\mathbf{x} = \lambda\mathbf{x}$, then $A^2\mathbf{x} = \lambda A\mathbf{x} = \lambda^2\mathbf{x}.$

6. Let A and B be square matrices such that $K^{-1}AK = B$. Show that
   (a) $K^{-1}A^2K = B^2.$
   (b) $K^{-1}A^mK = B^m$, where m is any positive integer.
   (c) $K^{-1}f(A)K = f(B)$, where f is any polynomial.

7. (a) Show that the Cayley-Hamilton theorem holds for any diagonal matrix.
   (b) Show that the Cayley-Hamilton theorem holds for any diagonalizable matrix. (Make use of Part (c) of the previous exercise.)

8. For the given matrix, find the characteristic polynomial p, observing that it is of degree 3. Then verify that the matrix satisfies the indicated polynomial m, of lower degree. (See the additional exercises at the end of the chapter for more about the polynomial of lowest degree satisfied by a matrix.)

   (a) $\begin{bmatrix} -2 & 2 & -2 \\ 0 & 0 & 0 \\ 6 & -3 & 5 \end{bmatrix},$
   $m(\lambda) = (\lambda - 2)(\lambda - 1)$

(b) $\begin{bmatrix} 5 & -3 & 1 \\ 4 & -3 & 2 \\ 2 & -3 & 4 \end{bmatrix}$,

$m(\lambda) = \lambda(\lambda - 3)$

9. Show (without using the Cayley-Hamilton theorem) that if A is an $n \times n$ matrix, there exists a polynomial g of degree less than n^2 such that $g(A) = 0$. Suggestion: the vector space of $n \times n$ matrices has dimension n^2.

10. A proof of the Cayley-Hamilton theorem can be obtained as follows. Recall from Section 2.12 that if we replace each element of a square matrix A by its cofactor and then take the transpose, we obtain a matrix B such that $AB = \det(A) I$. If we do this with the matrix $\lambda I - A$, we obtain a matrix $K(\lambda)$ whose elements are polynomials of degree $n - 1$ in λ, and which may therefore be written as

$$K = K_0 + \lambda K_1 + \cdots + \lambda^{n-1} K_{n-1}.$$

Substitute this expression into the equation

$$(\lambda I - A)K = \det(\lambda I - A)I.$$

Equate coefficients of like powers of λ, and use these relations to prove the theorem.

Additional Exercises for Chapter 4

1. For each matrix A, find the characteristic values and vectors. If the matrix is diagonalizable, find a matrix K such that $K^{-1} AK$ is diagonal.

   (a) $\begin{bmatrix} -4 & 2 \\ -5 & 3 \end{bmatrix}$

   (b) $\begin{bmatrix} -3 & 2 \\ -4 & 3 \end{bmatrix}$

   (c) $\begin{bmatrix} -1 & 4 & 2 \\ -1 & 3 & 1 \\ -1 & 2 & 2 \end{bmatrix}$

   (d) $\begin{bmatrix} -1 & 1 & -1 \\ 6 & -3 & 5 \\ 6 & -2 & 4 \end{bmatrix}$

2. For each matrix A, find a real matrix K such that $K^{-1}AK$ is either diagonal or has one of the forms (4.8) or (4.9).

   (a) $\begin{bmatrix} -2 & 2 \\ -4 & 2 \end{bmatrix}$ (b) $\begin{bmatrix} 5 & -2 \\ 4 & -1 \end{bmatrix}$

   (c) $\begin{bmatrix} 1 & 1 \\ -4 & 5 \end{bmatrix}$ (d) $\begin{bmatrix} -2 & 1 \\ -2 & -4 \end{bmatrix}$

3. Show that the characteristic values of a triangular matrix are the diagonal elements.

4. True or false?
   (a) If two rows of a matrix are interchanged, the characteristic values of the resulting matrix are the same as those of the original matrix.
   (b) The number λ is a characteristic value of the matrix A if and only if $-\lambda$ is a characteristic value of $-A$.

5. (a) If $A^k = 0$ for some positive integer k, then A is called *nilpotent*. Show that A is nilpotent if and only if all its characteristic values are zero.
   (b) If a symmetric matrix A is nilpotent, is A necessarily the zero matrix? Justify your answer.

6. Show that the matrix

$$\begin{bmatrix} a & b \\ 0 & a \end{bmatrix}$$

is similar to a diagonal matrix if and only if $b = 0$.

7. Let $p(\lambda) = \lambda^n + a_1 \lambda^{n-1} + \cdots + a_n$, where p is the characteristic polynomial of a matrix A. Show that
   (a) $a_n = (-1)^n \det A$
   (b) $a_1 = -(a_{11} + a_{22} + \cdots + a_{nn})$

8. If A is any real matrix show that $A^T A$ is a real symmetric matrix with nonnegative characteristic values.

9. Let A be real and nonsingular. Show that A can be written as $A = PQ$, where P is orthogonal and Q is real symmetric. Suggestion: $A^T A$ is symmetric and there is an orthogonal matrix K such that $K^T A^T A K = D^2$, where D is diagonal. Let $Q = KDK^T$.

10. A matrix A with complex elements is called an *Hermitian* matrix if $A^* = A$. (Here $A^* = \bar{A}^T$ is the transposed conjugate of A.)
    (a) Show that every real symmetric matrix is Hermitian.
    (b) Show that the diagonal elements of an Hermitian matrix are real.
    (c) Show that the characteristic values of an Hermitian matrix are real.
    (d) If $\mathbf{x}$ and $\mathbf{y}$ are characteristic vectors of an Hermitian matrix that correspond to different characteristic values, show that $\mathbf{x}$ and $\mathbf{y}$ are orthogonal.

11. According to the Cayley-Hamilton theorem, $p(A) = 0$, where p is the characteristic polynomial of A. Let q be *any* polynomial such that $q(A) = 0$. Prove that every root of p is also a root of q. Suggestion: use the result of Exercise 5, Section 4.5, to show that if c is a characteristic value of A, then $q(c)$ is a characteristic value of the zero matrix and hence is zero.

12. (a) Show that there is a smallest positive integer k such that $I, A, A^2, \ldots, A^k$ are linearly independent.
    (b) Show that there is a nonzero polynomial m, of degree k, such that $m(A) = 0$, and that no polynomial of lower degree has this property.
    (c) Show that the polynomial m of Part (b) is unique (except for a constant factor). The polynomial of lowest degree with unity as the coefficient of the term of degree k is called the *minimal polynomial* of A.

13. Show that the minimal polynomial m of the previous exercise and the characteristic polynomial p have the same roots (although not necessarily with the same multiplicities).

Suggestion: we have

$$p(\lambda) = q(\lambda)m(\lambda) + r(\lambda)$$

where the degree of r is less than the degree of m. Show that r must be the zero polynomial, and then show that $m(c) = 0$ implies $p(c) = 0$. Thus every root of m is also a root of p. Next, suppose that $p(c) = 0$. We have $m(\lambda) = (\lambda - c)g(\lambda) + r$ where r is a constant. Hence

$$m(A) = (A - cI)g(A) + rI = 0.$$

Show by consideration of determinants that $r = 0$. But this implies that c is a root of m.

14. If f is any polynomial such that $f(A) = 0$, show that f is divisible by the minimal polynomial (Exercise 12) of A.

15. (a) If A is a 2×2 matrix whose minimal polynomial (Exercise 12) is of lower degree than its characteristic polynomial, show that A is a diagonal matrix.
    (b) Find the minimal polynomials for the matrices of Exercises 1(c) and 1(d).

16. Let A be an $n \times n$ matrix with distinct characteristic values $\lambda_1, \lambda_2, \ldots, \lambda_n$ and characteristic polynomial $p(\lambda) = (\lambda - \lambda_1)(\lambda - \lambda_2) \cdots (\lambda - \lambda_n)$.
    (a) Using partial fractions, show that

$$\frac{1}{p(\lambda)} = \frac{a_1}{\lambda - \lambda_1} + \frac{a_2}{\lambda - \lambda_2} + \cdots + \frac{a_n}{\lambda - \lambda_n}$$

and hence that

$$1 \equiv a_1 p_1(\lambda) + a_2 p_2(\lambda) + \cdots + a_n p_n(\lambda)$$

where

$$p_i(\lambda) = \prod_{\substack{j=1 \\ j \neq i}}^{n} (\lambda - \lambda_j).$$

Show that

$$I = a_1 p_1(A) + a_2 p_2(A) + \cdots + a_n p_n(A).$$

(b) Let g be any polynomial. Then

$$g(x) = g(\lambda_i) + (x - \lambda_i)g_i(x)$$

and

$$g(A) = g(\lambda_i)I + (A - \lambda_i I)g_i(A)$$

for each i. Show, using the Cayley-Hamilton theorem, that

$$g(A)a_i p_i(A) = g(\lambda_i)a_i p_i(A)$$

for each i. Summing, show that

$$g(A) = \sum_{i=1}^{n} g(\lambda_i)a_i p_i(A).$$

(c) Let

$$A = \begin{bmatrix} -1 & -3 \\ 2 & 4 \end{bmatrix}.$$

Show that, for any polynomial g, $g(A) = g(2)(A - I) - g(1)(A - 2I)$. Find A^{100}.

17. Continuing the previous exercise, let f be any function that is defined on an a set that con-tains all the characteristic values of A. Let us define

$$f(A) = \sum_{i=1}^{n} f(\lambda_i)a_i p_i(A).$$

(a) If A is the 2×2 matrix of the previous exercise, then find $\sqrt{A}$ and verify that $(\sqrt{A})^2 = A$.

(b) Find $\sin A$, where A is the 2×2 matrix of the previous exercise.

(c) Find e^A, where A is the 2×2 matrix of the previous exercise.

18. Using the terminology of Exercise 16, let $E_i = a_i p_i(A)$. Show that $E_i E_j = 0$ if $i \neq j$, $E_i^2 = E_i$,

$$I = E_1 + E_2 + \cdots + E_n,$$

$$A = \lambda_1 E_1 + \lambda_2 E_2 + \cdots + \lambda_n E_n.$$

If x is any element of R^n, show that it can be written in the form $x = v_1 + v_2 + \cdots + v_n$ where $v_i = E_i x$ and $Ax = \lambda_1 v_1 + \lambda_2 v_2 + \cdots + \lambda_n v_n$.

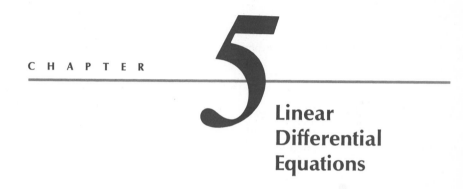

CHAPTER 5

Linear Differential Equations

5.1
INTRODUCTION

A linear differential equation of order n is of the form

$$a_0 y^{(n)} + a_1 y^{(n-1)} + \cdots + a_{n-1} y' + a_n y = F,$$

where the functions a_i and F are specified on some interval $\mathscr{I}$. The functions a_i are called the *coefficients* of the differential equation. If F is the zero function, the equation is said to be *homogeneous*; otherwise it is said to be *nonhomogeneous*. A homogeneous equation always has the zero function as one of its solutions. This solution is sometimes called the *trivial solution*.

Associated with the equation is the linear differential operator

$$L = a_0 D^n + a_1 D^{n-1} + \cdots + a_{n-1} D + a_n.$$

Such operators were discussed in Section 3.10. If $\mathscr{F}^n$ is the space of all functions that possess at least n derivatives on $\mathscr{I}$, then L maps $\mathscr{F}^n$ into $\mathscr{F}^0$, or $\mathscr{F}$. A function f in $\mathscr{F}^n$ for which $Lf = 0$ is a solution of the homogeneous equation

$$Ly = 0$$

on $\mathscr{I}$. The set of all solutions of this equation on $\mathscr{I}^1$ constitutes a vector

---

[1] Suppose that L satisfies the conditions in Theorem 5.1. Then if g is a solution on an interval $\mathscr{J}$ that is contained in $\mathscr{I}$, it can be shown that there is a solution f on $\mathscr{I}$ such that $g(x) = f(x)$ for all x in $\mathscr{J}$. For this reason we consider only solutions that exist throughout $\mathscr{I}$.

space, namely the kernel of the operator L. Consequently, if f is a solution and c is a number, then cf is also a solution. If f and g are both solutions, then so is $f + g$. Finally, if $f_1, f_2, \ldots, f_m$ are solutions and if $c_1, c_2, \ldots, c_m$ are numbers, then $c_1 f_1 + c_2 f_2 + \cdots + c_m f_m$ is again a solution.

These properties of solutions do not occur for nonlinear differential equations. For example, let us consider the equation

$$y' = -y^2,$$

which is nonlinear and separable. The solutions of this equation consist of the zero function and those functions of the form

$$y = \frac{1}{x + c},$$

where c is an arbitrary constant. Taking $c = 0$ and $c = 1$, we see that the functions

$$f_1(x) = \frac{1}{x}, \qquad f_2(x) = \frac{1}{x + 1}$$

are both solutions on the interval $(0, \infty)$. Setting $f = f_1 + f_2$, we see that

$$f'(x) = -\frac{1}{x^2} - \frac{1}{(x + 1)^2}$$

and

$$-[f(x)]^2 = -\frac{1}{x^2} - \frac{1}{(x + 1)^2} - \frac{2}{x(x + 1)}.$$

Since $f' \neq -f^2$, $f_1 + f_2$ is not a solution.

The space $\mathcal{F}^n$ is not finite-dimensional, as was shown in Section 3.5. However, the *kernel* of the operator L is a finite-dimensional subspace of $\mathcal{F}^n$. To be precise, the following theorem can be proved.

Theorem 5.1   Let the functions $a_0, a_1, \ldots, a_n$ be continuous, with a_0 never zero on an interval $\mathcal{I}$. Then the kernel of the linear operator L is a vector space of dimension n.

Since the facts stated in this theorem will be used repeatedly throughout the remainder of this chapter, let us elaborate. The theorem says that the set of all solutions of the equation $Ly = 0$ on $\mathcal{I}$ forms a vector space of dimension n. If we can find a set of n linearly independent solutions $y_1, y_2, \ldots, y_n$ of this equation, then $\{y_1, y_2, \ldots, y_n\}$ is a basis for the space of solutions. Every solution of the equation $Ly = 0$ is of the form

$$c_1 y_1 + c_2 y_2 + \cdots + c_n y_n.$$

Thus by finding a particular set of n linearly independent solutions we also find the general solution.

In order to find the specific solution that satisfies the initial conditions

$$y(x_0) = k_0, \qquad y'(x_0) = k_1, \ldots, \qquad y^{(n-1)}(x_0) = k_{n-1}$$

we must determine the constants c_i from the requirements

$$c_1 f_1(x_0) + c_2 f_2(x_0) + \cdots + c_n f_n(x_0) = k_0,$$
$$c_1 f_1'(x_0) + c_2 f_2'(x_0) + \cdots + c_n f_n'(x_0) = k_1,$$
$$\cdots \cdots \cdots \cdots \cdots \cdots \cdots \cdots \cdots \cdots \cdots \cdots$$
$$c_1 f_1^{(n-1)}(x_0) + c_2 f_2^{(n-1)}(x_0) + \cdots + c_n f_n^{(n-1)}(x_0) = k_{n-1}.$$

Example To illustrate, we consider the second-order differential equation

$$y'' - 3y' + 2y = 0$$

on the interval $(-\infty, \infty)$. It is easy to verify that the functions f_1 and f_2, where

$$f_1(x) = e^x, \qquad f_2(x) = e^{2x}$$

are solutions and hence belong to the kernel of the operator

$$L = D^2 - 3D + 2.$$

Furthermore, we see that these functions are linearly independent, since their Wronskian

$$W(x) = \begin{vmatrix} e^x & e^{2x} \\ e^x & 2e^{2x} \end{vmatrix} = e^{3x}$$

does not vanish for all x. Hence the general solution of the equation is

$$y = c_1 e^x + c_2 e^{2x}.$$

If we wish to find a specific solution that satisfies a set of initial conditions, such as

$$y(0) = 4, \qquad y'(0) = 1,$$

we need only determine the constants c_1 and c_2 in such a way that the conditions are met. Since

$$y' = c_1 e^x + 2c_2 e^{2x}$$

the initial conditions require that

$$c_1 + c_2 = 4, \qquad c_1 + 2c_2 = 1.$$

We find that $c_1 = 7$ and $c_2 = -3$. Hence the desired solution is

$$y = 7e^x - 3e^{2x}.$$

If a set of functions is linearly independent on an interval $\mathcal{I}$, the Wronskian of the functions may still vanish at every point of $\mathcal{I}$. An example was given in Section 3.4. However, if the functions $f_1, f_2, \ldots, f_n$ are linearly independent solutions of the nth-order linear differential equation $Ly = 0$, this cannot happen.

We state without proof the following important result. (See, however, Exercise 7.)

Theorem 5.2   Let $f_1, f_2, \ldots, f_n$ be n solutions of the nth-order equation

$$a_0 y^{(n)} + a_1 y^{(n-1)} + \cdots + a_n y = 0$$

on an interval where the functions a_i are continuous and a_0 is never zero. If the functions are linearly dependent, their Wronskian is zero at every point of the interval, but if they are linearly independent their Wronskian is nowhere zero on the interval.

Because of this theorem, we know that the system of equations for $c_1, c_2, \ldots, c_n$ in the system preceding the example has a nonzero determinant. By Theorem 2.10, we know that this system has a unique solution.

The main problem, of course, is how to find the linearly independent solutions of the differential equation. In the next few sections we shall investigate certain classes of linear equations for which this can readily be accomplished.

A theory for the nonhomogeneous equation $Ly = F$ can be based on the theory for the associated homogeneous equation $Ly = 0$. Nonhomogeneous equations will be discussed in Sections 5.6–5.8.

Exercises for Section 5.1
---

1. Find at least one solution of the equation $y'' + e^x y = 0$.

2. Verify that the given functions form a basis for the space of solutions of the given differential equation.

   (a) $y'' + y = 0$, $f_1(x) = \cos x$, $f_2(x) = \sin x$

   (b) $y' - 2xy = 0$, $f_1(x) = \exp(x^2)$

   (c) $x^2 y'' - 2xy' + 2y = 0$, $f_1(x) = x$, $f_2(x) = x^2$, $x > 0$

   (d) $y''' - y' = 0$, $f_1(x) = 1$, $f_2(x) = e^x$, $f_3(x) = e^{-x}$

3. Find the solution of the corresponding differential equation in Exercise 2 that satisfies the given initial conditions.

   (a) $y(\pi) = 0$, $y'(\pi) = -2$

(b) $y(0) = 5$

(c) $y(1) = 0, \quad y'(1) = 0$

(d) $y(0) = 3, \quad y'(0) = 1, \quad y''(0) = 1$

4. If $a_0, a_1, \ldots, a_n$ possess derivatives of all orders on $\mathscr{I}$ and if a_0 is never zero on $\mathscr{I}$, show that every solution of $Ly = 0$ possesses derivatives of all orders on $\mathscr{I}$. Suggestion: write

$$y^{(n)} = -\frac{a_1}{a_0} y^{(n-1)} - \cdots - \frac{a_n}{a_0} y.$$

5. What can be said about the solution of the differential equation

$$y''' + xy' - (\sin x)y = 0, \qquad \text{for all } x,$$

if $y(1) = y'(1) = y''(1) = 0$?

6. If u and v are both solutions of the equation $Ly = F$ show that $u - v$ is a solution of the equation $Ly = 0$.

7. (a) If f_1 and f_2 are solutions of the second order equation

$$a_0 y'' + a_1 y' + a_2 y = 0,$$

then show that the Wronskian $W = f_1 f_2' - f_2 f_1'$ satisfies the equation $W' + (a_1/a_0)W = 0$, and hence that

$$W(x) = c \exp\left[-\int \frac{a_1(x)}{a_0(x)} \, dx \right].$$

This formula is known as *Abel's formula.* Observe that W is everywhere zero or nowhere zero, depending on whether the constant is zero or not.

(b) If $f_1, f_2, \ldots, f_n$ are solutions of the nth-order equation

$$a_0 y^{(n)} + a_1 y^{(n-1)} + \cdots + a_n y = 0,$$

show that the Wronskian of the functions is given by the same formula (Abel's formula) as in Part (a).

8. The hypotheses of Theorem 5.1 include the condition that the coefficient a_0 of the highest order derivative not be zero at any point.

(a) Consider the first-order equation $a_0 y' + a_1 y = 0$ with $a_0(x) = x$ and $a_1(x) = 1$. Show that no solution (other than the trivial solution) exists on an interval that contains the point $x = 0$ (where a_0 vanishes).

(b) Consider the second order equation $x^2 y'' + 4xy' + 2y = 0$, where $a_0(x) = x^2$, $a_1(x) = 4x$, and $a_2(x) = 2$. Show that on any interval not containing $x = 0$, the nontrivial solutions are $y = c_1 x^{-1} + c_2 x^{-2}$, but that no nontrivial solutions exist throughout any interval that contains $x = 0$.

9. (a) Let f_1 and f_2 be functions such that their Wronskian $W(x; f_1, f_2)$ does not vanish on an interval. Show that the equation $W(x; f_1, f_2, y) = 0$ is a second-order linear differential equation whose solutions on the interval are $y = c_1 f_1 + c_2 f_2$.

(b) Find a second order linear differential equation whose solutions on the interval $(0, \infty)$ are $y = c_1(x + 1) + c_2 e^x$.

5.2 POLYNOMIAL OPERATORS

Linear differential equations whose coefficients are constant functions are of particular interest. One reason for this is that they are easy to solve. Another is that important problems in mechanics and electric circuits can be described by such equations. Examples of such problems are discussed in Sections 5.9 and 5.10.

An nth-order linear equation with constant coefficients has the form

$$a_0 y^{(n)} + a_1 y^{(n-1)} + \cdots + a_{n-1} y' + a_n y = F,$$

where the a_i are constants, with $a_0 \neq 0$. This equation can be written as $Ly = F$, where

$$L = a_0 D^n + a_1 D^{n-1} + \cdots + a_{n-1}D + a_n.$$

We say that L is a linear differential operator with constant coefficients. Associated with this operator in an obvious way is the polynomial P, where

$$P(r) = a_0 r^n + a_1 r^{n-1} + \cdots + a_{n-1}r + a_n.$$

We therefore write

$$P(D) = a_0 D^n + a_1 D^{n-1} + \cdots + a_{n-1}D + a_n$$

and call $P(D)$ a *polynomial operator*.

We recall that the sum $L_1 + L_2$ and the product $L_1 L_2$ of two operators are defined according to the rules

$$(L_1 + L_2)u = L_1 u + L_2 u,$$

$$(L_1 L_2)v = L_1(L_2 v),$$

where u and v are any functions such that the indicated operations on the right are defined.

Example 1   Let $P(D) = D^2 - D + 2$ and $Q(D) = D + 1$. Then

$$[P(D) + Q(D)]u = P(D)u + Q(D)u$$
$$= (u'' - u' + 2u) + (u' + u)$$
$$= u'' + 3u.$$

Thus $P(D) + Q(D) = D^2 + 3$. Since

$$P(r) + Q(r) = (r^2 - r + 2) + (r + 1) = r^2 + 3,$$

we see that $P(D) + Q(D)$ is the polynomial operator whose associated polynomial is $P + Q$.

Next we have

$$[P(D)Q(D)]v = (D^2 - D + 2)(v' + v)$$
$$= v''' + v'' - v'' - v' + 2v' + 2v$$
$$= v''' + v' + 2v.$$

Hence

$$P(D)Q(D) = D^3 + D + 2.$$

Since

$$P(r)Q(r) = (r^2 - r + 2)(r + 1) = r^3 + r + 2,$$

we see that $P(D)Q(D)$ is the polynomial operator whose associated polynomial is PQ.

This example illustrates the following general principles.

Theorem 5.3   Let P and Q be polynomials. Then $P(D) + Q(D)$ and $P(D)Q(D)$ are polynomial operators whose associated polynomials are $P + Q$ and PQ, respectively.

The proof is omitted.

There is associated with every polynomial operator one and only one ordinary polynomial. For example, associated with the operator $D^2 - 3D + 2$ is the polynomial P, with $P(r) = r^2 - 3r + 2$. According to Theorem 5.3, when we add (or multiply) two polynomial operators, the result is another polynomial operator whose associated polynomial is the sum (or product) of the polynomials associated with the original operators. Suppose we consider the product $P(D)Q(D)$ of two operators. The associated polynomial is $P(r)Q(r)$. But $P(r)Q(r) = Q(r)P(r)$ for all r, since ordinary multiplication is commutative. This means that the operators $P(D)Q(D)$ and $Q(D)P(D)$ have the same associated polynomial, and hence must be the same. Thus

$$P(D)Q(D) = Q(D)P(D).$$

Also, in view of the associative property $P(r)[Q(r)R(r)] = [P(r)Q(r)]R(r)$ for polynomials, we have

$$P(D)[Q(D)R(D)] = [P(D)Q(D)]R(D).$$

If P is any polynomial that can be written in the factored form

$$P(r) = P_1(r)P_2(r) \cdots P_k(r),$$

it follows that

$$P(D) = P_1(D)P_2(D) \cdots P_k(D).$$

Because of the commutative and associative properties of polynomial operators under multiplication, the order and manner of grouping of the operators in this product do not matter.

If P is any polynomial,

$$P(r) = a_0 r^n + a_1 r^{n-1} + \cdots + a_{n-1}r + a_n,$$

then P can be written as the product of linear factors,

$$P(r) = a_0(r - r_1)(r - r_2) \cdots (r - r_n).$$

The numbers $r_1, r_2, \ldots, r_n$, which may not be distinct, are the zeros of P and the roots of the equation $P(r) = 0$. Some of these roots may be complex. If P has real coefficients and if $r_1 = a + ib$, $b \neq 0$, is a complex root, then $r_2 = a - ib$ is also a root. The second-degree polynomial

$$(r - r_1)(r - r_2) = [r - (a + ib)][r - (a - ib)] = (r - a)^2 + b^2$$

has real coefficients. Therefore any polynomial operator $P(D)$ with real coefficients can be written as the product of first- and second-order polynomial operators with real coefficients. For example, let

$$P(D) = D^3 - 5D^2 + 9D - 5.$$

Since

$$P(r) = (r - 1)(r - 2 - i)(r - 2 + i) = (r - 1)(r^2 - 4r + 5),$$

we may write

$$P(D) = (D - 1)(D^2 - 4D + 5) = (D^2 - 4D + 5)(D - 1).$$

If a function is to be a solution of the equation

$$P(D)y = 0, \tag{5.1}$$

where

$$P(D) = a_0 D^n + a_1 D^{n-1} + \cdots a_n,$$

a certain linear combination of that function and its first n derivatives must vanish. If f is an exponential function, of the form $f(x) = e^{rx}$, the derivatives of f are multiples of f itself. We have

$$D^m e^{rx} = r^m e^{rx}$$

for every nonnegative integer m. We therefore attempt to find solutions of Eq. (5.1) that are of this form. In view of the last formula we have

$$(a_0 D^n + a_1 D^{n-1} + \cdots + a_n)e^{rx} = (a_0 r^n + a_1 r^{n-1} + \cdots + a_n)e^{rx}$$

or

$$P(D)e^{rx} = P(r)e^{rx}.$$

The polynomial P is called the *auxiliary polynomial* associated with the differential equation (5.1) and the equation $P(r) = 0$ is called the *auxiliary equation*. We may write

$$P(D)e^{rx} = a_0(r - r_1)(r - r_2) \cdots (r - r_n)e^{rx}.$$

If r_i is a real zero of P, the function $e^{r_i x}$ is a solution of Eq. (5.1). If $r_1, r_2, \ldots, r_n$ are real and distinct, each of the functions

$$e^{r_1 x}, \quad e^{r_2 x}, \ldots, \quad e^{r_n x}$$

is a solution. These solutions are linearly independent, as was shown in Section 3.4. In this case the general solution of Eq. (5.1) is

$$y = c_1 e^{r_1 x} + c_2 e^{r_2 x} + \cdots + c_n e^{r_n x}.$$

Example 2   Consider the equation

$$y'' - y' - 2y = 0,$$

which may be written as

$$(D^2 - D - 2)y = 0.$$

The auxiliary equation is

$$r^2 - r - 2 = (r + 1)(r - 2) = 0.$$

The roots are $r_1 = -1$ and $r_2 = 2$. Hence e^{-x} and e^{2x} are linearly independent solutions and the general solution is

$$y(x) = c_1 e^{-x} + c_2 e^{2x}.$$

In the general case of Eq. (5.1), the polynomial P may not have n distinct zeros. Even if it does, some of the zeros may be complex. The next two examples illustrate these possibilities.

Example 3
$$y'' - 4y' + 4y = 0.$$

The auxiliary equation

$$r^2 - 4r + 4 = (r - 2)^2$$

has only one distinct root, $r_1 = 2$. Thus e^{2x} is a solution, but we need still another solution to be able to describe the general solution.

Example 4
$$y'' - 2y' + 5y = 0.$$

The auxiliary polynomial equation

$$r^2 - 2r + 5 = 0$$

has the complex roots $r_1 = 1 + 2i$ and $r_2 = 1 - 2i$. Thus we are unable (as yet) to write down any nontrivial solution.

The difficulties caused by the appearance of complex zeros of the auxiliary polynomial will be removed in the next section.

Exercises for Section 5.2
---

1. Write the differential equation in the form $P(D)y = 0$.

   (a) $y'' - 3y' + 2y = 0$

   (b) $y' + 4y = 0$

   (c) $y''' - 3y'' - y' + y = 0$

   (d) $y^{(4)} - y'' + y = 0$

2. Write the differential equation in factored form, in terms of real polynomial operators of first- and second-order.

   (a) $(D^2 + D - 2)y = 0$

   (b) $(D^4 - 1)y = 0$

   (c) $(D^3 - 3D + 2)y = 0$

   (d) $(D^3 + D^2 - 4D + 6)y = 0$

3. Find a differential equation of the form $P(D)y = 0$, with real coefficients, whose associated polynomial P has the given numbers among its zeros. The order of P should be as low as possible.

   (a) $r_1 = 2, \quad r_2 = -1$

   (b) $r_1 = 3 - 2i$

   (c) $r_1 = r_2 = 2, \quad r_3 = 0$

   (d) $r_1 = 2i, \quad r_2 = -1, \quad r_3 = 2$

4. If the function u is a solution of the equation $P(D)y = 0$ and v is a solution of the equation $Q(D)y = 0$, show that u and v are both solutions of the equation $P(D)Q(D)y = 0$.

In Exercises 5–10, express the general solution of the given differential equation in terms of exponential functions, if possible.

5. $y'' - 5y' + 6y = 0$

6. $2y'' + 5y' - 3y = 0$

7. $y''' + 5y'' - y' - 5y = 0$

8. $y'' + 4y = 0$

9. $y'' - 6y' + 9y = 0$

10. $y''' - 4y' = 0$

11. Show that no differential equation of the form $P(D)y = 0$ has the function f, where $f(x) = 1/x$, $x > 0$, as a solution.

12. (a) If $P(D)$ is any polynomial operator and $Q(D) = cD^k$, show that $Q(D)P(D)$ is a polynomial operator whose associated polynomial is QP.

    (b) Use the result of part (a) to show that if P and Q are any two polynomials, then $Q(D)P(D)$ is a polynomial operator whose corresponding polynomial is QP.

13. Show that a function f is a solution of an equation of the type $P(D)y = 0$ if and only if there is a positive integer k such that $f, f^{(1)}, f^{(2)}, \ldots, f^{(k)}$ are linearly dependent. (The superscripts denote differentiation.)

5.3

COMPLEX SOLUTIONS

In applications that give rise to differential equations, we are almost always concerned with real solutions of the equations. However, it is sometimes convenient to extract a desired real solution from a complex solution, as will be seen later.

A complex function w of a real variable can be regarded as an ordered pair of real functions (u, v). We write

$$w(x) = u(x) + iv(x),$$

where i is the imaginary unit with the property that $i^2 = -1$. We call u the real part and v the imaginary part of w. The derivative of the complex function w is defined by the relation

$$w'(x) = u'(x) + iv'(x),$$

provided that $u'(x)$ and $v'(x)$ both exist. For example, if $w(x) = \cos 2x + i \sin 2x$, then $w'(x) = -2 \sin 2x + 2i \cos 2x$.

In this chapter we are concerned with linear differential equations with *real* coefficients. Let

$$L = a_0 D^n + a_1 D^{n-1} + \cdots + a_{n-1}D + a_n,$$

where the functions a_i are real valued. Since

$$a_k D^{n-k}w = a_k D^{n-k}u + ia_k D^{n-k}v$$

for $k = 0, 1, \ldots, n$, it follows that

$$Lw = Lu + iLv,$$

where Lu and Lv are both real functions. To illustrate, let $L = D^2 + 2D - 3$, and let $w = u + iv$. Then

$$Lw = (D^2 + 2D - 3)(u + iv)$$
$$= D^2(u + iv) + 2D(u + iv) - 3(u + iv)$$
$$= D^2u + iD^2v + 2Du + 2iDv - 3u - 3iv.$$

Collecting real and imaginary parts, we have

$$Lw = (D^2u + 2Du - 3u) + i(D^2v + 2Dv - 3v)$$
$$= (D^2 + 2D - 3)u + i(D^2 + 2D - 3)v$$
$$= Lu + iLv.$$

In general, if w is a complex solution of the homogeneous equation $Ly = 0$, then

$$Lu + iLv = 0.$$

Since a complex number can be zero only if its real and imaginary parts are both zero, it follows that $Lu = 0$ and $Lv = 0$. Consequently the real and imaginary parts, u and v, of a complex solution w are real solutions of the differential equation.

We shall also be interested in nonhomogeneous linear equations of the form

$$Ly = F, \qquad (5.2)$$

where F is a complex function,

$$F(x) = f(x) + ig(x).$$

If $w = u + iv$ is a solution of Eq. (5.2), then

$$Lw = F$$

or

$$Lu + iLv = f + ig.$$

But this means that

$$Lu = f, \qquad Lv = g.$$

Thus by finding a complex solution of Eq. (5.2) we obtain real solutions for the two equations

$$Ly = f, \qquad Ly = g.$$

We shall be particularly concerned with a class of complex functions known as complex exponential functions. In order to define these functions we first define the complex number e^{a+ib}, where e is the base of natural logarithms, as

$$e^{a+ib} = e^a \cos b + ie^a \sin b. \qquad (5.3)$$

When $a + ib$ happens to be real ($b = 0$) we see that e^a has its usual real value. Other special cases of interest are

$$e^{ib} = \cos b + i \sin b, \qquad e^{-ib} = \cos b - i \sin b. \qquad (5.4)$$

From these relations, we obtain the formulas

$$\cos b = \frac{1}{2}(e^{ib} + e^{-ib}), \qquad \sin b = \frac{1}{2i}(e^{ib} - e^{-ib}). \qquad (5.5)$$

The laws of exponents

$$e^{z_1} \cdot e^{z_2} = e^{z_1 + z_2},$$
$$e^{z_1}/e^{z_2} = e^{z_1 - z_2}, \qquad (5.6)$$

where z_1 and z_2 are arbitrary complex numbers, can be derived from the definition (5.3). We shall consider the first rule here, leaving the second to the exercises. Writing $z_1 = a_1 + ib_1$ and $z_2 = a_2 + ib_2$, we have

$$e^{z_1} \cdot e^{z_2} = e^{a_1}(\cos b_1 + i \sin b_1) \cdot e^{a_2}(\cos b_2 + i \sin b_2)$$
$$= e^{a_1 + a_2}[(\cos b_1 \cos b_2 - \sin b_1 \sin b_2)$$
$$+ i(\cos b_1 \sin b_2 + \sin b_1 \cos b_2)]$$
$$= e^{a_1 + a_2}[\cos(b_1 + b_2) + i \sin(b_1 + b_2)]$$
$$= e^{(a_1 + a_2) + i(b_1 + b_2)}$$
$$= e^{z_1 + z_2}.$$

We now consider a complex function F of the form

$$F(x) = e^{h(x)},$$

where $h(x) = u(x) + iv(x)$. We seek a formula for the derivative of this function. Writing

$$F(x) = e^{u(x)} \cos v(x) + ie^{u(x)} \sin v(x)$$

and using the definition of the derivative of a complex function, we have

$$F' = u'(e^u \cos v + ie^u \sin v) + v'(-e^u \sin v + ie^u \cos v)$$
$$= (u' + iv')(e^u \cos v + ie^u \sin v).$$

Hence

$$\frac{de^{h(x)}}{dx} = h'(x)e^{h(x)}. \tag{5.7}$$

A function w of the form

$$w(x) = e^{cx},$$

where $c = a + ib$ is a complex constant, is called a *complex exponential function*. Using the rule (5.7), we see that

$$\frac{de^{cx}}{dx} = ce^{cx}$$

for all x. From Eqs. (5.4) and (5.5) we obtain the formulas

$$e^{ibx} = \cos bx + i \sin bx,$$
$$e^{-ibx} = \cos bx - i \sin bx,$$

and

$$\cos bx = \frac{1}{2}(e^{ibx} + e^{-ibx}),$$

$$\sin bx = \frac{1}{2i}(e^{ibx} - e^{-ibx}),$$

which relate the real trigonometric functions and the complex exponential functions. As an example of their use, let us consider the homogeneous differential equation $(D^2 + 4)y = 0$. Seeking a solution of the form $y = e^{cx}$, we find that

$$(D^2 + 4)e^{cx} = (c^2 + 4)e^{cx} = (c + 2i)(c - 2i)e^{cx}.$$

Since this is identically zero when $c = \pm 2i$ the functions e^{2ix} and e^{-2ix} are complex solutions. But then the real and imaginary parts, $\cos 2x$ and $\sin 2x$, are real solutions.

Next we consider the nonhomogeneous differential equation

$$Ly = Ae^{ibx},$$

where L has real coefficients and A is a real constant. We may write

$$Ae^{ibx} = A \cos bx + iA \sin bx.$$

If we can find a complex solution of the given equation, then the real and imaginary parts of this solution will be real solutions of the equations

$$Ly = A \cos bx,$$

$$Ly = A \sin bx,$$

respectively. The reason for preferring to deal with complex exponential functions rather than real trigonometric functions is that the differentiation formulas for the former are simpler. A number of examples are presented in Sections 5.4 and 5.7.

If $c = a + ib$ is any complex number, and if α is any positive real number, we define

$$\alpha^c = e^{c \ln \alpha}. \tag{5.8}$$

Notice that when c is real ($b = 0$) this formula agrees with the definition of α^c given in calculus. The laws of exponents

$$\alpha^{z_1} \cdot \alpha^{z_2} = \alpha^{z_1 + z_2},$$

$$\alpha^{z_1}/\alpha^{z_2} = \alpha^{z_1 - z_2}, \tag{5.9}$$

follow easily from the laws (5.6). Their derivations are left to the exercises.

If x is a positive real number we have

$$x^c = e^{c \ln x}$$

according to the definition (5.8). Application of the differentiation formula (5.7) yields

$$\frac{dx^c}{dx} = \frac{c}{x} e^{c \ln x} = \frac{c}{x} x^c$$

or

$$\frac{dx^c}{dx} = cx^{c-1}, \qquad x > 0.$$

Notice that we have not defined α^β when α and β are both complex, or even when α is real and negative. To give proper definitions would take us further into the theory of complex variables than is necessary for our study of linear differential equations.

─────────────── Exercises for Section 5.3 ───────────────

1. (a) If $w = u + iv$ is a complex function and $c = a + ib$ is a complex constant, show that

$$\frac{d[cw(x)]}{dx} = c\frac{dw(x)}{dx}.$$

(b) If $w_1 = u_1 + iv_1$ and $w_2 = u_2 + iv_2$ are complex functions, show that $(w_1 + w_2)' = w_1' + w_2'$

2. Let $w_1 = u_1 + iv_1$ and $w_2 = u_2 + iv_2$ be complex functions. Show that
   (a) $(w_1 w_2)' = w_1' w_2 + w_1 w_2'$
   (b) $(w_1/w_2)' = (w_1' w_2 - w_1 w_2')/w_2^2$

3. Express each complex exponential function in terms of trigonometric functions and express each trigonometric function in terms of complex exponential functions.
   (a) e^{3ix} (b) e^{-2ix}
   (c) $e^{(2-3i)x}$ (d) $e^{(-2+i)x}$
   (e) $\cos 2x$ (f) $\sin 5x$
   (g) $\sin x$ (h) $\cos 5x$

4. Show that $e^{z_1}/e^{z_2} = e^{z_1 - z_2}$ when z_1 and z_2 are arbitrary complex numbers.

5. The *modulus* of a complex number $c = a + ib$, written $|c|$, is defined to be $|c| = (a^2 + b^2)^{1/2}$. Show that $|e^{a+ib}| = e^a$.

6. For every real number θ and every integer m show that

$$(e^{i\theta})^m = e^{im\theta}$$

and hence that

$$(\cos \theta + i \sin \theta)^m = \cos m\theta + i \sin m\theta.$$

This is known as DeMoivre's formula. Suggestion: first let m be nonnegative and use induction.

7. Show that the complex function $e^{(-1+2i)x}$ is a solution of the differential equation

$$(D^2 + 2D + 5)y = 0.$$

Use this fact to find two linearly independent real solutions.

8. For each of the following differential equations find all numbers r, real and complex,

such that e^{rx} is a solution. Find two linearly independent real solutions.

(a) $(D^2 + 9)y = 0$

(b) $(D^2 - 3D + 2)y = 0$

(c) $(D^2 - 4D + 5)y = 0$

(d) $(D^2 + 4D + 5)y = 0$

9. Let $P(D) = D^2 - D + 5$. Find a complex solution of the equation

$$P(D)y = 10e^{2ix}$$

of the form $y = Ae^{2ix}$, where A is a complex constant. Use your answer to find a real solution of each of the equations

$$P(D)y = 10 \cos 2x, \qquad P(D)y = 10 \sin 2x.$$

10. Derive the laws of exponents (5.9) from the laws (5.6).

11. Find all solutions of the differential equation

$$x^2 y'' + xy' + 4y = 0$$

on the interval $(0, \infty)$ that are of the form $y = x^c$, where c may be complex. Use your answer to find two linearly independent real solutions.

12. Show that the set of all complex functions defined on an interval forms a vector space over the complex numbers.

5.4

EQUATIONS WITH CONSTANT COEFFICIENTS

By making use of complex exponential functions we can now solve any differential equation of the type

$$P(D)y = 0, \qquad (5.10)$$

provided that the auxiliary polynomial P does not have multiple zeros. If $a + ib$ and $a - ib$ are complex zeros of P, then

$$e^{(a + ib)x} = e^{ax}(\cos bx + i \sin bx),$$

$$e^{(a - ib)x} = e^{ax}(\cos bx - i \sin bx).$$

are complex solutions of Eq. (5.10). The real and imaginary parts

$$e^{ax} \cos bx, \qquad e^{ax} \sin bx$$

are real solutions. Thus, to every pair of complex conjugate zeros of P there corresponds a pair of real solutions of Eq. (5.10), and to each real zero c of P there corresponds a real solution e^{cx}. It will be shown in this section later on that the set of all real solutions so obtained is linearly independent. First we consider some examples.

Example 1

$$y'' + a^2 y = 0. \qquad (5.11)$$

The auxiliary equation, $r^2 + a^2 = 0$, has the roots ai and $-ai$. Hence

$$e^{iax} = \cos ax + i \sin ax$$

is a complex solution. The real and imaginary parts, $\cos ax$ and $\sin ax$, are real solutions. The general solution is

$$y = c_1 \cos ax + c_2 \sin ax.$$

Equation (5.11) is important for many applications. We rewrite the last formula as

$$y = (c_1^2 + c_2^2)^{1/2}[p \cos ax + q \sin ax],$$

where

$$p = \frac{c_1}{(c_1^2 + c_2^2)^{1/2}}, \qquad q = \frac{c_2}{(c_1^2 + c_2^2)^{1/2}}.$$

Since $p^2 + q^2 = 1$, there exist angles α and β such that

$$\cos \alpha = p, \qquad \sin \alpha = -q,$$
$$\sin \beta = p, \qquad \cos \beta = q.$$

Consequently the general solution is also described by either of the formulas

$$y = A \cos(ax + \alpha), \qquad y = B \sin(ax + \beta),$$

where A, B, α, and β are arbitrary constants. (See Exercise 28.)

Example 2
$$(D^3 - 5D^2 + 9D - 5)y = 0.$$

The polynomial

$$P(r) = r^3 - 5r^2 + 9r - 5 = (r - 1)(r^2 - 4r + 5)$$

has the zeros $r_1 = 1$, $r_2 = 2 + i$, and $r_3 = 2 - i$. Thus

$$e^{(2 + i)x} = e^{2x}(\cos x + i \sin x)$$

is a complex solution and $e^{2x} \cos x$, $e^{2x} \sin x$ are real solutions. The general solution is

$$y = c_1 e^x + c_2 e^{2x} \cos x + c_3 e^{2x} \sin x.$$

Let us now consider the general case of an nth-order equation (5.10) where the polynomial P,

$$P(r) = a_0(r - r_1)(r - r_2) \cdots (r - r_n),$$

has n distinct zeros. The functions

$$e^{r_1 x}, \quad e^{r_2 x}, \ldots, \quad e^{r_n x},$$

some of which may be complex, belong to the vector space of complex functions that are defined on $(-\infty, \infty)$. The linear independence of these functions follows from a consideration of their Wronskian. We shall now show that the corresponding set of n real solutions is a linearly independent set of elements of the space of real functions that are defined on $(-\infty, \infty)$. Suppose that $c_1, c_2, \ldots, c_n$ are real numbers such that

$$c_1 e^{ax} \cos bx + c_2 e^{ax} \sin bx + \cdots = 0$$

for all x. Then

$$c_1' e^{(a+ib)x} + c_2' e^{(a-ib)x} + \cdots = 0$$

for all x, where

$$c_1 = c_1' + c_2', \qquad c_2 = i(c_1' - c_2').$$

But then $c_1' = c_2' = \cdots = 0$ so $c_1 = c_2 = \cdots = 0$. Hence the n real solutions are linearly independent.

In order to treat the case where the polynomial P has multiple zeros, we need the following result.

Lemma   If r is any number, real or complex, and if the function w, which may be complex, has at least m derivatives, then

$$(D - r)^m [e^{rx} w(x)] = e^{rx} D^m w(x).$$

This formula evidently holds when $m = 0$. When $m = 1$, we have

$$(D - r) [e^{rx} w(x)] = e^{rx} w'(x) + r e^{rx} w(x) - r e^{rx} w(x)$$
$$= e^{rx} D w(x).$$

The reader who understands mathematical induction should be able to establish that the formula holds for every nonnegative integer m.

Now suppose that r_1 is a zero of P of multiplicity k, so that

$$P(r) = a_0 (r - r_1)^k (r - r_2) \cdots (r - r_{n-k+1}).$$

Then

$$P(D) = Q(D)(D - r_1)^k,$$

where $Q(D)$ is a polynomial operator of order $n - k$. We shall show that each of the k functions

$$x^j e^{r_1 x}, \qquad 0 \le j \le k - 1,$$

is a solution of the equation $P(D)\,y = 0$. Using the lemma, we have

$$P(D)(x^j e^{r_1 x}) = Q(D)(D - r_1)^k x^j e^{r_1 x}$$
$$= Q(D)e^{r_1 x}D^k x^j$$
$$= 0$$

since $D^k x^j = 0$ when $j < k$. This proves our assertion. Of course if r_1 is complex, the functions will be complex.

If $r_1 = a + ib$ is a complex zero of P of multiplicity k, then $r_1 = a - ib$ is also a zero of multiplicity k. Then each of the $2k$ functions

$$x^j e^{(a+ib)x}, \qquad x^j e^{(a-ib)x}, \qquad 0 \le j \le k - 1$$

is a complex solution of the differential equation $P(D)y = 0$. But since the real and imaginary parts of a complex solution are both real solutions, we know that each of the $2k$ real functions

$$x^j e^{ax} \cos bx, \qquad x^j e^{ax} \sin bx, \qquad 0 \le j \le k - 1$$

is a real solution.

Thus even when P has complex and multiple zeros, we can still find n seemingly distinct real solutions of the nth-order equation $P(D)y = 0$. Proofs that these n solutions are linearly independent are given in the references. We summarize our results as follows.

Theorem 5.4 If r_1 is a real root of multiplicity k of the nth-degree polynomial equation $P(r) = 0$, then each of the k functions

$$x^j e^{r_1 x}, \qquad 0 \le j \le k - 1$$

is a solution of the nth-order differential equation $P(D)y = 0$. If $a + ib$, $b \ne 0$, is a complex root of multiplicity k (in which case $a - ib$ is also a root of multiplicity k) then each of the $2k$ real functions

$$x^j e^{ax} \cos bx, \qquad x^j e^{ax} \sin bx, \qquad 0 \le j \le k - 1$$

is a solution. The n real solutions of the differential equation (corresponding to the n zeros of the polynomial) that are described here are linearly independent on every interval.

Example 3
$$y'' + 4y' + 4y = 0.$$

Since $r^2 + 4r + 4 = (r + 2)^2$, we have $r_1 = r_2 = -2$. According to Theorem 5.4, e^{-2x} and xe^{-2x} are solutions. The general solution is

$$y = c_1 e^{-2x} + c_2 x e^{-2x}.$$

Example 4
$$D^2(D + 2)^3(D - 3)y = 0.$$

The auxiliary equation $r^2(r + 2)^3(r - 3) = 0$ has roots $0, 0, -2, -2, -2,$ and 3. Each of the functions

$$1, \quad x, \quad e^{-2x}, \quad xe^{-2x}, \quad x^2e^{-2x}, \quad e^{3x}$$

is a solution of the differential equation. (Remember that $e^{0x} = 1$.) The general solution is

$$y = c_1 + c_2x + (c_3 + c_4x + c_5x^2)e^{-2x} + c_6e^{3x}.$$

Example 5
$$(D^4 + 8D^2 + 16)y = 0.$$

The auxiliary equation is

$$r^4 + 8r^2 + 16 = (r^2 + 4)^2 = 0,$$

with roots $2i, 2i, -2i,$ and $-2i$. Thus

$$e^{2ix}, \quad xe^{2ix}, \quad e^{-2ix}, \quad xe^{-2ix}$$

are complex solutions. The functions

$$\cos 2x, \quad x \cos 2x, \quad \sin 2x, \quad x \sin 2x$$

are real solutions and the general solution is

$$y = (c_1 + c_2x) \cos 2x + (c_3 + c_4x) \sin 2x.$$

The behavior of solutions as $x \to \infty$ depends on the roots of the polynomial equation $P(r) = 0$. We leave the simple proof of the next theorem as an exercise.

Theorem 5.5   Every solution of the equation $P(D)y = 0$ approaches 0 as $x \to \infty$ if and only if every root of the polynomial equation $P(r) = 0$ has a negative real part.

The criterion provided by the next theorem[2] (the Routh-Hurwitz criterion) does not require finding the roots of the polynomial equation. In order to state the theorem, we need a definition. The *principal leading minor* of order r of an $n \times n$ matrix A is the determinant of the $r \times r$ submatrix of A formed by deleting all but the first r rows and the first r columns.

---

[2] A proof may be found in W. Kaplan, *Operational Methods for Linear Systems* (Reading, Mass.: Addison-Wesley, 1962).

Theorem 5.6   Given the polynomial

$$P(r) = a_0 r^n + a_1 r^{n-1} + a_2 r^{n-2} + \cdots + a_{n-1} r + a_n$$

with $a_0 > 0$, form the $n \times n$ matrix

$$A = \begin{bmatrix} a_1 & a_3 & a_5 & a_7 & \cdots \\ a_0 & a_2 & a_4 & a_6 & \cdots \\ 0 & a_1 & a_3 & a_5 & \cdots \\ 0 & a_0 & a_2 & a_4 & \cdots \\ 0 & 0 & a_1 & a_3 & \cdots \\ 0 & 0 & a_0 & a_2 & \cdots \\ \cdots & \cdots & \cdots & \cdots & \cdots \\ 0 & 0 & 0 & 0 & \cdots \end{bmatrix}$$

where it is understood that $a_i = 0$ if $i > n$. Then all the roots of the equation $P(r) = 0$ have negative real parts if and only if all the principal leading minors of A are positive; that is, if and only if

$$|a_1| > 0, \qquad \begin{vmatrix} a_1 & a_3 \\ a_0 & a_2 \end{vmatrix} > 0, \qquad \begin{vmatrix} a_1 & a_3 & a_5 \\ a_0 & a_2 & a_4 \\ 0 & a_1 & a_3 \end{vmatrix} > 0, \cdots, \qquad \det(A) > 0.$$

In the formation of A, the odd-numbered rows contain only coefficients with odd subscripts and the even numbered rows contain only those with even subscripts. The first two rows are repeated as often as necessary, but are shifted to the right one position each time, filling in the leftmost positions with zeros.

Example 6   Consider the differential equation $y''' + y'' + 2y' + 3y = 0$, with associated polynomial $r^3 + r^2 + 2r + 3$. We have $a_0 = 1$, $a_1 = 1$, $a_2 = 2$, $a_3 = 3$, and $a_i = 0$ for $i > 3$. The matrix of Theorem 5.6 is

$$A = \begin{bmatrix} 1 & 3 & 0 \\ 1 & 2 & 0 \\ 0 & 1 & 3 \end{bmatrix}$$

and the minors of interest are

$$|1| = 1 > 0, \qquad \begin{vmatrix} 1 & 3 \\ 1 & 2 \end{vmatrix} = -1 < 0, \qquad \det(A) = -3 < 0.$$

Since the minors are not all positive, the roots of the polynomial equation do not all have negative real parts, and not every solution of the differential equation approaches 0 as $x \to \infty$.

   Differential equations of certain classes can be transformed into equations with constant coefficients by means of a change of variable. One such class is considered in the next section.

───────────────── **Exercises for Section 5.4** ─────────────────

In Exercises 1–20, find the general solution of the differential equation.

1. $y'' - y' - 6y = 0$

2. $2y'' - 5y' + 2y = 0$

3. $y'' + 2y' = 0$

4. $y''' + 2y'' - y' - 2y = 0$

5. $y''' + 3y'' - 4y' = 0$

6. $y^{(4)} - 10y'' + 9y = 0$

7. $y'' + 2y' + y = 0$

8. $y'' - 6y' + 9y = 0$

9. $y''' - 6y'' + 12y' - 8y = 0$

10. $y''' + 5y'' + 3y' - 9y = 0$

11. $y''' + y'' = 0$

12. $(D - 1)^3(D + 2)^2 y = 0$

13. $y'' + 9y = 0$

14. $y'' + 2y' + 10y = 0$

15. $y'' - 6y' + 13y = 0$

16. $y''' + 2y'' + y' + 2y = 0$

17. $y^{(4)} + 2y'' + y = 0$

18. $(D^2 - 2D + 5)^2 y = 0$

19. $(D - 2)^2(D^2 + 2)y = 0$

20. $y^{(5)} + 4y''' = 0$

In Exercises 21–25, find the solution of the initial value problem.

21. $y'' - 4y' + 3y = 0, \quad y(0) = -1, \quad y'(0) = 3$

22. $y'' - 4y' + 4y = 0, \quad y(0) = 2, \quad y'(0) = 1$

23. $y'' + 4y = 0, \quad y(\pi) = 1, \quad y'(\pi) = -4$

24. $y'' + 2y' + 2y = 0, \quad y(0) = 2, \quad y'(0) = -3$

25. $y''' + y'' = 0, \quad y(0) = 2, \quad y'(0) = 1,$
    $y''(0) = -1$

26. Prove Theorem 5.5.

27. Show that the general solution of the equation $y'' - a^2 y = 0$, where a is a constant, can be written either as

$$y = c_1 e^{ax} + c_2 e^{-ax}$$

or as

$$y = C_1 \cosh ax + C_2 \sinh ax.$$

28. Show that the general solution of the equation $y'' + a^2 y = 0$, where a is a constant, can be written in the forms shown in Example 1.

29. Find a linear homogeneous differential equation with real constant coefficients, whose order is as low as possible, that has the given function as a solution.
    (a) xe^{-2x} (b) $x - e^{3x}$
    (c) $\cos 2x$ (d) $e^x \sin 2x$
    (e) $x \sin 3x$ (f) $\cos 2x + 3e^{-x}$

30. Use Theorem 5.6 to determine whether all solutions of the given equation approach 0 as x approaches ∞.
    (a) $(2D^3 + 3D^2 + 4D + 5)y = 0$
    (b) $(5D^3 + 7D^2 + 2D + 4)y = 0$
    (c) $(2D^4 + D^3 + 2D^2 + 6D + 3)y = 0$
    (d) $(3D^4 + 5D^3 + 8D^2 + D + 1)y = 0$

31. (a) Use Theorem 5.6 to show that all the roots of the second degree equation $r^2 + a_1 r + a_2 = 0$ have negative real parts if and only if $a_1 > 0$ and $a_2 > 0$.
    (b) Show that a *necessary* condition for all the roots of the equation

$$a_0 r^n + a_1 r^{n-1} + \cdots + a_{n-1}r + a_n, \quad a_0 > 0$$

to have negative real parts is that all the

coefficients be positive. Suggestion: the polynomial can be factored into first and second degree polynomials with real coefficients. Make use of Part (a).

(c) Show by means of an example that the condition of Part (b) is not *sufficient*.

5.5

CAUCHY-EULER EQUATIONS

A linear differential equation of the form

$$(b_0 x^n D^n + b_1 x^{n-1} D^{n-1} + \cdots + b_{n-1} xD + b_n)y = 0, \qquad (5.12)$$

where $b_0, b_1, \ldots, b_n$ are constants, is known as a *Cauchy-Euler equation*, or as an *equidimensional equation*. Examples of Cauchy-Euler equations are

$$x^2 y'' - 3xy' + 4y = 0, \qquad x^3 y''' + 2xy' = 0.$$

Equations of this type can be transformed into equations with constant coefficients by means of a change of independent variable.

Let

$$x = e^t, \qquad t = \ln x,$$

where as t varies over the set of all real numbers, x varies over the interval $(0, \infty)$. In what follows we assume that $x > 0$. The case $x < 0$ is treated in Exercise 19.

Let us write

$$y(e^t) = Y(t).$$

Then, by the chain rule for differentiation,

$$x \, Dy(x) = x \frac{dy(x)}{dx} = e^t \frac{dY(t)}{dt} \frac{dt}{dx}.$$

Since $dt/dx = 1/x = e^{-t}$ we have

$$x \, Dy(x) = DY(t).$$

Next, using the chain rule again,

$$x^2 D^2 y(x) = x^2 \frac{d}{dx} \left[\frac{dy(x)}{dx} \right]$$

$$= e^{2t} \frac{dt}{dx} \frac{d}{dt} \left[e^{-t} \frac{dY(t)}{dt} \right]$$

$$= e^t \left[e^{-t} \frac{d^2 Y(t)}{dt^2} - e^{-t} \frac{dY(t)}{dt} \right]$$

$$= \frac{d^2 Y(t)}{dt^2} - \frac{dY(t)}{dt}.$$

Thus

$$x^2D^2y(x) = (D^2 - D)Y(t) = D(D - 1)Y(t).$$

By the use of mathematical induction, it can be shown that

$$x^mD^my(x) = D(D - 1)(D - 2) \cdots (D - m + 1)Y(t)$$

for every positive integer m.

Using this formula we see that the original equation (5.12) becomes

$$[b_0D(D - 1) \cdots (D - n + 1) + b_1D(D - 1) \cdots (D - n + 2)$$
$$+ \cdots + b_{n-1}D + b_n]Y = 0. \tag{5.13}$$

This equation with constant coefficients has the auxiliary polynomial Q, where

$$Q(r) = b_0r(r - 1) \cdots (r - n + 1) + b_1r(r - 1) \cdots (r - n + 2)$$
$$+ \cdots + b_{n-1}r + b_n. \tag{5.14}$$

This polynomial is of degree n. If r_1 is a real zero of multiplicity k, then each of the k functions

$$t^je^{r_1t}, \quad 0 \le j \le k - 1$$

is a solution of Eq. (5.13) and each of the functions

$$(\ln x)^jx^{r_1}, \quad 0 \le j \le k - 1$$

is a solution of the original equation (5.12). If $a + ib$ and $a - ib$ are zeros of Q of multiplicity k, the functions

$$t^je^{at} \cos bt, \quad t^je^{at} \sin bt, \quad 0 \le j \le k - 1$$

are solutions of Eq. (5.13). The corresponding solutions of the original equation are

$$x^a(\ln x)^j \cos(b \ln x), \quad x^a(\ln x)^j \sin(b \ln x), \quad 0 \le j \le k - 1.$$

Example 1
$$2x^2y'' - 5xy' + 3y = 0.$$

The change of variable $x = e^t$ leads to the equation

$$[2D(D - 1) - 5D + 3]Y = 0$$

or

$$(2D - 1)(D - 3)Y = 0.$$

handwritten at top: $2r - 1 = 0 \quad r_1 = \frac{1}{2}$
$r - 3 = 0 \quad r_2 = 3$

Hence

$$Y(t) = c_1 e^{t/2} + c_2 e^{3t}$$

or *handwritten:* $y(x) = c_1 e^{\frac{1}{2}\ln x} + c_2 e^{3\ln x}$

$$y(x) = c_1 x^{1/2} + c_2 x^3, \qquad x > 0.$$

Example 2

$$x^2 y'' - xy' + 5y = 0.$$

handwritten: $\theta(\theta - 1) - \theta(\theta + 5)$

This equation becomes

$$[D(D - 1) - D + 5]Y = 0$$

or

$$(D^2 - 2D + 5)Y = 0.$$

The roots of the auxiliary equation are $1 + 2i$ and $1 - 2i$. Thus

$$Y(t) = c_1 e^t \cos 2t + c_2 e^t \sin 2t$$

or

$$y(x) = c_1 x \cos(2 \ln x) + c_2 x \sin(2 \ln x).$$

The equation

$$Ly \equiv (b_0 x^n D^n + b_1 x^{n-1} D^{n-1} + \cdots + b_{n-1} xD + b_n)y = 0$$

can be solved more directly by attempting to find solutions of the form $y = x^r$, without any change of variable. Observing that

$$D^k x^r = r(r - 1)(r - 2) \cdots (r - k + 1)x^{r-k}$$

and

$$x^k D^k x^r = r(r - 1)(r - 2) \cdots (r - k + 1)x^r,$$

we have

$$L(x^r) = [b_0 r(r - 1) \cdots (r - n + 1) + b_1 r(r - 1) \cdots (r - n + 2) + \cdots + b_{n-1} r + b_n]x^r$$

or

$$L(x^r) = Q(r)x^r,$$

where Q is the polynomial (5.14). If r_1 is a real zero of Q of multiplicity k, the functions

$$x^{r_1}(\ln x)^j, \qquad 0 \le j \le k - 1$$

are solutions. If $a + ib$ and $a - ib$ are zeros of multiplicity k, then the functions

$$x^a(\ln x)^j \cos(b \ln x), \qquad x^a(\ln x)^j \sin(b \ln x), \qquad 0 \le j \le k - 1$$

are solutions.

Example 3

$$x^3y''' - x^2y'' + xy' = 0.$$

$D(D-2)(D-1) - D(D-1) + D = 0$

Setting $y = x^r$, we must have

$$r(r - 1)(r - 2) - r(r - 1) + r = 0$$

or

$$r(r - 2)^2 = 0.$$

The general solution is

$$y = c_1 + (c_2 + c_3 \ln x)x^2.$$

───────────────────── **Exercises for Section 5.5** ─────────────────────

In Exercises 1–12, find the general solution of the differential equation if x is restricted to the interval $(0, \infty)$.

1. $x^2y'' - 2y = 0$

2. $x^2y'' + 3xy' - 3y = 0$

3. $3xy'' + 2y' = 0$

4. $x^3y''' + x^2y'' - 2xy' + 2y = 0$

5. $4x^2y'' + y = 0$

6. $x^2y'' - 3xy' + 4y = 0$

7. $xy''' + 2y'' = 0$

8. $x^3y''' + 6x^2y'' + 7xy' + y = 0$

9. $x^2y'' + xy' + 4y = 0$

10. $x^2y'' - 5xy' + 13y = 0$

11. $x^3y''' + 2x^2y'' + xy' - y = 0$

12. $x^4y^{(4)} + 6x^3y''' + 15x^2y'' + 9xy' + 16y = 0$

In Exercises 13–15, find the solution of the initial value problem on the interval $(0, \infty)$.

13. $x^2y'' + 4xy' + 2y = 0$, $y(1) = 1$, $y'(1) = 2$

14. $x^2y'' - 3xy' + 4y = 0$, $y(1) = 2$, $y'(1) = 1$

15. $x^2y'' + xy' + 4y = 0$, $y(1) = 1$, $y'(1) = 4$

In Exercises 16–18, find all solutions on the interval $(0, \infty)$ that have a finite limit as x tends to zero.

16. $4x^2y'' + 4xy' - y = 0$

17. $x^2y'' + 2xy' - 2y = 0$

18. $x^2y'' + 6xy' + 6y = 0$

19. If the function f is a solution of Eq. (5.12) on the interval $(0, \infty)$ show that the function g, where $g(x) = f(-x)$, is a solution on $(-\infty, 0)$.

20. Show that the change of variable $t = ax + b$ transforms the equation

$$b_0(ax + b)^2y'' + b_1(ax + b)y' + b_2y = 0$$

into a Cauchy-Euler equation.

21. Use the result of Exercise 20 to find the general solution of the given equation.

    (a) $(x - 3)^2 y'' + 3(x - 3)y' + y = 0$, $x > 3$

    (b) $(2x + 1)^2 y'' + 4(2x + 1)y' - 24y = 0$, $x > -\frac{1}{2}$

22. What conditions must the zeros of the polynomial Q satisfy in order that every solution of Eq. (5.12) tend to zero as

    (a) x approaches zero through positive values?

    (b) x becomes positively infinite?

23. If the functions $f_1, f_2, \ldots, f_n$ are linearly independent on the set of all real numbers and if $g_i(x) = f_i(\ln x)$ for $x > 0$, show that the functions g_i are linearly independent on $(0, \infty)$.

5.6
NON-HOMOGENEOUS EQUATIONS

The linear equation

$$Ly = F, \tag{5.15}$$

where

$$L = a_0 D^n + a_1 D^{n-1} + \cdots + a_{n-1} D + a_n$$

(the functions a_i need not be constants) is called *nonhomogeneous* if F is not the zero function. Associated with the nonhomogeneous equation is the homogeneous equation

$$Ly = 0. \tag{5.16}$$

It turns out that we can solve the nonhomogeneous equation if we can solve the homogeneous equation and if we can also find just one particular solution of the nonhomogeneous equation.

Theorem 5.7   Let $u_1, u_2, \ldots, u_n$ be linearly independent solutions of the homogeneous equation (5.16) on an interval $\mathscr{I}$[3] and let y_p be any particular solution of the nonhomogeneous equation (5.15) on $\mathscr{I}$. Then the set of all solutions of equation (5.15) on $\mathscr{I}$ consists of all functions of the form

$$c_1 u_1 + c_2 u_2 + \cdots + c_n u_n + y_p, \tag{5.17}$$

where $c_1, c_2, \ldots, c_n$ are constants.

Proof   First let us verify that every function of the form (5.17) is a solution of Eq. (5.15). Since $Lu_i = 0$, $1 \le i \le n$, and $Ly_p = F$, we have

$$L(c_1 u_1 + c_2 u_2 + \cdots + c_n u_n + y_p) = c_1 L u_1 + c_2 L u_2 + \cdots + c_n L u_n + L y_p$$

$$= 0 + 0 + \cdots + 0 + F$$

$$= F$$

---

[3] The functions a_i and F are assumed to be continuous on $\mathscr{I}$ with a_0 never zero on $\mathscr{I}$.

which we wished to show. Next we must show that every solution of Eq. (5.15) is of the form (5.17). Let u be any solution. Then $Lu = F$. Since also $Ly_p = F$ we have

$$L(u - y_p) = Lu - Ly_p = F - F = 0.$$

Thus the function $u - y_p$ is a solution of the homogeneous equation $Ly = 0$. By Theorem 5.1, $u - y_p$ must be of the form

$$u - y_p = c_1 u_1 + c_2 u_2 + \cdots + c_n u_n.$$

Then

$$u = c_1 u_1 + c_2 u_2 + \cdots + x_n u_n + y_p,$$

which we wished to show.

The general solution of the homogeneous equation $Ly = 0$, written

$$y_c = c_1 u_1 + c_2 u_2 + \cdots + c_n u_n,$$

is often called the *complementary function* associated with the nonhomogeneous equation $Ly = F$. Thus, according to Theorem 5.7, the general solution of the nonhomogeneous equation is

$$y = y_c + y_p,$$

the sum of the complementary function y_c and a particular solution y_p.

To illustrate the use of this theorem, let us consider the equation

$$y'' - 4y' + 4y = 9e^{-x}.$$

It is easy to verify that a particular solution is $y_p(x) = e^{-x}$. The associated homogeneous equation has e^{2x} and xe^{2x} as linearly independent solutions. Hence the complementary function is

$$y_c(x) = c_1 e^{2x} + c_2 xe^{2x}$$

and the general solution of the original equation is

$$y = c_1 e^{2x} + c_2 xe^{2x} + e^{-x}.$$

If the associated homogeneous equation has constant coefficients, or is of the Cauchy-Euler type, we can solve it. There remains the problem of finding one solution u_p of the nonhomogeneous equation. A method that applies in certain cases is described in the next section. A more general method is discussed in Section 5.8. In finding particular solutions, the following result is often useful. It is known as the *superposition principle.*

Theorem 5.8   If u_p and v_p are solutions of the equations $Ly = f$ and $Ly = g$, respectively, then $u_p + v_p$ is a solution of the equation $Ly = f + g$.

Proof   Since $Lu_p = f$ and $Lv_p = g$, we have

$$L(u_p + v_p) = Lu_p + Lv_p = f + g.$$

For example, suppose that u_p and v_p are solutions of the equations

$$Ly = 3e^x, \qquad Ly = -2\sin x,$$

respectively. Then $u_p + v_p$ is a solution of the equation

$$Ly = 3e^x - 2\sin x.$$

Theorem 5.7 can be regarded as a special case of a more general result. Let T be a linear transformation from a vector space V into a vector space W. For a given element $\mathbf{w}$ in W consider the nonhomogeneous equation

$$T\mathbf{v} = \mathbf{w}.$$

Associated with this equation is the homogeneous equation

$$T\mathbf{v} = \mathbf{0}.$$

If $\mathbf{v}_p$ is any particular solution of the nonhomogeneous equation and if $\mathbf{z}$ is any solution of the homogeneous equation, then $\mathbf{z} + \mathbf{v}_p$ is a solution of the nonhomogeneous equation. To see this, observe that

$$T(\mathbf{z} + \mathbf{v}_p) = T\mathbf{z} + T\mathbf{v}_p = \mathbf{0} + \mathbf{w} = \mathbf{w}.$$

It can also be shown (Exercise 8) that *every* solution of the nonhomogeneous equation is of the form $\mathbf{z} + \mathbf{v}_p$, where $\mathbf{z}$ is a solution of the homogeneous equation. If the kernel of T is finite dimensional, and if $\mathbf{z}_1, \mathbf{z}_2, \ldots, \mathbf{z}_n$ form a basis for the kernel, then the general solution of the equation $T\mathbf{v} = \mathbf{w}$ consists of those elements of V that are of the form

$$c_1\mathbf{z}_1 + c_2\mathbf{z}_2 + \cdots + c_n\mathbf{z}_n + \mathbf{v}_p.$$

This is because every solution is of this form, and every element of this form is a solution.

Exercises for Section 5.6

1. Verify that y_p, where $y_p(x) = \sin 2x$, is a solution of the differential equation

$$y'' - y = -5 \sin 2x.$$

Use this fact to find the general solution of the equation.

2. Show that the nonhomogeneous equation $y' - ay = ke^{bx}$ has a solution of the form $y_p = Ae^{bx}$ if $b \neq a$. Show that in this case the solution is $ke^{bx}/(b - a)$.

3. Use the result of Exercise 2 to find the general solution of the given equation.

(a) $y' - 2y = 6e^{5x}$ (b) $y' + 3y = 4e^{-x}$

4. If a_n is a nonzero constant and c is a constant, show that a solution of the equation

$$a_0 y^{(n)} + a_1 y^{(n-1)} + \cdots + a_n y = c$$

is $y = c/a_n$.

5. Show that the equation

$$b_0 x^2 y'' + b_1 xy' + b_2 y = cx^a$$

has a solution of the form $y = Ax^a$ provided that

$$b_0 a(a - 1) + b_1 a + b_2 \neq 0.$$

6. If F belongs to the kernel of the operator $Q(D)$, show that every solution of the nonhomogeneous equation $P(D)y = F$ is a solution of the homogeneous equation $Q(D)P(D)y = 0$.

7. Use the result of Exercise 2 and Theorem 5.8 to solve the given differential equation.

(a) $y' + 3y = -2e^{2x} + 8e^{-x}$
(b) $y' - y = 4e^{-x} - 2e^{-2x}$

8. Let T be a linear transformation from a vector space V into a vector space W. For a given element $\mathbf{w}$ of W consider the equation $T\mathbf{v} = \mathbf{w}$. If $\mathbf{v}_p$ is a solution, show that every solution is of the form $\mathbf{z} + \mathbf{v}_p$, where $\mathbf{z}$ belongs to the kernel of T.

5.7

THE METHOD OF UNDETERMINED COEFFICIENTS

In this section we describe a method that yields a particular solution of the nonhomogeneous equation $Ly = F$ when the following two conditions are both met.

(a) The operator L has constant coefficients.

(b) The function F is itself a solution of some linear homogeneous differential equation with constant coefficients.

Here F must consist of a linear combination of functions of the types

$$x^j, \quad x^j e^{cx}, \quad x^j e^{ax} \cos bx, \quad x^j e^{ax} \sin bx. \qquad (5.18)$$

These functions can be characterized in another way. If F is one of the types (5.18), then there is a *finite* set S of functions such that F and each of its derivatives can be expressed as a linear combination of the functions in S. For example, if $F(x) = x \cos 3x$, then F and every derivative of F is a linear combination of functions of the set

$$S = \{\cos 3x, \sin 3x, x \cos 3x, x \sin 3x\}.$$

(For a justification of this assertion the reader is referred to Exercise 20 in the set of additional exercises at the end of this chapter.)

Actually, according to Theorem 5.8, we can concentrate on the case where F is a constant multiple of just *one* of these functions. For instance, to find a solution of the equation

$$Ly = 5e^x \sin 2x - 4x^2 e^{-x} \tag{5.19}$$

we first find solutions u and v of the equations

$$Ly = 5e^x \sin 2x, \qquad Ly = -4x^2 e^{-x},$$

respectively. Then $u + v$ will be a solution of Eq. (5.19).

We therefore consider an nth-order equation

$$P(D)y = F, \tag{5.20}$$

where $F(x)$ is $kx^j e^{ax}$, $kx^j e^{ax} \cos bx$, or $kx^j e^{ax} \sin bx$. (In particular, a may be zero.) Then there exists a polynomial operator $Q(D)$, with real coefficients, such that

$$Q(D)F = 0.$$

We say that the operator $Q(D)$ *annihilates* F. If $F(x) = kx^j e^{ax}$ we may take

$$Q(D) = (D - a)^{j+1}$$

while if $F(x) = kx^j e^{ax} \cos bx$ or $F(x) = kx^j e^{ax} \sin bx$ we may choose

$$Q(D) = \{[D - (a + ib)][D - (a - ib)]\}^{j+1} = [(D - a)^2 + b^2]^{j+1}.$$

With these choices, the order of $Q(D)$ is as low as possible.

Suppose that the order of $Q(D)$ is m. If we operate on both members[4] of Eq. (5.20) with $Q(D)$, we see that every solution of Eq. (5.20) is also a solution of the homogeneous equation

$$Q(D)P(D)y = 0. \tag{5.21}$$

(However, not every solution of Eq. (5.21) need be a solution of Eq. (5.20).) The order of this equation is $m + n$. Every solution of the equation

$$P(D)y = 0 \tag{5.22}$$

---

[4] Every solution of Eq. (5.20) possesses derivatives of all orders, so the derivatives of y in $Q(D)P(D)y$ all exist.

is a solution of Eq. (5.21), but the latter equation also possesses additional solutions. Let the general solution of Eq. (5.21) be

$$A_1 u_1 + \cdots + A_m u_m + B_1 v_1 + \cdots + B_n v_n,$$

where the functions u_i and v_i are of the types (5.18) and the functions v_i are linearly independent solutions of Eq. (5.22).

Since every solution of the nonhomogeneous equation (5.20) is also of the above form, it must be possible to choose the constants A_i and B_i in such a way that

$$P(D)(A_1 u_1 + \cdots + A_m u_m + B_1 v_1 + \cdots + B_n v_n) = F.$$

Since

$$P(D)(B_1 v_1 + \cdots + B_n v_n) = 0$$

for every choice of the B_i, it must be possible to find constants A_i such that

$$A_1 u_1 + \cdots + A_m u_m \tag{5.23}$$

is a solution of Eq. (5.20). If the functions u_i are known, the constants A_i can be determined by substituting the expression (5.23) into the differential equation (5.20) and requiring that the latter be satisfied identically. The expression (5.23) is called a *trial solution* for Eq. (5.20).

Let us pause to consider some specific cases.

Example 1
$$y'' - y' - 2y = 20e^{4x}.$$

This equation may be written as

$$(D + 1)(D - 2)y = 20e^{4x}. \tag{5.24}$$

The operator $D - 4$ annihilates e^{4x}. Operating on both sides of this equation with $D - 4$, we see that every solution is also a solution of the homogeneous equation

$$(D - 4)(D + 1)(D - 2)y = 0.$$

The solutions of this equation are of the form

$$y(x) = Ae^{4x} + B_1 e^{-x} + B_2 e^{2x},$$

and hence every solution of Eq. (5.24) is of this form. Thus there exist constants A, B_1, and B_2 such that

$$(D + 1)(D - 2)(Ae^{4x} + B_1 e^{-x} + B_2 e^{2x}) = 20e^{4x}.$$

Since e^{-x} and e^{2x} are solutions of the homogeneous equation

$$(D + 1)(D - 2)y = 0,$$

it must be possible to choose A so that

$$(D + 1)(D - 2)(Ae^{4x}) = 20e^{4x}.$$

This yields the requirement

$$(D^2 - D - 2)(Ae^{4x}) = 20e^{4x},$$

$$10Ae^{4x} = 20e^{4x},$$

or

$$A = 2.$$

Hence a particular solution of Eq. (5.24) is $y_p(x) = 2e^{4x}$. The general solution is

$$y(x) = c_1e^{-x} + c_2e^{2x} + 2e^{4x},$$

where c_1 and c_2 are arbitrary constants.

Example 2 $$y'' - 4y' + 4y = 5 \sin x. \qquad (5.25)$$

The operator $(D - i)(D + i) = D^2 + 1$ annihilates $\sin x$. Operating on both members of Eq. (5.25) with $D^2 + 1$, we see that every solution of Eq. (5.25) is a solution of the equation

$$(D^2 + 1)(D - 2)^2 y = 0.$$

The solutions of this equation that are not solutions of the homogeneous equation $(D - 2)^2 y = 0$ are $\cos x$ and $\sin x$. Hence Eq. (5.25) has a solution of the form

$$y_p(x) = A \cos x + B \sin x.$$

Substituting in Eq. (5.25), we see that

$$-A \cos x - B \sin x - 4(-A \sin x + B \cos x)$$
$$+ 4(A \cos x + B \sin x) = 5 \sin x,$$

or, upon collecting terms,

$$(3A - 4B) \cos x + (4A + 3B) \sin x = 5 \sin x.$$

Thus we must have

$$3A - 4B = 0, \qquad 4A + 3B = 5$$

and so $A = \frac{4}{5}$, $B = \frac{3}{5}$. A particular solution of Eq. (5.25) is

$$y_p(x) = \frac{1}{5}(4 \cos x + 3 \sin x)$$

and the general solution is

$$y = (c_1 + c_2 x)e^{2x} + \frac{1}{5}(4 \cos x + 3 \sin x).$$

We now return to the general case and seek to determine the nature of the functions u_i in the trial solution (5.23). These functions are the solutions of the equation $Q(D)P(D)y = 0$ that are not solutions of the equation $P(D)y = 0$. If P and Q have no common zeros, the functions u_i are simply the solutions of the equation $Q(D)y = 0$. Hence our trial solution for the equation

$$P(D)y = kx^j e^{ax}$$

is of the form

$$y_p(x) = (A_0 + A_1 x + \cdots + A_j x^j)e^{ax}. \tag{5.26}$$

However, if a is a zero of P of multiplicity m, then

$$e^{ax}, \quad xe^{ax}, \ldots, \quad x^{m-1}e^{ax} \tag{5.27}$$

are solutions of the homogeneous equation $P(D)y = 0$. Since a is a zero of the polynomial QP of multiplicity $m + j + 1$, the functions

$$e^{ax}, \quad xe^{ax}, \ldots, \quad x^{m+j}e^{ax} \tag{5.28}$$

are solutions of the equation $Q(D)P(D)y = 0$. Selecting those functions in the set (5.28) that are not in the set (5.27), we see that our trial solution takes the form

$$y_p(x) = x^m(A_0 + A_1 x + \cdots + A_j x^j)e^{ax}. \tag{5.29}$$

Similarly, if the equation is one of the types

$$P(D)y = kx^j e^{ax} \cos bx, \qquad P(D)y = kx^j e^{ax} \sin bx,$$

and if $a + ib$ is not a zero of P, then

$$y_p(x) = e^{ax}[(A_0 \cos bx + B_0 \sin bx) + x(A_1 \cos bx + B_1 \sin bx)$$
$$+ \cdots + x^j(A_j \cos bx + B_j \sin bx)]. \tag{5.30}$$

However, if $a + ib$ is a zero of P of multiplicity m, then a particular solution is of the form

$$y_p(x) = x^m e^{ax}[(A_0 \cos bx + B_0 \sin bx) + x(A_1 \cos bx + B_1 \sin bx)$$
$$+ \cdots + x^j(A_j \cos bx + B_j \sin bx)]. \tag{5.31}$$

We may now formulate the following rule to be used in determining the form of a particular solution y_p. A tentative trial solution of the nonhomogeneous equation $P(D)y = F$ is of the form (5.26) or (5.30), depending on the form of F. But if a term in formula (5.26) or (5.30) is a solution of the homogeneous equation, the trial solution must be modified by multiplying through by x^m where m is the smallest positive integer such that none of the terms in formula (5.29) or (5.31) is a solution of the homogeneous equation.

We now examine several more examples.

Example 3

$$y'' - 3y' + 2y = 6e^{3x}.$$

The complementary function is

$$c_1 e^x + c_2 e^{2x}.$$

The tentative trial solution is

$$y_p(x) = Ae^{3x}.$$

Since no function of the form $x^m e^{3x}$ is a solution of the homogeneous equation, the trial solution need not be modified. The constant A is determined by the requirement that

$$(D^2 - 3D + 2)(Ae^{3x}) = 6e^{3x}$$

or

$$2Ae^{3x} = 6e^{3x}.$$

Thus $A = 3$ and a particular solution of the nonhomogeneous equation is

$$y_p(x) = 3e^{3x}.$$

The general solution of the nonhomogeneous equation is

$$y(x) = c_1 e^x + c_2 e^{2x} + 3e^{3x}.$$

Example 4

$$y'' - 4y' + 4y = 12xe^{2x}.$$

The complementary function is

$$c_1 e^{2x} + c_2 x e^{2x}.$$

The tentative trial solution is

$$y_p(x) = (A + Bx)e^{2x}.$$

But since xe^{2x} is a solution of the homogeneous equation, we must take

$$y_p(x) = x^2(A + Bx)e^{2x} = (Ax^2 + Bx^3)e^{2x}.$$

Differentiation shows that

$$y_p'(x) = [2Ax + (2A + 3B)x^2 + 2Bx^3]e^{2x}$$
$$y_p''(x) = [2A + (8A + 6B)x + (4A + 12B)x^2 + 4Bx^3]e^{2x}.$$

Calculation shows that

$$(D^2 - 4D + 4)y_p = (2A + 6Bx)e^{2x}$$

and this must be equal to $12xe^{2x}$ if y_p is to be a solution of the original equation. Hence we take $A = 0$ and $B = 2$. Then

$$y_p(x) = 2x^3 e^{2x}$$

and the general solution is

$$y(x) = (c_1 + c_2 x)e^{2x} + 2x^3 e^{2x}.$$

Example 5

$$y'' + y' - 2y = 4 \sin 2x.$$

The complementary function is

$$y_c = c_1 e^x + c_2 e^{-2x}$$

and our tentative trial solution is

$$y_p = A \cos 2x + B \sin 2x.$$

Since none of the terms in y_p appear in y_c, no modification of y_p is necessary. Substituting y_p in the differential equation, we find that

$$(D^2 + D - 2)y_p = (-6A + 2B)\cos 2x + (-2A - 6B)\sin 2x,$$

so we require that

$$-6A + 2B = 0, \qquad -2A - 6B = 4,$$

or $A = -\frac{1}{5}$, $B = -\frac{3}{5}$. Then

$$y_p = -\frac{1}{5}\cos 2x - \frac{3}{5}\sin 2x$$

and the general solution is

$$y = y_c + y_p = c_1 e^x + c_2 e^{-2x} - \frac{1}{5}\cos 2x - \frac{3}{5}\sin 2x.$$

Example 6
$$y''' + y' = 4\cos x.$$

The complementary function is

$$y_c = c_1 + c_2 \cos x + c_3 \sin x$$

and the tentative form of a particular solution is

$$y_p = A\cos x + B\sin x.$$

However, since terms in y_p also appear in y_c, the form of y_p must be modified according to the rule formulated earlier. We must take

$$y_p = x(A\cos x + B\sin x) = Ax\cos x + Bx\sin x.$$

Calculation shows that

$$Dy_p = A(-x\sin x + \cos x) + B(x\cos x + \sin x)$$
$$D^3 y_p = A(x\sin x - 3\cos x) + B(-x\cos x - 3\sin x).$$

Substituting in the differential equation, we find that

$$(D^3 + D)y_p = -2A\cos x - 2B\sin x.$$

Hence we must choose $-2A = 4$, $-2B = 0$, or $A = -2$, $B = 0$. Then

$$y_p = -2x \cos x$$

and the general solution is

$$y = c_1 + c_2 \cos x + c_3 \sin x - 2x \cos x.$$

Example 7 $x^2 y'' - 2xy' + 2y = 6 \ln x, \qquad x > 0.$

The change of variable $x = e^t$ leads to the equation

$$(D - 1)(D - 2)Y = 6t,$$

where $Y(t) = y(e^t)$. The complementary function is

$$Y(t) = c_1 e^t + c_2 e^{2t}.$$

The tentative trial solution is

$$Y_p(t) = A + Bt.$$

Since the homogeneous equation has no solution of the form t^m, this is correct as it stands. Calculation shows that $A = \frac{9}{2}$, $B = 3$. The general solution of the equation with constant coefficients is

$$Y(t) = c_1 e^t + c_2 e^{2t} + 3t + \frac{9}{2}.$$

The general solution of the original equation, obtained by setting $t = \ln x$, is

$$y(x) = c_1 x + c_2 x^2 + 3 \ln x + \frac{9}{2}.$$

Exercises for Section 5.7

In Exercises 1–20, find the general solution of the differential equation. If initial conditions are given, also find the solution that satisfies those conditions.

1. $y'' + 2y' - 3y = 5e^{2x}, \quad y(0) = 5, \quad y'(0) = 2$

2. $y'' + 4y' + 4y = 3e^{-x}, \quad y(0) = 3, \quad y'(0) = 1$

3. $y'' + 3y' + 2y = 36xe^x$

4. $y'' + y' - 2y = 6e^{-x} + 4e^{-3x}, \quad y(0) = -1, \\ y'(0) = 1$

5. $y'' + 3y' + 2y = 20 \cos 2x, \quad y(0) = -1, \\ y'(0) = 6$

6. $y'' + y = 5e^x \sin x$

7. $y'' - y' - 6y = 2$

8. $y'' + 3y' + 2y = 4x^2$

9. $(D - 1)^2(D + 1)y = 10 \cos 2x$

10. $y'' + 2y' + y = -4e^{-3x} \sin 2x$

11. $y'' + 3y' + 2y = 5e^{-2x}$

12. $y'' - y = 4e^x - 3e^{2x}$

13. $y'' - y' - 2y = -6xe^{-x}$

14. $y'' + 4y' + 4y = 4e^{-2x}$

15. $y'' + 4y' + 3y = 6x^2e^{-x}$

16. $y'' + 2y' = -4$

17. $y'' + y' = 3x^2$

18. $y'' + y = 4 \sin x$

19. $y'' + 2y' + 5y = 4e^{-x} \cos 2x$

20. $y'' + 4y = 16x \sin 2x$

21. Show that a particular solution of the non-homogeneous equation

$$P(D)y = ke^{cx}$$

is

$$y_p = \frac{k}{P(c)} e^{cx}$$

provided that c is not a root of the auxiliary equation $P(r) = 0$.

22. Use the result of Exercise 21 to find the general solution of the differential equation.

(a) $y'' - 3y' + 2y = 6e^{-x}$
(b) $y'' - 2y' + y = -3e^{2x}$
(c) $y'' - y' - 2y = 10 \cos x$
(d) $y'' - 3y' + 2y = 4e^{3x} + 6e^{-x}$
(e) $(D - 1)^2(D + 1)y = 9e^{2x}$
(f) $y'' + y' = 6 \sin 2x$

23. If the number c is an m-fold root of the polynomial equation $P(r) = 0$ then $P(r) = Q(r)(r - c)^m$, where $Q(c) \neq 0$. Show that in this case the differential equation

$$P(D)y = ke^{cx}$$

possesses the solution

$$y = \frac{k}{m!Q(c)} x^m e^{cx}.$$

24. Use the result of Exercise 23 to find the general solution of the differential equation.

(a) $y'' - y' - 2y = 6e^{2x}$
(b) $y'' - 4y' + 4y = 4e^{2x}$
(c) $y'' + y = 4 \cos x$
(d) $(D + 1)(D - 2)^3 y = 6e^{2x}$

In Exercises 25–30, find the general solution of the differential equation. Assume that the independent variable is restricted to the interval $(0, \infty)$.

25. $x^2y'' - 6y = 6x^4$

26. $x^2y'' + xy' - y = 9x^2 \ln x$

27. $x^2y'' - 3xy' + 3y = -6$

28. $x^2y'' - xy' = -4$

29. $x^2y'' + 2xy' - 2y = 6x$

30. $x^2y'' - xy' + y = 6x \ln x$

5.8
VARIATION OF
PARAMETERS

The method of undetermined coefficients discussed in the last section allows us to find a particular solution of the nonhomogeneous equation $Ly = F$ only in special cases. The operator L must have constant coefficients, and F must belong to a certain class of functions. The method of variation of parameters is more general. It gives a formula for a particular solution provided only that the general solution of the homogeneous equation $Ly = 0$ is known. The operator L need not have constant coefficients and there is no restriction (other than continuity) on F. However, the method of undetermined coefficients is usually easier to use when that method applies. The method of variation of parameters is valuable for theoretical purposes, as well as for actually finding solutions of equations. Some examples of its use are given in the exercises at the end of this section.

In the method of variation of parameters, we assume that n linearly independent solutions $u_1, u_2, \ldots, u_n$ of the homogeneous equation

$$a_0(x)y^{(n)} + a_1(x)y^{(n-1)} + \cdots + a_{n-1}(x)y' + a_n(x)y = 0 \qquad (5.32)$$

are known. We attempt to find a solution of the nonhomogeneous equation

$$a_0(x)y^{(n)} + a_1(x)y^{(n-1)} + \cdots + a_{n-1}(x)y' + a_n(x)y = F(x) \qquad (5.33)$$

that is of the form

$$y = C_1(x)u_1(x) + C_2(x)u_2(x) + \cdots + C_n(x)u_n(x), \qquad (5.34)$$

where the functions $C_1, C_2, \ldots, C_n$ are to be determined. If we simply calculate the first n derivatives of this expression and substitute them into the differential equation (5.33), we shall obtain one condition to be satisfied by the n functions. We shall impose $n - 1$ other conditions *en route*. Differentiating once, we have

$$y' = (C_1u_1' + \cdots + C_nu_n') + (C_1'u_1 + \cdots + C_n'u_n).$$

We now impose the requirement

$$C_1'u_1 + \cdots + C_n'u_n = 0.$$

This simplifies the expression for the first derivative since it now becomes

$$y' = C_1u_1' + \cdots + C_nu_n'.$$

We have also obtained one condition to be satisfied by the functions C_i. Differentiating again, we have

$$y'' = (C_1u_1'' + \cdots + C_nu_n'') + (C_1'u_1' + \cdots + C_n'u_n').$$

This time we require that

$$C_1'u_1' + \cdots + C_n'u_n' = 0.$$

The second derivative simplifies to

$$y'' = C_1 u_1'' + \cdots + C_n u_n''.$$

We have now imposed two conditions on the functions C_i. Continuing in this way through $n - 1$ differentiations, we impose the $n - 1$ conditions

$$C_1' u_1^{(k)} + C_2' u_2^{(k)} + \cdots + C_n' u_n^{(k)} = 0, \qquad 0 \le k \le n - 2,$$

on the functions C_i, and the derivatives of y are given by the formula

$$y^{(k)} = C_1 u_1^{(k)} + C_2 u_2^{(k)} + \cdots + C_n u_n^{(k)}, \qquad 0 \le k \le n - 1. \tag{5.35}$$

Then

$$y^{(n)} = C_1 u_1^{(n)} + \cdots + C_n u_n^{(n)} + C_1' u_1^{(n-1)} + \cdots + C_n' u_n^{(n-1)}. \tag{5.36}$$

To obtain a final nth condition on the functions C_i, we substitute the expressions (5.35) and (5.36) into the differential equation (5.33). We find that

$$a_0[C_1 u_1^{(n)} + \cdots + C_n u_n^{(n)} + C_1' u_1^{(n-1)} + \cdots + C_n' u_n^{(n-1)}]$$
$$+ a_1[C_1 u_1^{(n-1)} + \cdots + C_n u_n^{(n-1)}] + \cdots + a_n[C_1 u_1 + \cdots + C_n u_n] = F,$$

or, upon regrouping terms,

$$a_0[C_1' u_1^{(n-1)} + \cdots + C_n' u_n^{(n-1)}] + C_1[a_0 u_1^{(n)} + \cdots + a_n u_1]$$
$$+ \cdots + C_n[a_0 u_n^{(n)} + \cdots + a_n u_n] = F.$$

This reduces to

$$C_1' u_1^{(n-1)} + \cdots + C_n' u_n^{(n-1)} = F/a_0$$

since the functions u_i are solutions of the homogeneous equation (5.32). We have now obtained the n conditions

$$C_1' u_1 + \cdots + C_n' u_n = 0,$$
$$C_1' u_1' + \cdots + C_n' u_n' = 0,$$
$$C_1' u_1'' + \cdots + C_n' u_n'' = 0,$$
$$\cdots\cdots\cdots\cdots\cdots\cdots\cdots\cdots \tag{5.37}$$
$$C_1' u_1^{(n-2)} + \cdots + C_n' u_n^{(n-2)} = 0,$$
$$C_1' u_1^{(n-1)} + \cdots + C_n' u_n^{(n-1)} = F/a_0$$

for $C_1', C_2', \ldots, C_n'$. The determinant of this system is the Wronskian of the functions $u_1, u_2, \ldots, u_n$. As mentioned in Theorem 5.2, this Wronskian is never zero on an interval where a_0 does not vanish. Consequently the system (5.37) possesses a unique solution. The solution functions are continuous, for reasons explained in Section 2.11. If we solve for the quantities C_i', the functions C_i can be found by integration.

In arriving at the conditions (5.37), we proceeded under the assumption that the Eq. (5.33) had a solution of the form (5.34). To make our argument rigorous, let the functions C_i be functions that satisfy the conditions (5.37). Then the corresponding function (5.34) has derivatives given by the formulas (5.35) and (5.36). It can now be verified that the function (5.34) is indeed a solution of the differential equation. We sum up our results as follows.

Theorem 5.9   Let the functions a_i and F be continuous on an interval I, with a_0 never zero on I. Let $u_1, u_2, \ldots, u_n$ be linearly independent solutions of the homogeneous equation (5.32) on I. If the functions C_i are such that their derivatives C_i' satisfy the system of equations (5.37), then the function

$$y_p = C_1 u_1 + C_2 u_2 + \cdots + C_n u_n$$

is a solution of the nonhomogeneous equation (5.33).

Example 1
$$y'' - 3y' + 2y = -\frac{e^{2x}}{e^x + 1}.$$

The functions u_1 and u_2, where

$$u_1(x) = e^x, \qquad u_2(x) = e^{2x},$$

are linearly independent solutions of the associated homogeneous equation

$$y'' - 3y' + 2y = 0.$$

We seek a solution of the nonhomogeneous equation of the form

$$y_p = C_1 e^x + C_2 e^{2x}.$$

The conditions (5.37) in this case are

$$C_1' e^x + C_2' e^{2x} = 0,$$

$$C_1' e^x + 2C_2' e^{2x} = -\frac{e^{2x}}{e^x + 1}$$

or, upon dividing through in each equation by e^x,

$$C_1' + C_2' e^x = 0,$$

$$C_1' + 2C_2' e^x = -\frac{e^x}{e^x + 1}.$$

Solving, we find that

$$C_1' = \frac{e^x}{e^x + 1}, \qquad C_2' = \frac{-1}{e^x + 1} = -\frac{e^{-x}}{1 + e^{-x}}.$$

Then we may take

$$C_1 = \ln(e^x + 1), \qquad C_2 = \ln(1 + e^{-x}).$$

A particular solution of the differential equation is

$$y_p = e^x \ln(e^x + 1) + e^{2x} \ln(1 + e^{-x}).$$

The general solution is

$$y = c_1 e^x + c_2 e^{2x} + e^x \ln(e^x + 1) + e^{2x} \ln(1 + e^{-x}).$$

Example 2

$$x^2 y'' + xy' - y = -2x^2 e^x, \quad x > 0.$$

Notice that if we make the change of variable $x = e^t$, the equation becomes

$$[D(D - 1) + D - 1]Y = -2e^{2t} \exp(e^t).$$

The nonhomogeneous term is one for which the method of undetermined coefficients does not apply; therefore, we must use the method of variation of parameters. The homogeneous equation

$$x^2 y'' + xy' - y = 0$$

possesses the solutions $u_1(x) = x$ and $u_2(x) = x^{-1}$. We seek a solution of the original equation that is of the form

$$y_p = C_1(x)x + C_2(x)x^{-1}.$$

Conditions (5.37) become

$$C_1' x + C_2' x^{-1} = 0,$$
$$C_1' - C_2' x^{-2} = -2e^x.$$

(Here $a_0(x) = x^2$, therefore $F(x)/a_0(x) = -2e^x$.) We find that

$$C_1' = -e^x, \qquad C_2' = x^2 e^x,$$

so we may take

$$C_1 = -e^x, \qquad C_2 = e^x(x^2 - 2x + 2).$$

Our general solution is

$$y = c_1 x + c_2 x^{-1} - xe^x + x^{-1}e^x(x^2 - 2x + 2)$$

or

$$y = c_1 x + c_2 x^{-1} + 2(x^{-1} - 1)e^x.$$

The method of variation of parameters requires that the associated homogeneous equation can be solved. Thus far we are able to do this when the homogeneous equation

(a)   has constant coefficients,

(b)   is of the Cauchy-Euler type, or

(c)   is a second-order equation with dependent or independent variable absent (Section 1.11).

One other situation might be mentioned here. If one nontrivial solution of the homogeneous equation is known, the problem of solving the equation can be reduced to that of solving an equation whose order is one less. If the original equation is of second order, we arrive at a first-order linear equation that can be treated by the method of Section 1.4.
    Suppose that the function u_1 is a nontrivial solution of the equation

$$a_0(x)y'' + a_1(x)y' + a_2(x)y = 0,$$

and that we wish to solve the equation

$$a_0(x)y'' + a_1(x)y' + a_2(x)y = F(x), \tag{5.38}$$

where F may or may not be the zero function. We introduce a new dependent variable v, where $y = u_1(x)v$. Then

$$y' = u_1 v' + u_1' v, \qquad y'' = u_1 v'' + 2u_1' v' + u_1'' v.$$

Substituting into Eq. (5.38), we have

$$a_0(u_1 v'' + 2u_1' v' + u_1'' v) + a_1(u_1 v' + u_1' v) + a_2 u_1 v = F$$

or

$$a_0 u_1 v'' + (2a_0 u_1' + a_1 u_1)v' + (a_0 u_1'' + a_1 u_1' + a_2 u_1)v = F.$$

Here the coefficient of v is zero since u_1 is a solution of the homogeneous equation. The equation

$$a_0 u_1 v'' + (2a_0 u_1' + a_1 u_1)v' = F$$

has the dependent variable missing. Setting $w = v'$, we obtain the first-order equation

$$a_0 u_1 w' + (2a_0 u_1' + a_1 u_1)w = F.$$

We solve this equation for w, find v by integration, and then multiply v by u_1 to obtain the solutions of the original equation (5.38).

Example 3
$$x^2(x + 1)y'' - 2xy' + 2y = 0.$$

We observe that a solution is $u_1(x) = x$. Setting $y = vx$, we have

$$x^2(x + 1)(xv'' + 2v') - 2x(xv' + v) + 2xv = 0$$

or

$$(x + 1)v'' + 2v' = 0.$$

This first-order equation for v' has the solutions

$$v' = -\frac{c_1}{(x + 1)^2};$$

therefore

$$v = \frac{c_1}{x + 1} + c_2.$$

The general solution of the original equation is

$$y = vx = c_1 \frac{x}{x + 1} + c_2 x.$$

Example 4
$$xy'' + 2(1 - x)y' + (x - 2)y = 2e^x.$$

It is easy to verify that $u_1(x) = e^x$ is a solution of the associated homogeneous equation. Setting $y = ve^x$, we find that

$$[x(v'' + 2v' + v) + (2 - 2x)(v' + v) + (x - 2)v]e^x = 2e^x$$

or

$$xv'' + 2v' = 2.$$

Then

$$v' = -\frac{c_1}{x^2} + 1$$

and

$$v = \frac{c_1}{x} + c_2 + x.$$

The general solution of the original equation is

$$y = [(c_1/x) + c_2 + x]e^x.$$

——————————— Exercises for Section 5.8 ———————————

In Exercises 1–10, find the general solution of the differential equation.

1. $y'' - y = \dfrac{2}{e^x + 1}$

2. $y'' - 2y' + y = \dfrac{e^x}{x}$

3. $y'' + 2y' + y = 4e^{-x} \ln x$

4. $9y'' - y = \dfrac{e^{x/3}}{x^2}$

5. $4y'' + y = \sec \dfrac{x}{2}$

6. $y'' + y = \tan x \sec x$

7. $y'' + 2y' + 2y = 2e^{-x} \tan^2 x$

8. $x^2 y'' - 2xy' + 2y = x^3 e^x$

9. $(D - 1)^3 y = 2\dfrac{e^x}{x^2}$

10. $(D - 1)(D + 1)(D + 2)y = \dfrac{6}{e^x + 1}$

11. Let F be defined and continuous on the interval $[0, \infty)$.
   (a) Show that the general solution of the equation

$$y'' + k^2 y = F(x)$$

   may be written as

$$y = A \sin(kx + \alpha) - \frac{1}{k} \int_{x_0}^{x} \sin k(t - x) F(t)\, dt,$$

where A and α are arbitrary constants and x_0 is any nonnegative number.
   (b) Suppose that there exist numbers M, x_1, and a, $a > 1$, such that

$$|F(x)| \le Mx^{-a}, \qquad x \ge x_1.$$

Show that every solution of the equation in part (a) is bounded on the interval $[0, \infty)$.
   (c) If $\int_0^\infty |F(x)|\, dx$ converges, show that every solution of the equation in part (a) is bounded on $[0, \infty)$.

12. Let a and b be positive real numbers with $a \ne b$. Let the function F be defined and continuous on $[0, \infty)$.
   (a) Show that the general solution of the equation

$$(D + a)(D + b)y = F(x)$$

   may be written as

$$y = c_1 e^{-ax} + c_2 e^{-bx}$$
$$+ \frac{1}{b - a} \int_{x_0}^{x} [e^{-a(x-t)} - e^{-b(x-t)}] F(t)\, dt,$$

where x_0 is any nonnegative number.
   (b) If F is bounded (that is, $|F(x)| \le M$ for some number M and $x \ge 0$), show that every solution of the equation in part (a) is bounded on $[0, \infty)$.
   (c) If $\int_0^\infty |F(x)|\, dx$ converges, show that every solution of the equation in part (a) is bounded.
   (d) If $\lim_{x \to \infty} F(x) = L$, show that every solution of the equation in part (a) tends to the limit $L/(ab)$ as x becomes infinite.

In Exercises 13–16, find the general solution of the differential equation, given one solution of the homogeneous equation.

13. $x^3y'' + xy' - y = 0,\ y = x$

14. $xy'' + (1 - 2x)y' + (x - 1)y = 0,\ y = e^x$

15. $2xy'' + (1 - 4x)y' + (2x - 1)y = e^x,\ y = e^x$

16. $x^2(x + 2)y'' + 2xy' - 2y = (x + 2)^2,\ y = x$

17. Suppose that u_1 and u_2 are linearly independent solutions of the third-order equation

$$a_0y''' + a_1y'' + a_2y' + a_3y = 0.$$

Show that the change of variable $y = u_1(x)v$ leads to a second-order equation for $v' = w$. Find a solution of this equation, in terms of u_2, and use it to reduce the equation to one of first order.

5.9 SIMPLE HARMONIC MOTION

Suppose that a spring hangs vertically from a support, as in Fig. 5.1a. Let L denote the length of the spring when it is at rest. When the spring is stretched or compressed a distance s by a force F applied at the ends, it is found by experiment that the magnitude of the force is approximately proportional to the distance s, at leeast when s is not too large. Thus

$$F = ks,$$

where k is a positive constant of proportionality known as the *spring constant*. For example, if a force of 30 lb is required to stretch a spring 2 in, then

$$30 = 2k;$$

therefore $k = 15$ lb/in.

When a body of mass m, and weight mg, is attached to the free end of the spring, the body will remain at rest in a position such that the spring

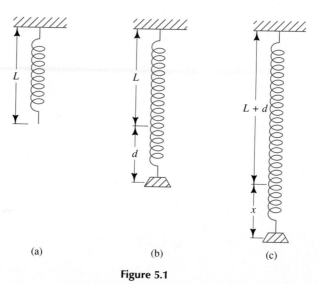

(a)   (b)   (c)

Figure 5.1

has length $L + d$ (Fig. 5.1b), where $mg = kd$. The downward force mg due to gravity is balanced by the restoring force kd of the spring. Let us denote by x the directed distance downward of the center of mass of the body from its position of rest, or equilibrium (Fig. 5.1c). Then the equation of motion $m\ddot{x} = F$ becomes

$$m\ddot{x} = mg - k(x + d)$$

or

$$m\ddot{x} + kx = 0. \qquad (5.39)$$

Example 1   A spring is stretched 2 in by a force of 16 lb. An object weighing 4 lb is attached to the end, is pulled down 3 in below the equilibrium position, and released. Find the subsequent motion.

We shall convert all distance measurements to feet. Then $w = mg = 32m$ so $m = w/32 = 4/32 = 1/8$ slug. From the relation $F = ks$ we find $16 = k/6$ or $k = 96$ lb/ft. Eq. (5.39) becomes

$$\frac{1}{8}\ddot{x} + 96x = 0 \qquad \text{or} \qquad \ddot{x} + 768x = 0.$$

The general solution of this equation is

$$x = c_1 \cos 16\sqrt{3}\,t + c_2 \sin 16\sqrt{3}\,t.$$

Since the initial conditions are $x(0) = 1/4$ ft, $\dot{x}(0) = 0$, we have $c_1 = 1/4$ and $c_2 = 0$. Hence the motion is described by the formula

$$x = \frac{1}{4} \cos 16\sqrt{3}\,t,$$

where t is in seconds and x in feet.

We now return to Eq. (5.39) and consider the general case. The general solution of this equation is

$$x = c_1 \cos \omega t + c_2 \sin \omega t,$$

where

$$\omega = (k/m)^{1/2}.$$

The constant ω is called the *angular frequency* of the motion.

If the body is held in the position $x = x_0$ and released from rest at time $t = 0$, the initial conditions are

$$x(0) = x_0, \qquad \dot{x}(0) = 0.$$

Using these conditions to determine the constants c_1 and c_2 in the formula for the general solution, we find that

$$x = x_0 \cos \omega t. \tag{5.40}$$

Thus the body oscillates about the equilibrium position $x = 0$ between the points $x = \pm|x_0|$ without ever coming to rest.

If the body is struck sharply when it is in the equilibrium position, giving it a velocity v_0, the initial conditions become

$$x(0) = 0, \qquad \dot{x}(0) = v_0.$$

This time we find that

$$x = \frac{v_0}{\omega} \sin \omega t. \tag{5.41}$$

In the more general case where

$$x(0) = x_0, \qquad \dot{x}(0) = v_0,$$

the solution is the sum of the solutions (5.40) and (5.41). We have

$$x = x_0 \cos \omega t + \frac{v_0}{\omega} \sin \omega t. \tag{5.42}$$

Let us write

$$A = [x_0^2 + (v_0/\omega)^2]^{1/2}.$$

Then there is an angle θ_1 such that

$$\cos \theta_1 = x_0/A, \qquad \sin \theta_1 = -v_0/(\omega A),$$

and formula (5.42) can be written

$$x = A \cos(\omega t + \theta_1).$$

If we put

$$\sin \theta_2 = x_0/A, \qquad \cos \theta_2 = v_0/(\omega A),$$

it becomes

$$x = A \sin(\omega t + \theta_2).$$

Straight-line motion that can be described by a function of the form

$$x = A \cos(\omega t + \theta) \qquad \text{or} \qquad x = A \sin(\omega t + \theta)$$

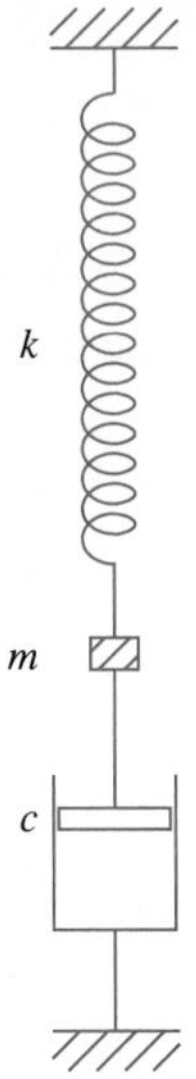

is called *simple harmonic motion.* The number $|A|$ is called the *amplitude* of the motion. Notice that x fluctuates between $-|A|$ and $|A|$ periodically. The period P of the motion is given by the formula $P = 2\pi/\omega$. This is the time required for the body to move through one cycle. The frequency f is the number of cycles per unit time. Thus $f = 1/P = \omega/(2\pi)$.

The presence of a damping force equal to c times the velocity may be indicated by means of a *dashpot*, as shown in Fig. 5.2. In this case, the body is said to exhibit *damped harmonic motion.* The equation of motion becomes

$$m\ddot{x} + c\dot{x} + kx = 0. \tag{5.43}$$

The auxiliary equation is

$$mr^2 + cr + k = 0,$$

with roots

$$r = \frac{1}{2m}\left[-c \pm (c^2 - 4mk)^{1/2}\right].$$

Figure 5.2

We consider separately the following cases:

(1) $c^2 < 4mk$ (two complex roots)

(2) $c^2 > 4mk$ (two distinct real roots)

(3) $c^2 = 4mk$ (equal real roots).

These cases are called the *underdamped* (or *oscillatory*), *overdamped*, and *critically damped* cases, respectively.

For case 1, we may write the general solution of the equation of motion (5.43) as

$$x = c_1 e^{-\alpha t}\cos \omega t + c_2 e^{-\alpha t}\sin \omega t,$$

where

$$\alpha = \frac{c}{2m}, \qquad \omega = \frac{1}{2m}(4mk - c^2)^{1/2}.$$

The solution that satisfies the initial conditions

$$x(0) = x_0, \qquad \dot{x}(0) = 0$$

is

$$x = x_0 e^{-\alpha t}\left(\cos \omega t + \frac{\alpha}{\omega}\sin \omega t\right).$$

This may be written

$$x = Ae^{-\alpha t}\cos(\omega t + \theta_2),$$

where

$$A = x_0\left[1 + \left(\frac{\alpha}{\omega}\right)^2\right]^{1/2}, \qquad \theta_1 = \tan^{-1}\frac{\omega}{\alpha},$$

or

$$x = Ae^{-\alpha t}\sin(\omega t + \theta_1),$$

where

$$\theta_2 = -\tan^{-1}\frac{\alpha}{\omega}.$$

In this case the body still oscillates back and forth across the equilibrium position, but the amplitude decreases exponentially with time. The situation is illustrated in Fig. 5.3.

In the overdamped case, the roots of the auxiliary equation are real, negative, and distinct. The solution has the form

$$x = c_1 e^{-\alpha t} + c_2 e^{-\beta t},$$

where α and β are positive and distinct. In the critically damped case, the auxiliary equation has equal negative roots, so the solution has the form

$$x = (c_1 + c_2 t)e^{-\alpha t},$$

where α is real and positive. From these last two formulas, it may be determined that the displacement $x(t)$ approaches 0 as t increases, and that

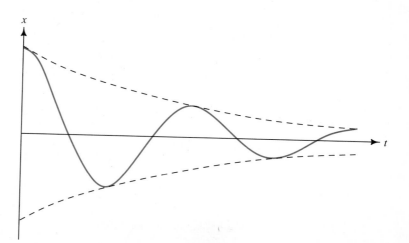

Figure 5.3

the equilibrium position is reached at most once, so there is no oscillation (see Exercises 8 and 9).

If an external force $F(t)$ is applied to the body on the spring, the equation becomes

$$m\ddot{x} + c\dot{x} + kx = F(t). \tag{5.44}$$

We shall consider here the case where F is a periodic force of the form

$$F(t) = A \sin \omega_1 t$$

and where damping is absent $(c = 0)$. Then Eq. (5.44) becomes

$$m\ddot{x} + kx = A \sin \omega_1 t. \tag{5.45}$$

The complementary function is

$$x_c = c_1 \cos \omega t + c_2 \sin \omega t,$$

where $\omega = (k/m)^{1/2}$ is the natural angular frequency. A particular solution is of the form

$$x_p = B \cos \omega_1 t + C \sin \omega_1 t,$$

provided $\omega_1 \neq \omega$. It turns out that

$$x_p = \frac{A}{m(\omega^2 - \omega_1^2)} \sin \omega_1 t.$$

The general solution of Eq. (5.44) is

$$x = c_1 \cos \omega t + c_2 \sin \omega t + \frac{A}{m(\omega^2 - \omega_1^2)} \sin \omega_1 t.$$

However, if $\omega_1 = \omega$, a particular solution is of the form

$$x_p = t(B \cos \omega t + C \sin \omega t),$$

and calculation shows that

$$x_p = -\frac{A}{2\omega m} t \cos \omega t.$$

The general solution of the equation is

$$x = c_1 \cos \omega t + c_2 \sin \omega t - \frac{A}{2\omega m} t \cos \omega t. \tag{5.46}$$

In this case, $\omega_1 = \omega$, the magnitude of the oscillations increases indefinitely because of the presence of the term $t \cos \omega t$. This phenomenon is called *resonance*. Actually, when the oscillations become sufficiently large the law $|F| = ks$ does not hold and our mathematical model no longer applies.

Example 2 Consider a rectangular box attached to a spring, and constrained (by frictionless walls) to move in a vertical direction. The situation is illustrated in Fig. 5.4. Inside the box is a wheel of radius r. The mass of the box and wheel is m. There is a weight of mass m_1 on the rim of the wheel. The wheel rotates with constant angular velocity ω_1, so that the x-coordinate of the weight at time t is $x + r \cos \omega_1 t$. Then the center of mass for the box and its contents has x-coordinate

$$\bar{x} = \frac{mx + m_1(x + r \cos \omega_1 t)}{m + m_1}.$$

The equation of motion is therefore

$$(m + m_1)\frac{d^2\bar{x}}{dt^2} = -kx.$$

This equation simplifies to

$$\frac{d^2x}{dt^2} + \omega^2 x = \frac{m_1}{m + m_1}\omega_1^2 r \cos \omega_1 t, \qquad \omega^2 = \frac{k}{m + m_1}.$$

In the case of resonance $(\omega_1 = \omega)$ a particular solution is found to be

$$x_p = \frac{1}{2}\frac{m_1}{m + m_1}\omega r t \sin \omega t.$$

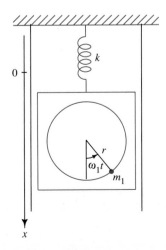

Figure 5.4

1. A weight of 4 lb is attached to a spring and stretches it 6 in. The weight is then pulled down 8 in from its equilibrium position and released from rest. Find a formula for the position of the weight at any time. Also find the period and frequency of the motion.

2. A weight of 8 lb stretches a spring 16 in. The weight is struck sharply, giving it an initial velocity of 18 in/sec downward from its equilibrium position. Find a formula for the position at any time. Also find the period and frequency of the motion.

3. A 5 lb weight is attached to a spring and set in motion. If it oscillates 4 times per second, what is the spring constant?

4. A mass of 30 gm stretches a spring 2 cm. An object of unknown mass is attached to the spring and set in motion. If the period is 0.4 sec, what is the mass of the object? ($g = 980$ cm/sec^2.)

5. A weight of 4 lb stretches a spring 6 in. The weight is pulled down 8 in from its equilibrium position and released from rest. There is a damping force of $2.5v$ lb when v is in ft/sec. Find a formula for the position at time t. Is the motion oscillatory, overdamped, or critically damped?

6. A force of 1300 dynes stretches a spring 4 cm. An object of mass 5 gm is attached to the spring. While at rest in its equilibrium position, the object is struck sharply, giving it a velocity of 24 cm/sec downward. There is a damping force of $10v$ dynes when v is in cm/sec. Find a formula for the position at time t, and find the number of oscillations per second. ($g = 980$ cm/sec^2.)

7. An object of mass 5 kg stretches a spring 2 m. The object is set in motion and the period observed to be 4 sec. If there is a damping force of cv newtons when v is in m/sec, what must be the value of the constant c? ($g = 9.8$ m/sec^2.)

8. Find the general solution of the equation of motion (5.43) in the overdamped case $c^2 > 4mk$. Show that every solution tends to zero as t becomes infinite, and show that the equilibrium position is reached at most once.

9. Find the general solution of the equation of motion (5.43) in the critically damped case $c^2 = 4mk$. Show that every solution tends to zero as t becomes infinite, and show that the equilibrium position is reached at most once.

10. The equation of motion of a particle that moves in a straight line is

$$m\ddot{x} + 2\dot{x} + x = 0.$$

For what values of m is the motion oscillatory?

11. The displacement $x(t)$ of a particle that moves in a straight line satisfies the equation of motion

$$\ddot{x} + c\dot{x} + 4x = 0.$$

For what values of the constant c is the motion oscillatory?

12. According to a physical principle, the buoyant force acting on a floating object is equal to the weight of the water displaced. A bobber in the shape of a right circular cylinder has mass m and radius r. It floats on water of density k with its axis perpendicular to the surface of the water. If the bobber is pulled down and released it bobs up and down. Show that its motion is periodic and find the period. Suggestion: find the differential equation for $x(t)$, the downward displacement from the equilibrium position at time t.

13. The position $x(t)$ of a body on a spring satisfies the differential equation

$$\ddot{x} + c\dot{x} + 4x = F(t),$$

where F is an applied force.

(a) If $c = 0$ and $F(t) = 5 \sin at$, for what values of a, if any, does resonance occur?

(b) If $c = 5$ and $F(t) = 8$, find $\lim_{t \to \infty} x(t)$.

14. The equation of motion of a particle is

$$\ddot{x} + 4x = F,$$

where $F(t) = 4 \cos 2t + 5 \cos 3t$. Show that F does not have period $\omega = \pi$, but that resonance still occurs.

15. If the applied force in Eq. (5.44) is constant, $F(t) = F_0$, find the limiting position of the body on the spring.

16. If an external force $F(t)$ is applied to the body on the spring, its equation of motion becomes

$$m\ddot{x} + c\dot{x} + kx = F(t).$$

Assume that $c > 0$ and that F is a periodic function of the form

$$F(t) = A \sin \omega_1 t.$$

(a) When t is large, show that every solution is approximately equal to

$$x_p(t) = B \sin(\omega_1 t - \theta),$$

where

$$B = \frac{A}{D}, \quad D = [(k - m\omega_1^2)^2 + c^2\omega_1^2]^{1/2},$$

$$\theta = \cos^{-1} \frac{k - m\omega_1^2}{D}.$$

This is called the *steady-state solution* of the equation.

(b) Show that when $c^2 \geq 2mk$, the amplitude $|B|$, considered as a function of ω_1, is a decreasing function.

(c) Show that when $c^2 < 2mk$ the amplitude $|B|$, considered as a function of ω_1, is a maximum when

$$\omega_1 = \left(\frac{k}{m} - \frac{c^2}{2m^2} \right)^{1/2}.$$

Show further that when $c^2 < 2mk$ the homogeneous equation has oscillatory solutions with angular frequency

$$\omega = \left(\frac{k}{m} - \frac{c^2}{4m^2} \right)^{1/2},$$

and that $|B|$ is largest when $\omega_1^2 = \omega^2 - \alpha^2$, where $\alpha = c/(2m)$. Thus $|B|$ does *not* have its largest value when ω_1 is equal to the natural angular frequency ω.

5.10 ELECTRIC CIRCUITS

Let us consider an electric circuit in which a resistance, capacitance, and inductance are connected in series with a voltage source, as in Fig. 5.5. When the switch is closed at $t = 0$, a current will flow in the loop. We denote the value of the current at time t by $I(t)$. The arrow in the figure

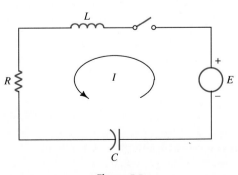

Figure 5.5

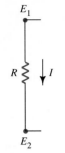

Figure 5.6

gives the loop a direction. We understand that I is positive when the flow is in the direction of the arrow and negative when in the opposite direction. In Fig. 5.6 we have isolated one circuit element, the resistance. The voltages at the terminals are denoted by E_1 and E_2. When the current is positive, in the direction of the arrow, $E_1 > E_2$ and the *voltage drop* across the element is $E_1 - E_2$, a positive quantity. Thus a positive current flows in the direction of decreasing voltage. This is a matter of convention, and the term "current" as used here should not be identified with a flow of electrons. Actually it may help the reader to think of the current as a flow of positively charged particles.

We shall use the following system of units: *amperes* for the current I, *volts* for the voltage, *ohms* for the resistance R, *henrys* for the inductance L, *farads* for the capacitance C, *coulombs* for the charge on the capacitance, and seconds for the time t. Using this system of units, the voltage drop across the resistance is RI and that across the inductance is $L\,dI/dt$. The voltage drop across the capacitance is Q/C, where Q is the charge on the capacitance. The charge and current are related by the equations

$$I = \frac{dQ}{dt}, \qquad Q(t) = \int_0^t I(s)\,ds + Q_0, \tag{5.47}$$

where Q_0 is the charge on the capacitance at $t = 0$.

According to one of Kirchhoff's laws the sum of the voltage drops around the loop must be equal to the applied voltage. Therefore the equality

$$L \frac{dI}{dt} + RI + \frac{1}{C} Q = E(t) \tag{5.48}$$

must hold for $t \geq 0$, provided that the sign of $E(t)$ is chosen in accordance with the $+$ and $-$ signs in Fig. 5.5. Upon differentiating with respect to t, and using relations (5.47), we arrive at the second-order differential equation

$$L \frac{d^2 I}{dt^2} + R \frac{dI}{dt} + \frac{1}{C} I = E'(t) \tag{5.49}$$

for I. We notice the resemblance of this equation to that for a damped harmonic oscillator,

$$m \frac{d^2 x}{dt^2} + c \frac{dx}{dt} + kx = F(t).$$

In particular, the term $L\,d^2 I/dt^2$ in the circuit equation corresponds to the inertia term $m\,d^2 x/dt^2$ in this equation. This means that the current passing through the inductance must be the same immediately before and after a sudden change[5] or jump in the voltage drop across it. Since the current

---

[5] Unless the change is infinite, as can happen in some idealized situations.

was zero before the switch was closed we must have

$$I(0) = 0.$$

Using this fact we can find the initial value of dI/dt from Eq. (5.48). Assuming that the initial charge on the capacitance is zero, we have

$$LI'(0) + RI(0) = E(0)$$

or

$$I'(0) = \frac{E(0)}{L}.$$

In many applications, the applied voltage is approximately constant, as in the case of a battery, or sinusoidal, as in the case of an alternating current generator. Let us consider the cases where the applied voltage has the constant value E_0. Then $E'(t) = 0$ and Eq. (5.49) becomes

$$L\frac{d^2I}{dt^2} + R\frac{dI}{dt} + \frac{1}{C}I = 0.$$

The initial conditions are

$$I(0) = 0, \qquad I'(0) = \frac{E_0}{L}.$$

The form of the solution depends on whether the quantity $R^2 - 4L/C$ is positive, negative, or zero. We consider here only the case where $R^2 - 4L/C > 0$. The solution of the initial value problem, as found by routine methods, is

$$I(t) = \frac{E_0}{(R^2 - 4L/C)^{1/2}}(e^{-\alpha t} - e^{-\beta t}),$$

where

$$\alpha = \frac{1}{2L}[R - (R^2 - 4L/C)^{1/2}], \qquad \beta = \frac{1}{2L}[R + (R^2 - 4L/C)^{1/2}].$$

Since α and β are both positive, $I(t)$ tends to zero as t becomes infinite.

We next consider the circuit of Fig. 5.7, in which the applied voltage is sinusoidal, of the form $E(t) = E_0 \cos \omega t$. No inductance is present. Kirchhoff's law leads to the equation

$$RI + \frac{1}{C}Q = E_0 \cos \omega t \qquad (5.50)$$

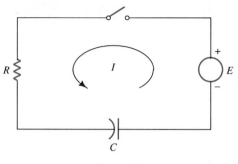

Figure 5.7

or

$$R\frac{dI}{dt} + \frac{1}{C}I = -\omega E_0 \sin \omega t. \qquad (5.51)$$

From Eq. (5.50) we obtain the initial condition

$$I(0) = \frac{E_0}{R},$$

again assuming that the initial charge on the capacitance is zero. The solution of the first-order equation (5.51) that satisfies the initial condition is found to be

$$I(t) = I_1(t) + I_2(t),$$

where

$$I_1(t) = \frac{E_0}{R}\left[1 + \frac{\omega^2}{(RC)^{-2} + \omega^2}\right]e^{-t/(RC)}$$

and

$$I_2(t) = \frac{E_0\omega}{R[(RC)^{-2} + \omega^2]}\left(\frac{1}{RC}\sin \omega t - \omega \cos \omega t\right).$$

The function I_1 dies out as t becomes infinite. It is called the *transient solution* of the initial-value problem. When t is large the solution I is very nearly equal to I_2; this is called the *steady-state* solution.

———————————————— Exercises for Section 5.10 ————————————————

1. A resistance and an inductance are connected in series with a battery of constant voltage E_0, as shown in Fig. 5.8. The switch is closed at $t = 0$. Assuming $L = 0.2$ henrys, $R = 2$ ohms, and $E_0 = 4$ volts, find

(a) a formula for the current as a function of t.

(b) the voltage drop across the resistance and that across the inductance.

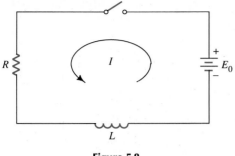

Figure 5.8

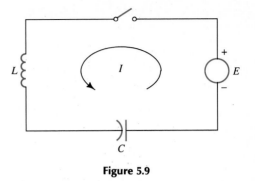

Figure 5.9

2. In the circuit of Fig. 5.5, assume that $L = 0.5$ henrys, $R = 2$ ohms, $C = 0.1$ farads, $E = 4$ volts, and $Q_0 = 0$. Find the current in the loop.

3. In the circuit of Fig. 5.5, assume that R, L, C, and Q_0 are as in Exercise 2 but that $E(t) = 2 \sin 4t$ volts. Find the steady-state and transient currents for the loop.

4. In the circuit of Fig. 5.9, suppose that $L = 0.25$ henrys, $C = 0.01$ farads, and $E(t) = 10 \cos \alpha t$ volts. Find a formula for the current at time t if
   (a) $\alpha \neq 20$ (b) $\alpha = 20$.

5. In the circuit of Fig. 5.10, a charge of $Q_0 = 3$ coulombs has been placed on the capacitance.

If $L = 0.2$ henrys and $C = 0.5$ farads, find a formula for the current at time t. The switch is closed at $t = 0$.

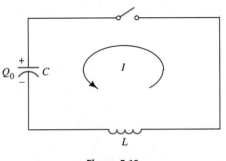

Figure 5.10

Additional Exercises for Chapter 5

In Exercises 1–10, find the general solution.

1. $y'' + 6y' + 10y = 0$

2. $y''' + 6y'' + 12y' + 8y = 0$

3. $x^2 y'' + 7xy' + 9y = 0$

4. $y'' + 3y' + 2y = 4e^{-x} + 6e^x$

5. $y'' - 4y' + 4y = 6 \cos 2x + 8x$

6. $y'' + 9y = 2 \sin 3x$

7. $y''' + 3y'' + 3y' + y = 4e^{-x}$

8. $y''' + y' = 6x$

9. $y'' + y = \csc x$

10. $y'' + 2y' + y = e^{-x} \sec^2 x$

11. $x^2 y'' - 2y = x^3 e^{-x}$

12. $(D^2 + 4D + 13)^2 y = 0$

13. $(D^4 + 3D^2 + 2)y = 0$

14. $(D^2 + 1)(D^2 + 4)y = \sec x$

15. A mass of 6 gm executes simple harmonic motion on the end of a spring. If the frequency is 4 cycles per second, what is the spring constant?

16. A weight of 2 lb stretches a spring 6 in. The weight is pulled down 24 in from the equilibrium position and released from rest.
    (a) If no damping is present, how long will it take for the object to reach the equilibrium position, and what is its velocity when it does so?
    (b) Suppose damping is present, such that the motion is critically damped. Does the weight reach the equilibrium position? If so, what is the time required?

17. An object of mass $m = 2$ gm is suspended from a spring with spring constant $k = 8$ dynes/cm. Damping is present, with $c = 4$ dyne-sec/cm. There is an external force $F(t) = 10 \cos \alpha t$. Initially, at $t = 0$, the displacement and velocity are zero. Find a formula for the steady-state displacement if
    (a) $\alpha = (k/m)^{1/2}$
    (b) $\alpha = \left(\dfrac{k}{m} - \dfrac{c^2}{4m^2}\right)^{1/2}$
    (c) $\alpha = \left(\dfrac{k}{m} - \dfrac{c^2}{2m^2}\right)^{1/2}$.

    What is the significance of these 3 values for α? (See Ex. 16, Section 5.9.)

18. An inductance of 0.5 henrys and a capacitance of 0.02 farads are connected in series with a 4 volt battery. There is zero charge on the capacitance initially. A switch is closed to complete the circuit at $t = 0$. Find the current.

19. A resistance of 4 ohms, an inductance of 0.4 henrys, and a capacitance of 0.02 farads are connected in series with a voltage source of $100 \sin(120\pi t)$ volts. Assume $I(0) = 0$ and $I'(0) = 0$.
    (a) Find the transient and steady-state currents.
    (b) If the capacitance can be varied, what should be its value for the amplitude of the steady-state current to be a maximum?

20. Suppose that F possesses derivatives of all orders, and that there is a finite set S of n functions such that F and each of its derivatives can be expressed as a linear combination of the functions of S. Show that F, F', F'', ..., $F^{(n)}$ are linearly dependent, and hence that F is a solution of a linear homogeneous differential equation with constant coefficients. Then show that F must be a linear combination of functions of the types (5.18).

21. Determine whether every solution approaches 0 as x approaches ∞.
    (a) $(D^3 + 3D^2 + 5D + 2)y = 0$
    (b) $(D^3 + 3D^2 + D + 8)y = 0$
    (c) $(D^4 + 4D^3 + D^2 + 3D + 1)y = 0$
    (d) $(D^4 + 2D^3 + 7D^2 + 3D + 5)y = 0$

22. Show that the change of dependent variable $y = f(x)v$ in the second-order linear equation
$$a_0(x)y'' + a_1(x)y' + a_2(x)y = 0$$
leads to a second-order equation for v. Show that if f is chosen appropriately, the equation for v has the form $v'' + b(x)v = 0$.

23. Show that a change of independent variable $t = g(x)$ in the second-order linear equation
$$a_0(x)\frac{d^2y}{dx^2} + a_1(x)\frac{dy}{dx} + a_2(x)y = 0$$
leads to a second-order linear equation. Show that g may be chosen so that the equation has one of the forms
$$\frac{d^2y}{dt^2} + c_1(t)\frac{dy}{dt} \pm y = 0.$$

24. Show that the change of dependent variable
$$y = e^{\int v\,dx}, \qquad v = y'/y$$
transforms a second-order linear homogeneous equation in y into a first-order nonlinear Riccati equation in v.

25. If y_1 is a solution of the equation $y'' + a(x)y = 0$, show that a second independent solution is
$$y = y_1(x)\int [y_1(x)]^{-2}\,dx.$$

CHAPTER 6

Systems of Differential Equations

6.1

INTRODUCTION

In this chapter we shall deal with sets of simultaneous equations that involve several unknown functions and their derivatives. Such a set of equations is called a *system of differential equations*. An example of a system is

$$\frac{d^2x_1}{dt^2} - 2\frac{dx_1}{dt} - \frac{dx_2}{dt} = -3e^{3t},$$

$$\frac{dx_2}{dt} - 6x_1 = 0. \tag{6.1}$$

A solution of this system consists of an ordered pair of functions (x_1, x_2) that satisfies both the equations. An example of a solution of the system is the pair of functions (f_1, f_2), where

$$f_1(t) = e^{3t}, \qquad f_2(t) = 2e^{3t}$$

for all t. To verify this assertion, we observe that

$$f''_1(t) - 2f'_1(t) - f'_2(t) = 9e^{3t} - 6e^{3t} - 6e^{3t} = -3e^{3t},$$

$$f'_2(t) - 6f_1(t) = 6e^{3t} - 6e^{3t} = 0.$$

We now present several examples of problems that give rise to systems of differential equations. More applications appear in Sections 6.12–6.14.

263

Example 1   Systems of differential equations commonly arise in problems of mechanics. For example, let $x_1(t)$, $x_2(t)$, and $x_3(t)$ denote the rectangular coordinates, at time t, of the center of mass of a moving object. Then, according to Newton's laws of motion we must have

$$m\mathbf{a} = \mathbf{F},$$

where m is the mass of the object, where

$$\mathbf{a} = \ddot{x}_1\mathbf{i} + \ddot{x}_2\mathbf{j} + \ddot{x}_3\mathbf{k}$$

is the acceleration, and where

$$\mathbf{F} = F_1\mathbf{i} + F_2\mathbf{j} + F_3\mathbf{k}$$

is the force. Equating corresponding components in the vector equation $m\mathbf{a} = \mathbf{F}$, we have

$$m\ddot{x}_1 = F_1, \qquad m\ddot{x}_2 = F_2, \qquad m\ddot{x}_3 = F_3.$$

In many instances the components F_1, F_2, and F_3 of the force are known functions of t, x_1, x_2, x_3, $\dot{x}_1$, $\dot{x}_2$, and $\dot{x}_3$. If this is the case, we have a system of three differential equations for x_1, x_2, and x_3.

Example 2   Suppose that two tanks each initially contain 100 gal of a solution of a chemical, with 20 lb of the chemical in the first tank and 10 lb in the second. At $t = 0$, clear water begins to flow into the first tank at the rate of 2 gal/min. The (stirred) mixture runs into the second tank at the same rate of 2 gal/min, and the (stirred) mixture in the second tank runs out at still the same rate. We wish to find formulas for the amounts of chemical in each tank at time t.

Let $x_1(t)$ and $x_2(t)$ denote the amounts in the first and second tanks, respectively. Then we must have

$$\frac{dx_1}{dt} = -\frac{2}{100}x_1,$$

$$\frac{dx_2}{dt} = \frac{2}{100}x_1 - \frac{2}{100}x_2,$$

(6.2)

since the rate of change of the amount of chemical in a tank must be equal to the rate at which it enters minus the rate at which it leaves. We also have

$$x_1(0) = 20, \qquad x_2(0) = 10.$$

(6.3)

Here the first equation of the system (6.2) involves only the one unknown

x_1. Solving this equation subject to the first of the conditions (6.3), we find

$$x_1(t) = 20e^{-t/50}. \tag{6.4}$$

Substituting for x_1 in the second equation of the system (6.2), we arrive at the equation

$$\frac{dx_2}{dt} + \frac{1}{50}x_2 = \frac{2}{5}e^{-t/50}$$

for x_2. The solution of this equation that satisfies the second of the conditions (6.3) is

$$x_2(t) = \left(\frac{2}{5}t + 10\right)e^{-t/50}. \tag{6.5}$$

Formulas (6.4) and (6.5) describe the desired solution of the system.

Example 3   Our final example is from economics. Suppose that the rate of change of the supply of a commodity is proportional to the difference between the demand and the supply. In the case of two commodities we have

$$\frac{dS_1}{dt} = h(D_1 - S_1), \qquad \frac{dS_2}{dt} = k(D_2 - S_2),$$

where h and k are positive constants. We assume that each commodity is used in the manufacture of the other. In addition, we assume that the first commodity is used in the manufacture of itself (for example, steel). Finally, we assume that there is no other significant contribution to the demand for either commodity. For our model we take

$$D_1 = aS_1 + bS_2, \qquad D_2 = cS_1,$$

where a, b, and c are positive constants. Thus we arrive at the system of differential equations

$$\frac{dS_1}{dt} = h(a-1)S_1 + hbS_2, \qquad \frac{dS_2}{dt} = kcS_1 - kS_2.$$

A technique for solving such a system will be developed in Section 6.3.

Exercises for Section 6.1

1. Verify that the ordered pair of functions (f_1, f_2), where

    $$f_1(t) = t^2, \qquad f_2(t) = 2t$$

    for all t, is a solution of the system

    $$x_1'' - x_1'x_2 + 4x_1 = 2,$$
    $$x_1 - x_2' + x_2^2 = 5t^2 - 2.$$

2. Verify that the ordered triple of functions (f_1, f_2, f_3), where

$$f_1(t) = e^t, \qquad f_2(t) = 2e^t, \qquad f_3(t) = -e^t,$$

for all t, constitutes a solution of the system

$$x_1' = x_2 + x_3, \qquad x_2' = x_1 - x_3,$$
$$x_3' = -x_2 - x_3.$$

Also find another solution by inspection.

3. Describe a procedure that could be used for solving a system of the form

$$F(t, x_1, x_1') = 0,$$
$$G(t, x_1, x_2, x_2') = 0,$$
$$H(t, x_1, x_2, x_3, x_3') = 0,$$

where t is the independent variable and x_1, x_2, and x_3 are the unknown functions. (Notice that the first equation involves only one unknown.)

4. Two tanks each contain 50 gal of a salt solution initially, with 10 lb of salt in the first tank and 20 lb in the second tank. A salt solution containing 2 lb of salt per gallon runs into the first tank at the rate of 1 gal/min. The mixture runs into the second tank at the rate of 1 gal/min, and the mixture in the second tank runs out at the same rate. Find the amount of salt in each tank after 30 min.

5. A radioactive substance A decays at a rate equal to k_1 times the amount of A present. Let k_2 be the proportion of A that goes into a second radioactive substance B, where B decays at a rate equal to k_3 times the amount of B present. Let $x_1(t)$ and $x_2(t)$ be the amounts of A and B present at time t, and let $x_1(0) = a$, $x_2(0) = b$.
   (a) Find a system of differential equations for x_1 and x_2.
   (b) Find formulas for x_1 and x_2. (Assume $k_1 \neq k_3$.)

6. Two bodies, of masses m_1 and m_2, are suspended from springs as shown in Fig. 6.1. The

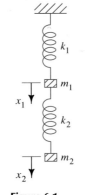

Figure 6.1

spring constants are k_1 and k_2. If x_1 and x_2 denote the directed distances downward of the bodies from their equilibrium positions, find a system of differential equations for x_1 and x_2.

7. Two bodies, with masses m_1 and m_2, respectively, move along a straight line as shown in Fig. 6.2. Let x_1 and x_2 denote the respective directed distances of the bodies from a fixed point on the line of motion. Assume that the bodies attract each other with a force equal to km_1m_2/r^2, where k is a positive constant and r is the distance between the bodies. Find a system of differential equations for x_1 and x_2.

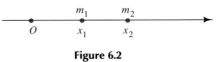

Figure 6.2

8. In the economic model of Example 3, suppose that $D_1 = aS_2 + b$ and $D_2 = cS_1 + d$, where a, b, c, and d are positive constants. Explain the economic meaning of these assumptions and write down the system of differential equations for S_1 and S_2.

6.2
FIRST-ORDER
SYSTEMS

A system of differential equations for unknown functions $x_1, x_2, \ldots, x_n$ that is of the form

$$\frac{dx_1}{dt} = f_1(t, x_1, x_2, \ldots, x_n),$$

$$\frac{dx_2}{dt} = f_2(t, x_1, x_2, \ldots, x_n),$$

$$\ldots\ldots\ldots\ldots\ldots\ldots\ldots\ldots\ldots$$

$$\frac{dx_n}{dt} = f_n(t, x_1, x_2, \ldots, x_n),$$

where the functions f_i are given, is called a *first-order system*. Notice that the number of equations is the same as the number of unknowns, and that no derivatives of order higher than 1 appear. A *solution* of the system is an ordered set of functions $(x_1, x_2, \ldots, x_n)$ that satisfies the system on some interval. The set of all solutions is called the *general solution*.

A set of auxiliary conditions of the form

$$x_1(t_0) = k_1, \qquad x_2(t_0) = k_2, \ldots, \qquad x_n(t_0) = k_n,$$

where the k_i are given numbers, is called a set of *initial conditions* for the system. The system together with the initial conditions is called an *initial value problem*. It can be shown that, with certain restrictions on the functions f_i, the initial value problem possesses a solution and that there is essentially only one solution. We shall usually be able to find explicit formulas for the solutions of the systems that we encounter in this chapter.

We notice that although the system (6.2) is a first-order system, the system (6.1) is not. In particular, this latter system involves a derivative of order higher than 1. However, most systems of importance that are not first-order can be rewritten as first-order systems, as we shall presently show. In much advanced work in differential equations, attention is centered on first-order systems. However, the results are still applicable to systems such as (6.1).

Let us show how the system (which is the same as (6.1))

$$x_1'' - 2x_1' - x_2' = -3e^{3t}, \qquad x_2' - 6x_1 = 0 \qquad (6.6)$$

can be replaced by a first-order system. We introduce new unknowns u_1, u_2, and u_3 by setting

$$u_1 = x_1, \qquad u_2 = x_2, \qquad u_3 = \frac{dx_1}{dt}.$$

Then we have immediately

$$\frac{du_1}{dt} = u_3,$$

and from the system (6.6) we also have

$$\frac{du_3}{dt} - 2u_3 - \frac{du_2}{dt} = -3e^{3t},$$

$$\frac{du_2}{dt} - 6u_1 = 0.$$

Notice that only *first* derivatives appear in these equations. Upon solving algebraically for du_2/dt and du_3/dt, we arrive at the first-order system

$$\frac{du_1}{dt} = u_3,$$

$$\frac{du_2}{dt} = 6u_1, \qquad\qquad (6.7)$$

$$\frac{du_3}{dt} = 6u_1 + 2u_3 - 3e^{3t}$$

for u_1, u_2, and u_3. Since initial conditions for the system (6.7) are of the form

$$u_1(t_0) = k_1, \qquad u_2(t_0) = k_2, \qquad u_3(t_0) = k_3,$$

appropriate initial conditions for the original system (6.6) are of the form

$$x_1(t_0) = k_1, \qquad x_2(t_0) = k_2, \qquad x_1'(t_0) = k_3.$$

More generally, suppose we have a system for two unknown functions x_1 and x_2 that is of the form

$$F[t, x_1, x_1', \ldots, x_1^{(m)}, x_2, x_2', \ldots, x_2^{(n)}] = 0,$$

$$G[t, x_1, x_1', \ldots, x_1^{(m)}, x_2, x_2', \ldots, x_2^{(n)}] = 0.$$

Here m and n are the orders of the highest order derivatives of x_1 and x_2 that appear in the two equations. Suppose that it is possible to solve algebraically for the highest derivatives $x_1^{(m)}$ and $x_2^{(n)}$, so that we can replace the system by the equivalent[1] system

$$x_1^{(m)} = f[t, x_1, x_1', \ldots, x_1^{(m-1)}, x_2, x_2', \ldots, x_2^{(n-1)}],$$

$$x_2^{(n)} = g[t, x_1, x_1', \ldots, x_1^{(m-1)}, x_2, x_2', \ldots, x_2^{(n-1)}].$$

---

[1] Two systems are said to be *equivalent* if they have the same solutions.

Let us set

$$u_1 = x_1, \qquad u_2 = x'_1, \qquad u_3 = x''_1, \ldots, \qquad u_m = x_1^{(m-1)},$$

$$u_{m+1} = x_2, \qquad u_{m+2} = x'_2, \qquad u_{m+3} = x''_2, \ldots, \qquad u_{m+n} = x_2^{(n-1)}.$$

Then we have

$$u'_1 = u_2,$$
$$u'_2 = u_3,$$
$$\cdots\cdots\cdots$$
$$u'_{m-1} = u_m,$$
$$u'_m = f(t, u_1, u_2, \ldots, u_m, u_{m+1}, \ldots, u_{m+n}),$$
$$u'_{m+1} = u_{m+2},$$
$$\cdots\cdots\cdots\cdots$$
$$u'_{m+n-1} = u_{m+n}.$$
$$u'_{m+n} = g(t, u_1, u_2, \ldots, u_m, u_{m+1}, \ldots, u_{m+n}),$$

This is a first-order system for the unknowns $u_1, u_2, \ldots, u_{m+n}$. Our discussion has been about a system with two equations and two unknowns, but the general case of k equations and k unknowns can be treated similarly.

An important special case is that of a single differential equation for one unknown function, of the form

$$x^{(n)} = F[t, x, x', \ldots, x^{(n-1)}]. \tag{6.8}$$

Setting

$$u_1 = x, \qquad u_2 = x', \qquad u_3 = x'', \ldots, \qquad u_n = x^{(n-1)}, \tag{6.9}$$

we have

$$u'_1 = u_2,$$
$$u'_2 = u_3,$$
$$\cdots\cdots\cdots \tag{6.10}$$
$$u'_{n-1} = u_n,$$
$$u'_n = F(t, u_1, u_2, \ldots, u_n).$$

The last equation of this first-order system comes from the differential equation (6.8). If x is a solution of Eq. (6.8) the relations (6.9) yield a solution $(u_1, u_2, \ldots, u_n)$ of the system (6.10). On the other hand, if a solution

$(u_1, u_2, \ldots, u_n)$ of the system (6.10) is known, the function u_1 is a solution of Eq. (6.8).

As an example, we consider the equation

$$\frac{d^3x}{dt^3} = 3tx - \frac{dx}{dt} + \frac{d^2x}{dt^2}.$$

Setting

$$u_1 = x, \qquad u_2 = x', \qquad u_3 = x'',$$

we have

$$u_1' = u_2,$$

$$u_2' = u_3,$$

$$u_3' = 3tu_1 - u_2 + u_3.$$

Notice that the initial conditions

$$x(t_0) = k_1, \qquad x'(t_0) = k_2, \qquad x''(t_0) = k_3$$

for the original equation correspond to the initial conditions

$$u_1(t_0) = k_1, \qquad u_2(t_0) = k_2, \qquad u_3(t_0) = k_3$$

for the first-order system.

We conclude this section with some remarks about appropriate restrictions on the functions f_i in the general first-order system. In Section 1.13 we discussed the single equation $dx/dt = f(t, x)$, with initial condition $x(t_0) = x_0$. If the function f and its partial derivative with respect to x are continuous in a region that contains the point (t_0, x_0), one and only one solution exists in some interval that contains t_0. In the case of a first-order system $x_i' = f_i(t, x_1, x_2, \ldots, x_n)$ with initial conditions $x_i(t_0) = k_i$, it can be shown that if the functions f_i and their partial derivatives with respect to $x_1, x_2, \ldots, x_n$ are continuous in a region of $n + 1$ dimensional space that contains the point $(t_0, k_1, k_2, \ldots, k_n)$, then one and only one solution exists in some interval that contains t_0.

Exercises for Section 6.2

In Exercises 1–8, rewrite the given system as a first-order system. Also, indicate what quantities must be specified in an appropriate set of auxiliary conditions for the given system at a point t_0.

1. $x_1' - x_2' - x_1 = \cos t$
    $x_2' - 3x_1 = e^t$

2. $x_1' - 2x_2' - x_2 = t^2$
    $x_1' - 3x_2' = 0$

3. $x_1' - x_2' = e^t$
    $x_2'' - x_1 - x_2 - x_2' = \sin t$

4. $x_1'' = 1$
    $x_2'' = 2$

5. $x_1''' - x_1' + x_1 x_2' = \sin t$
    $x_1'' - x_2 x_1' - x_2'' = \cos t$

6. $x_1'' = x_1 \sin t + x_2 x_3$
    $x_2'' = x_1^2 x_2' + x_1'$
    $x_3' = x_1 x_2 x_3$

7. $m_1 x'' = -k_1 x_1 + c(x_2' - x_1')$
    $m_2 x_2'' = -k_2 x_2 - c(x_2' - x_1')$
    (m_1, m_2, k_1, k_2 and c are constants)

8. $LI_1' + (R_1 + R_2)I_1 - R_2 I_2 = 0$
    $R_2 I_2' - R_2 I_1' + \dfrac{1}{C} I_2 = 0$
    (L, C, R_1 and R_2 are constants)

In Exercises 9–14, rewrite the differential equation as a first-order system.

9. $x'' - tx' + x^2 = \sin t$

10. $x'' + x = 0$

11. $x''' - x'' + x = e^t$

12. $x^{(4)} + x'''x - (x'')^2 = 0$

13. $\theta'' + \dfrac{g}{L} \sin \theta = 0$ (g and L are constants)

14. $mx'' + cx' + kx = A \cos \omega t$
    (m, c, k, A, and ω are constants)

15. Show that the linear system

$$P_{11}(D)x_1 + P_{12}(D)x_2 = b_1(t),$$
$$P_{21}(D)x_1 + P_{22}(D)x_2 = b_2(t),$$

where

$$P_{ij}(D) = a_{ij}D^2 + b_{ij}D + c_{ij},$$
$$i = 1, 2, \qquad j = 1, 2,$$

can be replaced by a first-order system with four equations if $a_{11}a_{22} - a_{21}a_{12} \neq 0$.

6.3
LINEAR SYSTEMS WITH CONSTANT COEFFICIENTS

A first-order system of the special form

$$\frac{dx_1}{dt} = a_{11}(t)x_1 + a_{12}(t)x_2 + \cdots + a_{1n}(t)x_n + b_1(t).$$

$$\frac{dx_2}{dt} = a_{21}(t)x_1 + a_{22}(t)x_2 + \cdots + a_{2n}(t)x_n + b_2(t),$$

. . . .

$$\frac{dx_n}{dt} = a_{n1}(t)x_1 + a_{n2}(t)x_2 + \cdots + a_{nn}(t)x_n + b_n(t),$$

where the functions a_{ij} and b_i are given, is called a first-order *linear* system. The functions a_{ij} are called the *coefficients* of the system. If each of the functions b_i is the zero function, the system is said to be *homogeneous*; otherwise it is said to be *nonhomogeneous*. The system can be written more briefly as

$$\frac{dx_i}{dt} = \sum_{j=1}^{n} a_{ij}(t)x_j + b_i(t), \qquad 1 \leq i \leq n.$$

In an initial value problem associated with the system we wish to find a solution that satisfies conditions of the form

$$x_1(t_0) = k_1, \qquad x_2(t_0) = k_2, \ldots, \qquad x_n(t_0) = k_n.$$

For example, when $n = 2$ (two equations and two unknowns) the initial value problem has the form

$$\frac{dx_1}{dt} = a_{11}(t)x_1 + a_{12}(t)x_2 + b_1(t),$$

$$\frac{dx_2}{dt} = a_{21}(t)x_1 + a_{22}(t)x_2 + b_2(t),$$

$$x_1(t_0) = k_1, \qquad x_2(t_0) = k_2.$$

In what follows, we restrict our attention to linear systems with *constant coefficients*. For convenience we use the operator notation

$$Df = f', \qquad Df(t) = f'(t).$$

A first-order linear system with constant coefficients has the form

$$
\begin{aligned}
Dx_1 &= a_{11}x_1 + a_{12}x_2 + \cdots + a_{1n}x_n + b_1(t), \\
Dx_2 &= a_{21}x_1 + a_{22}x_2 + \cdots + a_{2n}x_n + b_2(t), \\
&\;\cdots\cdots\cdots\cdots\cdots\cdots\cdots\cdots\cdots\cdots\cdots\cdots \\
Dx_n &= a_{n1}x_1 + a_{n2}x_2 + \cdots + a_{nn}x_n + b_n(t),
\end{aligned}
\tag{6.11}
$$

where the a_{ij} are constants. We shall also be concerned with more general linear systems with constant coefficients, of the form

$$
\begin{aligned}
P_{11}(D)x_1 + P_{12}(D)x_2 + \cdots + P_{1n}(D)x_n &= b_1(t), \\
P_{21}(D)x_1 + P_{22}(D)x_2 + \cdots + P_{2n}(D)x_n &= b_2(t), \\
&\quad\cdots\cdots\cdots\cdots\cdots\cdots\cdots\cdots\cdots\cdots \\
P_{n1}(D)x_1 + P_{n2}(D)x_2 + \cdots + P_{nn}(D)x_n &= b_n(t),
\end{aligned}
\tag{6.12}
$$

where the $P_{ij}(D)$ are polynomial operators.[2] Every system of the form (6.11) is of the form (6.12); but not every system of the form (6.12) is a first-order system. Usually, however, a system such as (6.12) can be rewritten as a first-order system, as was shown in Section 6.2.

One procedure that can be used to solve systems of the type (6.12) is called the *method of elimination*. The theory and technique are reminiscent of the elimination method for solving systems of linear algebraic equations. Two systems of differential equations are said to be *equivalent* if they have

---

[2] See Section 5.2.

the same solutions. In the method of elimination, we replace a given system by an equivalent, but simpler, system that is relatively easier to solve.

The reduction to a simpler system is carried out by performing a sequence of operations, each of which leads to an equivalent system. These operations are of three types.

First, we can simply interchange two equations of the system. For example, the two systems

$$(a) \quad Dx_1 = 1, \qquad (b) \quad Dx_2 = 2,$$
$$Dx_2 = 2, \qquad\qquad Dx_1 = 1,$$

are equivalent. Notice, however, that the two systems

$$(c) \quad Dx_1 = 1, \qquad (d) \quad Dx_2 = 1,$$
$$Dx_2 = 2, \qquad\qquad Dx_1 = 2,$$

are *not* equivalent, because a solution consists of an *ordered* pair of functions.

The second type of operation consists of simply multiplying through in one equation of the system by a nonzero constant.

In the third type of operation we operate on both members of one equation of the system, say the ith with a polynomial operator $Q(D)$ and add the result to another equation of the system, say the jth. In the new system, only the jth equation has changed. To illustrate, suppose that the ith and jth equations of the original system are

$$P_{i1}(D)x_1 + \cdots + P_{in}(D)x_n = b_i(t),$$
$$P_{j1}(D)x_1 + \cdots + P_{jn}(D)x_n = b_j(t). \tag{6.13}$$

Operating on both members of the ith equation with $Q(D)$ and adding the result to the jth equation, we have

$$P_{i1}(D)x_1 + \cdots + P_{in}(D)x_n = b_i(t),$$
$$[P_{j1}(D) + Q(D)P_{i1}(D)]x_1 + \cdots + [P_{jn}(D) + Q(D)P_{in}(D)]x_n$$
$$= b_j(t) + Q(D)b_i(t) \tag{6.14}$$

Notice that the equation

$$Q(D)P_{i1}(D)x_1 + \cdots + Q(D)P_{in}(D) = Q(D)b_i(t)$$

does not appear in either system. It is not hard to verify that if an ordered

n-tuple of functions $(x_1, x_2, \ldots, x_n)$ satisfies the pair of equations (6.13) it also satisfies the pair (6.14) and vice versa.[3]

Example 1   In order to illustrate the method, we solve the system

$$-(D + 2)x_1 + (D^2 - 4)x_2 = 4t,$$
$$(D + 3)x_1 + (D + 7)x_2 = 0. \tag{6.15}$$

Let us eliminate x_1. Adding the second equation to the first, we obtain

$$x_1 + (D^2 + D + 3)x_2 = 4t,$$
$$(D + 3)x_1 + (D + 7)x_2 = 0. \tag{6.16}$$

Next we operate on both sides of the first equation with $(D + 3)$, obtaining the equation

$$(D + 3)x_1 + (D + 3)(D^2 + D + 3)x_2 = 4 + 12t.$$

Subtracting this equation from the second equation of the system (6.16), we obtain the system

$$x_1 + (D^2 + D + 3)x_2 = 4t,$$
$$[-(D + 3)(D^2 + D + 3) + (D + 7)]x_2 = -4 - 12t. \tag{6.17}$$

The second equation, which involves only the one unknown x_2, can be written as

$$(D + 1)^2(D + 2)x_2 = 4 + 12t.$$

Using the methods of Chapter 5 to solve this equation, we have

$$x_2 = c_1e^{-t} + c_2te^{-t} + c_3e^{-2t} + 6t - 13. \tag{6.18}$$

From the first equation of the system (6.17) we have

$$x_1 = -(D^2 + D + 3)x_2 + 4t$$

or

$$x_1 = -3c_1e^{-t} + c_2(1 - 3t)e^{-t} - 5c_3e^{-2t} - 14t + 33. \tag{6.19}$$

Formulas (6.18) and (6.19) describe the general solution of the system (6.15).

---

[3] The functions $b_i, x_1, x_2, \ldots, x_n$ must be sufficiently differentiable, so that all indicated derivatives exist. This is usually the case, in practice.

Notice that three arbitrary constants are involved in the description of the general solution, even though the system involves only two unknowns. This is in accordance with the fact that the system (6.15) can be replaced by a *first-order* system with three unknown functions.

In the general case where we have a system of the form (6.12), we attempt to find an equivalent system of the form[4]

$$Q_{11}(D)x_1 + Q_{12}(D)x_2 + \cdots + Q_{1,n-1}(D)x_{n-1} + Q_{1n}(D)x_n = f_1(t),$$

$$Q_{22}(D)x_2 + \cdots + Q_{2,n-1}(D)x_{n-1} + Q_{2n}(D)x_n = f_2(t),$$

. . . .

$$Q_{n-1,n-1}(D)x_{n-1} + Q_{n-1,n}(D)x_n = f_{n-1}(t),$$

$$Q_{nn}(D)x_n = f_n(t),$$

where none of the operators $Q_{ii}(D)$, $1 \le i \le n$, is the zero operator. We can solve the nth equation for x_n, then find x_{n-1} from the $(n-1)$st equation, and so on.

Example 2   As a final example, we consider the first-order homogeneous system for three unknowns,

$$Dx_1 = 3x_1 + x_2 - 2x_3,$$

$$Dx_2 = -x_1 + 2x_2 + x_3,$$

$$Dx_3 = 4x_1 + x_2 - 3x_3.$$

This system can be written as

$$(D - 3)x_1 - x_2 + 2x_3 = 0,$$

$$x_1 + (D - 2)x_2 - x_3 = 0,$$

$$-4x_1 - x_2 + (D + 3)x_3 = 0.$$

The second equation can be used to eliminate x_1 from the first and third equations. First we multiply through in the second equation by 4 and add the result to the third equation. Next we operate on the second equation with $(D - 3)$ and subtract the result from the first equation. In this way we arrive at the equivalent system

$$(-D^2 + 5D - 7)x_2 + (D - 1)x_3 = 0,$$

$$x_1 + (D - 2)x_2 - x_3 = 0,$$

$$(4D - 9)x_2 + (D - 1)x_3 = 0.$$

---

[4] The unknowns may have to be renumbered.

We next eliminate x_3 between the first and third equations. Subtracting the third equation from the first, we have

$$(-D^2 + D + 2)x_2 = 0,$$
$$x_1 + (D - 2)x_2 - x_3 = 0, \qquad (6.20)$$
$$(4D - 9)x_2 + (D - 1)x_3 = 0.$$

The first equation of this system, which involves only the one unknown x_2, can be written as

$$(D - 2)(D + 1)x_2 = 0.$$

Hence we have

$$x_2(t) = c_1 e^{-t} + c_2 e^{2t}. \qquad (6.21)$$

From the third equation of the system (6.20) we have

$$(D - 1)x_3 = (9 - 4D)x_2 = 13c_1 e^{-t} + c_2 e^{2t}.$$

This is a first-order equation for x_3 and we find that

$$x_3(t) = -\frac{13}{2} c_1 e^{-t} + c_2 e^{2t} + c_3 e^{t}. \qquad (6.22)$$

The unknown x_1 can now be found from the second equation of the system (6.20). We have

$$x_1 = (2 - D)x_2 + x_3$$

or

$$x_1(t) = -\frac{7}{2} c_1 e^{-t} + c_2 e^{2t} + c_3 e^{t}. \qquad (6.23)$$

The formulas (6.22) and (6.23) become slightly simpler if we set $c_1 = 2c_1'$, where c_1' is an arbitrary constant. With this change the formulas (6.21), (6.22), and (6.23) become

$$x_1(t) = -7c_1' e^{-t} + c_2 e^{2t} + c_3 e^{t},$$
$$x_2(t) = 2c_1' e^{-t} + c_2 e^{2t},$$
$$x_3(t) = -13c_1' e^{-t} + c_2 e^{2t} + c_3 e^{t}.$$

Cramer's rule, a set of formulas for the solutions of a set of linear algebraic equations, was described in Chapter 2, Section 11. There is an analogous rule for a system of linear differential equations with constant coefficients. We shall consider the case of two equations, but the method applies to any number. Suppose the system is

$$L_1 x_1 + L_2 x_2 = f_1$$
$$L_3 x_1 + L_4 x_2 = f_2,$$

where the differential operators L_i have constant coefficients. The analogy to Cramer's rule says that

$$\begin{vmatrix} L_1 & L_2 \\ L_3 & L_4 \end{vmatrix} x_1 = \begin{vmatrix} f_1 & L_2 \\ f_2 & L_4 \end{vmatrix}, \qquad \begin{vmatrix} L_1 & L_2 \\ L_3 & L_4 \end{vmatrix} x_2 = \begin{vmatrix} L_1 & f_1 \\ L_3 & f_2 \end{vmatrix}$$

or

$$(L_1 L_4 - L_2 L_3)x_1 = L_4 f_1 - L_2 f_2$$
$$(L_1 L_4 - L_2 L_3)x_2 = L_1 f_2 - L_3 f_1.$$

The number of arbitrary constants in the solution is the order of the operator $L_1 L_4 - L_2 L_3$.[5]
We illustrate with the system of Example 1,

$$-(D + 2)x_1 + (D^2 - 4)x_2 = 4t$$
$$(D + 3)x_1 + (D + 7)x_2 = 0,$$

in which

$$L_1 = -(D + 2), \qquad L_2 = D^2 - 4, \qquad L_3 = D + 3, \qquad L_4 = D + 7,$$
$$f_1(t) = 4t, \qquad f_2(t) = 0.$$

Then the main differential operator becomes

$$L_1 L_4 - L_2 L_3 = -(D + 2)(D + 7) - (D + 3)(D^2 - 4)$$
$$= -(D^3 + 4D^2 + 5D + 2)$$
$$= -(D + 1)^2(D + 2),$$

and

$$L_4 f_1(t) - L_2 f_2(t) = (D + 7)(4t) = 4 + 28t$$
$$L_1 f_2(t) - L_3 f_1(t) = -(D + 3)(4t) = -4 - 12t.$$

The uncoupled system[6] is

$$-(D + 1)^2(D + 2)x_1 = 4 + 28t$$
$$-(D + 1)^2(D + 2)x_2 = -4 - 12t.$$

---

[5] See E. L. Ince, *Ordinary Differential Equations* (New York: Dover, 1956), 144.
[6] An uncoupled system is one in which each equation involves only one known function.

The number of arbitrary constants in the system is 3, since 3 is the order of the operator $-(D + 1)^2(D + 2)$. However, the formulas for x_1 and x_2 obtained from the uncoupled system will each involve 3 constants, for a total of 6. Hence the expressions for x_1 and x_2 must be substituted back into one of the original equations to determine the relationships among the 6 constants. The Cramer's rule analogy is easy to remember and leads immediately to an uncoupled system for the unknowns. It also yields some information about the nature of the solutions. In practice, however, determination of the relationships among the constants is often tedious.

––––––––––––––––––––––––––––––– Exercises for Section 6.3 –––––––––––––––––––––––––––––––

In Exercises 1–15, find the general solution of the system. If initial conditions are given, also find the solution that satisfies the conditions.

1. $Dx_1 = -4x_1 - 6x_2 + 9e^{-3t}$
    $Dx_2 = x_1 + x_2 - 5e^{-3t}$
    $x_1(0) = -9, \quad x_2(0) = 4$

2. $Dx_1 = -2x_1 + x_2$
    $Dx_2 = -3x_1 + 2x_2 + 2 \sin t$
    $x_1(0) = 3, \quad x_2(0) = 4$

3. $Dx_1 = -x_1 - 3e^{-2t}$
    $Dx_2 = -2x_1 - x_2 - 6e^{-2t}$

4. $D^2x_1 + (D + 2)x_2 = 2e^{-2t}$
    $Dx_1 - (D + 2)x_2 = 0$
    $x_1(0) = 4, \quad x_2(0) = 1, \quad x_1'(0) = -2$

5. $(2D + 1)x_1 + (D^2 - 4)x_2 = -7e^{-t}$
    $Dx_1 - (D + 2)x_2 = -3e^{-t}$

6. $(D + 2)x_1 + (D^2 + 2D)x_2 = 5e^{-t}$
    $(D + 1)x_1 - (D + 2)x_2 = 0$

7. $(D^2 + 1)x_1 + 2Dx_2 = 0$
    $-3(D^2 + 1)x_1 + 2(D^2 + 2)x_2 = 0$
    $x_1(0) = 1, \quad x_2(0) = 1, \quad x_1'(0) = 0, \quad x_2'(0) = -1$

8. $(D^3 - 2D^2 + 3D)x_1 + (2D^2 - 8)x_2 = 4 - 6t$
    $Dx_1 - (D + 2)x_2 = -2t$

9. $Dx_1 = x_1 - 3x_2 + 2x_3$
    $Dx_2 = -x_2$
    $Dx_3 = -x_2 - 2x_3$
    $x_1(0) = -3, \quad x_2(0) = 0, \quad x_3(0) = 3$

10. $Dx_1 = -x_1 + x_2$
    $Dx_2 = 2x_1 - 2x_2 + 2x_3$
    $Dx_3 = -x_2 - x_3$

11. $Dx_1 = x_1 - 2x_2 - t^2$
    $Dx_2 = x_1 + x_3 - 1 - t^2$
    $Dx_3 = -2x_1 + 2x_2 - x_3 + 2t^2 + 2t$

12. $Dx_1 = x_1 + x_2$
    $Dx_2 = -2x_1 + x_2 - 2x_3 + e^{2t}$
    $Dx_3 = -x_2 + x_3$

13. $Dx_1 = x_1 - 3x_2 + 2x_3$
    $Dx_2 = -2x_2 + 2x_3$
    $Dx_3 = x_1 - 5x_2 + 2x_3$

14. $Dx_1 = -x_1 + x_3$
    $Dx_2 = 2x_3$
    $Dx_3 = x_1 - 2x_2 - 3x_3$

15. $Dx_1 = -x_1 - x_3 + \cos t$
    $Dx_2 = -x_2 - x_3 + \sin t$
    $Dx_3 = -2x_3 + \cos t + 2 \sin t$

16. A tank initially contains 100 gal of water; a second tank initially contains 100 gal of a solution containing 20 lb of a solute. Liquid is pumped through a pipe from the first tank to the second at the rate of 2 gal per min, and liquid is pumped from the second tank to the first through another pipe at the same rate. Find formulas that express the amounts of solute in the two tanks as functions of time.

17. The contents of two tanks are initially as in the previous exercise. Water runs into the first tank at the rate of 3 gal per min, liquid is

pumped from the first tank into the second at the rate of 4 gal per min, liquid is pumped from the second tank into the first at the rate of 1 gal per min, and liquid is drained from the second tank at the rate of 3 gal per min. Find formulas that express the amounts of solute in the two tanks as functions of time.

18. Initially the first of three tanks contains 100 gal of water, the second 100 gal of a solution containing 14 lb of a solute, and the third 100 gal of a solution containing 7 lb of the solute. Liquid is pumped from tank 1 into tank 2 at the rate of 1 gal per min, from tank 2 into tank 3 at the rate of 1 gal per minute, and from tank 3 into tank 1 at the rate of 1 gal per minute. (a) Formulate a system of differential equations for the amounts of solutes in these three tanks. (b) Find formulas that express the amounts of solutes in the three tanks as functions of time. Suggestion: let $a = 1/100$.

6.4

MATRIX FORMULATION OF LINEAR SYSTEMS

In this and the next several sections we shall develop a theory for linear systems of differential equations that relies heavily on matrix algebra. This approach yields compact formulas that are convenient for theoretical purposes. We shall also present a practical method of solution for linear systems with constant coefficients that uses matrices. In the examples of the previous section, we dealt with systems involving only two or three unknowns and the coefficients of these systems were integers. In a practical problem, the solver may be faced with a system having a large number of unknowns and coefficients with several significant digits. For such systems, it is desirable to have a method of solution in which the computational work is facilitated.

We shall begin by showing how a first order linear system can be written in compact form by the use of matrices. To do this, we need the concept of a *matrix function*. An $m \times n$ matrix function is a rule that assigns to each number of an interval an $m \times n$ matrix. For example, a 2×3 matrix function A is defined by the formula

$$A(t) = \begin{bmatrix} 2t & t^2 & e^t \\ -1 & \cos t & t+2 \end{bmatrix}, \qquad \text{for all } t.$$

In general, if the quantities a_{ij} are ordinary (scalar) functions of a real variable defined on a common interval then the formula

$$A(t) = \begin{bmatrix} a_{11}(t) & \cdots & a_{1n}(t) \\ \vdots & & \vdots \\ a_{m1}(t) & \cdots & a_{mn}(t) \end{bmatrix}.$$

defines an $m \times n$ matrix function A. Thus, for each t, $A(t)$ is just an $m \times n$ matrix of scalars. We call the function a_{ij} the *elements* of A and write

$$A = \begin{bmatrix} a_{11} & \cdots & a_{1n} \\ \vdots & & \vdots \\ a_{m1} & \cdots & a_{mn} \end{bmatrix}.$$

A matrix function is said to be *continuous* if each of its elements is continuous.

Two $m \times n$ matrix functions A and B are said to be *equal* if they have the same domain and if $A(t) = B(t)$ for each t in the domain. Addition and multiplication of matrix functions are defined by

$$(A + B)(t) = A(t) + B(t),$$

$$(AB)(t) = A(t)B(t).$$

If f is a function whose domain is that of the $m \times n$ matrix function A, then the product $fA = Af$ is defined to be the $m \times n$ matrix function with elements fa_{ij}. If c is a number, the product $cA = Ac$ is defined by the equation $(cA)(t) = cA(t)$. It is not hard to show (Exercise 8) that the set of all $m \times n$ matrix functions on an interval forms a vector space. In particular, the set of all $n \times 1$ matrix functions, called *vector functions*, defined on an interval forms a vector space. The space is real or complex, depending on whether we restrict the scalars to be real or allow them to be complex numbers. However, the set of all n-dimensional column vector functions is not finite-dimensional, as was R^n.

The *derivative* A' of a matrix function A is defined by

$$A' = \begin{bmatrix} a'_{11} & \cdots & a'_{1n} \\ \vdots & & \vdots \\ a'_{m1} & \cdots & a'_{mn} \end{bmatrix}$$

provided that each of the functions a_{ij} is differentiable. In particular, a column vector function $\mathbf{x}$ and its derivative $\mathbf{x}'$ are written as

$$\mathbf{x} = \begin{bmatrix} x_1 \\ x_2 \\ \vdots \\ x_n \end{bmatrix}, \qquad \mathbf{x}' = \begin{bmatrix} x'_1 \\ x'_2 \\ \vdots \\ x'_n \end{bmatrix},$$

where $x_1, x_2, \ldots, x_n$ are functions defined on an interval. Verification of the differentiation rules

$$(A + B)' = A' + B', \qquad (AB)' = A'B + AB'$$

for the sum and product of matrix functions is left to the reader.

Let us now consider the linear system

$$x'_1 = a_{11}x_1 + a_{12}x_2 + \cdots + a_{1n}x_n + b_1,$$

$$x'_2 = a_{21}x_2 + a_{22}x_2 + \cdots + a_{2n}x_n + b_2,$$

$$\cdots\cdots\cdots\cdots\cdots\cdots\cdots\cdots\cdots\cdots\cdots\cdots$$

$$x'_n = a_{n1}x_1 + a_{n2}x_2 + \cdots + a_{nn}x_n + b_n,$$

where the functions a_{ij} and b_i are defined on a common interval. If $\mathbf{x}$ and $\mathbf{b}$ are the column vector functions

$$\mathbf{x} = \begin{bmatrix} x_1 \\ x_2 \\ \vdots \\ x_n \end{bmatrix}, \qquad \mathbf{b} = \begin{bmatrix} b_1 \\ b_2 \\ \vdots \\ b_n \end{bmatrix},$$

then the linear system of differential equations can be written as

$$\mathbf{x}' = A\mathbf{x} + \mathbf{b}.$$

For example, the system

$$x_1' = 2x_1 - 3x_2 + 3e^t,$$
$$x_2' = -x_1 + x_2 - e^t,$$

can be written as $\mathbf{x}' = A\mathbf{x} + \mathbf{b}$, with

$$A(t) = \begin{bmatrix} 2 & -3 \\ -1 & 1 \end{bmatrix}, \qquad \mathbf{b}(t) = \begin{bmatrix} 3e^t \\ -e^t \end{bmatrix}$$

for all t.

A set of initial conditions

$$x_1(t_0) = k_1, \qquad x_2(t_0) = k_2, \dots, \qquad x_n(t_0) = k_n$$

can be described by the vector equation $\mathbf{x}(t_0) = \mathbf{k}$, where $\mathbf{k}$ is the constant vector with components $k_1, k_2, \dots, k_n$.

We now consider the special case of a homogeneous equation

$$\mathbf{x}' = A\mathbf{x}, \tag{6.24}$$

where the matrix function A is continuous on an interval $\mathscr{I}$. The set of all real solutions of this equation on $\mathscr{I}$[7] constitutes a real vector space. To verify this fact, we first observe that if $\mathbf{u}$ is a solution on $\mathscr{I}$ then $c\mathbf{u}$, where c is any real number, is also a solution. For we have

$$(c\mathbf{u})' = c\mathbf{u}' = c(A\mathbf{u}) = A(c\mathbf{u}).$$

Also, if $\mathbf{u}$ and $\mathbf{v}$ are both solutions then $\mathbf{u} + \mathbf{v}$ is also a solution, because since

---

[7] If A is continuous on $\mathscr{I}$ and if $\mathbf{u}$ is a solution on some interval $\mathscr{J}$ that is contained in $\mathscr{I}$, it can be shown that there is a solution $\mathbf{v}$ on $\mathscr{I}$ such that $\mathbf{u}(t) = \mathbf{v}(t)$ for t in $\mathscr{J}$. For this reason we consider only solutions that exist throughout $\mathscr{I}$.

$\mathbf{u}' = A\mathbf{u}$ and $\mathbf{v}' = A\mathbf{v}$ we have

$$(\mathbf{u} + \mathbf{v})' = \mathbf{u}' + \mathbf{v}' = A\mathbf{u} + A\mathbf{v} = A(\mathbf{u} + \mathbf{v}).$$

Thus the set of all solutions is a subspace of the space of all vector functions defined on $\mathscr{I}$. The zero vector function $\mathbf{0}$ is evidently a solution. It is called the *trivial solution*. From properties of a vector space, we see that if $\mathbf{u}_1, \mathbf{u}_2, \ldots, \mathbf{u}_m$ are solutions and if $c_1, c_2, \ldots, c_m$ are real numbers, then $c_1\mathbf{u}_1 + c_2\mathbf{u}_2 + \cdots + c_m\mathbf{u}_m$ is a solution.

We shall also have occasion to consider complex solutions of a real vector differential equation. Let $\mathbf{w}$ be a *complex vector function* of the form

$$\mathbf{w} = \mathbf{u} + i\mathbf{v},$$

where $\mathbf{u}$ and $\mathbf{v}$ are real vector functions with components $u_1, u_2, \ldots, u_n$ and $v_1, v_2, \ldots, v_n$, respectively. The components of $\mathbf{w}$ are the complex functions $w_j = u_j + iv_j$, $1 \le j \le n$. The derivative of $\mathbf{w}$ is defined as $\mathbf{w}' = \mathbf{u}' + i\mathbf{v}'$. If $\mathbf{w}$ is a complex solution of Eq. (6.24), so that $\mathbf{w}' = A\mathbf{w}$, then $\mathbf{u}$ and $\mathbf{v}$ are real solutions of that equation. To see this, observe that

$$(\mathbf{u} + i\mathbf{v})' = A(\mathbf{u} + i\mathbf{v})$$

or

$$\mathbf{u}' + i\mathbf{v}' = A\mathbf{u} + iA\mathbf{v}.$$

Equating real and imaginary parts, we have

$$\mathbf{u}' = A\mathbf{u}, \qquad \mathbf{v}' = A\mathbf{v}.$$

Hence $\mathbf{u}$ and $\mathbf{v}$ are both real solutions. It is left to the reader (Exercise 9) to verify that the set of all complex solutions of Eq. (6.24) is a complex vector space. This space contains all the real solutions of the equation.

Exercises for Section 6.4

1. Write out the equations of the system that corresponds to the vector equation $\mathbf{x}' = A\mathbf{x} + \mathbf{b}$ if A and $\mathbf{b}$ are as given.

   (a) $A(t) = \begin{bmatrix} 2 & -1 \\ 3 & 0 \end{bmatrix}$, $\mathbf{b}(t) = \begin{bmatrix} e^t \\ 3e^{-2t} \end{bmatrix}$

   (b) $A(t) = \begin{bmatrix} t^2 & -e^t \\ 1 & \cos t \end{bmatrix}$, $\mathbf{b}(t) = \begin{bmatrix} 1-t \\ 0 \end{bmatrix}$

   (c) $A(t) = \begin{bmatrix} 2 & -1 & 0 \\ 0 & 1 & 1 \\ 3 & 2 & 0 \end{bmatrix}$, $\mathbf{b}(t) = \begin{bmatrix} e^{2t} \\ 0 \\ e^t \end{bmatrix}$

   (d) $A(t) = \begin{bmatrix} t+1 & 0 & t \\ -1 & 2-t & 0 \\ t & 0 & 0 \end{bmatrix}$, $\mathbf{b} = \mathbf{0}$.

2. Find A and $\mathbf{b}$ if the given system is written as $\mathbf{x}' = A\mathbf{x} + \mathbf{b}$.

   (a) $x_1' = -2x_1 + x_2 + \cos 2t$
       $x_2' = -x_1 - x_2 - 2\sin 2t$

   (b) $x_1' = e^t x_1 - e^{-t} x_2$
       $x_2' = 2e^{-t} x_1 + 3e^t x_2$

(c) $x_1' = 2x_1 + x_2 - x_3 + 2e^{-t}$
 $x_2' = x_1 - x_2 - e^{-t}$
 $x_3' = x_2 + 2x_3$

(d) $x_1' = te^t x_1 - x_2 + 1$
 $x_2' = x_3 - e^t$
 $x_3' = x_1 + e^t x_2$

3. Write out the vector-matrix form of the linear system that corresponds to the higher order equation:
(a) $y'' + a_1 y' + a_2 y = f$
(b) $y^{(n)} + a_1 y^{(n-1)} + \cdots + a_n y = f$

4. If
$$A(t) = \begin{bmatrix} -2 & 1 \\ -3 & 2 \end{bmatrix}, \qquad b(t) = \begin{bmatrix} 0 \\ 2 \sin t \end{bmatrix},$$
verify that a solution of $x' = Ax + b$ is
$$u(t) = \begin{bmatrix} -\sin t \\ -\cos t - 2 \sin t \end{bmatrix}.$$

5. If
$$A(t) = \begin{bmatrix} 1 & -3 & 2 \\ 0 & -1 & 0 \\ 0 & -1 & -2 \end{bmatrix},$$
verify that a solution of $x' = Ax$ is
$$u(t) = \begin{bmatrix} -2e^{-2t} \\ 0 \\ 3e^{-2t} \end{bmatrix}.$$

6. Verify the differentiation rules
$$(A + B)' = A' + B', \qquad (AB)' = A'B + AB'$$
for matrix functions.

7. If $w = u + iv$ is a complex solution of the equation $x' = Ax + b$, where $b = c + id$, show that u and v are real solutions of $x' = Ax + c$ and $x' = Ax + d$, respectively. (Here u, v, A, c, and d are assumed to be real.)

8. Show that the set of all n-dimensional real (complex) vector functions defined on an interval forms a vector space over the real (complex) numbers.

9. Show that the set of all complex solutions of the equation $x' = Ax$ on an interval $\mathscr{I}$ is a complex vector space.

10. Let F_n denote the set of all n-dimensional vector functions defined on an interval $\mathscr{I}$ and let $F_n^{(m)}$ be the set of all those vector functions in F_n that possess at least m derivatives on $\mathscr{I}$.
(a) Show that $F_n^{(m)}$ is a subspace of F_n.
(b) Let $Tx = x' - Ax$ for x in $F_n^{(1)}$, where A is an $n \times n$ matrix function defined on $\mathscr{I}$. Show that T is a linear transformation from $F_n^{(1)}$ into F_n.
(c) Show that a vector function is a solution of the equation $x' = Ax$ if and only if it belongs to the kernel of the operator T of part (b).

6.5 FUNDAMENTAL SETS OF SOLUTIONS

The solutions of the vector differential equation

$$x' = Ax \tag{6.25}$$

on an interval form a vector space, as we remarked in the last section. This space is finite-dimensional. Although we shall not prove this fact, we emphasize its importance by stating it as a theorem.

Theorem 6.1 Let A be an $n \times n$ matrix function, continuous on an interval $\mathscr{I}$. Then the set of all solutions of Eq. (6.25) is a vector space of dimension n.

In order to solve Eq. (6.25), we must find a basis for its solution space. If $u_1, u_2, \ldots, u_n$ are linearly independent solutions, the general solution consists of all vector functions of the form

$$c_1\mathbf{u}_1 + c_2\mathbf{u}_2 + \cdots + c_n\mathbf{u}_n,$$

where $c_1, c_2, \ldots, c_n$ are real numbers. A basis for the solution space is called a *fundamental set of solutions* for the equation. The next theorem gives a criterion for determining if a set of n vector functions of dimension n is linearly independent.

Theorem 6.2   Let $U = [\mathbf{u}_1, \mathbf{u}_2, \ldots, \mathbf{u}_n]$ be an $n \times n$ matrix function defined on an interval $\mathscr{I}$. If the column vector functions $\mathbf{u}_1, \mathbf{u}_2, \ldots, \mathbf{u}_n$ are linearly dependent then $\det U(t) = 0$ for all t in $\mathscr{I}$. Hence if there exists even one point t_1 in $\mathscr{I}$ such that $\det U(t_1) \neq 0$, the vector functions are linearly independent.

Proof   Suppose that the vector functions are linearly dependent. Then there exist numbers $c_1, c_2, \ldots, c_n$, not all zero, such that

$$c_1\mathbf{u}_1(t) + c_2\mathbf{u}_2(t) + \cdots + c_n\mathbf{u}_n(t) = \mathbf{0}$$

for all t in $\mathscr{I}$. This equation may be written as

$$U(t)\mathbf{c} = \mathbf{0},$$

where $\mathbf{c}$ is the column vector with components $c_1, c_2, \ldots, c_n$. Since $\mathbf{c}$ is not the zero vector, $U(t)$ must be singular for every t in $\mathscr{I}$. If there is a point t_1 such that $U(t_1)$ is nonsingular, the vector functions cannot be linearly dependent. Hence they must be linearly independent.

Theorem 6.2 does *not* say that if the determinant, $\det U(t)$, of a set of vector functions is zero on an interval then the vector functions are linearly dependent. The vector functions $\mathbf{u}_1$ and $\mathbf{u}_2$, where

$$\mathbf{u}_1(t) = \begin{bmatrix} t \\ 1 \end{bmatrix}, \qquad \mathbf{u}_2(t) = \begin{bmatrix} t^2 \\ t \end{bmatrix}$$

for all t, provide an example. The condition

$$c_1\mathbf{u}_1(t) + c_2\mathbf{u}_2(t) = \mathbf{0}$$

is satisfied when $t = 1$ and $t = -1$ only if $c_1 = c_2 = 0$, so the vector functions are linearly independent. However, $\det U(t) = 0$ for all t. If the vector functions $\mathbf{u}_1, \mathbf{u}_2, \ldots, \mathbf{u}_n$ are linearly independent and *in addition are solutions of Eq. (6.25)*, it can be shown that $\det U(t)$ is *never* zero.

As an application of the above theory, let us consider the equation $\mathbf{x}' = A\mathbf{x}$, where

$$A = \begin{bmatrix} 3 & 1 & -2 \\ -1 & 2 & 1 \\ 4 & 1 & -3 \end{bmatrix}.$$

(This is the system of Example 2, Section 6.3, formulated in terms of matrices.) The reader may verify directly that each of the vector functions $\mathbf{u}_1$, $\mathbf{u}_2$, $\mathbf{u}_3$, where

$$\mathbf{u}_1(t) = e^{-t}\begin{bmatrix} -7 \\ 2 \\ -13 \end{bmatrix}, \qquad \mathbf{u}_2(t) = e^{2t}\begin{bmatrix} 1 \\ 1 \\ 1 \end{bmatrix}, \qquad \mathbf{u}_3(t) = e^{t}\begin{bmatrix} 1 \\ 0 \\ 1 \end{bmatrix}$$

for all t, is a solution of the equation. If $U(t)$ is the matrix whose column vectors are $\mathbf{u}_1(t)$, $\mathbf{u}_2(t)$, $\mathbf{u}_3(t)$, then

$$\det U(0) = \begin{vmatrix} -7 & 1 & 1 \\ 2 & 1 & 0 \\ -13 & 1 & 1 \end{vmatrix} = 6 \neq 0.$$

By Theorem 6.2, the vectors are linearly independent. By Theorem 6.1, the general solution of the equation may be written as

$$\mathbf{x} = c_1 e^{-t}\begin{bmatrix} -7 \\ 2 \\ -13 \end{bmatrix} + c_2 e^{2t}\begin{bmatrix} 1 \\ 1 \\ 1 \end{bmatrix} + c_3 e^{t}\begin{bmatrix} 1 \\ 0 \\ 1 \end{bmatrix}$$

where c_1, c_2, and c_3 are arbitrary constants.

If A is an $n \times n$ matrix function defined on an interval $\mathscr{I}$, we can seek to determine an $n \times n$ matrix function X such that

$$X' = AX \tag{6.26}$$

on $\mathscr{I}$. If the column vectors of X are $\mathbf{x}_1, \mathbf{x}_2, \ldots, \mathbf{x}_n$, then the column vectors of X' are $\mathbf{x}_1', \mathbf{x}_2', \ldots, \mathbf{x}_n'$. The product AX is an $n \times n$ matrix function whose column vectors are $A\mathbf{x}_1, A\mathbf{x}_2, \ldots, A\mathbf{x}_n$. Hence a matrix function X satisfies Eq. (6.26) if and only if each of its column vectors is a solution of the vector equation (6.25). A matrix function whose column vectors form a basis for the solution space of Eq. (6.25) is called a *fundamental matrix* for that equation. Thus a matrix function is a fundamental matrix if and only if its column vectors constitute a fundamental set of solutions.

In dealing with an initial value problem

$$\mathbf{x}' = A\mathbf{x}, \qquad \mathbf{x}(t_0) = \mathbf{k},$$

one particular fundamental matrix is especially convenient. Suppose that $\mathbf{u}_1, \mathbf{u}_2, \ldots, \mathbf{u}_n$ are the solutions of the differential equation for which

$$\mathbf{u}_1(t_0) = \begin{bmatrix} 1 \\ 0 \\ 0 \\ \vdots \\ 0 \end{bmatrix}, \qquad \mathbf{u}_2(t_0) = \begin{bmatrix} 0 \\ 1 \\ 0 \\ \vdots \\ 0 \end{bmatrix}, \ldots, \qquad \mathbf{u}_n(t_0) = \begin{bmatrix} 0 \\ 0 \\ 0 \\ \vdots \\ 1 \end{bmatrix}.$$

(It can be shown that such solutions exist.) If U is the matrix function that has these vector functions as its column vectors, we see that

$$U(t_0) = I,$$

where I is the $n \times n$ identity matrix. Since I is nonsingular, U is a fundamental matrix, according to Theorem 6.2. Let $\mathbf{u}$ be defined by the relation

$$\mathbf{u} = U\mathbf{k}.$$

Making use of Theorem 2.4, we have

$$\mathbf{u} = k_1\mathbf{u}_1 + k_2\mathbf{u}_2 + \cdots + k_n\mathbf{u}_n;$$

thus $\mathbf{u}$ is a solution of the differential equation. Since $\mathbf{u}(t_0) = \mathbf{k}$, $\mathbf{u}$ is the solution of the initial value problem. A method for finding such a fundamental matrix for a system with constant coefficients will be developed in Sections 6.9 and 6.10.

In order to illustrate the ideas discussed here, let us consider the equation $\mathbf{x}' = A\mathbf{x}$, where

$$A = \begin{bmatrix} -2 & 1 \\ -4 & 3 \end{bmatrix}, \qquad \mathbf{x} = \begin{bmatrix} x_1 \\ x_2 \end{bmatrix}.$$

It may be verified that $\mathbf{u}_1$ and $\mathbf{u}_2$, where

$$\mathbf{u}_1(t) = \begin{bmatrix} e^{-t} \\ e^{-t} \end{bmatrix}, \qquad \mathbf{u}_2(t) = \begin{bmatrix} e^{2t} \\ 4e^{2t} \end{bmatrix},$$

are linearly independent solutions. Then the matrix function U, where

$$U(t) = [\mathbf{u}_1(t), \mathbf{u}_2(t)] = \begin{bmatrix} e^{-t} & e^{2t} \\ e^{-t} & 4e^{2t} \end{bmatrix},$$

is a fundamental matrix for the equation. It may also be verified that the matrix function V, where

$$V = \frac{1}{3}[4\mathbf{u}_1 - \mathbf{u}_2, -\mathbf{u}_1 + \mathbf{u}_2]$$

and

$$V(t) = \frac{1}{3}\begin{bmatrix} 4e^{-t} - e^{2t} & -e^{-t} + e^{2t} \\ 4e^{-t} - 4e^{2t} & -e^{-t} + 4e^{2t} \end{bmatrix},$$

is also a fundamental matrix, with $V(0) = I$. The solution of the equation that satisfies $\mathbf{x}(0) = \mathbf{k}$ is given by

$$\mathbf{x}(t) = V(t)\mathbf{k}.$$

———————————————————— **Exercises for Section 6.5** ————————————————————

In Exercises 1–3, determine whether or not the given set of vector functions is linearly dependent. The interval of definition is assumed to be the set of all real numbers.

1. (a) $\mathbf{u}_1(t) = \begin{bmatrix} 2t-1 \\ -t \end{bmatrix}$, $\mathbf{u}_2(t) = \begin{bmatrix} -t+1 \\ 2t \end{bmatrix}$

   (b) $\mathbf{u}_1(t) = \begin{bmatrix} \cos t \\ \sin t \end{bmatrix}$, $\mathbf{u}_2(t) = \begin{bmatrix} \sin t \\ \cos t \end{bmatrix}$

   (c) $\mathbf{u}_1(t) = \begin{bmatrix} t-2t^2 \\ -t \end{bmatrix}$, $\mathbf{u}_2(t) = \begin{bmatrix} -2t+4t^2 \\ 2t \end{bmatrix}$

   (d) $\mathbf{u}_1(t) = \begin{bmatrix} te^t \\ t \end{bmatrix}$, $\mathbf{u}_2(t) = \begin{bmatrix} e^t \\ 1 \end{bmatrix}$

2. (a) $\mathbf{u}_1(t) = \begin{bmatrix} 2-t \\ t \\ -2 \end{bmatrix}$, $\mathbf{u}_2(t) = \begin{bmatrix} t \\ -1 \\ 2 \end{bmatrix}$,

   $\mathbf{u}_3(t) = \begin{bmatrix} 2+t \\ t-2 \\ 2 \end{bmatrix}$

   (b) $\mathbf{u}_1(t) = \begin{bmatrix} \cos t \\ \sin t \\ 0 \end{bmatrix}$, $\mathbf{u}_2(t) = \begin{bmatrix} \cos t \\ 0 \\ \sin t \end{bmatrix}$,

   $\mathbf{u}_3(t) = \begin{bmatrix} 0 \\ \cos t \\ \sin t \end{bmatrix}$

   (c) $\mathbf{u}_1(t) = \begin{bmatrix} e^t \\ -e^t \\ e^t \end{bmatrix}$, $\mathbf{u}_2(t) = \begin{bmatrix} -e^t \\ 2e^t \\ -e^t \end{bmatrix}$,

   $\mathbf{u}_3(t) = \begin{bmatrix} 0 \\ e^t \\ 0 \end{bmatrix}$

   (d) $\mathbf{u}_1(t) = \begin{bmatrix} e^t \\ 0 \\ 0 \end{bmatrix}$, $\mathbf{u}_2(t) = \begin{bmatrix} 0 \\ \cos t \\ \cos t \end{bmatrix}$,

   $\mathbf{u}_3(t) = \begin{bmatrix} 0 \\ \sin t \\ \sin t \end{bmatrix}$

3. (a) $\mathbf{u}_1(t) = \begin{bmatrix} 2-t \\ t \end{bmatrix}$, $\mathbf{u}_2(t) = \begin{bmatrix} t+1 \\ -2 \end{bmatrix}$,

   $\mathbf{u}_3(t) = \begin{bmatrix} t \\ t+2 \end{bmatrix}$

   (b) $\mathbf{u}_1(t) = \begin{bmatrix} e^t \\ 0 \end{bmatrix}$, $\mathbf{u}_2(t) = \begin{bmatrix} 0 \\ 0 \end{bmatrix}$,

   $\mathbf{u}_3(t) = \begin{bmatrix} 0 \\ e^t \end{bmatrix}$

   (c) $\mathbf{u}_1(t) = \begin{bmatrix} t^2 \\ t^4 \end{bmatrix}$, $\mathbf{u}_2(t) = \begin{bmatrix} t|t| \\ t|t^3| \end{bmatrix}$

   (d) $\mathbf{u}_1(t) = \begin{bmatrix} \cos(t+\pi/4) \\ 0 \\ 0 \\ 0 \end{bmatrix}$, $\mathbf{u}_2(t) = \begin{bmatrix} \cos t \\ 0 \\ 0 \\ e^t \end{bmatrix}$,

   $\mathbf{u}_3(t) = \begin{bmatrix} \sin t \\ 0 \\ 0 \\ e^t \end{bmatrix}$

4. Show that any set of vector functions that contains the zero function is linearly dependent.

5. Let $\mathbf{u}$ and $\mathbf{v}$ be vector functions, where

$$\mathbf{u} = \begin{bmatrix} u_1 \\ u_2 \end{bmatrix}, \qquad \mathbf{v} = \begin{bmatrix} v_1 \\ v_2 \end{bmatrix}.$$

   (a) If u_1 and v_1 are linearly independent functions, is it necessarily true that $\mathbf{u}$ and $\mathbf{v}$ are linearly independent?
   (b) Suppose that u_1 and v_1 are linearly dependent and also that u_2 and v_2 are linearly dependent. Is it necessarily true that $\mathbf{u}$ and $\mathbf{v}$ are linearly dependent?

6. For all t, let

$$\mathbf{u}_1(t) = \begin{bmatrix} e^{2t} \\ e^{2t} \end{bmatrix}, \qquad \mathbf{u}_2(t) = \begin{bmatrix} 3e^{3t} \\ 2e^{3t} \end{bmatrix},$$

$$A(t) = \begin{bmatrix} 5 & -3 \\ 2 & 0 \end{bmatrix}.$$

   (a) Verify that $\mathbf{u}_1$ and $\mathbf{u}_2$ form a fundamental set of solutions for the equation $\mathbf{x}' = A\mathbf{x}$.
   (b) If U is the matrix function with $\mathbf{u}_1$ and $\mathbf{u}_2$ as its column vectors, verify that $U' = AU$.

(c) Write down a formula that describes the set of all solutions of the equation $x' = Ax$.

(d) Find the solution of $x' = Ax$ for which

$$x(0) = \begin{bmatrix} 1 \\ 0 \end{bmatrix}.$$

7. Let the matrix function U be defined by the equation

$$U(t) = \begin{bmatrix} \cos 2t & \sin 2t \\ \sin 2t & -\cos 2t \end{bmatrix}$$

for all t. Verify that U is a fundamental matrix for the equation $x' = Ax$, where

$$A = \begin{bmatrix} 0 & -2 \\ 2 & 0 \end{bmatrix}.$$

Find the solution of $x' = Ax$ for which

$$x(0) = \begin{bmatrix} 2 \\ 3 \end{bmatrix}.$$

8. Let

$$A(t) = \begin{bmatrix} -1 & 4 & -4 \\ 0 & -1 & 1 \\ 0 & 0 & 0 \end{bmatrix},$$

$$U(t) = \begin{bmatrix} 0 & 4te^{-t} & e^{-t} \\ 1 & e^{-t} & 0 \\ 1 & 0 & 0 \end{bmatrix}$$

for all t.

(a) Verify that U is a fundamental matrix for the equation $x' = Ax$.

(b) Find the solution of $x' = Ax$ for which

$$x(0) = \begin{bmatrix} 0 \\ 1 \\ 2 \end{bmatrix}.$$

9. Let U and A be $n \times n$ matrix functions such that $U'(t) = A(t) U(t)$ for t in an interval J.

(a) Show that

$$\frac{d}{dt} \det U(t)$$

$$= [a_{11}(t) + a_{22}(t) + \cdots + a_{nn}(t)] \det U(t)$$

and hence that

$$\det U(t)$$

$$= (\det U(t_0)) \exp\left\{ \int_{t_0}^{t} [a_{11}(s) + \cdots + a_{nn}(s)] \, ds \right\},$$

where t_0 is any point in J. Suggestion: differentiate $\det U$ by rows.

(b) Use the result of part (a) to show either $\det U(t) = 0$ for all t in J or $\det U(t)$ is not zero for any t in J.

10. Let U be a fundamental matrix for the equation $x' = Ax$, and let $V = UC$ where C is a square constant matrix.

(a) Show that $V' = AV$.

(b) Show that V is a fundamental matrix for the equation $x' = Ax$ if and only if C is nonsingular.

(c) If $W = CU$, show that $W' \neq AW$ in general.

(d) Show that if $V(t) = U(t)U^{-1}(t_0)$, then V is a fundamental matrix such that $V(t_0) = I$.

6.6
SOLUTIONS BY CHARACTERISTIC VALUES

In Chapter 5 we attacked the equation with constant coefficients $P(D)y = 0$ by attempting to find solutions of the form $y = e^{rx}$. Let us consider here the system with constant coefficients,

$$x_1' = x_1 - 3x_2,$$
$$x_2' = -2x_1 + 2x_2. \tag{6.27}$$

We shall seek solutions of the form

$$x_1(t) = k_1 e^{\lambda t}, \qquad x_2(t) = k_2 e^{\lambda t}, \tag{6.28}$$

where λ, k_1, and k_2 are numbers to be determined. Substituting in the system (6.27), we obtain the requirements

$$\lambda k_1 e^{\lambda t} = (k_1 - 3k_2)e^{\lambda t}$$
$$\lambda k_2 e^{\lambda t} = (-2k_1 + 2k_2)e^{\lambda t}$$

or

$$(\lambda - 1)k_1 + 3k_2 = 0,$$
$$2k_1 + (\lambda - 2)k_2 = 0.$$

Thus a nontrivial solution of the form (6.28) exists if and only if λ is a characteristic value of the matrix

$$A = \begin{bmatrix} 1 & -3 \\ -2 & 2 \end{bmatrix},$$

which is the coefficient matrix of the system (6.27). The vector $\mathbf{k} = (k_1, k_2)$ must be a corresponding characteristic vector. The characteristic values of the matrix are $\lambda_1 = -1$ and $\lambda_2 = 4$. Corresponding characteristic vectors are $(3, 2)$ and $(1, -1)$, respectively. Thus the vector functions $\mathbf{x}_1$ and $\mathbf{x}_2$, where

$$\mathbf{x}_1(t) = e^{-t}\begin{bmatrix} 3 \\ 2 \end{bmatrix}, \qquad \mathbf{x}_2(t) = e^{4t}\begin{bmatrix} 1 \\ -1 \end{bmatrix},$$

are nontrivial solutions of the equation $\mathbf{x}' = A\mathbf{x}$. If $X = [\mathbf{x}_1, \mathbf{x}_2]$, we see that $\det X(0) = -5 \neq 0$, so these solutions form a fundamental set. The general solution of the system (6.27) is $\mathbf{x} = c_1\mathbf{x}_1 + c_2\mathbf{x}_2$, or

$$x_1 = 3c_1 e^{-t} + c_2 e^{4t}, \qquad x_2 = 2c_1 e^{-t} - c_2 e^{4t}.$$

Let us now consider the general case of an equation

$$\mathbf{x}' = A\mathbf{x}, \tag{6.29}$$

where A is an $n \times n$ constant matrix. A vector function $\mathbf{x}$ of the form

$$\mathbf{x}(t) = e^{\lambda t}\mathbf{k}, \tag{6.30}$$

where $\mathbf{k}$ is a constant vector, is a solution if and only if

$$\lambda e^{\lambda t}\mathbf{k} = e^{\lambda t}A\mathbf{k}$$

or

$$(\lambda I - A)\mathbf{k} = \mathbf{0}.$$

Thus a vector function of the form (6.30) is a nontrivial solution when and only when λ is a characteristic value of A and $\mathbf{k}$ is a corresponding characteristic vector. The question now is whether it is possible to find a fundamental set of solutions of the form (6.30).

Theorem 6.3   Let the $n \times n$ matrix A possess n real linearly independent characteristic vectors $\mathbf{k}_1, \mathbf{k}_2, \ldots, \mathbf{k}_n$, and let λ_i be the real characteristic value that corresponds to $\mathbf{k}_i$. (The numbers $\lambda_1, \lambda_2, \ldots, \lambda_n$ need not all be distinct.) Then the n vector functions $\mathbf{u}_1, \mathbf{u}_2, \ldots, \mathbf{u}_n$, where

$$\mathbf{u}_i(t) = e^{\lambda_i t}\mathbf{k}_i \tag{6.31}$$

for all t, constitute a fundamental set of solutions for Eq. (6.29).

Proof   We have already shown that each of the vector functions (6.31) is a solution. If $U = [\mathbf{u}_1, \mathbf{u}_2, \ldots, \mathbf{u}_n]$, then $U(0) = [\mathbf{k}_1, \mathbf{k}_2, \ldots, \mathbf{k}_n]$. By Theorem 3.4, $U(0)$ is nonsingular. By Theorem 6.2, U is a fundamental matrix for Eq. (6.29).

In Chapter 4, we learned of two situations in which a matrix possessed a complete set of characteristic vectors. One was the case where an $n \times n$ matrix possessed n distinct characteristic values (Corollary to Theorem 4.3). The other was the case of a real symmetric matrix (Theorem 4.5). If a matrix does not fall into either of these categories, it may or may not possess a complete set of characteristic vectors. (There is no easy criterion to determine whether it does.) If the matrix A does not possess a complete set of characteristic vectors, the methods of the next section can be used to find a fundamental set of solutions of the system $\mathbf{x}' = A\mathbf{x}$.

It may happen that A possesses a set of n linearly independent characteristic vectors, some of which are complex. In this case it is still possible to find a fundamental set of real solutions by using the method of characteristic values. If $\lambda_1 = \alpha + i\beta$, $\beta \neq 0$, is a characteristic value of A, then by Theorem 4.1, so is $\lambda_2 = \alpha - i\beta$. If $\mathbf{k}_1 = \mathbf{r} + i\mathbf{s}$ is a characteristic vector corresponding to λ_1 then $\mathbf{k}_2 = \mathbf{r} - i\mathbf{s}$ is a characteristic vector corresponding to λ_2. The vector functions

$$\mathbf{w}_1(t) = (\mathbf{r} + i\mathbf{s})e^{(\alpha + i\beta)t} = \mathbf{u}(t) + i\mathbf{v}(t),$$
$$\mathbf{w}_2(t) = (\mathbf{r} - i\mathbf{s})e^{(\alpha - i\beta)t} = \mathbf{u}(t) - i\mathbf{v}(t),$$

are complex solutions of Eq. (6.29). The real and imaginary parts,

$$\mathbf{u}(t) = e^{\alpha t}(\mathbf{r} \cos \beta t - \mathbf{s} \sin \beta t),$$
$$\mathbf{v}(t) = e^{\alpha t}(\mathbf{r} \sin \beta t + \mathbf{s} \cos \beta t),$$

are real solutions. Thus to each pair of complex conjugate characteristic vectors corresponds a pair of real vector solutions of the differential equation. It can be shown (Exercise 12) that the n real solutions obtained by this process are linearly independent.

To consider an example, let

$$A = \begin{bmatrix} -3 & -1 \\ 2 & -1 \end{bmatrix}.$$

Then

$$p(\lambda) = \begin{vmatrix} \lambda + 3 & 1 \\ -2 & \lambda + 1 \end{vmatrix} = \lambda^2 + 4\lambda + 5;$$

therefore the characteristic values of A are

$$\lambda_1 = -2 + i, \qquad \lambda_2 = -2 - i.$$

The equation $(\lambda_1 I - A)\mathbf{k} = \mathbf{0}$ corresponds to the system

$$(1 + i)k_1 + k_2 = 0$$
$$-2k_1 + (-1 + i)k_2 = 0.$$

A characteristic vector is

$$\begin{bmatrix} -1 \\ 1 + i \end{bmatrix} = \begin{bmatrix} -1 \\ 1 \end{bmatrix} + i \begin{bmatrix} 0 \\ 1 \end{bmatrix}.$$

Then

$$e^{(-2 + i)t} \begin{bmatrix} -1 \\ 1 + i \end{bmatrix} = e^{-2t} \left(\begin{bmatrix} -1 \\ 1 \end{bmatrix} + i \begin{bmatrix} 0 \\ 1 \end{bmatrix} \right)(\cos t + i \sin t)$$

is a complex solution of the equation $\mathbf{x}' = A\mathbf{x}$. The real and imaginary parts are

$$\mathbf{u}(t) = e^{-2t}(\cos t) \begin{bmatrix} -1 \\ 1 \end{bmatrix} - e^{-2t}(\sin t) \begin{bmatrix} 0 \\ 1 \end{bmatrix}$$

and

$$\mathbf{v}(t) = e^{-2t}(\sin t) \begin{bmatrix} -1 \\ 1 \end{bmatrix} + e^{-2t}(\cos t) \begin{bmatrix} 0 \\ 1 \end{bmatrix},$$

respectively. The functions $\mathbf{u}$ and $\mathbf{v}$ are linearly independent real solutions.

As mentioned earlier, it is not possible to find a fundamental set of solutions of the form (6.31) if the matrix A does not have a complete set of characteristic vectors. A procedure for finding a fundamental set in this case is developed in the next section.

Exercises for Section 6.6

In Exercises 1–11, using the method of this section, find the general solution of the equation $\mathbf{x}' = A\mathbf{x}$, where A is the given matrix. If an initial condition is given, also find the solution that satisfies the condition.

1. $\begin{bmatrix} -2 & 4 \\ 1 & 1 \end{bmatrix}$, $\mathbf{x}(0) = \begin{bmatrix} -2 \\ 3 \end{bmatrix}$

2. $\begin{bmatrix} -3 & 2 \\ 1 & -2 \end{bmatrix}$

3. $\begin{bmatrix} 2 & 4 \\ -2 & -2 \end{bmatrix}$, $\mathbf{x}(0) = \begin{bmatrix} 1 \\ 3 \end{bmatrix}$

4. $\begin{bmatrix} -1 & 2 \\ -1 & -3 \end{bmatrix}$

5. $\begin{bmatrix} -2 & 0 \\ 0 & -2 \end{bmatrix}$

6. $\begin{bmatrix} 3 & 0 & -1 \\ -2 & 2 & 1 \\ 8 & 0 & -3 \end{bmatrix}$, $\mathbf{x}(0) = \begin{bmatrix} -1 \\ 2 \\ -8 \end{bmatrix}$

7. $\begin{bmatrix} -2 & 2 & 1 \\ 0 & -1 & 0 \\ 2 & -2 & -1 \end{bmatrix}$

8. $\begin{bmatrix} 3 & -4 & 4 \\ 4 & -5 & 4 \\ 4 & -4 & 3 \end{bmatrix}$, $\mathbf{x}(0) = \begin{bmatrix} 2 \\ 1 \\ -1 \end{bmatrix}$

9. $\begin{bmatrix} -3 & 0 & -3 \\ 1 & -2 & 3 \\ 1 & 0 & 1 \end{bmatrix}$

10. $\begin{bmatrix} 0 & 4 & 0 \\ -1 & 0 & 0 \\ 1 & 4 & -1 \end{bmatrix}$

11. $\begin{bmatrix} 5 & -5 & -5 \\ -1 & 4 & 2 \\ 3 & -5 & -3 \end{bmatrix}$

12. Let $\mathbf{w}_1 = \mathbf{u} + i\mathbf{v}$ and $\mathbf{w}_2 = \mathbf{u} - i\mathbf{v}$ be linearly independent elements of the space of complex vector functions.
    (a) Show that $\mathbf{u}$ and $\mathbf{v}$ are linearly independent elements of the space of real vector functions.
    (b) Show that any linearly independent set of complex vector functions that contains $\mathbf{w}_1$ and $\mathbf{w}_2$ remains linearly independent when $\mathbf{w}_1$ and $\mathbf{w}_2$ are replaced by $\mathbf{u}$ and $\mathbf{v}$.

13. Let $P(r) = r^n + a_1 r^{n-1} + \cdots + a_{n-1} r + a_n$. If the differential equation $P(D)y = 0$ is replaced by a first-order system, show that the characteristic polynomial of the matrix of that system is P.

6.7

REPEATED CHARACTERISTIC VALUES

Let us review the case of a single differential equation whose auxiliary polynomial has a multiple zero. We recall that the auxiliary polynomial of the equation

$$\frac{d^n x}{dt^n} + a_1 \frac{d^{n-1} x}{dt^{n-1}} + \cdots + a_{n-1} \frac{dx}{dt} + a_n x = 0$$

is

$$P(\lambda) = \lambda^n + a_1 \lambda^{n-1} + \cdots + a_{n-1} \lambda + a_n$$
$$= (\lambda - \lambda_1)^{m_1}(\lambda - \lambda_2)^{m_2} \cdots (\lambda - \lambda_r)^{m_r},$$

where $\lambda_1, \lambda_2, \ldots, \lambda_r$ are the distinct zeros and $m_1, m_2, \ldots, m_r$ are their respective multiplicities. If λ_1 has multiplicity 2, then $e^{\lambda_1 t}$ and $t e^{\lambda_1 t}$ are independent solutions. If λ_1 has multiplicity 3, then $e^{\lambda_1 t}$, $t e^{\lambda_1 t}$, and $t^2 e^{\lambda_1 t}$ are independent solutions, and so on.

In the case of a system,

$$\mathbf{x}' = A\mathbf{x},$$

where $\mathbf{x}$ is an n-dimensional vector, the situation is slightly different. Suppose that the characteristic polynomial of A is

$$p(\lambda) = \det(\lambda I - A) = (\lambda - \lambda_1)^{m_1}(\lambda - \lambda_2)^{m_2} \cdots (\lambda - \lambda_r)^{m_r},$$

where $\lambda_1, \lambda_2, \ldots, \lambda_r$ are the distinct characteristic values of A. Concentrating on λ_1, we know (from the previous section) that there is a solution of the form $e^{\lambda_1 t}\mathbf{k}_1$, where $\mathbf{k}_1$ is a characteristic vector of A that corresponds to λ_1. Now if λ_1 has multiplicity 2 ($m_1 = 2$), it turns out that there are two independent solutions

$$\mathbf{x}_1 = e^{\lambda_1 t}\mathbf{k}_1, \qquad \mathbf{x}_2 = e^{\lambda_1 t}(\mathbf{k}_2 + \mathbf{k}_3 t)$$

corresponding to λ_1. The vector $\mathbf{k}_3$ will not be zero if the dimension of the space V_1 of characteristic vectors associated with λ_1 is less than 2.

If λ_1 has multiplicity 3 ($m_1 = 3$), there are three independent solutions:

$$\mathbf{x}_1 = e^{\lambda_1 t}\mathbf{k}_1, \qquad \mathbf{x}_2 = e^{\lambda_1 t}(\mathbf{k}_2 + \mathbf{k}_3 t), \qquad \mathbf{x}_3 = e^{\lambda_1 t}(\mathbf{k}_4 + \mathbf{k}_5 t + \mathbf{k}_6 t^2).$$

Analogous results hold for higher multiplicities.

Let us consider a system of two equations. Then A is a 2×2 matrix with characteristic polynomial $p(\lambda) = (\lambda - \lambda_1)^2$. If A has two independent characteristic vectors $\mathbf{k}_1$ and $\mathbf{k}_2$ corresponding to the characteristic value λ_1, then

$$\mathbf{x}_1 = e^{\lambda_1 t}\mathbf{k}_1, \qquad \mathbf{x}_2 = e^{\lambda_1 t}\mathbf{k}_2$$

are linearly independent solutions and, according to Theorem 6.3, form a fundamental set. Suppose, however, that all characteristic vectors are multiples of $\mathbf{k}_1$. Then $\mathbf{x}_1 = e^{\lambda_1 t}\mathbf{k}_1$ is a nontrivial solution, but we must find a second independent solution. We make a change of dependent variable $\mathbf{x} = K\mathbf{y}$, where K is a nonsingular 2×2 matrix. Then $K\mathbf{y}' = AK\mathbf{y}$, or

$$\mathbf{y}' = K^{-1}AK\mathbf{y}.$$

As was shown in Section 4.3, K can be chosen so that

$$K^{-1}AK = \begin{bmatrix} \lambda_1 & 1 \\ 0 & \lambda_1 \end{bmatrix}.$$

Then our system for $\mathbf{y}$ becomes

$$y_1' = \lambda_1 y_1 + y_2, \qquad y_2' = y_2.$$

We find the solutions of this system to be

$$y_1 = c_1 t e^{\lambda_1 t} + c_2 e^{\lambda_1 t}, \qquad y_2 = c_1 e^{\lambda_1 t}$$

or, in vector form,

$$\mathbf{y} = \begin{bmatrix} c_2 \\ c_1 \end{bmatrix} e^{\lambda_1 t} + \begin{bmatrix} c_1 \\ 0 \end{bmatrix} t e^{\lambda_1 t}.$$

All solutions of the original system $\mathbf{x}' = A\mathbf{x}$ are of the form

$$\mathbf{x} = K\mathbf{y} = K \begin{bmatrix} c_2 \\ c_1 \end{bmatrix} e^{\lambda_1 t} + K \begin{bmatrix} c_1 \\ 0 \end{bmatrix} t e^{\lambda_1 t}$$

or

$$\mathbf{x} = e^{\lambda_1 t}(\mathbf{k}_2 + \mathbf{k}_3 t).$$

The general case of a system of n equations can be analyzed in similar fashion by making use of the Jordan canonical form (Section 4.3) of the matrix of the system.

Example   The system $\mathbf{x}' = A\mathbf{x}$, with

$$A = \begin{bmatrix} -3 & 1 \\ -1 & -1 \end{bmatrix},$$

has the characteristic polynomial

$$p(\lambda) = \begin{vmatrix} \lambda + 3 & -1 \\ 1 & \lambda + 1 \end{vmatrix} = (\lambda + 2)^2.$$

Thus $\lambda_1 = -2$ is a characteristic value with multiplicity 2. If $\mathbf{k}_1 = (a_1, a_2)$ is a characteristic vector, then its components satisfy the system

$$a_1 - a_2 = 0$$
$$a_1 - a_2 = 0.$$

All characteristic vectors are of the form $a(1, 1)$, with $a \neq 0$. Thus the dimension of the space of characteristic vectors associated with λ_1 is 1. One solution of the system is

$$\mathbf{x}_1 = \begin{bmatrix} 1 \\ 1 \end{bmatrix} e^{-2t}.$$

From the previous discussion, we know that there is a second independent solution of the form

$$\mathbf{x}_2 = (\mathbf{k}_2 + \mathbf{k}_3 t)e^{\lambda_1 t}.$$

The requirement that $\mathbf{x}_2' = A\mathbf{x}_2$ yields the condition

$$(\lambda_1 \mathbf{k}_2 + \lambda_1 \mathbf{k}_3 t + \mathbf{k}_3)e^{\lambda_1 t} = (A\mathbf{k}_2 + A\mathbf{k}_3 t)e^{\lambda_1 t}.$$

Equating terms in $e^{\lambda_1 t}$ and $te^{\lambda_1 t}$, we obtain the relations

$$\lambda_1 \mathbf{k}_2 + \mathbf{k}_3 = A\mathbf{k}_2$$
$$\lambda_1 \mathbf{k}_3 = A\mathbf{k}_3.$$

From the second equation we see that $\mathbf{k}_3$ must be a characteristic vector of A that corresponds to λ_1. The first equation becomes

$$(\lambda_1 I - A)\mathbf{k}_2 = -\mathbf{k}_3;$$

it corresponds to a nonhomogeneous equation for $\mathbf{k}_2$ once $\mathbf{k}_3$ has been chosen.[8] Taking $\mathbf{k}_3 = (1, 1)$ and $\mathbf{k}_2 = (b_1, b_2)$, we have

$$b_1 - b_2 = -1$$
$$b_1 - b_2 = -1.$$

Thus b_2 can be chosen arbitrarily and $b_1 = b_2 - 1$. A particularly simple choice is $b_2 = 1$, $b_1 = 0$. Then $\mathbf{k}_2 = (0, 1)$ and a second independent solution of the system of differential equations is

$$\mathbf{x}_2 = \begin{bmatrix} 1 \\ 1 \end{bmatrix} te^{-2t} + \begin{bmatrix} 0 \\ 1 \end{bmatrix} e^{-2t}.$$

The general solution is

$$\mathbf{x} = c_1 \mathbf{x}_1 + c_2 \mathbf{x}_2.$$

In component form,

$$x_1 = (c_1 + c_2 t)e^{-2t}$$
$$x_2 = (c_1 + c_2)e^{-2t} + c_2 te^{-2t}.$$

The truth of the following theorem follows from the results of our investigations in this section.

---

[8] In this particular example there is essentially only one possible choice for $\mathbf{k}_3$. In other examples $\mathbf{k}_3$ may have to be chosen with some care so that the equation for $\mathbf{k}_2$ has a solution.

Theorem 6.4   Every solution of the system $\mathbf{x}' = A\mathbf{x}$ approaches zero as t becomes infinite if and only if all the characteristic values of A have negative real parts.

Once the characteristic polynomial of A has been found (often no easy task)[9], the Routh-Hurwitz criterion (Theorem 5.6) can be used to determine whether the characteristic values all have negative real parts. Examples are presented in the exercises.

In the next three sections, we shall develop an alternative method for solving linear systems with constant coefficients. This method does not require a separate treatment for systems with multiple characteristic values. Also, the method is particularly convenient for finding the specific solution of an initial value problem. Although the derivations may appear formidable, the actual use of the formulas for the solutions is straightforward. A system with multiple characteristic values can be solved more easily by this method than by the method of this section.

───────────────── **Exercises for Section 6.7** ─────────────────

In Exercises 1–10, find a fundamental set of solutions for the problem $\mathbf{x}' = A\mathbf{x}$, where A is as specified.

1. $\begin{bmatrix} -1 & 1 \\ -4 & 3 \end{bmatrix}$ 2. $\begin{bmatrix} 2 & 1 \\ -9 & -4 \end{bmatrix}$

3. $\begin{bmatrix} 5 & 2 \\ -2 & 1 \end{bmatrix}$ 4. $\begin{bmatrix} -3 & 1 \\ -9 & 3 \end{bmatrix}$

5. $\begin{bmatrix} 3 & 2 \\ -8 & -5 \end{bmatrix}$ 6. $\begin{bmatrix} -1 & 1 \\ -4 & -5 \end{bmatrix}$

7. $\begin{bmatrix} 0 & 1 & 1 \\ 1 & 1 & -1 \\ -2 & 1 & 3 \end{bmatrix}$

8. $\begin{bmatrix} -2 & 1 & -1 \\ 3 & -3 & 4 \\ 3 & -1 & 2 \end{bmatrix}$

9. $\begin{bmatrix} -3 & 3 & 2 \\ -4 & 5 & 4 \\ 4 & -6 & -5 \end{bmatrix}$

10. $\begin{bmatrix} -3 & 1 & 0 \\ -6 & 2 & 1 \\ 2 & -1 & -2 \end{bmatrix}$

11. The second order equation $y'' + 2y' + y = 0$ posseses the linearly independent solutions $y_1 = e^{-t}$, $y_2 = te^{-t}$. Find the first-order system $\mathbf{x}' = A\mathbf{x}$ associated with the system and show that it does not possess a solution of the form $\mathbf{x} = te^{-t}\mathbf{k}$.

12. Use Theorem 6.4 and the Routh-Hurwitz criterion to determine whether all solutions of the system $\mathbf{x}' = A\mathbf{x}$ approach zero as t becomes infinite, if A is the given matrix.

(a) $\begin{bmatrix} -2 & 1 & 1 \\ 3 & 2 & 5 \\ -7 & 0 & -2 \end{bmatrix}$

(b) $\begin{bmatrix} -4 & 4 & 3 \\ -7 & 8 & 7 \\ 6 & -7 & -6 \end{bmatrix}$

(c) $\begin{bmatrix} 1 & 2 & 2 & 8 \\ 1 & 1 & 3 & 7 \\ 1 & 1 & 2 & 7 \\ -1 & -1 & -2 & -6 \end{bmatrix}$

────────────────────

[9] Numerical methods for finding the characteristic polynomial are described in several of the references. See, e.g., Cullen (1991) and Fraleigh (1990).

6.8
SERIES OF
MATRICES

In Chapter 1 we saw that the solution of the initial value problem

$$x' = ax, \qquad x(0) = k$$

was

$$x(t) = e^{at}k.$$

We shall presently show that the solution of the problem

$$\mathbf{x}' = A\mathbf{x}, \qquad \mathbf{x}(0) = \mathbf{k}$$

can be written as

$$\mathbf{x}(t) = e^{tA}\mathbf{k},$$

where, of course, A is a matrix. First, however, it is necessary to define what is meant by e with a matrix exponent.

In Section 4.5 we defined polynomial functions with square matrix arguments. Now we shall try to define quantities such as e^A and $\sin A$ by means of the formulas

$$e^A = I + A + \frac{1}{2!} A^2 + \cdots + \frac{1}{k!} A^k + \cdots,$$

$$\sin A = A - \frac{1}{3!} A^3 + \frac{1}{5!} A^5 - \cdots + (-1)^k \frac{1}{(2k-1)!} A^{2k-1} + \cdots.$$

In this approach we attempt to replace the number x in the relations

$$e^x = \sum_{k=0}^{\infty} \frac{1}{k!} x^k, \qquad \sin x = \sum_{k=1}^{\infty} \frac{(-1)^k}{(2k-1)!} x^{2k-1}$$

by a square matrix A. First, however, we must define what we mean by an infinite series of matrices.

We define the *norm* of an $m \times n$ matrix A, written $|A|$, as

$$|A| = \max_{\substack{1 \le i \le m \\ 1 \le j \le n}} |a_{ij}|.$$

Thus the norm of A is simply the largest of the absolute values of all the elements of A. It follows from this definition that

$$|cA| = |c|\,|A|$$

for every number c. If A and B are matrices of the same size it can be shown (Exercise 1) that

$$|A + B| \le |A| + |B|.$$

Also, if A is an $m \times n$ matrix and B is an $n \times q$ matrix, then

$$|AB| \leq n|A||B|.$$

To see this, let $C = AB$. Then

$$c_{ij} = a_{i1}b_{1j} + a_{i2}b_{2j} + \cdots + a_{in}b_{nj}$$

and

$$
\begin{aligned}
|c_{ij}| &\leq |a_{i1}||b_{1j}| + |a_{i2}||b_{2j}| + \cdots + |a_{in}||b_{nj}| \\
&\leq |A||B| + |A||B| + \cdots + |A||B| \\
&\leq n|A||B|
\end{aligned}
$$

for $1 \leq i \leq m$ and $1 \leq j \leq q$. Hence $|C| \leq n|A||B|$.

Let

$$(A_1, A_2, A_3, \ldots)$$

be an infinite sequence of $m \times n$ matrices. This sequence is said to converge to the $m \times n$ matrix A if

$$\lim_{k \to \infty} |A_k - A| = 0.$$

If the sequence converges to A we write

$$\lim_{k \to \infty} A_k = A.$$

Notice that, for each positive integer k, $|A_k - A|$ is a real number. Let the element in the ith row and jth column of A_k be denoted by $a_{ij}^{(k)}$. Since

$$|A_k - A| = \max_{i,j} |a_{ij}^{(k)} - a_{ij}|,$$

it can be seen that the sequence converges to A if and only if the sequence of numbers

$$(a_{ij}^{(1)}, a_{ij}^{(2)}, a_{ij}^{(3)}, \ldots)$$

converges to a_{ij} for $1 \leq i \leq m$ and $1 \leq j \leq n$. For example, the sequence of matrices whose first few terms are

$$\begin{bmatrix} 2 & 1 \\ 3 & 4 \end{bmatrix}, \quad \begin{bmatrix} 0 & 1 \\ 2 & -2 \end{bmatrix}, \quad \begin{bmatrix} -1 & 3 \\ 2 & 5 \end{bmatrix}, \ldots$$

converges to the matrix

$$\begin{bmatrix} a & b \\ c & d \end{bmatrix}$$

if and only if the sequences

$$(2, 0, -1, \ldots), (1, 1, 3, \ldots), (3, 2, 2, \ldots), (4, -2, 5, \ldots)$$

converge to a, b, c, and d, respectively.

In order to define convergence for an infinite series of $m \times n$ matrices

$$\sum_{k=1}^{\infty} B_k,$$

we form the sequence

$$(S_1, S_2, S_3, \ldots),$$

where

$$S_1 = B_1, \qquad S_2 = B_1 + B_2, \qquad S_3 = B_1 + B_2 + B_3,$$

and so on. We say that the series converges to the $m \times n$ matrix B if the sequence of partial sums converges to B. We see that

$$s_{ij}^{(k)} = b_{ij}^{(1)} + b_{ij}^{(2)} + \cdots + b_{ij}^{(k)},$$

where $s_{ji}^{(k)}$ and $b_{ji}^{(p)}$ are the elements in the ith row and jth column of S_k and B_p, respectively. Consequently the series converges to B if and only if the series of numbers

$$\sum_{k=1}^{\infty} b_{ij}^{(k)}$$

converges to b_{ij} for $1 \le i \le m$ and $1 \le j \le n$. Thus the series

$$\begin{bmatrix} 3 & 5 \\ -1 & 1 \end{bmatrix} + \begin{bmatrix} 2 & 0 \\ 4 & 6 \end{bmatrix} + \begin{bmatrix} -2 & 3 \\ 2 & 4 \end{bmatrix} + \cdots$$

converges to

$$\begin{bmatrix} a & b \\ c & d \end{bmatrix}$$

if and only if

$$3 + 2 - 2 + \cdots = a, \qquad 5 + 0 + 3 + \cdots = b,$$
$$-1 + 4 + 2 + \cdots = c, \qquad 1 + 6 + 4 + \cdots = d.$$

Using the notion of the norm of a matrix, we can formulate the following test for the convergence of an infinite series of matrices.

Theorem 6.5   Let $B_1, B_2, B_3, \ldots$ be $m \times n$ matrices. If the series of numbers

$$\sum_{k=1}^{\infty} |B_k| \tag{6.32}$$

converges, then the series of matrices

$$\sum_{k=1}^{\infty} B_k \tag{6.33}$$

converges.

Proof   Fixing i and j, we compare the series

$$\sum_{k=1}^{\infty} b_{ij}^{(k)} \tag{6.34}$$

with the series (6.32). Since

$$|b_{ij}^{(k)}| \le |B_k|, \qquad k = 1, 2, 3, \ldots$$

the series (6.34) converges absolutely for each i and j. Hence the series (6.33) converges.

We now define e^A for a square $n \times n$ matrix A. We first observe that

$$|A^2| \le n|A|\,|A| = n|A|^2,$$
$$|A^3| \le n|A|\,|A^2| \le n^2|A|^3,$$
$$\cdots\cdots\cdots\cdots\cdots\cdots\cdots\cdots$$
$$|A^k| \le n|A|\,|A^{k-1}| \le n^{k-1}|A|^k.$$

Since the series of numbers

$$1 + \sum_{k=1}^{\infty} \frac{1}{k!} n^{k-1}|A|^k$$

converges (as can be shown by the ratio test), the series of matrices

$$I + \sum_{k=1}^{\infty} \frac{1}{k!} A^k$$

converges, by Theorem 6.5. We may now define

$$e^A = I + \sum_{k=1}^{\infty} \frac{1}{k!} A^k$$

for any square matrix A. Note that e^A is a square matrix of the same size as A. Matrices of this form are important in the study of certain systems of differential equations, as we shall see.

Exercises for Section 6.8

1. Prove the inequality $|A + B| \leq |A| + |B|$.

2. Let

$$\sum_{k=1}^{\infty} B_k$$

be a series of matrices that converges to the matrix B. If A and C are matrices such that AB and BC are defined, and if b is a number, prove that the series

$$\sum_{k=1}^{\infty} AB_k, \qquad \sum_{k=1}^{\infty} B_k C, \qquad \sum_{k=1}^{\infty} bB_k$$

converge to AB, BC, and bB, respectively.

3. If

$$\sum_{k=1}^{\infty} A_k = A, \qquad \sum_{k=1}^{\infty} B_k = B,$$

where A and B are of the same size, show that

$$\sum_{k=1}^{\infty} (A_k + B_k) = A + B.$$

4. Show that the series

$$\sum_{k=1}^{\infty} \frac{(-1)^{k-1}}{(2k-1)!} A^{2k-1}$$

converges for every square matrix A.

5. Let A be an $n \times n$ matrix. Show that

$$(I - A)^{-1} = I + A + A^2 + \cdots + A^k + \cdots$$

provided that $|A| < 1/n$. (Use the results of Exercises 2 and 3.)

6. If $D = \text{diag}(d_1, d_2, \ldots, d_n)$ show that

$$e^D = \text{diag}(e^{d_1}, e^{d_2}, \ldots, e^{d_n}).$$

7. Given

$$A = \begin{bmatrix} 0 & 1 & 0 \\ 0 & 0 & 4 \\ 0 & 0 & 0 \end{bmatrix},$$

verify that $A^3 = 0$ and find e^A.

8. If A is similar to a diagonal matrix D and $K^{-1}AK = D$, show that

$$e^A = Ke^D K^{-1}.$$

9. Use the result of Exercise 8 to find e^A if

$$A = \begin{bmatrix} 2 & -1 \\ 3 & -2 \end{bmatrix}.$$

10. Let

$$f(x) = \sum_{k=0}^{\infty} c_k x^k$$

for all x. (The power series converges, and hence converges absolutely, for all x.) If A is any square matrix, show that the series of matrices in the definition

$$f(A) = \sum_{k=0}^{\infty} c_k A^k$$

is convergent.

6.9

THE EXPONENTIAL MATRIX FUNCTION

In the previous section we defined e^A, for a square matrix A, by the formula[10]

$$e^A = I + \sum_{k=1}^{\infty} \frac{1}{k!} A^k. \tag{6.35}$$

In a few cases we can find simple expressions for the elements of e^A directly from this definition. If A is the zero matrix we evidently have

$$e^0 = I.$$

If

$$A = \begin{bmatrix} 2 & 0 \\ 0 & -3 \end{bmatrix},$$

then

$$e^A = \begin{bmatrix} 1 & 0 \\ 0 & 1 \end{bmatrix} + \begin{bmatrix} 2 & 0 \\ 0 & -3 \end{bmatrix} + \frac{1}{2!}\begin{bmatrix} 2^2 & 0 \\ 0 & (-3)^2 \end{bmatrix} + \frac{1}{3!}\begin{bmatrix} 2^3 & 0 \\ 0 & (-3)^3 \end{bmatrix} + \cdots$$

$$= \begin{bmatrix} e^2 & 0 \\ 0 & e^{-3} \end{bmatrix}.$$

More generally, if $D = \operatorname{diag}(d_1, d_2, \ldots, d_n)$ then (Exercise 6, Section 6.8)

$$e^D = \operatorname{diag}(e^{d_1}, e^{d_2}, \ldots, e^{d_n}).$$

Another special case of interest is that where A is a scalar matrix, $A = dI$. It is left as an exercise to show that

$$e^{dI} = e^d I.$$

If some power of A is the zero matrix, then the series (6.27) is finite. For example, if

$$A = \begin{bmatrix} 4 & 1 & -2 \\ -6 & 0 & 3 \\ 8 & 2 & -4 \end{bmatrix},$$

then

$$A^2 = \begin{bmatrix} -6 & 0 & 3 \\ 0 & 0 & 0 \\ -12 & 0 & 6 \end{bmatrix}, \quad A^3 = \begin{bmatrix} 0 & 0 & 0 \\ 0 & 0 & 0 \\ 0 & 0 & 0 \end{bmatrix},$$

---

[10] The formula (6.35) is also used to define e^A when A has complex elements. In this case an understanding of the formula requires a knowledge of the theory of sequences and series of complex numbers.

and $A^n = 0$ if $n \geq 3$. Hence

$$e^A = I + A + \frac{1}{2}A^2 = \begin{bmatrix} 2 & 1 & -\frac{1}{2} \\ -6 & 1 & 3 \\ 2 & 2 & 0 \end{bmatrix}.$$

Usually it is not possible to find the components of e^A directly from Eq. (6.35), and other methods must be used. One such method is described in the next section. For now we seek to discover general properties of matrices of the form e^A.

If A and B are square matrices of the same size, it is natural to inquire whether e^{A+B} is equal to $e^A e^B$. We know that

$$e^A = I + A + \frac{1}{2}A^2 + \cdots$$

$$e^B = I + B + \frac{1}{2}B^2 + \cdots$$

$$e^{A+B} = I + (A + B) + \frac{1}{2}(A + B)^2 + \cdots.$$

If we multiply the series for e^A and e^B in the manner employed for the multiplication of ordinary power series (a procedure that we have certainly not justified) we find that

$$e^A e^B = I + (A + B) + \frac{1}{2}(A^2 + 2AB + B^2) + \cdots.$$

Comparing the terms of second degree here with those in the formula for e^{A+B}, we see that the matrices

$$A^2 + 2AB + B^2$$

and

$$(A + B)^2 = (A + B)(A + B) = A^2 + AB + BA + B^2$$

are not equal unless $BA = AB$. It can be shown, however (See Cullen (1991)), that

$$e^{A+B} = e^A e^B$$

if A and B commute.

In particular, since A and $-A$ commute, we have

$$e^A e^{-A} = e^{A-A} = e^0 = I.$$

Hence e^A is nonsingular and

$$(e^A)^{-1} = e^{-A}.$$

If A is a square matrix then tA is a square matrix for every real number t. Then the formula

$$e^{tA} = I + tA + \frac{1}{2!}\, t^2 A^2 + \cdots + \frac{1}{k!}\, t^k A^k + \cdots \tag{6.36}$$

defines a matrix function. By using facts about the termwise differentiation of ordinary power series, it is possible to show that termwise differentiation of the series in Eq. (6.36) is valid. Then

$$\frac{de^{tA}}{dt} = A + \frac{2}{2!}\, tA^2 + \cdots + \frac{k}{k!}\, t^{k-1} A^k + \cdots$$

$$= A\left[I + tA + \cdots + \frac{1}{(k-1)!}\, t^{k-1} A^{k-1} + \cdots \right]$$

or

$$\frac{de^{tA}}{dt} = A e^{tA}.$$

The importance of the exponential matrix function (6.36) in the theory of linear systems of differential equations with constant coefficients is given by the following theorem.

Theorem 6.6   Let the matrix function U be defined by the formula

$$U(t) = e^{tA}$$

for all t. Then U is a fundamental matrix for the vector differential equation $\mathbf{x}' = A\mathbf{x}$, and the solution of the initial value problem

$$\mathbf{x}' = A\mathbf{x}, \qquad \mathbf{x}(0) = \mathbf{k}$$

is given by the formula

$$\mathbf{x}(t) = U(t)\mathbf{k} = e^{tA}\mathbf{k}.$$

Proof   It follows from the relation

$$U'(t) = A e^{tA} = A U(t)$$

that the column vectors of U are solutions of the equation $\mathbf{x}' = A\mathbf{x}$. Since

$$\det U(0) = \det I = 1,$$

we know by Theorem 6.2 that U is a fundamental matrix. If U has column vectors $\mathbf{u}_1, \mathbf{u}_2, \ldots, \mathbf{u}_n$ and $\mathbf{k}$ has components $k_1, k_2, \ldots, k_n$, then

$$U\mathbf{k} = k_1\mathbf{u}_1 + k_2\mathbf{u}_2 + \cdots + k_n\mathbf{u}_n.$$

Hence $\mathbf{x} = U\mathbf{k}$ is a solution of the equation $\mathbf{x}' = A\mathbf{x}$ and

$$\mathbf{x}(0) = U(0)\mathbf{k} = I\mathbf{k} = \mathbf{k}.$$

Thus $U\mathbf{k}$ is the solution of the initial value problem.

It is interesting to compare the formulas for the solutions of the initial value problems

$$x' = ax, \qquad x(0) = k$$

and

$$\mathbf{x}' = A\mathbf{x}, \qquad \mathbf{x}(0) = \mathbf{k}.$$

The solution of the first is the scalar function $x(t) = e^{at}k$, while the solution of the second is the vector function $\mathbf{x}(t) = e^{tA}\mathbf{k}$. We now have a symbol and a series definition for the solution of the vector problem. In order to make use of these and to learn something of the properties of the solution, we need to know more about the exponential matrix function. This is the purpose of the next section.

The next corollary gives a formula for the solution of the initial value problem when the initial condition is imposed at some other point than $t = 0$. Its proof is left to the reader.

Corollary   The solution of the initial value problem

$$\mathbf{x}' = A\mathbf{x}, \qquad \mathbf{x}(t_0) = \mathbf{k}$$

is given by the formula

$$\mathbf{x}(t) = e^{(t-t_0)A}\mathbf{k}.$$

───────── **Exercises for Section 6.9** ─────────

1. Show that $e^{dI} = e^d I$, where d is a number and I is the identity matrix. Show that $e^0 = I$.

2. If D is a diagonal matrix, show that $e^D e^{-D} = I$.

3. Find e^{tA} if A is

(a) $\begin{bmatrix} 4 & 0 \\ 0 & -1 \end{bmatrix}$.   (b) $\begin{bmatrix} 0 & 0 & 0 \\ 0 & 3 & 0 \\ 0 & 0 & 2 \end{bmatrix}$.

4. Show that $(e^A)^2 = e^{2A}$, and that $(e^A)^m = e^{mA}$ for every positive integer m.

5. Use properties of ordinary power series to show that $de^{tA}/dt = Ae^{tA}$.

6. If $\mathbf{u}$ is a solution of the equation $\mathbf{x}' = A\mathbf{x}$, where A is a constant matrix, show that $\mathbf{v}$, where $\mathbf{v}(t) = \mathbf{u}(t - t_0)$, is also a solution.

7. Prove the Corollary to Theorem 6.6. (Use the result of Exercise 6.)

8. Use Theorem 6.6 to solve the initial value problem $\mathbf{x}' = A\mathbf{x}$, $\mathbf{x}(0) = \mathbf{k}$, where A and $\mathbf{k}$ are as given.

(a) $A = \begin{bmatrix} 1 & 0 \\ 0 & 2 \end{bmatrix}$, $\mathbf{k} = \begin{bmatrix} 3 \\ 5 \end{bmatrix}$

(b) $A = \begin{bmatrix} 2 & 0 & 0 \\ 0 & -1 & 0 \\ 0 & 0 & -1 \end{bmatrix}$, $\mathbf{k} = \begin{bmatrix} 4 \\ 5 \\ 6 \end{bmatrix}$

9. Find the solution of the problem $\mathbf{x}' = A\mathbf{x}$, $\mathbf{x}(t_0) = \mathbf{k}$, where A and $\mathbf{k}$ are as in Exercise 8.

10. Find the elements of e^{tA} if

$$A = \begin{bmatrix} 0 & 1 & -1 \\ 0 & 0 & 2 \\ 0 & 0 & 0 \end{bmatrix}.$$

Suggestion: $A^3 = 0$.

6.10
A MATRIX
METHOD

As explained in the last section, the matrix function $U(t) = e^{tA}$ plays an important role in the study of linear systems of differential equations with constant coefficients. Our object now is to derive a formula for e^{tA} that expresses the matrix function in terms of ordinary (scalar-valued) polynomials and exponential functions. The same techniques can be applied to other matrix functions, as explained in Smiley (1965), but our interest is limited to exponential matrix functions. It turns out that a knowledge of the characteristic values of A is what is necessary to obtain our formula for e^{tA}.

Let $\lambda_1, \lambda_2, \ldots, \lambda_k$ be the *distinct* characteristic values of the $n \times n$ matrix A, with corresponding multiplicities $m_1, m_2, \ldots, m_k$. Then the characteristic polynomial of A is

$$p(\lambda) = \det(\lambda I - A) = (\lambda - \lambda_1)^{m_1}(\lambda - \lambda_2)^{m_2} \cdots (\lambda - \lambda_k)^{m_k}.$$

We pause for a moment to introduce the product notation

$$\prod_{j=1}^{m} c_j = c_1 c_2 \cdots c_m.$$

We also use the notation

$$\prod_{\substack{j=1 \\ j \neq i}}^{m} c_j = c_1 c_2 \cdots c_{i-1} c_{i+1} \cdots c_m.$$

We now define polynomials p_i, $1 \leq i \leq k$, by means of the formula

$$p_i(\lambda) = \prod_{\substack{j=1 \\ j \neq i}}^{k} (\lambda - \lambda_j)^{m_j}.$$

(If $k = 1$, we define $p_1(\lambda) = 1$.) Thus

$$p_1(\lambda) = (\lambda - \lambda_2)^{m_2}(\lambda - \lambda_3)^{m_3} \cdots (\lambda - \lambda_k)^{m_k},$$

$$p_2(\lambda) = (\lambda - \lambda_1)^{m_1}(\lambda - \lambda_3)^{m_3} \cdots (\lambda - \lambda_k)^{m_k},$$

and so on.

If we expand $1/p(\lambda)$ by partial fractions, the terms corresponding to λ_i are

$$\frac{c_{i1}}{\lambda - \lambda_i} + \frac{c_{i2}}{(\lambda - \lambda_i)^2} + \cdots + \frac{c_{im_i}}{(\lambda - \lambda_i)^{m_i}} = \frac{a_i(\lambda)}{(\lambda - \lambda_i)^{m_i}},$$

where the c_{ij} are constants and a_i is a polynomial of degree less than m_i. Then

$$\frac{1}{p(\lambda)} = \frac{a_1(\lambda)}{(\lambda - \lambda_1)^{m_1}} + \frac{a_2(\lambda)}{(\lambda - \lambda_2)^{m_2}} + \cdots + \frac{a_k(\lambda)}{(\lambda - \lambda_k)^{m_k}}.$$

Multiplying through in this equation by $p(\lambda)$, we have

$$1 = a_1(\lambda)p_1(\lambda) + a_2(\lambda)p_2(\lambda) + \cdots + a_k(\lambda)p_k(\lambda).$$

Hence we have

$$I = a_1(A)p_1(A) + a_2(A)p_2(A) + \cdots + a_k(A)p_k(A). \tag{6.37}$$

For each fixed integer i, $1 \le i \le k$, we may write[11]

$$e^{tA} = e^{\lambda_i t}e^{t(A - \lambda_i I)}$$

and hence

$$e^{tA} = e^{\lambda_i t}\sum_{j=0}^{\infty} \frac{1}{j!} t^j(A - \lambda_i I)^j.$$

Premultiplying both members of this equation by $a_i(A)p_i(A)$, and observing that

$$p_i(A)(A - \lambda_i I)^{m_i} = p(A) = 0,$$

and hence that

$$p_i(A)(A - \lambda_i I)^j = 0$$

---

[11] The formulas here are valid when λ_i is complex. See the footnote at the beginning of the previous section.

for all $j \geq m_i$, we have

$$a_i(A)p_i(A)e^{tA} = e^{\lambda_i t}a_i(A)p_i(A) \sum_{j=0}^{m_i-1} \frac{t^j}{j!}(A - \lambda_i I)^j.$$

Summing from $i = 1$ to $i = k$, and using the relation (6.37), we arrive at the formula

$$e^{tA} = \sum_{i=1}^{k} e^{\lambda_i t}a_i(A)p_i(A) \sum_{j=0}^{m_i-1} \frac{t^j}{j!}(A - \lambda_i I)^j. \qquad (6.38)$$

At this point we pause to summarize our results.

Theorem 6.7   Let A be an $n \times n$ matrix with characteristic polynomial.

$$p(\lambda) = (\lambda - \lambda_1)^{m_1}(\lambda - \lambda_2)^{m_2} \cdots (\lambda - \lambda_k)^{m_k},$$

where $\lambda_1, \lambda_2, \ldots, \lambda_k$ are the distinct characteristic values of A. Then e^{tA} is given by formula (6.38), with

$$p_i(\lambda) = \prod_{\substack{j=1 \\ j \neq i}}^{k} (\lambda - \lambda_j)^{mj}$$

and where a_i is the polynomial of degree less than m_i in the expansion

$$\frac{1}{p(\lambda)} = \frac{a_1(\lambda)}{(\lambda - \lambda_1)^{m_1}} + \frac{a_2(\lambda)}{(\lambda - \lambda_2)^{m_2}} + \cdots + \frac{a_k(\lambda)}{(\lambda - \lambda_k)^{m_k}}.$$

Before working some examples, we point out that the formula (6.38) simplifies in the case where A has n distinct characteristic values.

Corollary   If A has distinct characteristic values $\lambda_1, \lambda_2, \ldots, \lambda_n$, then

$$e^{tA} = \sum_{i=1}^{n} e^{\lambda_i t}a_i p_i(A), \qquad (6.39)$$

where p_i is as defined in Theorem 6.7 and a_i is the constant

$$a_i = \frac{1}{p_i(\lambda_i)}, \qquad 1 \leq i \leq n.$$

Proof   Since $k = n$ we have $m_1 = m_2 = \cdots = m_k = 1$. Since the degree of a_i is less than m_i, a_i must be a constant. The equation preceding (6.37) becomes

$$1 = a_1 p_1(\lambda) + a_2 p_2(\lambda) + \cdots + a_n p_n(\lambda).$$

Setting $\lambda = \lambda_i$ in this equation, we see that $1 = a_i p_i(\lambda_i)$, and the formula for a_i follows.

Example 1   We consider the initial value problem

$$x_1' = -x_1 - 2x_2, \qquad x_1(0) = 2$$
$$x_2' = 3x_1 + 4x_2, \qquad x_2(0) = -1.$$

Here

$$A = \begin{bmatrix} -1 & -2 \\ 3 & 4 \end{bmatrix}$$

and

$$p(\lambda) = \begin{vmatrix} \lambda + 1 & 2 \\ -3 & \lambda - 4 \end{vmatrix} = (\lambda - 1)(\lambda - 2).$$

Since $n = k = 2$ we may use the corollary. Let $\lambda_1 = 1$ and $\lambda_2 = 2$. Since

$$p_1(\lambda) = \lambda - 2, \qquad p_2(\lambda) = \lambda - 1,$$

we have

$$a_1 = \frac{1}{p_1(\lambda_1)} = -1, \qquad a_2 = \frac{1}{p_2(\lambda_2)} = 1$$

and

$$p_1(A) = A - 2I, \qquad p_2(A) = A - I.$$

Then

$$e^{tA} = -(A - 2I)e^t + (A - I)e^{2t}.$$

Since

$$-(A - 2I) = \begin{bmatrix} 3 & 2 \\ -3 & -2 \end{bmatrix}, \qquad A - I = \begin{bmatrix} -2 & -2 \\ 3 & 3 \end{bmatrix},$$

we have

$$e^{tA} = \begin{bmatrix} 3 & 2 \\ -3 & -2 \end{bmatrix} e^t + \begin{bmatrix} -2 & -2 \\ 3 & 3 \end{bmatrix} e^{2t}.$$

The solution of the initial value problem is given by

$$\mathbf{x}(t) = e^{tA}\mathbf{k},$$

where

$$\mathbf{k} = \begin{bmatrix} 2 \\ -1 \end{bmatrix}.$$

It is a matter of routine calculation to show that

$$\mathbf{x}(t) = \begin{bmatrix} 4 \\ -4 \end{bmatrix} e^t + \begin{bmatrix} -2 \\ 3 \end{bmatrix} e^{2t}.$$

Example 2   This time we consider the system $\mathbf{x}' = A\mathbf{x}$, where

$$A = \begin{bmatrix} 3 & 1 & -1 \\ -1 & 2 & 1 \\ 2 & 1 & 0 \end{bmatrix}.$$

Here

$$p(\lambda) = \begin{vmatrix} \lambda - 3 & -1 & 1 \\ 1 & \lambda - 2 & -1 \\ -2 & -1 & \lambda \end{vmatrix} = (\lambda - 1)(\lambda - 2)^2.$$

Setting $\lambda_1 = 1$ and $\lambda_2 = 2$ we have $m_1 = 1$ and $m_2 = 2$. Then

$$p_1(\lambda) = (\lambda - 2)^2 = \lambda^2 - 4\lambda + 4, \qquad p_2(\lambda) = \lambda - 1.$$

Using partial fractions, we have

$$\frac{1}{p(\lambda)} = \frac{1}{\lambda - 1} + \frac{-\lambda + 3}{(\lambda - 2)^2}$$

so

$$a_1(\lambda) = 1, \qquad a_2(\lambda) = -\lambda + 3.$$

From the general formula (6.38), with $k = 2$, we have

$$e^{tA} = e^t(A - 2I)^2 + e^{2t}(-A + 3I)(A - I)[I + t(A - 2I)]$$

or

$$e^{tA} = (A^2 - 4A + 4I)e^t + (-A^2 + 4A - 3I)e^{2t} + (A^2 - 3A + 2I)te^{2t}.$$

Here we have used the Cayley-Hamilton theorem to simplify the coefficient of te^{2t}. Since

$$A^2 = \begin{bmatrix} 6 & 4 & -2 \\ -3 & 4 & 3 \\ 5 & 4 & -1 \end{bmatrix}$$

some calculation yields the formula

$$e^{tA} = \begin{bmatrix} -2 & 0 & 2 \\ 1 & 0 & -1 \\ -3 & 0 & 3 \end{bmatrix} e^t + \begin{bmatrix} 3 & 0 & -2 \\ -1 & 1 & 1 \\ 3 & 0 & -2 \end{bmatrix} e^{2t}$$
$$+ \begin{bmatrix} -1 & 1 & 1 \\ 0 & 0 & 0 \\ -1 & 1 & 1 \end{bmatrix} te^{2t}.$$

Suppose that $\lambda_1 = \alpha + i\beta$, $\lambda_2 = \alpha - i\beta$, $\beta \neq 0$, are a pair of complex conjugate characteristic values of the matrix A. Then in the expansion for $1/p(\lambda)$ it turns out that

$$a_2(\lambda) = \overline{a_1(\lambda)}$$

for all real λ. Since A is real,

$$a_2(A) = \overline{a_1(A)}.$$

Consequently the sum of the terms in formula (6.39) that correspond to λ_1 and λ_2 is equal to twice the real part of the sum of the terms that correspond to λ_1.

Example 3   We examine the equation $\mathbf{x}' = A\mathbf{x}$, where

$$A = \begin{bmatrix} -1 & -1 & 0 \\ 0 & -4 & -1 \\ 0 & 5 & 0 \end{bmatrix}.$$

We find that

$$p(\lambda) = \det(\lambda I - A)$$
$$= (\lambda + 1)(\lambda^2 + 4\lambda + 5)$$
$$= (\lambda + 1)[\lambda - (-2 + i)][\lambda - (-2 - i)].$$

Setting $\lambda_1 = -1$, $\lambda_2 = -2 + i$, $\lambda_3 = -2 - i$, we have

$$p_1(\lambda) = \lambda^2 + 4\lambda + 5,$$
$$p_2(\lambda) = (\lambda + 1)(\lambda + 2 + i),$$
$$p_3(\lambda) = (\lambda + 1)(\lambda + 2 - i)$$

and

$$a_1 = \frac{1}{2}, \qquad a_2 = -\frac{1 - i}{4}, \qquad a_3 = \overline{a_2} = \frac{1 + i}{4}.$$

In formula (6.39), the expression corresponding to λ_3 will be the complex conjugate of that corresponding to λ_2. Then

$$e^{tA} = \frac{1}{2}(A^2 + 4A + 5I)e^{-t}$$

$$+ 2 \operatorname{Re}\left\{ -\frac{1}{4}(1 - i)(A + I)[A + (2 + i)I]e^{(-2+i)t} \right\}$$

or

$$e^{tA} = \frac{1}{2}(A^2 + 4A + 5I)e^{-t} - \frac{1}{2}(A^2 + 4A + 3I)e^{-2t}\cos t$$

$$- \frac{1}{2}(A^2 + 2A + I)e^{-2t}\sin t.$$

Exercises for Section 6.10

In Exercises 1–12, find (a) e^{tA}, and (b) the solution of the initial value problem $\mathbf{x}' = A\mathbf{x}$, $\mathbf{x}(0) = \mathbf{k}$.

1. $A = \begin{bmatrix} 0 & 1 \\ 1 & 0 \end{bmatrix}$, $\quad \mathbf{k} = \begin{bmatrix} 3 \\ 1 \end{bmatrix}$

2. $A = \begin{bmatrix} 1 & -2 \\ -2 & 1 \end{bmatrix}$, $\quad \mathbf{k} = \begin{bmatrix} 2 \\ 0 \end{bmatrix}$

3. $A = \begin{bmatrix} 0 & -1 \\ 1 & 2 \end{bmatrix}$, $\quad \mathbf{k} = \begin{bmatrix} 2 \\ -3 \end{bmatrix}$

4. $A = \begin{bmatrix} -1 & -1 \\ 1 & -3 \end{bmatrix}$, $\quad \mathbf{k} = \begin{bmatrix} 2 \\ 1 \end{bmatrix}$

5. $A = \begin{bmatrix} 1 & -5 \\ 2 & -1 \end{bmatrix}$, $\quad \mathbf{k} = \begin{bmatrix} 3 \\ 0 \end{bmatrix}$

6. $A = \begin{bmatrix} 3 & -2 \\ 1 & 1 \end{bmatrix}$, $\quad \mathbf{k} = \begin{bmatrix} 4 \\ 1 \end{bmatrix}$

7. $A = \begin{bmatrix} 2 & 1 & 0 \\ -2 & -1 & 2 \\ 1 & 1 & 1 \end{bmatrix}$, $\quad \mathbf{k} = \begin{bmatrix} 3 \\ 0 \\ -3 \end{bmatrix}$

8. $A = \begin{bmatrix} 1 & 2 & -1 \\ -1 & 1 & 1 \\ 2 & 2 & -2 \end{bmatrix}$, $\quad \mathbf{k} = \begin{bmatrix} 0 \\ 1 \\ 0 \end{bmatrix}$

9. $A = \begin{bmatrix} 3 & -2 & 1 \\ 2 & -1 & 1 \\ -4 & 4 & 1 \end{bmatrix}$, $\quad \mathbf{k} = \begin{bmatrix} 1 \\ 1 \\ 0 \end{bmatrix}$

10. $A = \begin{bmatrix} -3 & 0 & 2 \\ 2 & -1 & -1 \\ -4 & 0 & 3 \end{bmatrix}$, $\quad \mathbf{k} = \begin{bmatrix} 1 \\ 0 \\ 1 \end{bmatrix}$

11. $A = \begin{bmatrix} -1 & -1 & 1 \\ 0 & 2 & 0 \\ -3 & -1 & 3 \end{bmatrix}$, $\quad \mathbf{k} = \begin{bmatrix} 0 \\ 2 \\ 2 \end{bmatrix}$

12. $A = \begin{bmatrix} -1 & 0 & 0 \\ -1 & 0 & 3 \\ 3 & -3 & 0 \end{bmatrix}$, $\quad \mathbf{k} = \begin{bmatrix} 0 \\ 1 \\ 1 \end{bmatrix}$

13. If $\lambda_2 = \bar{\lambda}_1$, show that $a_2(\lambda) = \overline{a_1(\lambda)}$ for all real λ in the formula for $1/p(\lambda)$.

In Exercises 14 and 15, solve the initial value problem by converting the differential equation to a first-order system and then using the method of this section.

14. $y'' - 2y' + y = 0$, $\quad y(0) = -2$, $\quad y'(0) = 3$

15. $y'' + 3y' + 2y = 0$, $\quad y(0) = 1$, $\quad y'(0) = 2$

6.11
NON-
HOMOGENEOUS
LINEAR SYSTEMS

The theory for nonhomogeneous linear systems parallels that for nonhomogeneous single linear equations. We write our nonhomogeneous linear system in vector form as

$$\mathbf{x}' = A\mathbf{x} + \mathbf{b}, \tag{6.40}$$

where $\mathbf{x}$ and $\mathbf{b}$ are vector functions with n components and A is an $n \times n$ matrix function. Associated with the nonhomogeneous equation is the homogeneous equation

$$\mathbf{x}' = A\mathbf{x}. \tag{6.41}$$

The connection between the two equations is provided by the following theorem.

Theorem 6.8   Let $\mathbf{u}_1, \mathbf{u}_2, \ldots, \mathbf{u}_n$ constitute a fundamental set of solutions for the homogeneous equation (6.41) on an interval $\mathscr{I}$, and let $\mathbf{u}_p$ be any particular solution of the nonhomogeneous equation (6.40) on $\mathscr{I}$. Then the general solution of the equation (6.40) on $\mathscr{I}$ consists of the set of all vector functions of the form

$$c_1\mathbf{u}_1 + c_2\mathbf{u}_2 + \cdots + c_n\mathbf{u}_n + \mathbf{u}_p, \tag{6.42}$$

where $c_1, c_2, \ldots, c_n$ are constants.

Proof   We first verify that every function of the form (6.42) is a solution of equation (6.40). Since

$$\mathbf{u}_p' = A\mathbf{u}_p + \mathbf{b} \quad \text{and} \quad \mathbf{u}_i' = A\mathbf{u}_i \quad \text{for} \quad 1 \le i \le n,$$

we have

$$(c_1\mathbf{u}_1 + \cdots + c_n\mathbf{u}_n + \mathbf{u}_p)' = c_1\mathbf{u}_1' + \cdots + c_n\mathbf{u}_n' + \mathbf{u}_p'$$
$$= c_1A\mathbf{u}_1 + \cdots + c_nA\mathbf{u}_n + A\mathbf{u}_p + \mathbf{b}$$
$$= A(c_1\mathbf{u}_1 + \cdots + c_n\mathbf{u}_n + \mathbf{u}_p) + \mathbf{b}.$$

Hence $c_1\mathbf{u}_1 + \cdots + c_n\mathbf{u}_n + \mathbf{u}_p$ is a solution.

We must next show that every solution of Eq. (6.40) is of the form (6.42). Let $\mathbf{u}$ be any solution. Then $\mathbf{u}' = A\mathbf{u} + \mathbf{b}$ and

$$(\mathbf{u} - \mathbf{u}_p)' = \mathbf{u}' - \mathbf{u}_p' = (A\mathbf{u} + \mathbf{b}) - (A\mathbf{u}_p + \mathbf{b}) = A(\mathbf{u} - \mathbf{u}_p).$$

Hence $\mathbf{u} - \mathbf{u}_p$ is a solution of the homogeneous equation (6.41). By Theorem 6.1, there exist numbers $c_1, c_2, \ldots, c_n$ such that

$$\mathbf{u} - \mathbf{u}_p = c_1\mathbf{u}_1 + c_2\mathbf{u}_2 + \cdots + c_n\mathbf{u}_n$$

or

$$\mathbf{u} = c_1\mathbf{u}_1 + c_2\mathbf{u}_2 + \cdots + c_n\mathbf{u}_n + \mathbf{u}_p.$$

Thus $\mathbf{u}$ is of the form (6.42).

The *method of variation of parameters* for systems, which we shall describe shortly, enables us to construct a particular solution of the nonhomogeneous equation whenever a fundamental matrix for the associated homogeneous equation is known. As in the case of a single linear equation, the method of variation of parameters is extremely important for theoretical purposes.

In what follows, we shall need the notion of the integral of a vector function. If $\mathbf{u}$ is a vector function with components $u_1, u_2, \ldots, u_n$ we define

$$\int_a^b \mathbf{u}(t)\, dt = \begin{bmatrix} \int_a^b u_1(t)\, dt \\ \int_a^b u_2(t)\, dt \\ \vdots \\ \int_a^b u_n(t)\, dt \end{bmatrix}.$$

Thus the integral of $\mathbf{u}$ is the vector whose components are the integrals of the corresponding components of $\mathbf{u}$. For example, if

$$\mathbf{u}(t) = \begin{bmatrix} 3t^2 \\ -2 \\ 2t + 1 \end{bmatrix},$$

then

$$\int_0^t \mathbf{u}(s)\, ds = \begin{bmatrix} t^3 \\ -2t \\ t^2 + t \end{bmatrix}, \qquad \int_0^1 \mathbf{u}(t)\, dt = \begin{bmatrix} 1 \\ -2 \\ 2 \end{bmatrix}.$$

We now proceed with a description of the method of variation of parameters.

Let $\mathbf{u}_1, \mathbf{u}_2, \ldots, \mathbf{u}_n$ constitute a fundamental set of solutions for the equation

$$\mathbf{x}' = A\mathbf{x} \tag{6.43}$$

on an interval $\mathscr{I}$. We attempt to find functions $c_1, c_2, \ldots, c_n$ such that

$$c_1\mathbf{u}_1 + c_2\mathbf{u}_2 + \cdots + c_n\mathbf{u}_n$$

is a solution of the nonhomogeneous equation

$$\mathbf{x}' = A\mathbf{x} + \mathbf{b} \tag{6.44}$$

on $\mathscr{I}$. Let U be the matrix function whose column vectors are $\mathbf{u}_1, \mathbf{u}_2, \ldots, \mathbf{u}_n$ and let $\mathbf{c}$ be the vector function with components $c_1, c_2, \ldots, c_n$. Since (by Theorem 2.4)

$$U\mathbf{c} = c_1 \mathbf{u}_1 + c_2 \mathbf{u}_2 + \cdots + c_n \mathbf{u}_n,$$

we are seeking to determine a vector function $\mathbf{c}$ such that

$$(U\mathbf{c})' = A(U\mathbf{c}) + \mathbf{b} \qquad \text{or} \qquad U'\mathbf{c} + U\mathbf{c}' = AU\mathbf{c} + \mathbf{b}.$$

Since $U' = AU$ we have $U'\mathbf{c} = AU\mathbf{c}$. The requirement for $\mathbf{c}$ becomes

$$U\mathbf{c}' = \mathbf{b}. \tag{6.45}$$

Since U is a fundamental matrix for Eq. (6.43) on $\mathscr{I}$, $U(t)$ is nonsingular for every t in $\mathscr{I}$. Then $U^{-1}(t)$ exists for each t and

$$\mathbf{c}'(t) = U^{-1}(t)\mathbf{b}(t).$$

We may choose

$$\mathbf{c}(t) = \int_{t_0}^{t} U^{-1}(s)\mathbf{b}(s) \, ds,$$

where t_0 is any point in $\mathscr{I}$. Then a particular solution $\mathbf{u}_p = U\mathbf{c}$ of Eq. (6.44) is defined by the formula

$$\mathbf{u}_p(t) = U(t) \int_{t_0}^{t} U^{-1}(s)\mathbf{b}(s) \, ds. \tag{6.46}$$

Note that $\mathbf{u}_p(t_0) = \mathbf{0}$.

If A is a constant matrix with n linearly independent characteristic vectors, we can find a fundamental matrix U by the method of the previous section. The equation $U\mathbf{c}' = \mathbf{b}$ can be solved for the components of $\mathbf{c}'$, and then the components of $\mathbf{c}$ can be found by integration. The vector function $\mathbf{u}_p = U\mathbf{c}$ is a particular solution of the equation $\mathbf{x}' = A\mathbf{x} + \mathbf{b}$.

Example Let us consider the equation $\mathbf{x}' = A\mathbf{x} + \mathbf{b}$, where

$$A = \begin{bmatrix} 4 & -3 \\ 2 & -1 \end{bmatrix}, \qquad \mathbf{b}(t) = \begin{bmatrix} t \\ e^t \end{bmatrix}.$$

The characteristic values of A are found to be $\lambda_1 = 1$ and $\lambda_2 = 2$. Corresponding characteristic vectors are

$$\mathbf{k}_1 = \begin{bmatrix} 1 \\ 1 \end{bmatrix}, \qquad \mathbf{k}_2 = \begin{bmatrix} 3 \\ 2 \end{bmatrix}.$$

Then the vector functions

$$\mathbf{u}_1(t) = \begin{bmatrix} 1 \\ 1 \end{bmatrix} e^t, \qquad \mathbf{u}_2(t) = \begin{bmatrix} 3 \\ 2 \end{bmatrix} e^{2t}$$

are solutions of the equation $\mathbf{x}' = A\mathbf{x}$, and

$$U(t) = \begin{bmatrix} e^t & 3e^{2t} \\ e^t & 2e^{2t} \end{bmatrix}$$

is a fundamental matrix. Then det $U(t) = 2e^{3t} - 3e^{3t} = -e^{3t}$ and, making use of Theorem 2.14, we find that

$$U^{-1}(t) = \begin{bmatrix} -2e^{-t} & 3e^{-t} \\ e^{-2t} & -e^{-2t} \end{bmatrix}.$$

Formula (6.46), with $t_0 = 0$, gives a particular solution:

$$\mathbf{u}_p(t) = \begin{bmatrix} e^t & 3e^{2t} \\ e^t & 2e^{2t} \end{bmatrix} \int_0^t \begin{bmatrix} -2e^{-s} & 3e^{-s} \\ e^{-2s} & -e^{-2s} \end{bmatrix} \begin{bmatrix} s \\ e^s \end{bmatrix} ds$$

$$= \begin{bmatrix} e^t & 3e^{2t} \\ e^t & 2e^{2t} \end{bmatrix} \int_0^t \begin{bmatrix} -2se^{-s} + 3 \\ se^{-2s} - e^{-s} \end{bmatrix} ds$$

$$= \begin{bmatrix} e^t & 3e^{2t} \\ e^t & 2e^{2t} \end{bmatrix} \begin{bmatrix} 2e^{-t}(t + 1) + 3t - 2 \\ e^{-2t}(-\frac{t}{2} - \frac{1}{4}) + e^{-t} - \frac{3}{4} \end{bmatrix}$$

$$= \begin{bmatrix} \frac{t}{2} + \frac{5}{4} + (3t + 1)e^t - \frac{9}{4}e^{2t} \\ t + \frac{3}{2} + 3te^t - \frac{3}{2}e^{2t} \end{bmatrix}.$$

Exercises for Section 6.11

In Exercises 1–8, use the method of variation of parameters to find a particular solution $\mathbf{u}_p$ of the equation $\mathbf{x}' = A\mathbf{x} + \mathbf{b}$, where A and $\mathbf{b}$ are as given.

1. $A = \begin{bmatrix} 2 & 1 \\ -3 & -2 \end{bmatrix}$, $\mathbf{b}(t) = \begin{bmatrix} 2e^t \\ 4e^t \end{bmatrix}$

2. $A = \begin{bmatrix} 0 & 1 \\ -1 & 2 \end{bmatrix}$, $\mathbf{b}(t) = \begin{bmatrix} e^t/(t + 1) \\ 0 \end{bmatrix}$

3. $A = \begin{bmatrix} 2 & 2 \\ -3 & -3 \end{bmatrix}$, $\mathbf{b}(t) = \begin{bmatrix} 1 \\ 2t \end{bmatrix}$

4. $A = \begin{bmatrix} 0 & -1 \\ 2 & 3 \end{bmatrix}$, $\mathbf{b}(t) = \begin{bmatrix} 0 \\ 2te^t \end{bmatrix}$

5. $A = \begin{bmatrix} 3 & 2 \\ -4 & -3 \end{bmatrix}$, $\mathbf{b}(t) = \begin{bmatrix} 2 \cos t \\ 2 \sin t \end{bmatrix}$

6. $A = \begin{bmatrix} 0 & 1 \\ -1 & 0 \end{bmatrix}$, $\mathbf{b}(t) = \begin{bmatrix} 4 \sin t \\ 0 \end{bmatrix}$

7. $A = \begin{bmatrix} 1 & -1 & 1 \\ 0 & 0 & 1 \\ 0 & -1 & 2 \end{bmatrix}$, $\mathbf{b}(t) = \begin{bmatrix} 0 \\ e^t \\ e^t \end{bmatrix}$

8. $A = \begin{bmatrix} 1 & 1 & 1 \\ 0 & -1 & 0 \\ -2 & -1 & -2 \end{bmatrix}$, $\mathbf{b}(t) = \begin{bmatrix} 1 \\ 0 \\ 2e^{-t} \end{bmatrix}$

9. In the case where A is constant, we may take $U(t) = e^{tA}$. Show that in this case Eq. (6.46) becomes

$$\mathbf{u}_p(t) = \int_{t_0}^{t} e^{(t-s)A} \mathbf{b}(s)\, ds.$$

10. Use the result of Exercise 9 to obtain particular solutions of the system $\mathbf{x}' = A\mathbf{x} + \mathbf{b}$. Take $t_0 = 0$.

(a) $A = \begin{bmatrix} 3 & -2 \\ 1 & 0 \end{bmatrix}$, $\mathbf{b}(t) = -2e^{-t} \begin{bmatrix} 1 \\ 1 \end{bmatrix}$

(b) $A = \begin{bmatrix} 3 & -1 \\ 3 & -1 \end{bmatrix}$, $\mathbf{b}(t) = e^{3t} \begin{bmatrix} 2 \\ 5 \end{bmatrix}$

11. (a) If $F(t)$ is an $n \times n$ matrix function and $a \le b$, show that

$$\left| \int_a^b F(t)\, dt \right| \le \int_a^b |F(t)|\, dt.$$

(b) If the distinct characteristic values of A are $\lambda_1, \lambda_2, \ldots, \lambda_k$, show that there exist polynomials q_i, with nonnegative coefficients, such that

$$|e^{tA}| \le \sum_{i=1}^{k} q_i(t) e^{\alpha_i t}$$

where α_i is the real part of λ_i.

12. If all the characteristic values of A have negative real parts, and if $|\mathbf{b}(t)| \le M$ for $t \ge t_0$, show that all solutions of $\mathbf{x}' = A\mathbf{x} + \mathbf{b}$ are bounded on $[t_0, \infty)$. Use the results of Exercise 9 and the previous exercise.

6.12 MECHANICAL SYSTEMS

In this section we consider some problems that involve the motion of one or more objects. The mathematical description of such a problem involves a system of differential equations in which the unknown functions are the coordinates of the centers of mass of the moving objects.

For our first example we consider the configuration of Fig. 6.3. Two bodies with masses m_1 and m_2, respectively, are suspended by springs as shown. Their displacements from their equilibrium positions are denoted by x_1 and x_2 respectively. At equilibrium, the forces due to gravity are balanced by forces due to a stretching of the springs. We denote by F_1 and F_2 the net forces acting on the moving bodies with masses m_1 and m_2, respectively. Suppose first that $x_1 > 0$ and that $x_2 > x_1$. Then the spring with constant k_2 is stretched by a distance $x_2 - x_1$. The net force acting on the object with mass m_1 (positive direction downward) is

$$F_1 = -k_1 x_1 + k_2(x_2 - x_1), \tag{6.47}$$

while the net force acting on the second body is

$$F_2 = -k_2(x_2 - x_1). \tag{6.48}$$

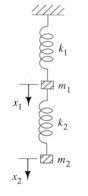

Figure 6.3

If $x_1 < 0$, the first spring is compressed and exerts a downward force on the first object. If $x_2 - x_1 < 0$, the second spring is compressed. It then exerts an upward force on the first object and a downward force on the second. But if either or both of these situations occur, it can be seen that the formulas (6.47) and (6.48) are still valid. Consequently our equations of motion become

$$m_1\ddot{x}_1 = -k_1 x_1 + k_2(x_2 - x_1),$$
$$m_2\ddot{x}_2 = -k_2(x_2 - x_1).$$

(6.49)

This is a linear system with constant coefficients. The nature of the solutions is discussed in Exercise 1.

We next consider the motion of a single body in a plane, subject to a constant gravitational force. It is convenient to work in a rectangular co-ordinate system in the plane. We choose the origin to be at the surface of the earth, with one axis vertical, as shown in Fig. 6.4. According to Newton's law of motion,

$$m\mathbf{a} = \mathbf{F},$$

(6.50)

where m is the mass of the body,

$$\mathbf{a} = \ddot{x}\mathbf{i} + \ddot{y}\mathbf{j}$$

is the acceleration, and

$$\mathbf{F} = 0\mathbf{i} - mg\mathbf{j}$$

is the force acting on the body. Then Eq. (6.50) becomes

$$m(\ddot{x}\mathbf{i} + \ddot{y}\mathbf{j}) = 0\mathbf{i} - mg\mathbf{j}.$$

Equating corresponding components, we arrive at the system

$$m\ddot{x} = 0, \qquad m\ddot{y} = -mg.$$

We denote by (a, b) the position of the body at $t = 0$. We assume that the initial velocity vector $\mathbf{v}_0$ has magnitude v_0 and is inclined at an angle α

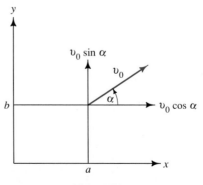

Figure 6.4

with the horizontal. Thus

$$\mathbf{v}_0 = v_0(\mathbf{i}\cos\alpha + \mathbf{j}\sin\alpha).$$

The initial conditions for the system are

$$x(0) = a, \qquad y(0) = b, \qquad \dot{x}(0) = v_0\cos\alpha, \qquad \dot{y}(0) = v_0\sin\alpha.$$

Some specific cases are considered in the exercises.

In some applications that involve the motion of a body in a plane, it is more convenient to work with the polar coordinates (r, θ) of the center of mass of the body. This is the case, for example, when attempting to describe the motion of a satellite about the earth. We introduce the unit vectors $\mathbf{e}_r$ and $\mathbf{e}_\theta$, where

$$\begin{aligned} \mathbf{e}_r &= \mathbf{i}\cos\theta + \mathbf{j}\sin\theta, \\ \mathbf{e}_\theta &= -\mathbf{i}\sin\theta + \mathbf{j}\cos\theta. \end{aligned} \qquad (6.51)$$

The vector $\mathbf{e}_r$ points away from the origin in the direction of increasing r. The vector $\mathbf{e}_\theta$ is perpendicular to $\mathbf{e}_r$ and points in the direction of increasing θ. The situation is illustrated in Fig. 6.5. From formulas (6.51), we deduce the relations

$$\frac{d}{d\theta}\mathbf{e}_r = \mathbf{e}_\theta, \qquad \frac{d}{d\theta}\mathbf{e}_\theta = -\mathbf{e}_r. \qquad (6.52)$$

We resolve the force $\mathbf{F}$ acting on the body into radial and circumferential components, writing

$$\mathbf{F} = F_r\mathbf{e}_r + F_\theta\mathbf{e}_\theta.$$

In many important cases, $F_\theta = 0$. For instance, when the origin is at the

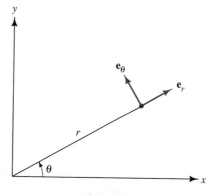

Figure 6.5

center of the earth the force exerted on a satellite of mass m by the earth is

$$\mathbf{F} = -\frac{mgR^2}{r^2}\,\mathbf{e}_r,$$ (6.53)

where R is the radius of the (spherical) earth.

Since

$$\mathbf{i}x + \mathbf{j}y = \mathbf{i}r\cos\theta + \mathbf{j}r\sin\theta = r\mathbf{e}_r,$$

the velocity of the moving body is (making use of the relations (6.52))

$$\mathbf{v} = \dot{r}\mathbf{e}_r + r\frac{d\mathbf{e}_r}{d\theta}\frac{d\theta}{dt} = \dot{r}\mathbf{e}_r + r\dot{\theta}\mathbf{e}_\theta$$

and the acceleration is

$$\mathbf{a} = \ddot{r}\mathbf{e}_r + \dot{r}\dot{\theta}\mathbf{e}_\theta + (r\ddot{\theta} + \dot{r}\dot{\theta})\mathbf{e}_\theta - r\dot{\theta}^2\mathbf{e}_r$$

or

$$\mathbf{a} = (\ddot{r} - r\dot{\theta}^2)\mathbf{e}_r + (r\ddot{\theta} + 2\dot{r}\dot{\theta})\mathbf{e}_\theta.$$

Then, by equating corresponding components in the equation

$$m\mathbf{a} = \mathbf{F},$$

we arrive at the system

$$m(\ddot{r} - r\dot{\theta}^2) = F_r,$$
$$m(r\ddot{\theta} + 2\dot{r}\dot{\theta}) = F_\theta.$$ (6.54)

In the special case where $\mathbf{F}$ is given by formula (6.53), we have

$$\ddot{r} - r\dot{\theta}^2 = -gR^2r^{-2},$$
$$r\ddot{\theta} + 2\dot{r}\dot{\theta} = 0.$$ (6.55)

Exercises for Section 6.12

1. (a) Show that the system (6.49) is equivalent to the system

$$P(D)x_1 = 0, \quad x_2 = \frac{1}{k_2}(m_1D^2 + k_1 + k_2)x_1,$$

where

$$P(D) = m_1m_2D^4 + [m_2(k_1 + k_2) + m_1k_2]D^2 + k_1k_2.$$

(b) Show that the polynomial P of part (a) has four distinct pure imaginary zeros, and hence that x_1 and x_2 are bounded functions.

2. (a) Find the equations of motion for the mechanical system of Fig. 6.6.

(b)   Solve the system of equations in Part (a) if $m_1 = m_2 = m$ and $k_1 = k_2 = k_3 = k$. Let $k/m = a^2$.

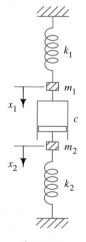

Figure 6.6

(a)   Find the equations of motion for the mechanical system of Fig. 6.7.
(b)   Solve the system of equations in Part (a) if $m_1 = m_2 = 1$, $c = 2$, and $k_1 = k_2 = 4$.

Figure 6.7

4.   Find the equations of motion for the mechanical system of Fig. 6.8.

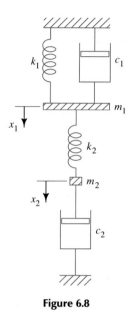

Figure 6.8

5.   A ball is thrown horizontally from the top of a tower of height h, with a velocity v_0. Neglect air resistance.
   (a)   How far from the base of the tower will the ball land?
   (b)   Find the xy equation of the path of the ball and show that it is a parabola.

6.   A ball is thrown horizontally from the top of a tower of height h, with a velocity v_0. Assume that the force due to air resistance is $\mathbf{F} = -c\mathbf{v}$, where c is a positive constant and $\mathbf{v}$ is the velocity vector.
   (a)   Find the coordinates, $x(t)$ and $y(t)$, of the ball.
   (b)   Find the xy equation of the path.

7.   A projectile is fired from a gun located at the origin of the coordinate system in Fig. 6.4. The gun is inclined at an angle α with the horizontal and its muzzle velocity is v_0. Neglect air resistance.
   (a)   How far from the gun will the projectile land?

(b) Find the xy equation of the path of the projectile and show that it is a parabola.

8. In Exercise 7, suppose that the force due to air resistance is $\mathbf{F} = -c\mathbf{v}$, where c is a positive constant and $\mathbf{v}$ is the velocity vector.
   (a) Find the coordinates, $x(t)$ and $y(t)$, of the projectile.
   (b) Find the xy equation of the path.

9. An earth satellite of mass m is attracted toward the center of the earth by a force of magnitude mkr^{-2}, where k is a positive constant and r is the distance to the earth's center. Using a rectangular xy coordinate system with origin at the center of the earth, find the equations of motion of the satellite.

6.13

THE TWO BODY PROBLEM

We consider the motion of two bodies of masses m_1 and m_2, which we refer to as the first body and second body, respectively. They are assumed to attract each other with a force of magnitude km_1m_2/r^2, where k is a positive constant and r is the distance between the centers of mass. We ignore all other forces, such as attractions of other bodies. An example of such a system is furnished by the sun and the earth, if the effects of the moon and other celestial objects are ignored. Another example consists of the earth and a satellite, assuming that the satellite stays sufficiently close to the earth so that attractions of the moon, sun, and other objects is negligible.

Let $\mathbf{P}_1$ and $\mathbf{P}_2$ be the position vectors of the centers of mass with respect to some fixed coordinate system. Then the equations of motion are

$$m_1\ddot{\mathbf{P}}_1 = \frac{km_1m_2}{r^2}\frac{\mathbf{P}_2 - \mathbf{P}_1}{r}, \tag{6.56}$$

$$m_2\ddot{\mathbf{P}}_2 = \frac{km_1m_2}{r^2}\frac{\mathbf{P}_1 - \mathbf{P}_2}{r}. \tag{6.57}$$

Notice that $(\mathbf{P}_2 - \mathbf{P}_1)/r$ is a unit vector directed away from the first body and toward the second.

The vector

$$\mathbf{P} = \frac{m_1\mathbf{P}_1 + m_2\mathbf{P}_2}{m_1 + m_2}$$

is called the position vector of the center of mass of the system. It is represented by an arrow from the origin to a point on the line segment joining the two centers of mass. One way to see this is to observe that

$$\mathbf{P} = \mathbf{P}_1 + \frac{m_2}{m_1 + m_2}(\mathbf{P}_2 - \mathbf{P}_1)$$

and note that $\mathbf{P}_2 - \mathbf{P}_1$ is parallel to the line through the centers of mass. Notice that if m_2 is very small compared with m_1, the center of mass of the system is very near the center of mass of the first body. This is the case if the first body is the sun and the second the earth, or if the first body is the earth and the second a satellite.

Addition of Eqs. (6.56) and (6.57) yields the relation

$$m_1\ddot{\mathbf{P}}_1 + m_2\ddot{\mathbf{P}}_2 = \mathbf{0},$$

from which it follows that $\ddot{\mathbf{P}} = \mathbf{0}$. Thus the center of mass of the system travels in a straight line with constant velocity.

Let us write $\mathbf{R} = \mathbf{P}_2 - \mathbf{P}_1$. The vector $\mathbf{R}$ describes the position of the second body relative to the first. The equations of motion may be written as

$$\ddot{\mathbf{P}}_1 = \frac{km_2}{r^3}\mathbf{R}, \qquad \ddot{\mathbf{P}}_2 = -\frac{km_1}{r^3}\mathbf{R}.$$

Subtracting, we find that

$$\ddot{\mathbf{P}}_2 - \ddot{\mathbf{P}}_1 = -k\frac{m_1 + m_2}{r^3}\mathbf{R}$$

or

$$\ddot{\mathbf{R}} = -K\frac{\mathbf{R}}{r^3}, \tag{6.58}$$

where K is the positive constant $K = k(m_1 + m_2)$.

If the first body does not move, it follows from the inverse square law that the equation of motion of the second body is

$$\ddot{\mathbf{R}} = -km_1 r^{-3}\mathbf{R},$$

but if the first body is allowed to move, Eq. (6.58) says that

$$\ddot{\mathbf{R}} = -k(m_1 + m_2)r^{-3}\mathbf{R}.$$

Thus the relative motion in the second case is the same as when the first body is fixed, but the mass of the first body has been increased from m_1 to $m_1 + m_2$. In the case of an earth satellite, m_2 is very small compared with m_1 and it matters little whether we regard the earth as fixed or moving.

If the first body is the earth, with radius $a = 3960$ miles, then when $r = a$ the force exerted on the second body is

$$k\frac{m_1 m_2}{a^2} = m_2 g,$$

where g is 32.174 ft/sec^2. Thus $km_1 = a^2 g$ and if m_2 is small compared with m_1 we have approximately

$$K = k(m_1 + m_2) \approx km_1 = a^2 g.$$

Some computation shows that

$$K = 1.41 \times 10^{16} \text{ ft}^3/\text{sec}^2 = 1.24 \times 10^{12} \text{ miles}^3/\text{hr}^2.$$

We have obtained a single vector equation for **R**, and by solving this equation we can describe the motion of the second body relative to the first. In what follows, we may think of a coordinate system with origin at the center of mass of the first body, with **R** as the position vector of the second body in this coordinate system. The second body moves in a plane that passes through the origin and contains the velocity vector, since the force of attraction has no component perpendicular to this plane. Using polar coordinates r and θ as described in the last section, we arrive at the system of differential equations

$$\ddot{r} - r\dot{\theta}^2 = -Kr^{-2}, \tag{6.59}$$

$$r\ddot{\theta} + 2\dot{r}\dot{\theta} = 0. \tag{6.60}$$

These equations follow from the vector equation (6.58), with $r = |\mathbf{R}|$.
Multiplying through in Eq. (6.60) by r gives

$$r^2\ddot{\theta} + 2r\dot{r}\dot{\theta} = 0$$

or

$$(r^2\dot{\theta})' = 0,$$

whence

$$r^2\dot{\theta} = c, \tag{6.61}$$

where c is a constant.
We can use Eq. (6.61) to eliminate θ from Eq. (6.59), thus obtaining a single differential equation for r. This equation, which turns out to be

$$\ddot{r} = cr^{-3} - Kr^{-2}, \tag{6.62}$$

does not yield a simple formula for r in terms of t. We shall instead seek the equation of the path of motion in polar coordinates. This equation has the form $r = f(\theta)$, where f is to be determined. Using the fact that $\dot{\theta} = cr^{-2}$, as seen from Eq. (6.61), we have

$$\frac{dr}{dt} = \frac{dr}{d\theta}\frac{d\theta}{dt} = cr^{-2}\frac{dr}{d\theta}$$

and

$$\frac{d^2r}{dt^2} = c^2r^{-4}\frac{d^2r}{d\theta^2} - 2c^2r^{-5}\left(\frac{dr}{d\theta}\right)^2.$$

Substitution of these expressions back into Eq. (6.59) yields the equation

$$\frac{d^2r}{d\theta^2} - 2r^{-1}\left(\frac{dr}{d\theta}\right)^2 - r = -Kc^{-2}r^{-2}.$$

This equation becomes greatly simplified if we make the change of dependent variable $r = 1/u$. Then

$$\frac{dr}{d\theta} = \frac{dr}{du}\frac{du}{d\theta} = -u^{-2}\frac{du}{d\theta},$$

$$\frac{d^2r}{d\theta^2} = -u^{-2}\frac{d^2u}{d\theta^2} + 2u^{-3}\left(\frac{du}{d\theta}\right)^2,$$

and the equation becomes

$$\frac{d^2u}{d\theta^2} + u = Kc^{-2}. \tag{6.63}$$

This linear equation with constant coefficients has the general solution

$$u(\theta) = A\cos(\theta - \alpha) + Kc^{-2},$$

where A and α are arbitrary constants. We shall assume $A \geq 0$. This is no restriction, for if $A < 0$ we can replace α by $\alpha + \pi$. Since $r = 1/u$ we have

$$r(\theta) = \frac{1}{Kc^{-2} + A\cos(\theta - \alpha)} = \frac{c^2K^{-1}}{1 + Ac^2K^{-1}\cos(\theta - \alpha)},$$

where A and α must be determined from initial conditions. Setting

$$\varepsilon = Ac^2K^{-1}, \qquad h = A^{-1},$$

we have

$$r(\theta) = \frac{\varepsilon h}{1 + \varepsilon\cos(\theta - \alpha)}, \tag{6.64}$$

where ε, h, and α are constants that depend on the initial conditions. Here ε and h are not independent, since $\varepsilon h = c^2K^{-1}$. Since $A \geq 0$ we have $\varepsilon \geq 0$.

We shall now determine the nature of the path of motion. Consider a particle that moves in such a way that the ratio of its distance r from the origin to its distance from a fixed line is ε. We know from analytic geometry that the curve described is a parabola, ellipse, or hyperbola, according as $\varepsilon = 1$, $\varepsilon < 1$, or $\varepsilon > 1$. The number ε is called the *eccentricity* of the curve. Suppose that the line is a distance h from the origin and inclined at an angle

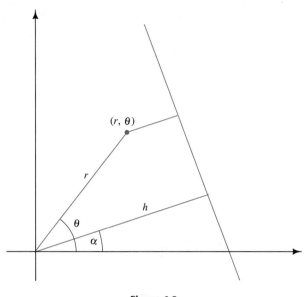

Figure 6.9

α with the vertical, as shown in Fig. 6.9. The distance from the particle to the line is $h - r \cos(\theta - \alpha)$, so

$$\frac{r}{h - r \cos(\theta - \alpha)} = \varepsilon.$$

Upon solving for r, we find that

$$r = \frac{\varepsilon h}{1 + \varepsilon \cos(\theta - \alpha)}.$$

This is Eq. (6.64). It must be remembered that this equation describes the motion of the second body relative to the first. The motion of the second body relative to a fixed coordinate system in space is more complicated.

─────────────── **Exercises for Section 6.13** ───────────────

1. A time $t = 0$ an earth satellite has position $r = a$, $\theta = 0$, and velocity $\dot{r} = 0$, $\dot{\theta} = v/a$. Show that the eccentricity ε of the orbit is given by $\varepsilon = (a^2 v^2 - a)/aK$. From this deduce that the path is an ellipse if and only if $v < [(K + 1)/a]^{1/2}$. Find this critical velocity if $a = 4060$ miles.

2. Show that a body in an elliptical orbit moves in such a way that its position vector from a focus of the ellipse sweeps out area at a constant rate. (This is Kepler's second law of planetary motion. The first law states that the earth moves in an elliptical orbit with the sun at one focus.)

3. A body moves in an elliptical orbit with major axis $2a$. If T is the time for one revolution, show that T is proportional to $a^{3/2}$. Use the result of Exercise 2. (This is Kepler's third law.)

4. If an earth satellite stays over a fixed point on earth, show that the orbit is circular and find the height above the earth. Suggestion: show that $\dot{\theta}$ is constant.

5. Given that the earth moves in an elliptical orbit about the sun, with the sun at one focus, and that the sun exerts a force $-F(r)\mathbf{e}_r$ on the earth, show that $F(r)$ must be proportional to r^{-2}. (Thus deduce the inverse square law from the assumption that the earth's orbit is an ellipse.) Suggestion: use Eqs. (6.64) and (6.54).

6. Derive Eq. (6.63).

7. Derive Eq. (6.62) by putting $r = 1/u$ in Eqs. (6.59) and (6.60) and then finding an equation for u as a function of θ.

8. A body moving under the influence of the earth's gravitational field is observed to have the position $r = a$, $\theta = 0$, and velocity $\dot{r} = v_1$, $\dot{\theta} = v_2/a$ at a certain instant. Show that the eccentricity of the path is given by

$$\varepsilon = \frac{1}{K}\left[(av_2^2 - K)^2 + (av_1v_2)^2\right]^{1/2}.$$

9. A rocket, with fuel burned, is observed to have the position $r = 4060$ miles, $\theta = 0$, and velocity $\dot{r} = 20{,}000$ mph, $r\dot{\theta} = 16{,}000$ mph. Is the path of the rocket an ellipse or a hyperbola? Use the result of Exercise 8.

6.14 ELECTRIC CIRCUITS   Figure 6.10 illustrates an electrical network that involves two loops. Our aim is to formulate a system of differential equations and initial conditions for the two unknown loop currents I_1 and I_2. We assume that the switch is closed at time $t = 0$, and that the charge on the capacitance is zero before this time.

Each loop has been oriented by assigning positive directions to the currents. The currents flowing from node 2 to node 1 must be $I_1 - I_2$. This follows from the law of Kirchhoff that says that the current leaving a node must be equal to the current entering it. Kirchhoff's other law says that the sum of the voltage drops around each loop must be equal to the applied voltage. Applying this law to each loop in turn, we arrive at the system of

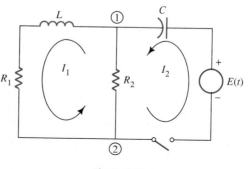

Figure 6.10

equations

$$L\frac{dI_1}{dt} + R_1 I_1 + R_2(I_1 - I_2) = 0, \qquad (6.65)$$

$$R_2(I_2 - I_1) + \frac{1}{C}Q = E(t). \qquad (6.66)$$

Differentiating through in Eq. (6.66) with respect to t and noting that $dQ/dt = I_2$, we obtain the system of differential equations

$$L\frac{dI_1}{dt} + (R_1 + R_2)I_1 - R_2 I_2 = 0,$$

$$R_2\frac{dI_2}{dt} - R_2\frac{dI_1}{dt} + \frac{1}{C}I_2 = E'(t)$$

for I_1 and I_2. It is not hard to see that this system is equivalent to a first-order system for I_1 and I_2. We must therefore specify $I_1(0)$ and $I_2(0)$ in our initial conditions. Because of the presence of the inductance in the loop for I_1, we must have

$$I_1(0) = 0.$$

From Eq. (6.66) (or by inspection of Fig. 6.10) we see that

$$I_2(0) = \frac{E(0)}{R_2}.$$

―――――――――――――― Exercises for Section 6.14 ――――――――――――――

In Exercises 1–6, formulate a system of differential equations and initial conditions for the loop currents, assuming that all initial charges are zero and that the switch is closed at $t = 0$.

1. The circuit of Fig. 6.11.

2. The circuit of Fig. 6.12.

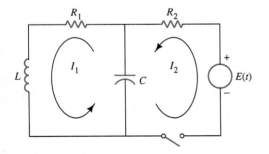

Figure 6.11

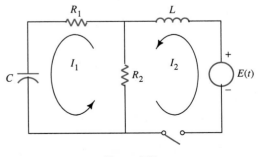

Figure 6.12

3. The circuit of Fig. 6.13. Find the loop currents if $E = 6$ volts, $R_1 = 2$ ohms, $R_2 = 1$ ohm, and $L = 4$ henrys.

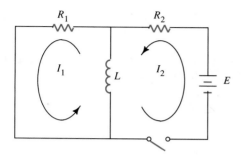

Figure 6.13

4. The circuit of Fig. 6.14. Find the *steady-state* loop currents if $R = 1.0$ ohms, $C = 0.5$ farad, $L = 0.5$ henry, and $E(t) = 2 \cos(t/2)$ volts.

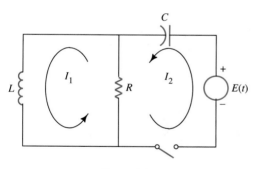

Figure 6.14

5. The circuit of Fig. 6.15.

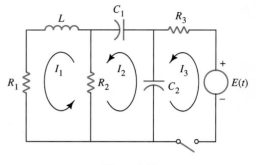

Figure 6.15

6. The circuit of Fig. 6.16.

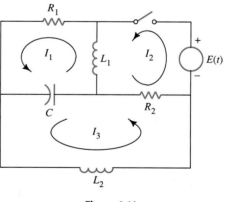

Figure 6.16

Additional Exercises for Chapter 6

In Exercises 1–16, solve the system.

1. $(-2D + 9)x_1 + Dx_2 = 0$
   $(3D - 7)x_1 - (D + 1)x_2 = 0$

2. $Dx_1 + (4D + 2)x_2 = e^{-3t}$
   $(D - 1)x_1 + (5D + 1)x_2 = e^{-3t}$

3. $(D - 1)x_1 + (D^2 + D - 2)x_2 = 0$
   $Dx_1 - (D + 2)x_2 = 0$

4. $Dx_1 - (2D - 1)x_2 = 0$
   $-(D^2 + D)x_1 + (D^3 + 3D^2 + 2D)x_2 = 0$

5. $x_1' = 5x_1 - 6x_2$
   $x_2' = 3x_1 - 4x_2$

6. $x_1' = x_1 + x_2 + 4e^{-t}$
   $x_2' = -8x_1 - 5x_2$

7. $x_1' = -2x_1 - 2x_2 + 3 \sin t$
   $x_2' = 4x_1 + 2x_2$

8. $x_1' = x_1 - 3x_2$
    $x_2' = 3x_1 - 5x_2$

9. $x_1' = -x_1 + 2x_2$
    $x_2' = -x_1 - 3x_2 + 4\cos t$

10. $x_1' = -7x_1 - 4x_2 + 9$
    $x_2' = 4x_1 + x_2 + 4$

11. $x_1' = -2x_1 + x_2 + x_3$
    $x_2' = -3x_1 + 2x_2 + x_3$
    $x_3' = 6x_1 - 6x_2 - x_3$

12. $x_1' = x_1 + x_2$
    $x_2' = -4x_1 + x_2 + x_3$
    $x_3' = -x_1 + x_2 + 2x_3$

13. $x_1' = -x_1 + x_2 + 3x_3$
    $x_2' = x_1 - x_2 - 2x_3$
    $x_3' = -x_1 - x_2$

14. $x_1' = -2x_1 + 2x_2 - x_3$
    $x_2' = x_1 + 2x_2 - 4x_3$
    $x_3' = 2x_2 - 3x_3$

15. $x_1' = -x_1 + 2x_2$
    $x_2' = -x_1 - 2x_2 + x_3$
    $x_3' = x_1 + x_2 - 2x_3$

16. $x_1' = -3x_1 + 3x_2 + 3x_3$
    $x_2' = -2x_1 - x_2 + 3x_3$
    $x_3' = -2x_1 + 2x_2 + 2x_3$

17. Rewrite the indicated system as a first-order system and solve by the method of characteristic values or the method of Section 6.10.
    (a) The system of Exercise 1.
    (b) The system of Exercise 2.
    (c) The system of Exercise 3.

18. Determine whether every solution of the system $x' = Ax$ approaches zero as t becomes infinite, if A is the given matrix. (It is not necessary to solve the system.)

    (a) $\begin{bmatrix} -1 & 1 & 1 \\ 1 & 2 & 7 \\ -2 & -1 & -5 \end{bmatrix}$

    (b) $\begin{bmatrix} 5 & 3 & 12 \\ 5 & 3 & 13 \\ -6 & -2 & -12 \end{bmatrix}$

    (c) $\begin{bmatrix} 0 & 1 & 0 & 1 \\ 1 & 4 & 8 & 15 \\ 1 & 4 & -7 & 7 \\ -1 & -4 & 7 & 8 \end{bmatrix}$

19. Given the system $x' = Ax + b$, where b is a constant vector and A is a nonsingular constant matrix, show that there is a change of variable $x = y + k$, where k is a constant vector, that yields a homogeneous system $y' = Ay$ for y. Show that $x = k$ is a constant solution of the original system.

20. Use the results of the previous exercise to find a constant solution of the given system. Determine whether every solution approaches this constant solution as t becomes infinite.
    (a) $x_1' = x_1 - x_2 - 1$
         $x_2' = 5x_1 - 4x_2 - 6$
    (b) $x_1' = 5x_1 + 7x_2 - 3$
         $x_2' = -5x_1 - 6x_2 - 2$
    (c) $x_1' = -4x_1 + 8x_2 + 7x_3 - 1$
         $x_2' = -4x_1 + 8x_2 + 8x_3 - 1$
         $x_3' = 3x_1 - 7x_2 - 7x_3 - 1$

21. Given the system of equations

    $$m_1 x_1'' = -k_1 x_1 + c(x_2' - x_1'),$$
    $$m_2 x_2'' = -k_2 x_2 - c(x_2' - x_1'),$$

    (a) describe, by means of a diagram, a mechanical system with two masses that is governed by such a system.
    (b) find a first-order system of four equations that is equivalent to the given system.
    (c) use the Routh-Hurwitz criterion (Theorem 5.6) to show that x_1 and x_2 both approach zero as t becomes infinite for all solutions if and only if $k_1/m_1 \neq k_2/m_2$.

22. Initially tank 1 contains 100 gal of a solution with 20 lb of dissolved solute, while tank 2 contains 25 gal of a solution with 10 lb of

the solute. Water flows into tank 1 at the rate of 1 gal/min, liquid is pumped from tank 1 into tank 2 at the rate of 2 gal/min, liquid is pumped from tank 2 into tank 1 at the rate of 1 gal/min, and liquid is pumped out of tank 2 at the rate of 1 gal/min. Find formulas that express the amount of solute in each tank as functions of time.

23. In the circuit of Fig. 6.17, the switch S is closed at time $t = 0$. Find the loop currents as functions of t. Assume $R = 2$, $L = 0.4$, $C = 0.2$ and $E = 6$.

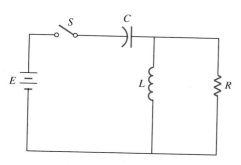

Figure 6.17

24. Suppose three masses connected by springs lie on a frictionless table, as shown in Fig. 6.18. Find the equations of motion, assuming that all masses move along the same straight line.

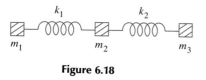

Figure 6.18

25. A homogeneous slab of mass m is mounted on two springs, as shown in Fig. 6.19, and then set in motion. Find a system of differential equations for x and θ, where x is the vertical displacement of the center of mass and θ is the angle shown. Let r be the radius of gyration of the slab about the appropriate axis through the center of mass. Assume that θ is small, so that $\sin \theta \approx \theta$.

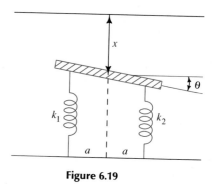

Figure 6.19

26. At an altitude of 250 miles above the earth, a satellite has a speed of 25,000 mph and its direction is 60 degrees from the vertical, away from the earth. Will the satellite go into orbit?

27. In Section 6.10 we derived a formula for e^{tA} (and for e^A, by setting $t = 1$). Let f be any polynomial, so that the series

$$f(A) = f(\lambda_i)I + \frac{f'(\lambda_i)}{1!}(A - \lambda_i I)$$

$$+ \frac{f''(\lambda_i)}{2!}(A - \lambda_i I)^2 + \cdots$$

is finite. Show (using the symbols introduced in Section 6.10) that

$$a_i(A)p_i(A)f(A)$$

$$= a_i(A)p_i(A)\sum_{j=0}^{m_i-1} \frac{f^{(j)}(\lambda_i)}{j!}(A - \lambda_i I)^j$$

for each i, and hence that

$$f(A) = \sum_{i=1}^{k} a_i(A)p_i(A)\sum_{j=0}^{m_i-1} \frac{f^{(j)}(\lambda_i)}{j!}(A - \lambda_i I)^j.$$

28. Continuing with Exercise 27, show that

$$I = \sum_{i=1}^{k} E_i, \qquad A = \sum_{i=1}^{k} (\lambda_i E_i + N_i)$$

where $E_i = a_i(A)p_i(A)$, $N_i = a_i(A)p_i(A)(A - \lambda_i I)$. Show that $E_i^2 = E_i$, $E_i E_j = 0$ if $i \neq j$, $N_i^{m_i} = 0$, $N_i N_j = 0$ if $i \neq j$, $N_i E_j = 0$ if $i \neq j$, and $N_i E_i = N_i$.

7

Series Solutions

7.1

POWER SERIES

The simplest functions in calculus are the polynomials, and those functions "built" from polynomials by a finite number of operations. These operations include the four arithmetic operations, raising to a positive integral power, and root-taking. Functions of this class are known as *algebraic functions*. Examples of such functions are f and g, where

$$f(x) = \left(\frac{x^2 - 3x + 1}{2x + 5} \right)^3, \qquad g(x) = \sqrt{x^3 + 1}.$$

Still other functions are too complicated to be defined by the procedures described above. Some can be defined in terms of integrals of simpler functions. For instance, the formulas

$$\ln x = \int_0^x \frac{1}{t}\, dt, \qquad x > 0,$$
$$\arcsin x = \int_0^x (1 - t^2)^{-1/2}\, dt, \qquad |x| \le 1$$

can be used to define the natural logarithm and inverse sine functions. Other functions can be defined by infinite sequences or series of simpler functions. Thus we may define

$$e^x = \sum_{n=0}^{\infty} \frac{x^n}{n!}, \qquad \sin x = \sum_{n=1}^{\infty} (-1)^{n+1} \frac{x^{2n-1}}{(2n-1)!}$$

333

for all x. In doing computational work with any of these functions, a calculator or computer with "built in" routines to calculate their values is convenient. A knowledge of the basic properties of the function is also useful. For instance, we know that the exponential function is nonnegative and increasing. The sine function is periodic with period 2π and its values vary between -1 and 1. Actually we may feel more comfortable in working with some of these functions than with many of the more complicated algebraic functions.

Even simple differential equations can have complicated solutions. Thus the solutions of the equation $y' = y$ are not algebraic functions but are multiples of the exponential function. The solutions of the equation $y'' = xy$ are not algebraic functions, nor can they be expressed in a simple way in terms of the standard exponential and trigonometric functions of our library of functions. Rather, they become new functions that can be added to our library if we find them sufficiently important.

In this chapter we consider the possibility of representing the solutions of certain linear differential equations in terms of power series. These "series solutions" form a starting point for the investigation of the general properties of the solutions and for the computation of their values. We begin with a review of power series.

A *power series* is a series of functions which at every real number x has the form

$$\sum_{n=0}^{\infty} a_n(x - x_0)^n. \tag{7.1}$$

Here x_0 is a fixed number, called the *center of expansion* (we speak of a power series *about the point* x_0) and the numbers a_i are the *coefficients* of the power series. The series (7.1) always converges at $x = x_0$ to the sum a_0. It may converge only at this point. It may converge for all values of x. If neither of these extreme situations occurs, there is a positive number R such that the series converges absolutely for $|x - x_0| < R$ and diverges when $|x - x_0| > R$. The number R is called the *radius of convergence* of the series. (If the series converges only at x_0 we set $R = 0$; if it converges everywhere we write $R = \infty$.) The interval $(x_0 - R, x_0 + R)$ is called the *interval of convergence* of the series.

The radius of convergence of a power series can sometimes be found by the following method, which is based on the ratio test for convergence of series. We leave the proof as an exercise.

Theorem 7.1   The radius of convergence of the series (7.1) is given by

$$R = \lim_{n \to \infty} \left| \frac{a_n}{a_{n+1}} \right|,$$

provided that the limit exists. (If $|a_n/a_{n+1}| \to \infty$, then $R = \infty$.)

Example 1   For the series

$$\sum_{n=0}^{\infty} \frac{2^n(x-1)^n}{n!}$$

we have $a_n = 2^n/n!$ and $a_{n+1} = 2^{n+1}/(n+1)!$ so $a_n/a_{n+1} = (n+1)/2$. Since this ratio becomes infinite as n increases, we have $R = \infty$. The series converges everywhere.

Example 2   The series

$$\sum_{n=0}^{\infty} n!x^n$$

converges only at the center of expansion $x = 0$ because

$$R = \lim_{n\to\infty}\left|\frac{a_n}{a_{n+1}}\right| = \lim_{n\to\infty}\frac{1}{n+1} = 0.$$

A power series defines a function f whose domain is the interval of convergence (plus perhaps one or both endpoints of the interval). The value $f(x)$ of the function f at each point x in this interval is the sum of the series at the point x.

In the next two theorems we state, without proof, some facts about functions defined by power series. Proofs are given in most books on advanced calculus.

Theorem 7.2   Let

$$f(x) = \sum_{n=0}^{\infty} a_n(x-x_0)^n, \qquad g(x) = \sum_{n=0}^{\infty} b_n(x-x_0)^n$$

for $|x - x_0| < r$. Then for $|x - x_0| < r$,

$$cf(x) = \sum_{n=0}^{\infty} ca_n(x-x_0)^n$$

for every real number c,

$$f(x) + g(x) = \sum_{n=0}^{\infty} (a_n + b_n)(x-x_0)^n,$$

and

$$f(x)g(x) = \sum_{n=0}^{\infty} c_n(x-x_0)^n,$$

where

$$c_n = a_0b_n + a_1b_{n-1} + \cdots + a_nb_0 = \sum_{k=0}^{n} a_kb_{n-k} = \sum_{k=0}^{n} a_{n-k}b_k.$$

Example 3   Let

$$f(x) = \sum_{n=0}^{\infty} (n + 1)x^n, \qquad g(x) = \sum_{n=0}^{\infty} x^n$$

for $|x| < 1$. Then

$$f(x)g(x) = \sum_{n=0}^{\infty} c_n x^n,$$

where

$$c_n = \sum_{k=0}^{n} (k + 1)(1) = 1 + 2 + \cdots + (n + 1) = \frac{(n + 1)(n + 2)}{2}.$$

The next theorem concerns facts about differentiation and integration of power series.

Theorem 7.3   Let

$$f(x) = \sum_{n=0}^{\infty} a_n(x - x_0)^n$$

for $|x - x_0| < R$. Then f is differentiable on the interval $(x_0 - R, x_0 + R)$, and

$$f'(x) = \sum_{n=1}^{\infty} na_n(x - x_0)^{n-1}$$

for $|x - x_0| < R$. If a and b are any points in $(x_0 - R, x_0 + R)$, then

$$\int_a^b f(x)\, dx = \sum_{n=0}^{\infty} \frac{a_n}{n + 1}\left[(b - x_0)^{n+1} - (a - x_0)^{n+1}\right].$$

In particular,

$$\int_{x_0}^x f(t)\, dt = \sum_{n=0}^{\infty} \frac{a_n}{n + 1}(x - x_0)^{n+1}, \qquad |x - x_0| < R.$$

The first part of the theorem says, roughly speaking, that the derivative of the sum of a power series is equal to the series of derivatives. Since the series of derivatives is again a power series, we may differentiate term-by-term again to find the second derivative. For instance, if

$$f(x) = \sum_{n=0}^{\infty} \frac{n + 1}{n + 2} x^n, \qquad |x| < 1,$$

then

$$f'(x) = \sum_{n=1}^{\infty} \frac{n(n+1)}{n+2} x^{n-1}, \qquad |x| < 1,$$

$$f''(x) = \sum_{n=2}^{\infty} \frac{n(n+1)(n-1)}{n+2} x^{n-2}, \qquad |x| < 1,$$

and so on. The series for f' (and the one for f'') can be rewritten in the form (7.1) by making a shift in the index of summation. If we set

$$k = n - 1$$

then as n runs over the sequence $(1, 2, 3, \ldots)$, k varies over $(0, 1, 2, \ldots)$. Thus

$$\sum_{n=1}^{\infty} \frac{n(n+1)}{n+2} x^{n-1} = \sum_{k=0}^{\infty} \frac{(k+1)(k+2)}{(k+3)} x^k.$$

This is also the same as

$$\sum_{n=0}^{\infty} \frac{(n+1)(n+2)}{n+3} x^n,$$

since the sum of the series depends only on x, and not on either the index of summation k or the index n. The procedure is much like making a "change of variable" in an integral. If in the series for f'' we set $n = k + 2$, we see that

$$\sum_{n=2}^{\infty} \frac{n(n+1)(n-1)}{n+2} x^{n-2} = \sum_{k=0}^{\infty} \frac{(k+2)(k+3)(k+1)}{k+4} x^k.$$

In working with power series with a center of expansion x_0 different from zero, it is often convenient to make a change of variable $z = x - x_0$. Then

$$\sum_{n=0}^{\infty} a_n(x - x_0)^n = \sum_{n=0}^{\infty} a_n z^n.$$

Exercises for Section 7.1

In Exercises 1–8, find the interval of convergence of the power series.

1. $\displaystyle\sum_{n=0}^{\infty} \frac{n+2}{n+1}(x-1)^n$

2. $\displaystyle\sum_{n=0}^{\infty} \frac{n+1}{2^n}(x+3)^n$

3. $\displaystyle\sum_{n=1}^{\infty} n^3(x-2)^n$

4. $\displaystyle\sum_{n=0}^{\infty} \frac{(n!)^2}{(2n)!} x^n$

5. $\displaystyle\sum_{n=0}^{\infty} \frac{1 \cdot 3 \cdot 5 \cdots (2n+1)}{(2n)!} x^n$

6. $\displaystyle\sum_{n=1}^{\infty} n^n (x+5)^n$

7. $\displaystyle\sum_{n=1}^{\infty} (-1)^n \frac{n}{\ln(n+1)} x^n$

8. $\displaystyle\sum_{n=1}^{\infty} \frac{n!}{n^n} x^n$

In Exercises 9–12, let f and g be defined by the indicated power series. Find power series expansions for $f + g$ and fg.

9. $f(x) = \displaystyle\sum_{n=0}^{\infty} (n+1)(x-2)^n$, $g(x) = \displaystyle\sum_{n=0}^{\infty} (x-2)^n$

10. $f(x) = \displaystyle\sum_{n=0}^{\infty} \frac{x^n}{n!}$, $g(x) = \displaystyle\sum_{n=1}^{\infty} nx^n$

11. $f(x) = \displaystyle\sum_{n=0}^{\infty} (n+1)x^n$, $g(x) = \displaystyle\sum_{n=0}^{\infty} \frac{x^n}{n+1}$

12. $f(x) = \displaystyle\sum_{n=0}^{\infty} \frac{n+1}{n+2} x^n$, $g(x) = \displaystyle\sum_{n=0}^{\infty} \frac{n+2}{n+3} x^n$

In Exercises 13–16, write the series in the form

$$\sum_{n=n_0}^{\infty} a_n x^n,$$

where n_0 and a_n are to be determined.

13. $\displaystyle\sum_{n=2}^{\infty} n(n-1)x^{n-2}$

14. $\displaystyle\sum_{n=3}^{\infty} \frac{n}{n+1} x^{n-1}$

15. $\displaystyle\sum_{n=0}^{\infty} (n^2+2)x^{n+1}$

16. $\displaystyle\sum_{n=1}^{\infty} (2n+1)(2n+3)x^{n+2}$

In Exercises 17–18, find power series that represent $f'(x)$ and $f''(x)$, and determine an interval on which these formulas are valid.

17. $f(x) = \displaystyle\sum_{n=0}^{\infty} \frac{(-1)^n x^n}{2^n(n+1)}$

18. $f(x) = \displaystyle\sum_{n=0}^{\infty} \frac{(x+1)^{2n}}{n!(n+1)!}$

19. Prove Theorem 7.1.

7.2
TAYLOR
SERIES

Let f be a function defined by a power series on its interval of convergence. Thus

$$f(x) = \sum_{n=0}^{\infty} a_n(x-x_0)^n, \qquad |x-x_0| < R. \qquad (7.2)$$

Differentiating and using Theorem 7.3, we have

$$f'(x) = \sum_{n=1}^{\infty} na_n(x-x_0)^{n-1},$$

$$f''(x) = \sum_{n=2}^{\infty} n(n-1)a_n(x-x_0)^{n-2},$$

. . . .

$$f^{(k)}(x) = \sum_{n=k}^{\infty} n(n-1)\cdots(n-k+1)a_n(x-x_0)^{n-k},$$

. . . .

for $|x - x_0| < R$. Setting $x = x_0$ in these formulas, we see that

$$f'(x_0) = a_1, \qquad f''(x_0) = 2 \cdot 1 a_2, \ldots, \qquad f^{(k)}(x_0) = k(k-1) \cdots 2 \cdot 1 a_k.$$

Thus if Eq. (7.2) holds we must have

$$a_k = \frac{f^{(k)}(x_0)}{k!} \tag{7.3}$$

for $k = 0, 1, 2, \ldots$. On the other hand, suppose that f is any function that possesses derivatives of all orders at a point x_0. Then we can form the series

$$\sum_{n=0}^{\infty} \frac{f^{(n)}(x_0)}{n!} (x - x_0)^n, \tag{7.4}$$

called the *Taylor series* of f at x_0. (In the special case when $x_0 = 0$, the series (7.4) is also referred to as the *Maclaurin series* for f.) If the series (7.4) converges to $f(x)$ for all x in some interval $(x_0 - r, x_0 + r)$, $r > 0$, we say that f is *analytic* at x_0.

As an example, let $f(x) = e^x$ for all x. Then $f^{(n)}(x) = e^x$ and $f^{(n)}(0) = 1$ for every positive integer n. The Maclaurin series for f is

$$\sum_{n=0}^{\infty} \frac{x^n}{n!}.$$

It is easily shown, by application of Theorem 7.1, that this series converges for all x. It is more difficult to show that the sum of the series is e^x. This can be established by the use of Taylor's formula with remainder, which is discussed in most books on calculus.

For convenience of reference we list a number of Maclaurin series whose validity has been established.

$$e^x = \sum_{n=0}^{\infty} \frac{x^n}{n!}, \qquad \text{for all } x, \tag{7.5}$$

$$\cos x = \sum_{n=0}^{\infty} \frac{(-1)^n}{(2n)!} x^{2n}, \qquad \text{for all } x, \tag{7.6}$$

$$\sin x = \sum_{n=1}^{\infty} \frac{(-1)^{n+1}}{(2n-1)!} x^{2n-1}, \qquad \text{for all } x, \tag{7.7}$$

$$\frac{1}{1-x} = \sum_{n=0}^{\infty} x^n, \qquad |x| < 1, \tag{7.8}$$

$$(1 + x)^m = 1 + \sum_{n=1}^{\infty} \frac{m(m-1) \cdots (m-n+1)}{n!} x^n, \qquad |x| < 1. \tag{7.9}$$

The last series is called the *binomial series*. If m is a nonnegative integer there are only a finite number of nonzero terms in the series and it is valid for all x. The series (7.8) is called the *geometric series*. It is a special case of Eq. (7.9).

By means of these formulas and Theorems 7.2 and 7.3, it is possible to establish the validity of many other Taylor series expansions rather easily. For example, we have

$$\ln(1 + x) = \int_0^x \frac{1}{1+t}\,dt, \qquad x > -1.$$

Using the geometric series (7.8) (with x replaced by $-t$) we have

$$\ln(1 + x) = \int_0^x \left[\sum_{n=0}^{\infty} (-1)^n t^n\right] dt, \qquad |x| < 1.$$

An application of Theorem 7.3 yields the formula

$$\ln(1 + x) = \sum_{n=0}^{\infty} \frac{(-1)^n}{n+1} x^{n+1}, \qquad |x| < 1.$$

As a second example, let us find the Maclaurin series for the function f, where

$$f(x) = \frac{x}{2x + 3}.$$

Writing

$$f(x) = x\,\frac{1}{3}\,\frac{1}{1 + \frac{2}{3}x}$$

and observing from formula (7.8) that

$$\frac{1}{1 + \frac{2}{3}x} = \sum_{n=0}^{\infty} (-1)^n \left(\frac{2}{3}x\right)^n, \qquad |x| < \frac{3}{2},$$

we have

$$f(x) = \frac{1}{2} \sum_{n=0}^{\infty} (-1)^n \left(\frac{2}{3}x\right)^{n+1}, \qquad |x| < \frac{3}{2}.$$

Exercises for Section 7.2

In Exercises 1–6, find the Taylor series for the function f about the point x_0 by calculating the derivatives of the function at x_0.

1. $f(x) = 2x^3 - 3x^2 + x - 3, \quad x_0 = 2$

2. $f(x) = -1/x, \quad x_0 = -1$

3. $f(x) = \ln x, \quad x_0 = 1$

4. $f(x) = \cos x, \quad x_0 = \pi/4$

5. $f(x) = (1 + x)^{1/2}, \quad x_0 = 0$

6. $f(x) = e^{3x}, \quad x_0 = 0$

In Exercises 7–10, use the geometric and binomial series to find the Taylor series for the given function f about the point x_0. Indicate an interval on which the series converges to the function.

7. $\dfrac{4}{4 + x}, \quad x_0 = 0$

8. $\dfrac{1}{x}, \quad x_0 = 2$

9. $\dfrac{-2}{(x - 1)(x + 2)}, \quad x_0 = 0$

10. $(1 - x^2)^{-1/2}, \quad x_0 = 0$

In Exercises 11–14, find the Maclaurin series of the given function by differentiation or integration of another series.

11. $f(x) = \dfrac{1}{(1 - x)^2}$

12. $f(x) = \tan^{-1} x$

13. $f(x) = \tanh^{-1} x$

14. $f(x) = \sin^{-1} x$

In Exercises 15–18, express the sum of the series in terms of elementary functions.

15. $\displaystyle\sum_{n=0}^{\infty} \frac{n + 1}{n!} x^{n+1}$

16. $\displaystyle\sum_{n=1}^{\infty} n^2 x^n$

17. $\displaystyle\sum_{n=0}^{\infty} \frac{n + 2}{n + 1} x^{n+1}$

18. $\displaystyle\sum_{n=1}^{\infty} n^2 x^{2n}$

7.3 ORDINARY POINTS

The linear homogeneous differential equation

$$y^{(n)} + a_1 y^{(n-1)} + \cdots + a_{n-1} y' + a_n y = 0 \tag{7.10}$$

is said to have an *ordinary point* at x_0 if each of the functions a_i is analytic at x_0. A point that is not an ordinary point is called a *singular point* for the equation. For the equation

$$y'' + \frac{2x}{(2x - 1)(x + 2)} y' + \frac{\cos x}{x^2} y = 0$$

the singular points are $\frac{1}{2}$, 0, and -2. All other points are ordinary points. The basic result about series solutions at an ordinary point is as follows. (A proof is given in Coddington (1961).)

Theorem 7.4   Let the functions $a_1, a_2, \ldots, a_n$ in Eq. (7.10) be analytic at x_0, and let each of these functions be represented by its Taylor series at x_0 on the interval $\mathcal{I} = (x_0 - R, x_0 + R)$. Then every solution of Eq. (7.10) on $\mathcal{I}$ is analytic at x_0 and is represented on $\mathcal{I}$ by its Taylor series at x_0.

Our interest is mainly in second-order equations, and we consider two examples.

Example 1   The equation

$$(2x + 1)y'' + y' + 2y = 0, \tag{7.11}$$

which may be written as

$$y'' + \frac{1}{2x + 1} y' + \frac{2}{2x + 1} y = 0, \tag{7.12}$$

has an ordinary point at $x = 0$. Here

$$a_1(x) = \frac{1}{1 + 2x} = \sum_{n=0}^{\infty} (-2)^n x^n, \qquad |x| < \frac{1}{2},$$

$$a_2(x) = \frac{2}{1 + 2x} = \sum_{n=0}^{\infty} 2(-2)^n x^n, \qquad |x| < \frac{1}{2}.$$

According to Theorem 7.4, the Maclaurin series for every solution converges at least for $|x| < \frac{1}{2}$. The solution for which

$$y(0) = A_0, \qquad y'(0) = A_1$$

has a Maclaurin series of the form

$$y = \sum_{n=0}^{\infty} A_n x^n, \tag{7.13}$$

where A_2, A_3, and so on, must be determined. Now

$$y' = \sum_{n=1}^{\infty} nA_n x^{n-1}, \qquad y'' = \sum_{n=2}^{\infty} n(n-1)A_n x^{n-2},$$

so if we substitute in Eq. (7.11) we obtain the requirement

$$2 \sum_{n=2}^{\infty} n(n-1)A_n x^{n-1} + \sum_{n=2}^{\infty} n(n-1)A_n x^{n-2}$$

$$+ \sum_{n=1}^{\infty} nA_n x^{n-1} + 2 \sum_{n=0}^{\infty} A_n x^n = 0.$$

(It is convenient to substitute in Eq. (7.11) instead of Eq. (7.12) because the coefficients of the former are polynomials.) The first series here starts with the first power of x while the last three start with x^0. Collecting the constant terms, we may write

$$(2A_2 + A_1 + 2A_0) + 2 \sum_{n=2}^{\infty} n(n-1)A_n x^{n-1} + \sum_{n=3}^{\infty} n(n-1)A_n x^{n-2}$$

$$+ \sum_{n=2}^{\infty} nA_n x^{n-1} + 2 \sum_{n=1}^{\infty} A_n x^n = 0. \tag{7.14}$$

Now all four series start with the first power of x. In order to combine the series, we shift the indices of summation in the first, third, and fourth, so that the exponent of x in each series is $n - 2$. We illustrate the procedure with the first series in Eq. (7.14),

$$2 \sum_{n=2}^{\infty} n(n - 1)A_n x^{n-1}.$$

Making the change of index $n - 1 = k - 2$ or $n = k - 1$, the series becomes

$$2 \sum_{k=3}^{\infty} (k - 1)(k - 2)A_{k-1} x^{k-2}.$$

The smallest value of k is 3, since $k = 3$ when $n = 2$. This series is the same as

$$2 \sum_{n=3}^{\infty} (n - 1)(n - 2)A_{n-1} x^{n-2},$$

since the sum of the series does not depend on the symbol used for the index. Similarly, we find for the remaining series in Eq. (7.14) that

$$\sum_{n=2}^{\infty} n A_n x^{n-1} = \sum_{n=3}^{\infty} (n - 1)A_{n-1} x^{n-2}$$

$$2 \sum_{n=1}^{\infty} A_n x^n = 2 \sum_{n=3}^{\infty} A_{n-2} x^{n-2}.$$

Equation (7.14) may now be written as

$$(2A_2 + A_1 + 2A_0) + \sum_{n=3}^{\infty} [n(n - 1)A_n + 2(n - 1)(n - 2)A_{n-1}$$

$$+ (n - 1)A_{n-1} + 2A_{n-2}]x^{n-2} = 0.$$

Since the coefficients in the Maclaurin series of the zero function are all zero, we must have[1]

$$2A_2 + A_1 + 2A_0 = 0 \qquad\qquad (7.15)$$

and

$$n(n - 1)A_n + (n - 1)(2n - 3)A_{n-1} + 2A_{n-2} = 0 \qquad\qquad (7.16)$$

---

[1] The power series representation of a function is unique. If a function f can be represented in the form (7.2) then the coefficients in the series must be given by formula (7.3).

for $n \geq 3$. This last relation is called a *recurrence relation* for the coefficients A_i. From Eq. (7.15) we can find A_2 in terms of A_0 and A_1. By using the recurrence relation (7.16) with $n = 3$ we may express A_3 in terms of A_0, A_1, and A_2 and hence in terms of A_0 and A_1. In fact, each coefficient A_i can be expressed in terms of A_0 and A_1.

From Eq. (7.15) we have

$$A_2 = -A_0 - \frac{1}{2} A_1. \tag{7.17}$$

Equation (7.16) may be written as

$$A_n = -\frac{2n-3}{n} A_{n-1} - \frac{2}{n(n-1)} A_{n-2}, \qquad n \geq 3. \tag{7.18}$$

For $n = 3$, we have

$$A_3 = -A_2 - \frac{1}{3} A_1.$$

Substituting from Eq. (7.17) for A_2 we find that

$$A_3 = -\left(A_0 - \frac{1}{2} A_1\right) - \frac{1}{3} A_1$$

or

$$A_3 = A_0 + \frac{1}{6} A_1.$$

Next, setting $n = 4$ in the relation (7.18), we see that

$$A_4 = -\frac{1}{6} A_2 - \frac{5}{4} A_3$$

$$= -\frac{1}{6}\left(-A_0 - \frac{1}{2} A_1\right) - \frac{5}{4}\left(A_0 + \frac{1}{6} A_1\right)$$

$$= -\frac{13}{12} A_0 - \frac{1}{8} A_1.$$

The first few terms in the Maclaurin series expansion of the solution are

$$y(x) = A_0 + A_1 x + \left(-A_0 - \frac{1}{2} A_1\right)x^2 + \left(A_0 + \frac{1}{6} A_1\right)x^3$$

$$+ \left(-\frac{13}{12} A_0 - \frac{1}{8} A_1\right)x^4 + \cdots . \tag{7.19}$$

Collecting terms that involve A_0 and A_1, we have

$$y(x) = A_0\left(1 - x^2 + x^3 - \frac{13}{12}x^4 + \cdots\right)$$

$$+ A_1\left(x - \frac{1}{2}x^2 + \frac{1}{6}x^3 - \frac{1}{8}x^4 + \cdots\right).$$

The series included in parentheses would be obtained by setting $A_0 = 1$ and $A_1 = 0$, or $A_0 = 0$ and $A_1 = 1$ in formula (7.19). Thus each converges at least for $|x| < \frac{1}{2}$. If we set

$$y_1(x) = 1 - x^2 + x^3 - \frac{13}{12}x^4 + \cdots,$$

$$y_2(x) = x - \frac{1}{2}x^2 + \frac{1}{6}x^3 - \frac{1}{8}x^4 + \cdots,$$

then the general solution on $\left(-\frac{1}{2}, \frac{1}{2}\right)$ consists of all functions of the form

$$y(x) = A_0 y_1(x) + A_1 y_2(x),$$

where A_0 and A_1 are arbitrary constants.

Example 2 $$y'' - 2(x - 1)y' - y = 0. \tag{7.20}$$

Suppose that we wish to find the Taylor series expansions of the solutions at $x = 1$. For convenience we make the change of variable

$$t = x - 1. \tag{7.21}$$

Then $x = 1$ corresponds to $t = 0$ and Eq. (7.20) becomes

$$\frac{d^2y}{dt^2} - 2t\frac{dy}{dt} - y = 0. \tag{7.22}$$

Seeking solutions of the form

$$y = \sum_{n=0}^{\infty} A_n t^n,$$

we obtain the requirement

$$\sum_{n=2}^{\infty} n(n-1)A_n t^{n-2} - 2\sum_{n=1}^{\infty} nA_n t^n - \sum_{n=0}^{\infty} A_n t^n = 0.$$

By proceeding as in the previous example, we see that this equation can be written as

$$(2A_2 - A_0) + \sum_{n=3}^{\infty} \{n(n-1)A_n - [2(n-2) + 1]A_{n-2}\}t^{n-2} = 0.$$

Then

$$A_2 = \frac{1}{2} A_0$$

and

$$A_n = \frac{2n-3}{(n-1)n} A_{n-2}, \qquad n \geq 3.$$

From these relations we see that

$$A_2 = \frac{1}{2} A_0,$$

$$A_4 = \frac{5}{3 \cdot 4} A_2 = \frac{1 \cdot 5}{2 \cdot 3 \cdot 4} A_0,$$

$$A_6 = \frac{9}{5 \cdot 6} A_4 = \frac{1 \cdot 5 \cdot 9}{6!} A_0,$$

$$\cdots\cdots\cdots\cdots\cdots\cdots\cdots\cdots$$

$$A_{2m} = \frac{1 \cdot 5 \cdot 9 \cdots (4m-3)}{(2m)!} A_0, \qquad m \geq 1$$

and

$$A_3 = \frac{3}{2 \cdot 3} A_1,$$

$$A_5 = \frac{7}{4 \cdot 5} A_3 = \frac{3 \cdot 7}{2 \cdot 3 \cdot 4 \cdot 5} A_1,$$

$$\cdots\cdots\cdots\cdots\cdots\cdots\cdots\cdots$$

$$A_{2m-1} = \frac{3 \cdot 7 \cdots (4m-5)}{(2m-1)!} A_1, \qquad m \geq 2.$$

The general solution of Eq. (7.21) (for all x) is

$$y(x) = A_0 \left[1 + \sum_{m=1}^{\infty} \frac{1 \cdot 5 \cdot 9 \cdots (4m-3)}{(2m)!} (x-1)^{2m} \right]$$

$$+ A_1 \left[(x-1) + \sum_{m=2}^{\infty} \frac{3 \cdot 7 \cdots (4m-5)}{(2m-1)!} (x-1)^{2m-1} \right].$$

In the general case of an equation

$$p(x)y'' + q(x)y' + r(x)y = 0,$$

where p, q, and r are analytic at x_0 and $p(x_0) \neq 0$, it can be shown (see Coddington (1961)) that substitution of the series

$$y = \sum_{n=0}^{\infty} A_n(x - x_0)^n$$

into the equation always leads to a recurrence relation that completely determines the coefficients A_i, $i \geq 2$, in terms of A_0 and A_1. Note that A_0 and A_1 are the values of $y(x_0)$ and $y'(x_0)$, respectively. In an initial value problem, the values of these constants would be specified.

Exercises for Section 7.3

1. Locate all singular points of the given differential equation.

   (a) $y'' + \dfrac{x}{(x-1)(x+2)} y' + \dfrac{1}{x(x-1)^2} y = 0$

   (b) $x(x+3)y'' + x^2 y' - y = 0$

   (c) $y'' + e^x y' + (\cos x)y = 0$

   (d) $(\sin x)y'' - y = 0$

In Exercises 2–10, verify that $x = 0$ is an ordinary point for the differential equation and express the general solution in terms of power series about this point. Discuss the interval of convergence of the series.

2. $y'' + xy' + y = 0$

3. $2y'' - xy' - 2y = 0$

4. $(1 - x^2)y'' - 5xy' - 3y = 0$

5. $(2 + x^2)y'' + 5xy' + 4y = 0$

6. $y'' - xy = 0$

7. $y'' - x^2 y' - 2xy = 0$

8. $y'' - (x+1)y' - y = 0$

9. $(1 + x)y'' - y = 0$

10. $y'' + e^x y' + y = 0$

In Exercises 11–14, express the general solution of the differential equation in terms of power series about the indicated point x_0. Suggestion: make the change of variable $t = x - x_0$.

11. $y'' + (x-1)y' + y = 0$, $x_0 = 1$

12. $(x^2 + 2x)y'' + (x+1)y' - 4y = 0$, $x_0 = -1$

13. $(3 - 4x + x^2)y'' - 6y = 0$, $x_0 = 2$

14. $y'' - (x^2 + 6x + 9)y' - 3(x+3)y = 0$, $x_0 = -3$

7.4 SINGULAR POINTS

The Cauchy-Euler equation

$$2x^2 y'' + 3xy' - y = 0$$

has a singular point at $x = 0$. Its general solution is

$$y = c_1 |x|^{1/2} + c_2 x^{-1},$$

and from this formula we see that no nontrivial solution is analytic at $x = 0$. On the other hand, the occurrence of a singular point may not preclude the existence of analytic solutions. The general solution of the equation

$$x^2 y'' - 2xy' + 2y = 0$$

is

$$y = c_1 x + c_2 x^2,$$

so *every* solution is analytic at $x = 0$.

We notice that every equation of the form

$$b_0 x^2 y'' + b_1 xy' + b_2 y = 0, \tag{7.23}$$

$b_0 \neq 0$, possesses at least one solution of the form

$$y = x^s,$$

where s is a number that may be complex. Our concern in this section is with a generalization of the class of equations (7.23). An equation that can be written in the form

$$(x - x_0)^2 y'' + (x - x_0)P(x)y' + Q(x)y = 0 \tag{7.24}$$

is said to have a *regular singular point* at x_0 if P and Q are analytic at x_0.[2] Such an equation may be written as

$$y'' + p(x)y' + q(x)y = 0, \tag{7.25}$$

where

$$p(x) = \frac{P(x)}{x - x_0}, \qquad q(x) = \frac{Q(x)}{(x - x_0)^2}.$$

Thus Eq. (7.25) has a regular singular point at x_0 if and only if the functions

$$(x - x_0)p(x), \qquad (x - x_0)^2 q(x)$$

are analytic at x_0. It turns out that Eq. (7.24) possesses at least one, and

---

[2] More generally, the nth-order equation

$$(x - x_0)^n y^{(n)} + (x - x_0)^{n-1} b_1(x)y^{(n-1)} + \cdots + (x - x_0)b_{n-1}(x)y' + b_n(x)y = 0$$

is said to have a regular singular point at x_0 if $b_1, b_2, \ldots, b_n$ are analytic at x_0.

sometimes two solutions of the form

$$y = (x - x_0)^s \sum_{n=0}^{\infty} A_n(x - x_0)^n, \qquad A_0 \neq 0, \tag{7.26}$$

where s is a number[3] that need not be an integer. The procedure for finding solutions of the type (7.26) is known as the *method of Frobenius*. We illustrate the method with some examples.

Example   The equation

$$2x^2y'' + 3xy' - (1 + x)y = 0 \tag{7.27}$$

has a regular singular point at $x = 0$. It can be written as

$$y'' + p(x)y' + q(x)y = 0,$$

with

$$p(x) = \frac{3}{2x}, \qquad q(x) = -\frac{1 + x}{2x^2},$$

and $xp(x)$ and $x^2q(x)$ are analytic at $x = 0$. If

$$y = x^s \sum_{n=0}^{\infty} A_n x^n = \sum_{n=0}^{\infty} A_n x^{n+s},$$

then

$$y' = \sum_{n=0}^{\infty} (n + s)A_n x^{n+s-1}$$

and

$$y'' = \sum_{n=0}^{\infty} (n + s)(n + s - 1)A_n x^{n+s-2}.$$

Substituting into the differential equation (7.27), we obtain the requirement

$$2 \sum_{n=0}^{\infty} (n + s)(n + s - 1)A_n x^{n+s} + 3 \sum_{n=0}^{\infty} (n + s)A_n x^{n+s}$$

$$- \sum_{n=0}^{\infty} A_n x^{n+s} - \sum_{n=0}^{\infty} A_n x^{n+s+1} = 0.$$

---

[3] It is possible that s may be complex, but this does not happen in the classical equations in which we are most interested.

The last sum in this equation begins with a term involving x^{s+1}; the remaining series start with a term involving x^s. In order to combine terms with like powers of x, we separate out those terms with x^s and make a shift of index $(n \rightarrow n-1)$ in the last series. The result is

$$[2s(s-1) + 3s - 1]A_0 x^s + \sum_{n=1}^{\infty} \{[2(n+s)(n+s-1)$$

$$+ 3(n+s) - 1]A_n - A_{n-1}\}x^{n+s} = 0.$$

Since $A_0 \neq 0$, we see that s must be a root of the quadratic equation

$$2s(s-1) + 3s - 1 = 0$$

or

$$(2s - 1)(s + 1) = 0. \tag{7.28}$$

Thus s must have one of the values $s_1 = \frac{1}{2}$ or $s_2 = -1$. In either case the coefficients A_i must satisfy the recurrence relation

$$[2(n+s)(n+s-1) + 3(n+s) - 1]A_n = A_{n-1}, \qquad n \geq 1. \tag{7.29}$$

When $s = \frac{1}{2}$ this becomes

$$n(2n + 3)A_n = A_{n-1}$$

or

$$A_n = \frac{1}{n(2n + 3)} A_{n-1}, \qquad n \geq 1.$$

Then

$$A_1 = \frac{1}{1 \cdot 5} A_0,$$

$$A_2 = \frac{1}{2 \cdot 7} A_1 = \frac{1}{(1 \cdot 2)(5 \cdot 7)} A_0,$$

$$\cdots\cdots\cdots\cdots\cdots\cdots\cdots\cdots$$

$$A_n = \frac{1}{n!5 \cdot 7 \cdots (2n + 3)} A_0.$$

Taking $A_0 = 1$, we arrive at the specific "series solution"

$$y_1(x) = x^{1/2}\left[1 + \sum_{n=1}^{\infty} \frac{x^n}{n!5 \cdot 7 \cdots (2n + 3)}\right]. \tag{7.30}$$

When $s = -1$, the recurrence relation (7.29) becomes

$$n(2n - 3)A_n = A_{n-1}$$

or

$$A_n = \frac{1}{n(2n - 3)} A_{n-1}, \qquad n \geq 1.$$

From this relation we find that

$$A_1 = \frac{1}{1 \cdot (-1)} A_0,$$

$$A_2 = \frac{1}{2 \cdot (1)} A_1 = \frac{1}{1 \cdot 2(-1)(1)} A_0,$$

$$A_3 = \frac{1}{3 \cdot 3} A_2 = \frac{1}{1 \cdot 2 \cdot 3(-1) \cdot 1 \cdot 3} A_0,$$

$$\cdots \cdots \cdots \cdots \cdots \cdots \cdots \cdots$$

$$A_n = \frac{1}{n!(-1)1 \cdot 3 \cdots (2n - 3)} A_0.$$

Thus a second series solution of Eq. (7.27) is

$$y_2(x) = x^{-1} \left[1 - x - \sum_{n=2}^{\infty} \frac{x^n}{n! 1 \cdot 3 \cdots (2n - 3)} \right]. \tag{7.31}$$

We have seen that a function of the form

$$f(x) = x^s \sum_{n=0}^{\infty} A_n x^n, \tag{7.32}$$

where the power series converges is some interval $(-R, R)$, is a solution of the differential equation (7.27) if and only if the exponent s and the coefficients A_i satisfy the relations (7.28) and (7.29). Now the power series in formulas (7.30) and (7.31) converge everywhere, as can be verified by the use of Theorem 7.1. Consequently the functions y_1 and y_2 are both solutions on the interval $(0, \infty)$. These solutions are linearly independent. For if

$$c_1 y_1(x) + c_2 y_2(x) = 0$$

for $x > 0$, then, letting $x \to 0$, we see that c_2 must be zero. This is because $y_2(x) \to \infty$ as $x \to 0$. We now have

$$c_1 y_1(x) = 0$$

for $x > 0$. But y_1 is not the zero function, so $c_1 = 0$ also. Hence y_1 and y_2 are linearly independent.

We now consider the general second-order equation with a regular singular point at zero. Such an equation has the form

$$Ly \equiv x^2 y'' + xP(x)y' + Q(x)y = 0, \qquad (7.33)$$

where

$$P(x) = \sum_{n=0}^{\infty} P_n x^n, \qquad Q(x) = \sum_{n=0}^{\infty} Q_n x^n,$$

for $|x| < R$. If y is a function of the form

$$y(x) = y^s \sum_{n=0}^{\infty} A_n x^n = \sum_{n=0}^{\infty} A_n x^{n+s}, \qquad (7.34)$$

then

$$y'(x) = \sum_{n=0}^{\infty} (n+s)A_n x^{n+s-1},$$

$$y''(x) = \sum_{n=0}^{\infty} (n+s)(n+s-1)A_n x^{n+s-2}.$$

Also

$$xP(x)y'(x) = x^s \left(\sum_{n=0}^{\infty} (n+s)A_n x^n \right) \left(\sum_{n=0}^{\infty} P_n x^n \right)$$

$$= \sum_{n=0}^{\infty} \left(\sum_{k=0}^{n} (k+s)A_k P_{n-k} \right) x^{n+s}$$

and

$$Q(x)y(x) = x^s \left(\sum_{n=0}^{\infty} A_n x^n \right) \left(\sum_{n=0}^{\infty} Q_n x^n \right)$$

$$= \sum_{n=0}^{\infty} \left(\sum_{k=0}^{n} A_k Q_{n-k} \right) x^{n+s}.$$

Here we have used Theorem 7.2 to multiply the power series. Upon substituting the various quantities into the differential equation (7.33) and combining like powers of x, we obtain the equation

$$\sum_{n=0}^{\infty} \left\{ (n+s)(n+s-1)A_n + \sum_{k=0}^{n} [(k+s)P_{n-k} + Q_{n-k}]A_k \right\} x^{n+s} = 0.$$

The coefficient of the lowest power of x (which corresponds to the value $n = 0$ for the summation index) is $f(s)A_0$, where

$$f(s) = s(s - 1) + sP_0 + Q_0$$
$$= s^2 + (P_0 - 1)s + Q_0.$$

Since $A_0 \neq 0$, the possible values of s are the roots of the *indicial equation*

$$f(s) = 0.$$

The two roots, which we denote by s_1 and s_2, are called the *exponents* of the differential equation at the regular singular point.

The coefficients A_i in the series for y must satisfy the recurrence relation

$$(n + s)(n + s - 1)A_n + \sum_{k=0}^{\infty} [(k + s)P_{n-k} + Q_{n-k}]A_k = 0$$

for $n \geq 1$. By collecting the terms that involve A_n, we can write this relation as

$$[(n + s)(n + s - 1) + (n + s)P_0 + Q_0]A_n = -\sum_{k=0}^{n-1} [(k + s)P_{n-k} + Q_{n-k}] A_k,$$

where $n \geq 1$. More briefly, we may write

$$f(s + n)A_n = \sum_{k=0}^{n-1} g_n(k, s)A_k, \qquad n \geq 1, \tag{7.35}$$

where

$$f(s) = (s - s_1)(s - s_2),$$
$$f(s + n) = (s + n - s_1)(s + n - s_2),$$

and the quantities $g_n(k, s)$ depend on the coefficients P_i and Q_i but not on the coefficients A_i. If, for a given value of s, say s_1 or s_2, the quantities $f(s + n)$, $n \geq 1$, do not vanish, then each of the coefficients $A_1, A_2, A_3, \ldots$, is uniquely determined, in terms of A_0, by the recurrence relation (7.35).

Let us first take up the case where the exponents s_1 and s_2 are real and distinct. We denote the larger of the two exponents by s_1. Since

$$f(s_1 + n) = (s_1 + n - s_1)(s_1 + n - s_2)$$
$$= n[n + (s_1 - s_2)],$$

we see that $f(s_1 + n) \neq 0$ for $n \geq 1$. Thus the differential equation always possesses a formal series solution y_1 of the form

$$y_1(x) = x^{s_1} \sum_{n=0}^{\infty} A_n x^n, \tag{7.36}$$

corresponding to the larger exponent s_1. It can be shown that the power series in this formula actually converges, at least for $|x| < R$, and that the function y_1 is a soluton of the differential equation, at least on the interval $(0, R)$. We refer the reader to Coddington (1961).

Considering now the smaller exponent s_2, we see that

$$f(s_2 + n) = (s_2 + n - s_1)(s_2 + n - s_2)$$
$$= n[n - (s_1 - s_2)].$$

If the difference $s_1 - s_2$ is not a positive integer, then $f(s_2 + n) \neq 0$ for $n \geq 1$, and we obtain a second series solution y_2 of the form

$$y_2(x) = x^{s_2} \sum_{n=0}^{\infty} A_n x^n \tag{7.37}$$

corresponding to the exponent s_2. The power series in this formula also converges for $|x| < R$ and y_2 is a solution, at least on the interval $(0, R)$.

However, if $s_1 - s_2 = N$, where N is a positive integer, then $f(s_2 + n)$ is zero when and only when $n = s_1 - s_2 = N$. In this case the recurrence relation (7.35) becomes, for $n = N$,

$$0 \cdot A_N = \sum_{k=0}^{N-1} g_N(k, s_2) A_k.$$

Unless it happens that the right-hand side of this equation is zero, it is impossible to find a number A_N that satisfies the equation, and no solution of the form (7.37) exists. If it does happen that the right-hand side of the equation is zero, then we have

$$0 \cdot A_N = 0$$

and *any* value for A_N will do. (In particular we can choose $A_N = 0$.) In this case we again obtain a series solution of the form (7.37). The case where $s_1 - s_2$ is a positive integer will be treated in a later section. If the exponents are equal, there is evidently only one solution of the type (7.34). We shall consider this case in the next section.

──────────── Exercises for Section 7.4 ────────────

1. Locate all regular singular points of the given differential equations.

   (a) $y'' + \dfrac{1-x}{x(x+1)(x+2)} y'$

   $+ \dfrac{x+3}{x^2(x+2)^3} y = 0$

   (b) $(x-1)^2(x-2)y'' + xy' + y = 0$

   (c) $(2x+1)(x-2)^2 y'' + (x+2)y' = 0$

   (d) $y'' + \dfrac{\sin x}{x^2} y' + \dfrac{e^x}{x+1} y = 0$

In Exercises 2–10, verify that $x = 0$ is a regular singular point of the differential equation. If possible, express the general solution in terms of series of the form (7.34).

2. $2x^2 y'' - 3xy' + (3-x)y = 0$

3. $2x^2 y'' + xy' - (x+1)y = 0$

4. $2x^2 y'' + (x - x^2)y' - y = 0$

5. $3xy'' + 2y' + y = 0$

6. $2xy'' + 3y' - xy = 0$

7. $3x^2 y'' + (5x + 3x^3)y' + (3x^2 - 1)y = 0$

8. $2x^2 y'' + 5xy' + (1 - x^3)y = 0$

9. $(2x^2 - x^3)y'' + (7x - 6x^2)y' + (3 - 6x)y = 0$

10. $(2x - 2x^2)y'' + (1 + x)y' + 2y = 0$

In Exercises 11 and 12, verify that x_0 is a regular singular point for the differential equation. If possible, express the general solution in terms of series of the form (7.26). *Suggestion*: make the change of variable $t = x - x_0$.

11. $9(x-1)^2 y'' + [9(x-1) - 3(x-1)^2]y'$

    $+ (4x - 5)y = 0, \quad x_0 = 1$

12. $2(x+1)y'' - (1 + 2x)y' + 7y = 0,$

    $x_0 = -1$

13. Show that the solutions (7.36) and (7.37) are linearly independent.

14. Let the functions P, Q, and F be analytic at $x = 0$. Show that the equation

    $$x^2 y'' + xP(x)y' + Q(x)y = x^\alpha F(x)$$

    possesses at least a formal solution of the form

    $$y(x) = x^\alpha \sum_{n=0}^{\infty} A_n x^n$$

    whenever the constant α is such that neither $\alpha - s_1$ nor $\alpha - s_2$ is a positive integer. Show, by means of an example, that the equation may still possibly have a solution of the indicated from even when α does not satisfy these conditions.

7.5

THE CASE OF EQUAL EXPONENTS

When the exponents s_1 and s_2 of the differential equation (7.33) are equal, we can find only one solution of the form (7.34). In order to get some idea as to how a second solution can be found, suppose we examine a Cauchy-Euler equation whose exponents are equal. Let the equation be

$$Ly \equiv x^2 y'' + b_1 xy' + b_2 y = 0,$$

where b_1 and b_2 are real constants. Seeking a solution of the form

$$y(x, s) = x^s,$$

we have
$$Ly(x, s) = [s(s - 1) + b_1 s + b_2]x^s.$$

The exponents are the roots of the indicial equation
$$(s - s_1)(s - s_2) = s(s - 1) + b_1 s + b_2 = 0.$$

If $s_1 = s_2$ then
$$Ly(x, s) = (s - s_1)^2 x^s. \tag{7.38}$$

Evidently $Ly(x, s_1) = 0$, so one solution is
$$y_1(x) = y(x, s_1) = x^{s_1}.$$

To obtain a second solution, let us differentiate both members of Eq. (7.38) with respect to s. We have
$$\frac{\partial Ly(x, s)}{\partial s} = L\frac{\partial y(x, s)}{\partial s}$$
$$= [2(s - s_1) + (s - s_1)^2 \ln x]x^s,$$

where the first equality involves merely the interchange of the order of differentiation between s and x. Upon setting $s = s_1$, we see that
$$L \left.\frac{\partial y(x, s)}{\partial s}\right|_{s=s_1} = 0$$

so that a second solution is
$$y_2(x) = \left.\frac{\partial y(x, s)}{\partial s}\right|_{s=s_1} = x^{s_1} \ln x.$$

We shall use this same technique to obtain a second solution of the equation
$$Ly = x^2 y'' + xP(x)y' + Q(x)y = 0. \tag{7.39}$$
Writing
$$y(x, s) = x^s \sum_{n=0}^{\infty} A_n(s)x^n, \tag{7.40}$$

where the coefficients A_i are functions to be determined, we find as in the previous section that
$$Ly(x, s) = f(s) A_0 x^s + \sum_{n=0}^{\infty} \left[f(s + n) A_n(s) - \sum_{k=0}^{n-1} g_n(k, s) A_k \right] x^{n+s}. \tag{7.41}$$

Here, since $s_2 = s_1$, we have

$$f(s) = (s - s_1)^2,$$
$$f(s + n) = (s + n - s_1)^2.$$

Hence $f(s + n) \neq 0$ for $n \geq 1$ and $|s - s_1| < 1$. The coefficients A_n can be determined in terms of A_0 (which we take to be a fixed nonzero constant, independent of s) by means of the recurrence relation

$$A_n(s) = \frac{1}{(s + n - s_1)^2} \sum_{k=0}^{n-1} g_n(k, s) A_k, \qquad n \geq 1. \tag{7.42}$$

The functions A_n that are so defined are rational functions of s, and hence possess derivatives of all orders for $|s - s_1| < 1$.

Let us assume that the coefficients $A_n(s)$ in the series (7.40) have been chosen in the manner just described. Then from Eq. (7.41) we have

$$Ly(x, s) = A_0(s - s_1)^2 x^s. \tag{7.43}$$

Evidently $Ly(x, s_1) = 0$, so the function

$$y_1(x) = y(x, s_1) = x^{s_1} \sum_{n=0}^{\infty} A_n(s_1) x^n \tag{7.44}$$

is a solution. Upon differentiating both sides of Eq. (7.43) with respect to s and setting $s = s_1$, we see that

$$\left. \frac{\partial y(x, s)}{\partial s} \right|_{s=s_1} = L \left. \frac{\partial y(x, s)}{\partial s} \right|_{s=s_1} = 0.$$

Thus a second solution (at least formally) is

$$y_2(x) = \left. \frac{\partial y(x, s)}{\partial s} \right|_{s=s_1}$$

$$= x^{s_1} \sum_{n=0}^{\infty} A_n(s_1) x^n \ln x + x^{s_1} \sum_{n=1}^{\infty} A_n'(s_1) x^n$$

or

$$y_2(x) = y_1(x) \ln x + x^{s_1} \sum_{n=1}^{\infty} A_n'(s_1) x^n, \tag{7.45}$$

where y_1 is the solution (7.44). It can be shown that the power series in this formula converges at least for $|x| < R$ and that y_2 is a solution, at least on the interval $(0, R)$.

Example   Let us consider the equation

$$Ly = x^2y'' + 3xy' + (1 - x)y = 0.$$

Setting

$$y(x, s) = \sum_{n=0}^{\infty} A_n(s)x^{n+s},$$

we find that

$$Ly(x, s) = (s + 1)^2 A_0 x^s + \sum_{n=1}^{\infty} [(n + s + 1)^2 A_n - A_{n-1}]x^{n+s}.$$

The indicial equation is $(s + 1)^2 = 0$ and the exponents are $s_1 = s_2 = -1$. We choose the coefficients $A_n(s)$ to satisfy the recurrence relation

$$(n + s + 1)^2 A_n(s) = A_{n-1}(s)$$

for $n \geq 1$. Then

$$A_1(s) = \frac{A_0}{(s + 2)^2},$$

$$A_2(s) = \frac{A_1(s)}{(s + 3)^2} = \frac{A_0}{(s + 2)^2(s + 3)^2},$$

and in general,

$$A_n(s) = \frac{A_0}{(s + 2)^2(s + 3)^2 \cdots (s + n + 1)^2}, \qquad n \geq 1. \qquad (7.46)$$

Setting $s = s_1 = -1$, we have

$$A_n(-1) = \frac{A_0}{1^2 \cdot 2^2 \cdots n^2} = \frac{A_0}{(n!)^2}.$$

Taking $A_0 = 1$, we obtain the solution

$$y_1(x) = x^{-1} \sum_{n=0}^{\infty} \frac{x^n}{(n!)^2}.$$

In order to obtain a second solution, we need to compute the derivatives $A_n'(-1)$. It is convenient to do this by logarithmic differentiation. From formula (7.46) we have

$$\ln A_n(s) = \ln A_0 - 2[\ln(s + 2) + \ln(s + 3) + \cdots + \ln(s + n + 1)].$$

Differentiating with respect to s, we have

$$\frac{A_n'(s)}{A_n(s)} = -2\left[\frac{1}{s+2} + \frac{1}{s+3} + \cdots + \frac{1}{s+n+1}\right].$$

Then

$$\frac{A_n'(-1)}{A_n(-1)} = -2\left[1 + \frac{1}{2} + \cdots + \frac{1}{n}\right]$$

or

$$A_n'(-1) = -2\frac{\phi(n)}{(n!)^2}, \qquad n \geq 1,$$

where we use the notation

$$\phi(n) = 1 + \frac{1}{2} + \frac{1}{3} + \cdots + \frac{1}{n}. \tag{7.47}$$

From the general formula (7.45) we see that a second solution of the differential equation is

$$y_2(x) = y_1(x) \ln x - 2x^{-1} \sum_{n=1}^{\infty} \frac{\phi(n)}{(n!)^2} x^n.$$

The second solution could also have been determined by substituting an expression of the form

$$y(x) = y_1(x) \ln x + x^{-1} \sum_{n=1}^{\infty} B_n x^n$$

into the differential equation. The coefficients B_n can be determined by collecting the like powers of x and equating the coefficient of each power of x to zero. This method, however, does not so readily yield a general formula for the coefficients B_n.

──────────── Exercises for Section 7.5 ────────────

In Exercises 1–10, verify that the exponents relative to $x = 0$ are equal, and find the general solution by using the method described in this section.

1. $(x^2 - x^3)y'' - 3xy' + 4y = 0$

2. $x^2y'' + 7xy' + (9 + 2x)y = 0$

3. $x^2y'' - xy' + (1 - x)y = 0$

4. $x^2y'' - (3x + x^2)y' + (4 - x)y = 0$

5. $xy'' + y' - 2xy = 0$

6. $(x^2 + x^4)y'' + (-x + 7x^3)y' + (1 + 9x^2)y = 0$

7. $x^2y'' + 5xy' + (4 - x)y = 0$

8. $x^2y'' + (3x + x^2)y' + y = 0$

9. $(x^2 + x^3)y'' - (x^2 + x)y' + y = 0$

10. $xy'' + (1 - x^2)y' + 4xy = 0$

7.6
THE CASE WHEN THE EXPONENTS DIFFER BY AN INTEGER

When $s_1 - s_2 = N$, N a positive integer, the equation

$$Ly = x^2y'' + xP(x)y' + Q(x)y = 0 \qquad (7.48)$$

may possess either one or two solutions of the form

$$y(x) = x^s \sum_{n=0}^{\infty} A_n x^n. \qquad (7.49)$$

(In the special case of a Cauchy-Euler equation, there are always two such solutions.) In any case, there is always a solution

$$y_1(s) = x^{s_1} \sum_{n=0}^{\infty} A_n x^n \qquad (7.50)$$

of this type, corresponding to s_1.

We now consider the case where there is only one solution of the form (7.49). Substituting the series

$$y(x, s) = x^s \sum_{n=0}^{\infty} A_n(s)x^n$$

into Eq. (7.48), we find that

$$Ly(x, s) = f(s)A_0 x^s + \sum_{n=1}^{\infty} \left[f(s + n)A_n - \sum_{k=0}^{n-1} g_n(k, s)A_k \right] x^{n+s}, \qquad (7.51)$$

where

$$f(s) = (s - s_1)(s - s_2) = (s - s_2 - N)(s - s_2)$$

and

$$f(s + n) = (s + n - s_2 - N)(s + n - s_2)$$

when $n \geq 1$. In particular, $f(s + N)$ is zero when $s = s_2$, because

$$f(s + N) = (s - s_2)(s + N - s_2).$$

This is the only one of the quantities $f(s + n)$ that vanishes when $s = s_2$. Let us choose the coefficient functions A_n (where A_0 is a fixed nonzero constant) so that they satisfy the recurrence relation

$$f(s + n)A_n(s) = \sum_{k=0}^{n-1} g_n(k, s) A_k(s). \tag{7.52}$$

Then the functions $A_1, A_2, \ldots, A_{N-1}$ are analytic at $s = s_2$, but for $n \geq N$, the quantities $A_n(s)$ contain the factor $s - s_2$ in the denominator. They may become infinite as s approaches s_2. However, the functions B_n, where

$$B_n(s) = (s - s_2)A_n(s),$$

are analytic at $s = s_2$ and satisfy the recurrence relation (7.52) not only for s near s_2 but also for $s = s_2$. (To see this, multiply through in Eq. (7.52) by $s - s_2$.)

Let

$$\tilde{y}(x, s) = (s - s_2)y(x, s)$$

$$= x^s \sum_{n=0}^{\infty} B_n(s)x^n.$$

Multiplying through in Eq. (7.51) by $s - s_2$, we see that

$$L\tilde{y}(x, s) = A_0(s - s_2)f(s)x^s$$

$$= A_0(s - s_1)(s - s_2)^2 x^s.$$

Because of the occurence of the factor $(s - s_2)^2$ in the last expression, it follows that each of the quantities

$$\tilde{y}_1(x) = \tilde{y}(x, s_2), \qquad y_2(x) = \frac{\partial \tilde{y}(x, s)}{\partial s}\bigg|_{s=s_2}$$

formally satisfies the differential equation.

We now examine the forms of these two formal solutions. Since $B_n(s_2) = 0$ for $0 \leq n \leq N - 1$, the solution $\tilde{y}_1$ has the form

$$\tilde{y}_1(x) = x^{s_2} \sum_{n=N}^{\infty} B_n(s_2)x^n$$

$$= x^{s_2 + N} \sum_{n=0}^{\infty} B_{n+N}(s_2)x^n$$

$$= x^{s_1} \sum_{n=0}^{\infty} B_{n+N}(s_2)x^n.$$

Thus $\tilde{y}_1$ must simply be a multiple of the solution y_1 given in formula (7.50). In fact,

$$\tilde{y}_1 = \frac{B_N(s_2)}{A_0} y_1.$$

The solution y_2 is obtained by setting $s = s_2$ in the formula

$$\frac{\partial \tilde{y}(x, s)}{\partial s} = \frac{\partial}{\partial s}\left[x^s \sum_{n=0}^{\infty} B_n(s)x^n \right]$$

$$= x^s \sum_{n=0}^{\infty} B_n(s)x^n \ln x + x^s \sum_{n=0}^{\infty} B'_n(s)x^n.$$

We have

$$y_2(x) = \tilde{y}_1(x) \ln x + x^{s_2} \sum_{n=0}^{\infty} B'_n(s_2)x^n$$

or

$$y_2(x) = \frac{B_N}{A_0} y_1(x) \ln x + x^{s_2} \sum_{n=0}^{\infty} B'_n(s_2)x^n. \qquad (7.53)$$

It can be shown that the power series in this formula converges, at least for $|x| < R$, and that the function y_2 is a solution of the differential equation, at least on the interval $(0, R)$.

Example 1   Let us consider the equation

$$Ly = xy'' + 2y' - y = 0.$$

Writing

$$y(x, s) = x^s \sum_{n=0}^{\infty} A_n(s)x^n,$$

we find that

$$Ly(x, s) = s(s + 1)A_0 x^{s-1} + \sum_{n=1}^{\infty} [(n + s)(n + s + 1)A_n - A_{n-1}]x^{n+s-1}.$$

The exponents for the equation are $s_1 = 0$ and $s_2 = -1$. The recurrence relation for the coefficients A_n is

$$(n + s)(n + s + 1)A_n(s) = A_{n-1}(s), \qquad n \geq 1.$$

From this relation we find that

$$A_1(s) = \frac{A_0}{(s + 1)(s + 2)}, \qquad A_2(s) = \frac{A_0}{(s + 1)(s + 2)^2(s + 2)},$$

and in general,

$$A_n(s) = \frac{A_0}{(s + 1)(s + 2)^2(s + 3)^2 \cdots (s + n)^2(s + n + 1)}$$

for $n \geq 2$. Setting $s = s_1 = 0$ in these formulas, we find that

$$A_n(s_1) = \frac{A_0}{1 \cdot 2^2 \cdot 3^3 \cdots n^2(n+1)} = \frac{A_0}{n!(n+1)!}, \qquad n \geq 0.$$

Therefore a solution of the equation that corresponds to the exponent $s_1 = 0$ is

$$y_1(x) = \sum_{n=0}^{\infty} \frac{x^n}{n!(n+1)!}.$$

The quantity $A_1(s)$ ($N = 1$ in this example) becomes infinite as s approaches $s_2 = -1$, because of the factor $(s+1)$ in its denominator. Hence the equation does not possess a second solution of the form (7.49) corresponding to the exponent s_2. The second solution is therefore logarithmic.
The functions

$$B_0(s) = (s+1)A_0,$$

$$B_1(s) = (s+1)A_1(s) = \frac{A_0}{s+2},$$

$$B_n(s) = (s+1)A_n(s) = \frac{A_0}{(s+2)^2(s+3)^2 \cdots (s+n)^2(s+n+1)}, \qquad n \geq 2,$$

are analytic at $s = s_2 = -1$. Routine calculation shows that

$$B_0'(-1) = A_0, \qquad B_1'(-1) = -A_0,$$

$$B_n'(-1) = -\frac{\phi(n-1) + \phi(n)}{(n-1)!n!} A_0, \qquad n \geq 2,$$

where $\phi(n)$ is defined by formula (7.47). Choosing $A_0 = 1$, we obtain the solution

$$y_2(x) = y_1(x) \ln x - x^{-1}\left[1 - x - \sum_{n=2}^{\infty} \frac{\phi(n-1) + \phi(n)}{(n-1)!n!} x^n\right],$$

where $y_1(x)$ is as above.
This second solution could also have been found by substituting an expression of the form

$$y(x) = y_1(x) \ln x + x^{-1} \sum_{n=0}^{\infty} C_n x^n$$

into the differential equation and determining the coefficients C_n. However, it is difficult to find a general formula for C_n using this method.

Example 2   The equation

$$xy'' + 3y' - x^2y = 0$$

has a regular singular point at $x = 0$. Seeking solutions of the form

$$y(x, s) = x^s \sum_{n=0}^{\infty} A_n x^n, \qquad A_0 \neq 0,$$

we find, after some calculation, that the indicial equation is

$$s(s + 2) = 0,$$

and that the coefficients A_i must satisfy the conditions

$$(s + 1)(s + 3)A_1 = 0,$$
$$(s + 2)(s + 4)A_2 = 0,$$

and

$$(n + s)(n + s + 2)A_n = A_{n-3}, \qquad n \geq 3.$$

The exponents are $s_1 = 0$ and $s_2 = -2$. For the larger exponent s_1, we have

$$A_1 = 0,$$
$$A_2 = 0,$$
$$A_n = \frac{A_{n-3}}{n(n + 2)}, \qquad n \geq 3.$$

All the coefficients A_i vanish except those whose subscripts are multiples of three. We have

$$A_3 = \frac{1}{3 \cdot 5} A_0,$$

$$A_6 = \frac{1}{6 \cdot 8} A_3 = \frac{1}{(3 \cdot 6)(5 \cdot 8)} A_0,$$

and in general,

$$A_{3m} = \frac{1}{(3 \cdot 6 \cdot 9 \cdots 3m)[5 \cdot 8 \cdot 11 \cdots (3m + 2)]} A_0$$

$$= \frac{1}{3^m m! [5 \cdot 8 \cdot 11 \cdots (3m + 2)]} A_0.$$

The solution that corresponds to the exponent s_1 is

$$y_1(x) = 1 + \sum_{m=1}^{\infty} \frac{x^{3m}}{3^m m! [5 \cdot 8 \cdot 11 \cdots (3m+2)]}.$$

For the smaller exponent $s_2 = -2$, we have

$$A_1 = 0,$$

$$0 \cdot A_2 = 0,$$

$$(n-2)nA_n = A_{n-3}, \qquad n \geq 3.$$

(Note that A_2 is the critical coefficient, since $N = 2$ in this example.) Here A_2 is arbitrary, and we may choose $A_2 = 0$. A solution that corresponds to the exponent s_2 is found to be

$$y_2(x) = x^{-2} \left[1 + \sum_{m=1}^{\infty} \frac{x^{3m}}{3^m m! [1 \cdot 4 \cdot 7 \cdots (3m-2)]} \right].$$

―――――――――――――――― Exercises for Section 7.6 ――――――――――――――――

In Exercises 1–14, express the general solution in terms of series about $x = 0$.

1. $xy'' - xy' - y = 0$

2. $xy'' + xy' + 2y = 0$

3. $(x^2 - x^4)y'' + 4xy' + (2 + 20x^2)y = 0$

4. $x^2 y'' + x^2 y' - 2y = 0$

5. $x^2 y'' - 2xy' + (2 - x)y = 0$

6. $(x^2 - x^3)y'' + (x - 5x^2)y' - (1 + 4x)y = 0$

7. $x^2 y'' + (3x - x^2)y' - xy = 0$

8. $x^2 y'' + (x + x^3)y' + (x^2 + 1)y = 0$

9. $xy'' - y = 0$

10. $xy'' - y' + y = 0$

11. $x^2 y'' + xy' - (2x + 1)y = 0$

12. $x^2 y'' + (2x - 2)y = 0$

13. $(x^2 - x^3)y'' + (3x - 5x^2)y' - 3y = 0$

14. $(x^2 + x^3)y'' + (x + x^2)y' - (4 + x)y = 0$

7.7
THE POINT
AT INFINITY

In some instances it may be desired to find the behavior of solutions of a differential equation as the independent variable x becomes infinite, rather than near some finite point. If we make the change of variable

$$x = \frac{1}{t}, \qquad (7.54)$$

then as t tends to zero through positive (negative) values, x becomes positively (negatively) infinite. Setting $Y(t) = y(1/t) = y(x)$, the chain rule shows that

$$\frac{dy}{dx} = \frac{dY}{dt}\frac{dt}{dx} = -\frac{1}{x^2}\frac{dY}{dt} = -t^2\frac{dY}{dt},$$

$$\frac{d^2y}{dx^2} = \frac{1}{x^4}\frac{d^2Y}{dt^2} + \frac{2}{x^3}\frac{dY}{dt} = t^4\frac{d^2Y}{dt^2} + 2t^3\frac{dY}{dt}.$$

The equation

$$\frac{d^2y}{dx^2} + P(x)\frac{dy}{dx} + Q(x)y = 0 \tag{7.55}$$

becomes

$$\frac{d^2Y}{dt^2} + p(t)\frac{dY}{dt} + q(t)Y = 0, \tag{7.56}$$

where

$$p(t) = \frac{2}{t} - \frac{1}{t^2}P\left(\frac{1}{t}\right), \qquad q(t) = \frac{1}{t^4}Q\left(\frac{1}{t}\right).$$

If Eq. (7.56) has an ordinary point at $t = 0$, then Eq. (7.55) is said to have an ordinary point at infinity. Similarly, if Eq. (7.56) has a regular singular point at $t = 0$, then Eq. (7.55) is said to have a regular singular point at infinity.

For purposes of illustration, let us attempt to find series solutions of the equation

$$2x^3\frac{d^2y}{dx^2} + 3x^2\frac{dy}{dx} - y = 0$$

that are valid for large values of $|x|$. After the transformation (7.54), our equation becomes

$$2t\frac{d^2Y}{dt^2} + \frac{dY}{dt} - Y = 0.$$

This equation has a regular singular point at $t = 0$. Applying the method of Frobenius, we find that the exponents at $t = 0$ are $s_1 = \frac{1}{2}$ and $s_2 = 0$, and that corresponding solutions are

$$Y_1(t) = t^{1/2}\sum_{n=0}^{\infty}\frac{2^n}{(2n+1)!}t^n,$$

$$Y_2(t) = \sum_{n=0}^{\infty}\frac{2^n}{(2n)!}t^n.$$

Replacing t by $1/x$ in these formulas, we obtain the solutions

$$y_1(x) = x^{-1/2} \sum_{n=0}^{\infty} \frac{2^n}{(2n+1)!} x^{-n},$$

$$y_2(x) = \sum_{n=0}^{\infty} \frac{2^n}{(2n)!} x^{-n}$$

of the original equation. Since the series for Y_1 and Y_2 converge for $|t| < \infty$, the series for y_1 and y_2 converge for $x \neq 0$.

--- Exercises for Section 7.7 ---

1. Find all singular points of the given differential equation and indicate which are regular singular points. Include any singularity at infinity.

(a) $x^4 y'' + x^3(x+2)y' + y = 0$

(b) $(x+1)^2 y'' + (x+1)y' - y = 0$

(c) $(x-2)y'' + y' - xy = 0$

(d) $y'' + ay' + by = 0$, a and b constants

(e) $y'' + x^{-2}y = 0$

(f) $y'' + e^x y = 0$

In Exercises 2–7, verify that the point at infinity is either an ordinary point or a regular singular point. Express the general solution in terms of series of powers of $1/x$.

2. $x^4 y'' + (2x^3 - x)y' + y = 0$

3. $x^4 y'' + (2x^3 + x)y' - y = 0$

4. $2x^3 y'' + x^2 y' - (x+1)y = 0$

5. $2x^3 y'' + (5x^2 - 2x)y' + (x+3)y = 0$

6. $x^3 y'' + (x^2 - x)y' + (2-x)y = 0$

7. $x^2(x^2 - 1)y'' + (x^3 + 5x)y' - 8y = 0$

8. Let us write $f(x) = O(x^m)$ whenever f is any function such that $f(x)/x^m$ is bounded when $|x|$ is sufficiently large. In particular, $f(x) = O(x^m)$ if $f(x)/x^m$ tends to a finite limit as $|x|$ becomes infinite. If the differential equation

$$y'' + P(x)y' + Q(x)y = 0$$

has a regular singular point at infinity, show that $P(x) = O(x^{-1})$ and that $Q(x) = O(x^{-2})$. If the equation has an ordinary point at infinity, show that $P(x) - 2/x = O(x^{-2})$ and that $Q(x) = O(x^{-4})$.

9. Verify that the equation $xy'' - (x+1)y = 0$ has a singular point at infinity, but that the singular point is not regular. Then show that the equation possesses *formal* solutions of the forms

$$y_1(x) = x^{1/2}e^x \sum_{n=0}^{\infty} A_n x^{-n},$$

$$y_2(x) = x^{-1/2}e^{-x} \sum_{n=0}^{\infty} B_n x^{-n},$$

but that both of the series involved *diverge* for all values of x.

7.8
LEGENDRE
POLYNOMIALS

In this and the next section, we consider two differential equations that are important in a variety of applications and that cannot be solved by elementary methods. Properties of the solutions of these equations have been extensively investigated and tables of values for some of the solutions have been compiled.

The first of these equations, *Legendre's equation*, is

$$(1 - x^2)y'' - 2xy' + \alpha y = 0, \tag{7.57}$$

where α is a constant. This equation has regular singular points at $x = 1$ and $x = -1$, and no other singular points. We shall seek series solutions that are valid near $x = 1$. Making the change of variable $t = x - 1$, we arrive at the equation

$$(2t + t^2)\frac{d^2y}{dt^2} + 2(1 + t)\frac{dy}{dt} - \alpha y = 0. \tag{7.58}$$

A function of the form

$$y = t^s \sum_{n=0}^{\infty} A_n t^n \tag{7.59}$$

is a solution only if the condition

$$2s^2 A_0 t^{s-1} + \sum_{n=1}^{\infty} \{2(n + s)^2 A_n + [(n + s)(n + s - 1) - \alpha]A_{n-1}\}t^{n+s-1} = 0$$

is satisfied. The exponents are $s_1 = s_2 = 0$, so there is only one solution of the form (7.59). The recurrence relation is

$$A_n = \frac{\alpha - (n - 1)n}{2n^2} A_{n-1}, \qquad n \geq 1.$$

From this relation we see that

$$A_1 = \frac{\alpha - 0}{2 \cdot 1^2} A_0,$$

$$A_2 = \frac{\alpha - 1 \cdot 2}{2 \cdot 2^2} A_1 = \frac{(\alpha - 0)(\alpha - 1 \cdot 2)}{2^2 \cdot 1^2 \cdot 2^2} A_0,$$

$$\cdots\cdots\cdots\cdots\cdots\cdots\cdots\cdots\cdots\cdots\cdots\cdots\cdots\cdots$$

$$A_n = \frac{\alpha(\alpha - 1 \cdot 2)(\alpha - 2 \cdot 3) \cdots [\alpha - (n - 1)n]}{2^n(n!)^2} A_0.$$

Thus a solution of the original equation (7.57) is

$$y_1(x) = 1 + \sum_{n=1}^{\infty} \frac{\alpha(\alpha - 1 \cdot 2)(\alpha - 2 \cdot 3) \cdots [\alpha - (n - 1)n]}{2^n(n!)^2}(x - 1)^n. \tag{7.60}$$

Notice that when α is of the form

$$\alpha = m(m + 1),$$

where m is a nonnegative integer, the series (7.60) terminates. The corresponding solution is denoted by P_m, and we have

$$P_m(x) =$$

$$1 + \sum_{n=1}^{m} \frac{m(m + 1)[m(m + 1) - 1 \cdot 2] \cdots [m(m + 1) - (n - 1)n]}{2^n (n!)^2} (x - 1)^n.$$

$$(7.61)$$

This solution P_m is a polynomial of degree m; it is known as the *Legendre polynomial* of degree m. The corresponding differential equation

$$(1 - x^2)y'' - 2xy' + m(m + 1)y = 0$$

is known as *Legendre's differential equation of order m*.
   From the definition (7.61), we find that

$$P_0(x) = 1, \qquad P_1(x) = x. \qquad (7.62)$$

It can be shown that the Legendre polynomials satisfy the recurrence relation

$$nP_n(x) = (2n - 1)xP_{n-1}(x) - (n - 1)P_{n-2}(x), \qquad n \geq 2. \qquad (7.63)$$

From this relation we find that

$$2P_2(x) = 3xP_1(x) - P_0(x)$$

or

$$P_2(x) = \frac{3}{2}x^2 - \frac{1}{2}.$$

Also, by repeated application of the recurrence relation, we obtain the formulas

$$P_3(x) = \frac{5}{2}x^3 - \frac{3}{2}x,$$

$$P_4(x) = \frac{35}{8}x^4 - \frac{15}{4}x^2 + \frac{3}{8},$$

$$P_5(x) = \frac{63}{8}x^5 - \frac{35}{4}x^3 + \frac{15}{8}x.$$

It follows from formulas (7.62) and (7.63) that $P_n(x)$ involves only even powers of x when n is even and only odd powers when n is odd.

Another important property of Legendre polynomials is the *orthogonality*[4] *relation*

$$\int_{-1}^{1} P_m(x)P_n(x)\, dx = 0, \qquad m \neq n. \tag{7.64}$$

To derive this relation, we start with the fact that P_n and P_m satisfy the equations

$$[(1 - x^2)P_n'(x)]' + n(n + 1)P_n(x) = 0,$$
$$[(1 - x^2)P_m'(x)]' + m(m + 1)P_m(x) = 0,$$

respectively. Multiplying through in the first equation by $P_m(x)$ and in the second by $P_n(x)$, and then subtracting, we find that

$$[n(n + 1) - m(m + 1)]P_m(x)P_n(x)$$
$$= P_n(x)[(1 - x^2)P_m'(x)]' - P_m(x)[(1 - x^2)P_n'(x)]'$$
$$= \{(1 - x^2)[P_m'(x)P_n(x) - P_n'(x)P_m(x)]\}'.$$

Integrating from -1 to 1, we see that

$$(n - m)(n + m + 1) \int_{-1}^{1} P_m(x)P_n(x)\, dx = 0.$$

Since $m \neq n$, the orthogonality relation (7.64) holds.

Two other useful properties of Legendre polynomials are

$$\int_{-1}^{1} [P_n(x)]^2\, dx = \frac{2}{2n + 1}, \qquad n \geq 0 \tag{7.65}$$

and

$$P_n(x) = \frac{1}{2^n n!} \frac{d^n}{dx^n} (x^2 - 1)^n, \qquad n \geq 0. \tag{7.66}$$

This last formula is known as *Rodrigues' formula.*

----

[4] Two functions f and g are said to be orthogonal on the interval $[a, b]$ if

$$\int_a^b f(x)g(x)\, dx = 0.$$

──────────── **Exercises for Section 7.8** ────────────

1. Given that $P_0(x) = 1$ and $P_1(x) = x$, find formulas for P_3 and P_4 from the recurrence relation (7.63).

2. Derive the formula (7.65) by using Rodrigues' formula (7.66) and repeated integration by parts.

3. Show that under the change of independent variable $x = \cos \phi$, Legendre's equation assumes the form

$$\frac{d^2 y}{d\phi^2} + \frac{dy}{d\phi} \cot \phi + \alpha(\alpha + 1)y = 0.$$

4. For nonnegative integers m and n, with $m \le n$, let

$$p_n^m(x) = \frac{d^m}{dx^m} P_n(x).$$

Show that the function p_n^m is a solution of the differential equation

$$(1 - x^2)y'' - 2(m + 1)xy' +$$
$$(n - m)(n + m + 1)y = 0.$$

5. Let f and g be solutions of the equations

$$[p(x)y']' + \lambda r(x)y = 0,$$
$$[p(x)y']' + \mu r(x)y = 0,$$

respectively, on an interval $[a, b]$. If $p(a) = p(b) = 0$, show that

$$\int_a^b r(x)f(x)g(x)\, dx = 0,$$

provided $\lambda \ne \mu$. (The functions f and g are said to be *orthogonal with respect to the weight function r on* $[a, b]$.)

6. The equation

$$xy'' + (1 - x)y' + ny = 0,$$

where n is a nonnegative integer, is known as Laguerre's equation of order n.

   (a) Show that the differential equation above possesses the polynomial solution

$$L_n(x) = \sum_{k=0}^{n} \frac{n!(-1)^k x^k}{(k!)^2(n - k)!}.$$

   (The function L_n is known as the Laguerre polynomial of degree n.)

   (b) Show that

$$\int_0^\infty e^{-x} L_n(x)L_m(x)\, dx = 0$$

   if $m \ne n$. Suggestion: observe that Laguerre's equation can be written as

$$(xe^{-x}y')' + ne^{-x}y = 0.$$

7. Show that the Hermite equation of order n,

$$y'' - 2xy' + 2ny = 0$$

where n is a nonnegative integer, possesses a polynomial solution of degree n. A certain constant multiple of this polynomial is known as the Hermite polynomial of degree n.

7.9 | Our study of Bessel functions, the main topic of this section, requires a
BESSEL | knowledge of another function, the gamma function. We define the *gamma*
FUNCTIONS | *function* Γ by means of the formula[5]

$$\Gamma(x) = \int_0^\infty t^{x-1} e^{-t}\, dt, \qquad x > 0. \tag{7.67}$$

─────────

[5] The improper integral diverges if $x \le 0$.

The two properties

$$\Gamma(1) = 1, \tag{7.68}$$

$$\Gamma(x + 1) = x\Gamma(x), \qquad x > 0 \tag{7.69}$$

are easily verified. From the definition (7.67), we see that

$$\Gamma(1) = \int_0^\infty e^{-t}\, dt = 1.$$

Verification of the other property involves an integration by parts. We have

$$\Gamma(x + 1) = \lim_{T \to \infty} \int_0^T t^x e^{-t}\, dx$$

$$= \lim_{T \to \infty} \left\{ [-t^x e^{-t}]_0^T + x \int_0^T t^{x-1}\, dx \right\}$$

$$= x\, \Gamma(x).$$

From the two properties (7.68) and (7.69) we see that

$$\Gamma(2) = 1 \cdot \Gamma(1) = 1,$$

$$\Gamma(3) = 2 \cdot \Gamma(2) = 2 \cdot 1,$$

$$\Gamma(4) = 3 \cdot \Gamma(3) = 3 \cdot 2 \cdot 1.$$

In general

$$\Gamma(n + 1) = n! \tag{7.70}$$

for every nonnegative integer n, as can be shown by mathematical induction.
   The relation

$$\Gamma(x) = \frac{\Gamma(x + 1)}{x}$$

is valid for $x > 0$. However, the right-hand side of this equation is defined for $x > -1$. We *define* $\Gamma(x)$ for $-1 < x < 0$ as

$$\Gamma(x) = \frac{\Gamma(x + 1)}{x}. \tag{7.71}$$

Replacing x by $x + 1$ in this formula, we have

$$\Gamma(x + 1) = \frac{\Gamma(x + 2)}{x + 1}, \qquad -2 < x < -1. \tag{7.72}$$

Then from formulas (7.71) and (7.72), we have

$$\Gamma(x) = \frac{\Gamma(x+2)}{x(x+1)}$$

for $x > -1$ and $x = 0$. This formula serves to define $\Gamma(x)$ for $-2 < x < -1$. Continuing in this way, we have

$$\Gamma(x) = \frac{\Gamma(x+k)}{x(x+1)(x+2)\cdots(x+k-1)} \tag{7.73}$$

for $x > -k$, $x \neq 0, -1, -2, \ldots, -k+1$. A graph of the gamma function is shown in Figure 7.1.

The differential equation

$$x^2 y'' + xy' + (x^2 - \alpha^2)y = 0, \tag{7.74}$$

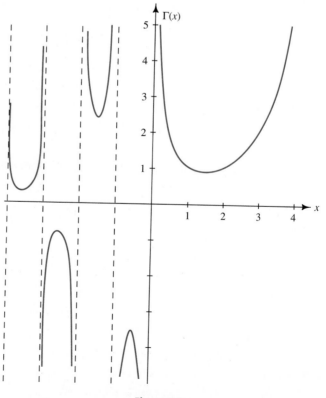

Figure 7.1

where α is a nonnegative real constant, is the main topic of this section. It is known as *Bessel's equation of order* α. It has a regular singular point at $x = 0$. We shall seek series solutions of the form

$$y(x) = x^s \sum_{n=0}^{\infty} A_n x^n. \qquad (7.75)$$

Some computation shows that a function of the form (7.75) is a solution if

$$[s(s - 1) + s - \alpha^2]A_0 x^s + [(s + 1)s + s + 1 - \alpha^2]A_1 x^{s-1}$$
$$+ \sum_{n=2}^{\infty} \{[(n + s)(n + s - 1) + (n + s) - \alpha^2]A_n + A_{n-2}\}x^{n+s} = 0.$$

Thus s must be a root of the equation

$$s^2 - \alpha^2 = 0, \qquad (7.76)$$

and the coefficients A_i must satisfy the relations

$$[(s + 1)^2 - \alpha^2]A_1 = 0, \qquad (7.77)$$
$$[(n + s)^2 - \alpha^2]A_n = -A_{n-2}, \qquad n \geq 2. \qquad (7.78)$$

The exponents of Bessel's equation at $x = 0$, as determined from Eq. (7.76), are $s_1 = \alpha$ and $s_2 = -\alpha$.

If $s = \alpha$, the relations (7.77) and (7.78) require that $A_1 = 0$ and

$$n(n + 2\alpha)A_n = -A_{n-2}, \qquad n \geq 2.$$

Thus we have $A_1 = A_3 = A_5 = \cdots = 0$ and

$$A_2 = -\frac{1}{2(2 + 2\alpha)} A_0,$$

$$A_4 = -\frac{1}{4(4 + 2\alpha)} A_2 = \frac{1}{2 \cdot 4(2 + 2\alpha)(4 + 2\alpha)} A_0,$$

$$\cdots\cdots\cdots\cdots\cdots\cdots\cdots\cdots$$

$$A_{2m} = \frac{(-1)^m}{2 \cdot 4 \cdots (2m)(2 + 2\alpha)(4 + 2\alpha) \cdots (2m + 2\alpha)} A_0$$

$$= \frac{(-1)^m}{2^{2m} m!(1 + \alpha)(2 + \alpha) \cdots (m + \alpha)} A_0.$$

A solution is

$$y_1(x) = A_0 x^\alpha \left[1 + \sum_{m=1}^{\infty} \frac{(-1)^m (x/2)^{2m}}{m!(1 + \alpha)(2 + \alpha) \cdots (m + \alpha)} \right],$$

where the power series converges for all x. Choosing

$$A_0 = \frac{1}{2^\alpha \Gamma(1 + \alpha)},$$

and making use of Eq. (7.73), we obtain a specific solution, known as the *Bessel function of the first kind* of order α. We denote it by J_α. Thus

$$J_\alpha(x) = \sum_{m=0}^{\infty} \frac{(-1)^m (x/2)^{2m+\alpha}}{m! \Gamma(m + \alpha + 1)}. \tag{7.79}$$

From this formula we see that $J_0(0) = 1$ and $J_\alpha(0) = 0$ for $\alpha > 0$. Graphs of J_0 and J_1 are shown in Figure 7.2.

The difference $s_1 - s_2 = 2\alpha$ is an integer whenever 2α is an integer. It turns out that Bessel's equation possesses a solution of the form (7.75) corresponding to $s_2 = -\alpha$ except when α is a nonnegative integer. A particular second solution is

$$J_{-\alpha}(x) = \sum_{m=0}^{\infty} \frac{(-1)^m (x/2)^{2m-\alpha}}{m! \Gamma(m - \alpha + 1)}. \tag{7.80}$$

Notice that $J_{-\alpha}(x)$ becomes infinite as x approaches zero through positive values. When α is an integer, there is a solution corresponding to $s_2 = -\alpha$ that is of the form

$$y_2(x) = J_\alpha(x) \ln x + x^{-\alpha} \sum_{m=0}^{\infty} B_m x^m.$$

A particular linear combination of y_2 and J_α that is denoted by Y_α,

$$Y_\alpha(x) = aJ_\alpha(x) + by_2,$$

is called *Weber's Bessel function of the second kind* of order α. The constants

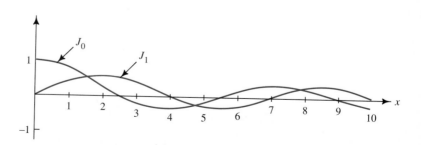

Figure 7.2

a and *b* are chosen so that the behavior of Y_α as $x \to \infty$ is similiar to that of J_α. It can be shown that

$$J_\alpha(x) = \sqrt{\frac{2}{\pi x}} \cos(x - \theta_\alpha)[1 + f_\alpha(x)],$$

$$Y_\alpha(x) = \sqrt{\frac{2}{\pi x}} \sin(x - \theta_\alpha)[1 + g_\alpha(x)],$$

where $\theta_\alpha = (2\alpha + 1)\pi/4$ and $f_\alpha(x) \to 0$, $g_\alpha(x) \to 0$ as $x \to \infty$. Thus the general solution of Bessel's equation on the interval $(0, \infty)$ may be written as

$$y = c_1 J_\alpha(x) + c_2 J_{-\alpha}(x)$$

when α is not an integer, but when $\alpha = N$, where N is a nonnegative integer, the general solution is

$$y = c_1 J_N(x) + c_2 Y_N(x).$$

The solution Y_N becomes infinite as x tends to zero through positive values. Thus the only solutions that are bounded near $x = 0$ are those that are multiples of J_α. Because of this fact, interest is centered on these solutions in many applications.

The Bessel functions of the first kind satisfy a number of recurrence relations. One of them is

$$\frac{d}{dx}[x^\alpha J_\alpha(x)] = x^\alpha J_{\alpha-1}(x). \tag{7.81}$$

Although this relation is not so easy to *discover* from the definition (7.79), it is easily verified. We have

$$\frac{d}{dx}[x^\alpha J_\alpha(x)] = \frac{d}{dx} \sum_{m=0}^{\infty} \frac{(-1)^m x^{2m+2\alpha}}{2^{2m+\alpha} m! \Gamma(m + \alpha + 1)}$$

$$= \sum_{m=0}^{\infty} \frac{(-1)^m x^{2m+2\alpha-1}}{2^{2m+\alpha-1} m! \Gamma(m + \alpha)}$$

$$= x^\alpha \sum_{m=0}^{\infty} \frac{(-1)^m (x/2)^{2m+\alpha-1}}{m! \Gamma(m + \alpha)}$$

$$= x^\alpha J_{\alpha-1}(x).$$

In similar fashion it can be shown that

$$\frac{d}{dx}[x^{-\alpha} J_\alpha(x)] = -x^{-\alpha} J_{\alpha+1}(x). \tag{7.82}$$

The relations (7.79) and (7.80) are equivalent to the relations

$$J_\alpha'(x) = J_{\alpha-1}(x) - \frac{\alpha}{x} J_\alpha(x), \qquad (7.83)$$

$$J_\alpha'(x) = -J_{\alpha+1}(x) + \frac{\alpha}{x} J_\alpha(x), \qquad (7.84)$$

respectively. Adding, we obtain the formula

$$J_\alpha'(x) = \frac{1}{2} \left[J_{\alpha-1}(x) - J_{\alpha+1}(x) \right]. \qquad (7.85)$$

Subtraction yields the relation

$$J_{\alpha+1}(x) = \frac{2\alpha}{x} J_\alpha(x) - J_{\alpha-1}(x). \qquad (7.86)$$

Because of this last relation, it is necessary to tabulate the functions J_α only for $0 \le \alpha < 2$. In particular, every function J_n, with n an integer, can be expressed in terms of J_0 and J_1. Extensive tables of these two functions have been compiled. See, for example, Jahnke and Emde (1945).

The solutions of some other second-order equations can be expressed in terms of Bessel functions. If we make the variable changes

$$t = ax^r, \qquad y = x^s u$$

in Bessel's equation

$$t^2 \frac{d^2u}{dt^2} + t \frac{du}{dt} + (t^2 - \alpha^2)u = 0,$$

it becomes (Exercise 9)

$$x^2 \frac{d^2y}{dx^2} + (1 - 2s)x \frac{dy}{dx} + [(s^2 - r^2\alpha^2) + a^2r^2x^{2r}]y = 0. \qquad (7.87)$$

A solution of this equation is

$$y = x^s J_\alpha(ax^r).$$

If α is not an integer, the general solution is

$$y = c_1 x^s J_\alpha(ax^r) + c_2 x^s J_{-\alpha}(ax^r), \qquad (7.88)$$

but if α is a nonnegative integer N it is

$$y = c_1 x^s J_N(ax^r) + c_2 x^s Y_N(ax^r). \tag{7.89}$$

Example   Let us attempt to solve the equation

$$x\frac{d^2 y}{dx^2} + 5\frac{dy}{dx} + 9xy = 0$$

in terms of Bessel functions. Multiplying through by x in an attempt (perhaps futile) to put the equation in the form (7.87), we have

$$x^2\frac{d^2 y}{dx^2} + 5x\frac{dy}{dx} + 9x^2 y = 0.$$

For this equation to agree with Eq. (7.87) we must have

$$1 - 2s = 5, \qquad s^2 - r^2\alpha^2 = 0, \qquad a^2 r^2 = 9, \qquad r = 1.$$

We find that these conditions are satisfied if

$$r = 1, \qquad a = 3, \qquad s = -2, \qquad \alpha = 2.$$

From formula (7.89) we conclude that

$$y = c_1 x^{-2} J_2(3x) + c_2 x^{-2} Y_2(3x)$$

is the general solution of the original equation.

Exercises for Section 7.9
---

1. Given that $\Gamma\left(\frac{1}{2}\right) = \sqrt{\pi}$, find:

    (a) $\Gamma\left(\frac{3}{2}\right)$ (b) $\Gamma\left(\frac{5}{2}\right)$

    (c) $\Gamma\left(-\frac{1}{2}\right)$ (d) $\Gamma\left(-\frac{3}{2}\right)$

2. If x is not zero or a negative integer, verify

    $$x(x + 1)(x + 2)\cdots(x + n)\Gamma(x)$$
    $$= \Gamma(x + n + 1)$$

    for every positive integer n.

3. Show that $J_\alpha(0) = 0$ if $\alpha > 0$, and also that $J_0(0) = 1$.

4. Use the series definition to calculate the following quantities, correct to three decimal places.

    (a) $J_0(0.2)$ (b) $J_1(0.2)$ (c) $J_2(0.2)$

5. Show that

    $$\frac{d}{dx}[x^{-\alpha} J_\alpha(x)] = -x^{-\alpha} J_{\alpha+1}(x).$$

6. (a) Express $J_3(x)$ in terms of $J_0(x)$ and $J_1(x)$.
    (b) Express $J_2'(x)$ in terms of $J_0(x)$ and $J_1(x)$.

7. Show that

(a) $\int x^{\alpha+1} J_\alpha(x)\, dx = x^{\alpha+1} J_{\alpha+1}(x) + c$,

(b) $\int x^{1-\alpha} J_\alpha(x)\, dx = -x^{1-\alpha} J_{\alpha-1}(x) + c$

8. (a) Show that the change of dependent variable $u = x^{1/2} y$ in Bessel's equation (7.74) leads to the equation

$$u'' + \left(1 + \frac{1 - 4\alpha^2}{x^2}\right) u = 0.$$

(b) Use the result of part (a) and the fact that $\Gamma\left(\dfrac{1}{2}\right) = \sqrt{\pi}$ to show that

$$J_{1/2}(x) = \sqrt{\frac{2}{\pi x}} \sin x,$$

$$J_{-1/2}(x) = \sqrt{\frac{2}{\pi x}} \cos x.$$

(c) Use the result of part (b) to express $J_{3/2}(x)$ in terms of elementary functions.

9. Show that the variable changes

$$t = ax^r, \qquad y = x^s u,$$

in Bessel's equation

$$t^2 \frac{d^2 u}{dt^2} + t \frac{du}{dt} + (t^2 - \alpha^2) u = 0,$$

lead to the equation (7.87).

10. Express the general solution of the given equation in terms of Bessel functions

(a) $y'' + x^2 y = 0$

(b) $x^2 y'' + 5xy' + (9x^6 - 12)y = 0$

(c) $3x^2 y'' + 5xy' + (3x^4 - 1)y = 0$

(d) $xy'' + y = 0$

(e) $4x^2 y'' + (1 + 4x^4)y = 0$

(f) $xy'' - 3y' + 3xy = 0$

Additional Exercises for Chapter 7

In Exercises 1–16, express the general solution in terms of series about $x = 0$.

1. $y'' - x^2 y' - 2xy = 0$

2. $(2 - x^2)y'' + 6y = 0$

3. $y'' + xy' + 2y = 0$

4. $(1 - x^2)y'' + 2y = 0$

5. $2x^2 y'' + (3x - x^2)y' - (1 + 2x)y = 0$

6. $2x^2 y'' + 5xy' + (1 - x)y = 0$

7. $xy'' + y' - y = 0$

8. $x^2 y'' + (3x - x^2)y' + y = 0$

9. $x^2 y'' + 4xy' + (2 - x)y = 0$

10. $x^2 y' + (x - x^2)y' - (1 - 2x)y = 0$

11. $x^2 y'' + x(5 - x)y' + (4 - 3x)y = 0$

12. $4x^2 y'' - 2x^2 y' + (1 - x)y = 0$

13. $xy'' - xy' - 3y = 0$

14. $x^2 y'' + 2xy' + (x - 2)y = 0$

15. $x(1 - x)y'' + 4y' + 2y = 0$

16. $(x^2 - x^3)y'' - 2y = 0$

17. Given the equation

$$x(1 - x)y'' + [a - (a + 2)x]y' - ay = 0,$$

determine the values of the constant a for which there are two independent solutions of the form (7.34) and find these solutions.

18. Express the general solution in terms of Bessel
Functions.

(a) $xy'' + 2y' + 4xy = 0$

(b) $9x^2y'' + 3xy' + (x^2 - 3)y = 0$

(c) $x^2y'' + 3xy' + (4x^4 - 3)y = 0$

(d) $x^2y'' + 5xy' + (9x^2 - 5)y = 0$

19. Show that a solution of the *modified Bessel's
equation* of order α,

$$x^2y'' + xy' - (x^2 + \alpha^2)y = 0, \qquad \alpha \geq 0,$$

is

$$I_\alpha(x) = \sum_{m=0}^{\infty} \frac{(x/2)^{2m+\alpha}}{m!\Gamma(m + \alpha + 1)}.$$

This particular solution is known as the *modified Bessel function of the first kind* of order α.

20. Use the result of Exercise 10 to show that

$$\frac{d}{dx}[x^\alpha I_\alpha(x)] = x^\alpha I_{\alpha-1}(x),$$

$$\frac{d}{dx}[x^{-\alpha}I_\alpha(x)] = x^{-\alpha}I_{\alpha+1}(x),$$

$$I_{\alpha-1}(x) - I_{\alpha+1}(x) = \frac{2\alpha}{x}I_\alpha(x).$$

21. For n a nonnegative integer, show that the
equation

$$(1 - x^2)y'' - xy' + n^2y = 0$$

possesses a polynomial solution of degree n,
and find a formula for the specific solution for
which $y(1) = 1$. This function is known as
Chebyshev's polynomial of degree n. It is important in approximation theory. Suggestion:
use powers of $x - 1$.

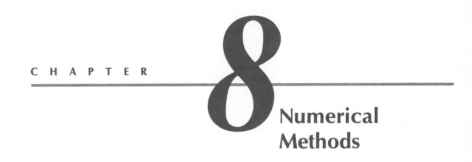

CHAPTER

8

Numerical
Methods

8.1

**THE EULER
METHOD**

In this chapter we shall describe some techniques that can be used to obtain approximate values of the solution of an initial value problem for a first-order differential equation. Such an initial value problem is of the form

$$y' = f(x, y), \qquad y(x_0) = y_0. \tag{8.1}$$

If h is a small positive number, we write

$$x_1 = x_0 + h, \qquad x_2 = x_1 + h = x_0 + 2h, \qquad x_3 = x_2 + h = x_0 + 3h,$$

and so on. In general, $x_n = x_0 + nh$. We refer to h as the *step size* to be used in the numerical method. We also use the abbreviations

$$y_1 = y(x_1), \qquad y_2 = y(x_2), \ldots, \qquad y_n = y(x_n)$$

for the exact values of the solution at the points $x_1, x_2, \ldots, x_n$. If a formula for the solution is not available, then these values are not known. For example, the problem

$$y' = x^2 + y^2, \qquad y(0) = 1$$

appears simple, but no explicit formula for its solution is known.

   In the methods to be investigated in this chapter, an approximation to y_1, denoted by w_1, is calculated. Then an approximate value w_2 for y_2 is

381

found, and so on. These approximate values are calculated step by step, with the value of w_2 depending on w_1 (and perhaps on w_0 also), while w_3 depends on w_2 (and possibly on w_1 and w_0 also). If w_n depends only on w_{n-1} (and not on w_{n-2} or other previously calculated values) the method is called a *single-step method* (as opposed to a *multi-step method*). We shall investigate several single-step methods and then discuss a multi-step technique.

All the single-step methods we shall discuss are based on Taylor's formula, which is derived in most calculus textbooks. This formula states that (under appropriate conditions[1])

$$y(x + h) = y(x) + hy'(x) + \frac{h^2}{2!} y''(x) + \cdots + \frac{h^k}{k!} y^{(k)}(x) + R_k(x, h),$$

where

$$R_k(x, h) = \frac{h^{k+1}}{(k+1)!} y^{(k+1)}(\xi)$$

with ξ between x and $x + h$. In particular, if $k = 1$, we have

$$y(x + h) = y(x) + hy'(x) + R_1(x, h),$$

and R_1 is proportional to h^2. If y is the solution of the initial value problem, then from the differential equation we see that $y'(x) = f(x, y(x))$, so the previous equation becomes

$$y(x + h) = y(x) + hf(x, y(x)) + R_1(x, h).$$

Setting $x = x_{n-1}$ and using the terminology previously introduced, we have the approximate relationship

$$y_n \approx y_{n-1} + hf(x_{n-1}, y_{n-1}). \tag{8.2}$$

Actually y_{n-1} is not known (when $n > 1$) and we must make an additional approximation by using w_{n-1} in place of y_{n-1} in the right-hand side of this last relationship. Thus we calculate approximate solution values step by step from the relation

$$w_n = w_{n-1} + hf(x_{n-1}, w_{n-1}). \tag{8.3}$$

This method of calculating approximate values of the solution at the points $x_1, x_2, \ldots$ is called the *Euler method*. In regard to accuracy, it is known as a *first-order method*, since it uses terms through those of first degree in h

---

[1] The function y is assumed to be $k + 1$ times differentiable on an open interval that contains x and $x + h$.

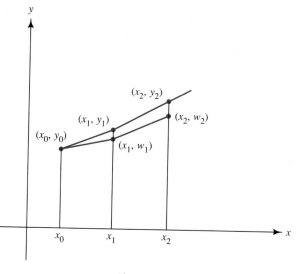

Figure 8.1

in Taylor's formula. Formula (8.2) has a simple geometrical interpretation. The tangent line to the solution curve at the point (x_{n-1}, y_{n-1}) has slope $f(x_{n-1}, y_{n-1})$. Hence y_n, as determined approximately by means of relation (8.2), is the value of y on the tangent line, rather than on the solution curve, that corresponds to $x = x_n$. Figure 8.1 illustrates the situation.

The choice of h in Eq. (8.3) is arbitrary, but in general the smaller the value of h the better, *up to a point*. Errors arise in the use of Eq. (8.3) for several reasons. First, the formula (8.3) is not exact, since the remainder term R_1 in Taylor's formula is omitted. Second, the exact value y_{n-1} is replaced by the approximation w_{n-1}. Finally, a computer cannot store numbers exactly, having only a finite number of binary or decimal places available. Each arithmetic calculation involves a small error, and many calculations can lead to a large error. In calculating a solution value at a particular point, say $x_0 + L$, use of a smaller value for h requires more calculations, since more intermediate points between x_0 and $x_0 + L$ are necessary. The subject of error analysis is a difficult one. The reader is referred to the chapter on ordinary differential equations in the reference on numerical analysis listed at the end of the book.

In order to illustrate the Euler method, we consider an initial value problem whose exact solution is known.

Example   We shall use the Euler method, first with step size $h = 0.2$, to obtain approximate values for the solution of the initial value problem

$$y' = -2x + 2y + 1, \qquad y(0) = 1,$$

on the interval $[0, 1]$. Since $f(x, y) = -2x + 2y + 1$, formula (8.3) becomes

$$w_n = w_{n-1} + h(-2x_{n-1} + 2w_{n-1} + 1).$$

Table 8.1
Comparison of Euler and Exact
Values for $h = 0.2$

x	y-Euler	y-exact
0.0	1.00000	1.00000
0.2	1.60000	1.69182
0.4	2.36000	2.62554
0.6	3.34400	3.92012
0.8	4.64160	5.75303
1.0	6.37824	8.38906

Proceeding step by step, we find that

$$w_0 = y_0 = y(0) = 1,$$

$$w_1 = 1 + 0.2[-(2)(0) + (2)(1) + 1] = 1.6,$$

$$w_2 = 1.6 + 0.2[-(2)(0.2) + (2)(1.6) + 1] = 2.36,$$

$$w_3 = 2.36 + 0.2[-(2)(0.4) + (2)(2.36) + 1] = 3.344,$$

$$w_4 = 3.344 + 0.2[-(2)(0.6) + (2)(3.344) + 1] = 4.6416,$$

$$w_5 = 4.6416 + 0.2[-(2)(0.8) + (2)(4.6416) + 1] = 6.37824.$$

The exact solution of the problem (the differential equation is first-order and linear) is $y = x + e^{2x}$. A comparison of the exact values (to 5 decimal places) and the approximate ones is shown in Table 8.1.

To show the effect of reducing the step size h, a computer was used to calculate approximations by the Euler method using $h = 0.1, 0.05$, and 0.01. The results are compared with the exact solution values in Table 8.2.

Table 8.2
Comparison of Euler and Exact Values for Various Step Sizes

x	$h = 0.1$	$h = 0.05$	$h = 0.01$	exact
0.0	1.00000	1.00000	1.00000	1.00000
0.1	1.30000	1.31000	1.31899	1.32140
0.2	1.64000	1.66410	1.68595	1.69182
0.3	2.02800	2.07156	2.11136	2.12212
0.4	2.47360	2.54359	2.60804	2.62554
0.5	2.98832	3.09374	3.19159	3.21828
0.6	3.58598	3.73843	3.88103	3.92012
0.7	4.28318	4.49750	4.69956	4.75520
0.8	5.09982	5.39497	5.67544	5.75303
0.9	6.05978	6.45992	6.84313	6.94965
1.0	7.19174	7.72750	8.24465	8.38906

The reader should keep several points in mind while examining this table. Formula (8.2) involves an error, known as the *local error*, that is proportional to h^2 if y_{n-1} is known exactly. Hence, the smaller the step size h the smaller the local error. However, the smaller the value of h, the more calculations required. Each calculation involves a roundoff or truncation error. It does not pay to decrease h beyond a certain point because the improvement in local error is at the expense of additional error due to additional computations.

The Euler method is relatively simple and unsophisticated, and is seldom used in practice. Better methods will be examined in the sections that follow. We conclude this section with a computer program, written in the BASIC language. This program uses the Euler method to calculate solution values for the example problem

$$y' = -2x + 2y + 1.$$

The reader who prefers another programming language such as FORTRAN, Pascal, or C can easily make the necessary conversion.

```
100 REM ***********************************************************
110 REM
120 REM This program illustrates the Euler method for a
130 REM particular equation. The significance of the
140 REM various variables is as follows:
150 REM
160 REM      X = starting value for x
170 REM      Y = starting value for y
180 REM      H = spacing
190 REM      N1 = number of iterations between printed values
200 REM      N2 = number of values printed
210 REM
220 REM ***********************************************************
230 DEF FNA(X, Y) = -2*X + 2*Y +1
240 F1$ = "##.##      ###.#####"
250 INPUT "Enter starting x and y values: "; X, Y
260 INPUT "Enter the spacing for x: "; H
270 INPUT "Enter number of iterations between printed values: "; N1
280 INPUT "Enter number of values to be printed: "; N2
290 PRINT : PRINT "    X              Y" : PRINT
300 PRINT USING F1$; X; Y
310 FOR I = 1 TO N2
320   FOR J = 1 TO N1
330     Y = Y + H * FNA(X, Y)
340     X = X + H
350   NEXT J
360   PRINT USING F1$; X; Y
370 NEXT I
380 END
```

Exercises for Section 8.1

1. Using the Euler method with step size $h = 0.1$, find approximate values for the solution of the initial value problem.

$$y' + 2y = 4x^2, \qquad y(0) = 2,$$

at $x = 0.1$, 0.2, and 0.3. Then find the exact solution and determine its values at the same points. Display the pairs of values in a table similar to Table 8.1.

2. Use the Euler method with $h = 0.1$ to find approximate values for the solution of the initial value problem

$$xy' + y = 3x^2, \qquad y(1) = -2,$$

at $x = 1.1$, 1.2, and 1.3. Then find the exact solution and determine its values at the same points.

3. Use the Euler method to find approximate values of the solution of the problem

$$y' = x + y^2, \qquad y(0) = 0,$$

at $x = 0.1, 0.2$, and 0.3, first using $h = 0.1$ and then using $h = 0.05$. Compare the two sets of values in a table.

4. Do as in Exercise 3 for the problem

$$y' = 2x + y^{-1}, \qquad y(0) = 1.$$

5. In the approximation (8.2), $f(x_{n-1}, y_{n-1})$ represents the slope of the solution curve at $x = x_{n-1}$. A modification of the Euler method is obtained by replacing this slope by the aver-

age of two slopes,

$$\frac{1}{2}[f(x_{n-1}, y_{n-1}) + f(x_n y_n)]$$

and then approximating $f(x_n, y_n)$ by $f(x_n, y_{n-1} + k)$, where $k = hf(x_{n-1}, y_{n-1})$. Show that this approach leads to the numerical scheme

$$w_n = w_{n-1} + \frac{1}{2}(k_k + k_1),$$

where

$$k_0 = hf(x_{n-1}, w_{n-1}),$$
$$k_1 = hf(x_{n-1} + h, w_{n-1} + k_0).$$

This scheme (an example of a class of methods to be described in Section 8.3) has a local error term proportional to h^3 (rather than h^2, as in the Euler method).

6. Compare values obtained by the Euler method, the method of Exercise 5, and the exact solution values, at $x = 0.1$, 0.2 using $h = 0.1$ for
   (a) the problem of Exercise 1.
   (b) the problem of Exercise 2.

7. Use a digital computer to print a table that compares the exact solution values and the approximate values from the Euler method for the initial value problem of Exercise 1 for $x = 0.0, 0.1, 0.2, \dots, 1.0$ using $h = 0.01$.

8. Do as in Exercise 7 for the initial value problem of Exercise 2.

8.2 TAYLOR SERIES METHODS

The Euler method, described in the previous section, can be regarded as a simple special case of a class of methods whose validity is based on Taylor's formula. This formula, mentioned earlier, says that (if y is sufficiently differentiable)

$$y(x + h) = y(x) + hy'(x) + \frac{h^2}{2!}y''(x) + \cdots + \frac{h^k}{k!}y^{(k)}(x) + R_k(x, h),$$

where

$$R_k(x, h) = \frac{h^{k+1}}{(k+1)!} \, y^{(k+1)}(\xi),$$

and ξ is between x and $x + h$. Thus we have the approximation

$$y(x + h) \approx y(x) + hy'(x) + \cdots + \frac{h^k}{k!} \, y^{(k)}(x), \tag{8.4}$$

and the local error is proportional to h^{k+1}.

If y is the solution of the initial value problem

$$y' = f(x, y), \qquad y(x_0) = y_0,$$

then the derivatives y', y'', and so on, can be expressed in terms of f and its partial derivatives. We have

$$y' = f,$$
$$y'' = f_x + f_y f,$$
$$y''' = f_{xx} + f_{yx} f + f_y f_x + f(f_{xy} + f_{yy} f + f_y^2),$$

with the calculation of higher derivatives becoming complicated. Formula (8.4) becomes

$$y(x + h) \approx y(x) + hf(x, y(x))$$
$$+ \frac{h^2}{2!} \left[f_x(x, y(x)) + f_y(x, y(x))f(x, y(x)) \right] + \cdots + \frac{h^k}{k!} \, y^{(k)}(x). \tag{8.5}$$

By taking $k = 1$ and setting $x = x_{n-1}$ in this relation, we arrive at Euler's formula,

$$w_n = w_{n-1} + hf(x_{n-1}, w_{n-1}).$$

By taking $k = 2$ in formula (8.5) we obtain a new approximation scheme,

$$w_n = w_{n-1} + hf(x_{n-1}, w_{n-1})$$
$$+ \frac{h^2}{2!} \left[f_x(x_{n-1}, w_{n-1}) + f(x_{n-1}, w_{n-1})f_y(x_{n-1}, w_{n-1}) \right]. \tag{8.6}$$

The method based on this formula is called a second-order method, since terms through those of second degree in h are retained in Taylor's formula. The local error is proportional to h^3; in the Euler method it was proportional to h^2. Other, more complicated approximation schemes can be derived by retaining more terms in Taylor's formula. In the following example, we compare the results obtained by using relation (8.6) with those obtained from Euler's method.

Example   We again consider the initial value problem

$$y' = -2x + 2y + 1, \qquad y(0) = 1,$$

in which

$$f(x, y) = -2x + 2y + 1, \qquad f_x(x, y) = -2, \qquad f_y(x, y) = 2.$$

Formula (8.6) becomes

$$w_n = w_{n-1} + hf(x_{n-1}, w_{n-1}) + \frac{h^2}{2}\left[-2 + 2f(x_{n-1}, w_{n-1})\right]$$

or

$$w_n = w_{n-1} + h(-2x_{n-1} + 2w_{n-1} + 1) + h^2(-2x_{n-1} + 2w_{n-1}).$$

The approximate values obtained from this formula are compared with those obtained by the Euler method, and also with the exact values of the solution in Table 8.3. A spacing of $h = 0.01$ was used in both approximation schemes.

The main drawbacks in using the second order Taylor method are, first, that expressions for the partial derivatives f_x and f_y must be derived (not always simple) and, second, that the values of the *three* functions f, f_x, and

Table 8.3
Comparison of Euler, Taylor, and Exact
Values, $h = 0.01$

x	y-Euler	2nd-order y-Taylor	y-exact
0.0	1.00000	1.00000	1.00000
0.1	1.31899	1.32139	1.32140
0.2	1.68595	1.69179	1.69182
0.3	2.11136	2.12205	2.12212
0.4	2.60804	2.62542	2.62554
0.5	3.19159	3.21810	3.21828
0.6	3.88103	3.91986	3.92012
0.7	4.69956	4.75483	4.75520
0.8	5.67544	5.75251	5.75303
0.9	6.84313	6.94893	6.94965
1.0	8.24465	8.38809	8.38906

f_y must be calculated at each step. As we shall see in the next section, formula (8.6) (and, indeed, any method based on Taylor's formula) can be replaced by one that gives the same degree of accuracy but is easier to use. However, formula (8.6) and its generalizations are essential for an understanding of the derivations of the more desirable methods that are commonly used in practice.

─────────────── **Exercises for Section 8.2** ───────────────

1. Use the second-order Taylor formula scheme (8.6) with $h = 0.1$ to find approximate solution values for the initial value problem

$$y' + 2y = 4x^2, \qquad y(0) = 2,$$

at $x = 0.1$, 0.2, and 0.3. Make a table comparing these values with those obtained by the Euler method in Exercise 1, Section 8.1, and with the exact solution values.

2. Use the second-order Taylor formula (8.6) with $h = 0.1$ to find approximate solution values for the problem

$$xy' + y = 3x^2, \qquad y(1) = -2,$$

at $x = 1.1$, 1.2, and 1.3. Compare these values with those obtained in Exercise 2, Section 8.1.

3. Use the second-order Taylor method with $h = 0.1$ to find approximate values of the solution of the problem.

$$y' = x + y^2, \qquad y(0) = 0,$$

at $x = 0.1$, 0.2, and 0.3. Compare these values with those obtained in Exercise 3, Section 8.1.

4. Do as in Exercise 3 for the problem

$$y' = 2x + y^{-1}, \qquad y(0) = 1,$$

but compare the results with those obtained in Exercise 4, Section 8.1.

5. Use the second-order Taylor method with $h = 0.1$ to find approximate solution values for the problem of the example of this section at $x = 0.1$ and 0.2.

6. If $y' = f(x, y)$, show that

$$y''' = f_{xx} + f_y f_x + f(f_{yx} + f_{xy} + f_{yy}f + f_y^2).$$

7. Using the result of Exercise 6, write out the third-order Taylor scheme for the initial value problem of Example 1. Then, using $h = 0.1$, obtain approximate solution values at $x = 0.1$ and 0.2 using this scheme. Compare with the results obtained in Exercise 5, and also with the exact solution values shown in Table 8.3.

8. Use a digital computer to print a table comparing the Euler, second-order Taylor, and exact solution values for the initial value problem of Exercise 1 for $x = 0.1, 0.2, 0.3, \ldots, 1.0$ using $h = 0.01$.

9. Do as in Exercise 8 for the initial value problem of Exercise 2.

8.3
RUNGE-KUTTA
METHODS

In the previous section we derived the formula

$$w_n = w_{n-1} + hf(x_{n-1}, w_{n-1})$$
$$+ \frac{h^2}{2} [f_x(x_{n-1}, w_{n-1}) + f(x_{n-1}, w_{n-1})f_y(x_{n-1}, w_{n-1})]$$

from the Taylor formula approximation

$$y_n \approx y_{n-1} + hy'_{n-1} + \frac{h^2}{2} y''_{n-1}.$$

The underlying idea of the Runge-Kutta methods is exemplified by the approach we now consider. The first formula of this section is to be replaced by one of the form

$$w_n = w_{n-1} + h[af(x_{n-1}, w_{n-1}) + bf(x'_{n-1}, w'_{n-1})],$$

where a, b, x'_{n-1}, and w'_{n-1} are to be chosen so that the right-hand members of the two equations differ only by a term proportional to h^3. Thus both formulas have second-order accuracy. The advantage of the last formula is that only one function, f, need be evaluated, although at two points. In particular, f_x and f_y are not involved.

The determination of a, b, x'_{n-1} and w'_{n-1} requires the use of the mean value theorem for functions of two variables. It is fairly elementary, but tedious. We content ourselves with the final result,[2] which may be stated as follows:

$$w_n = w_{n-1} + \frac{1}{2}(k_0 + k_1), \tag{8.7}$$

where

$$k_0 = hf(x_{n-1}, w_{n-1}), \qquad k_1 = hf(x_{n-1} + h, w_{n-1} + k_0).$$

The method (8.7) is called a second-order Runge-Kutta method. In some instances a Runge-Kutta method and Taylor method of the same order may give identical results (see Exercise 9), but this does not usually happen.

Example   We illustrate the use of formula (8.7) by calculating w_1 for a step size $h = 0.01$ for the initial value problem

$$y' = -x^2 + y, \qquad y(0) = 1.$$

Since $x_0 = 0$ and $w_0 = y_0 = 1$, we have

$$k_0 = 0.01(-0 + 1) = 0.01,$$

$$k_1 = 0.01[-(0.01)^2 + 1.01] = 0.010099,$$

$$w_1 = 1 + 0.5(0.01 + 0.010099) = 1.0100495.$$

---

[2] The quantities a, b, x'_{n-1}, and w'_{n-1} are not uniquely determined, and other choices are possible.

If the calculations are continued until $x = 1.0$, the approximate value obtained for $y(1)$ is 2.28168. The value obtained by the second-order Taylor method is 2.28176, and the exact value is 2.28172.

Other, higher-order, formulas can be derived. The Taylor formula

$$y_n \approx y_{n-1} + hy'_{n-1} + \frac{h^2}{2} y''_{n-1} + \frac{h^3}{6} y'''_{n-1}$$

can be replaced by

$$w_n = w_{n-1} + \frac{1}{6}(k_0 + 4k_1 + k_2), \qquad (8.8)$$

with

$$k_0 = hf(x_{n-1}, w_{n-1}),$$

$$k_1 = hf\left(x_{n-1} + \frac{1}{2}h, w_{n-1} + \frac{1}{2}k_0\right),$$

$$k_2 = hf(x_{n-1} + h, w_{n-1} + 2k_1 - k_0),$$

and the fourth-order Taylor approximation

$$y_n \approx y_{n-1} + hy'_{n-1} + \cdots + \frac{h^4}{24} y^{(4)}_{n-1}$$

leads to the fourth-order Runge-Kutta formula

$$w_n = w_{n-1} + \frac{1}{6}(k_0 + 2k_1 + 2k_2 + k_3) \qquad (8.9)$$

with

$$k_0 = hf(x_{n-1}, w_{n-1}),$$

$$k_1 = hf\left(x_{n-1} + \frac{1}{2}h, w_{n-1} + \frac{1}{2}k_0\right),$$

$$k_2 = hf\left(x_{n-1} + \frac{1}{2}h, w_{n-1} + \frac{1}{2}k_1\right),$$

$$k_3 = hf(x_{n-1} + h, w_{n-1} + k_2).$$

This last formula is widely used. For the problem

$$y' = -2x + 2y + 1, \qquad y(0) = 1$$

used to illustrate other methods, the scheme (8.9), with a step size $h = 0.01$, yields values that agree with the exact solution values to at least seven significant digits on the interval $[0, 3]$. A computer program in BASIC follows. It uses the fourth-order Runge-Kutta method to obtain solution values for the example just mentioned.

```
100 REM ************************************************************
110 REM
120 REM This program illustrates the fourth-order Runge-Kutta
130 REM method for a particular equation. The significance of the
140 REM various variables is as follows:
150 REM
160 REM      X = starting value for x
170 REM      Y = starting value for Y
180 REM      H = spacing
190 REM      N1 = number of iterations between printed values
200 REM      N2 = number of values printed
210 REM
220 REM ************************************************************
230 DEF FNA(X, Y) = - 2*X + 2*Y + 1
240 F1$ = "##.##        ##.#####"
250 INPUT "Enter starting x and y values: "; X, Y
260 INPUT "Enter the spacing for x: "; H
270 INPUT "Enter the number of iterations between printed values: "; N1
280 INPUT "Enter the number of values to be printed: "; N2
290 PRINT : PRINT"    X                Y" : PRINT
300 PRINT USING F1$; X; Y
310 FOR I = 1 TO N2
320    FOR J = 1 TO N1
330       K0 = H * FNA(X,Y)
340       K1 = H * FNA(X + .5*H,Y+.5*K0)
350       K2 = H * FNA(X + .5*X,Y + .5*K1)
360       K3 = H * FNA(X + H,Y + K2)
370       Y = Y + (K0 + 2*K1 + 2*K2 + K3) / 6
380       X = X + H
390    NEXT J
400    PRINT USING F1$; X; Y
410 NEXT I
420 END
```

Exercises for Section 8.3

1. Use the second-order Runge-Kutta method (8.7) with $h = 0.1$ to calculate approximate solution values for the problem

$$y' + 2y = 4x^2, \qquad y(0) = 2,$$

at $x = 0.1, 0.2$, and 0.3. Compare your results with those of Exercise 1, Section 8.2.

2. Use the second-order Runge-Kutta method (8.7) with $h = 0.1$ to calculate approximate solution values for the problem

$$xy' + y = 3x^2, \qquad y(1) = -2,$$

at $x = 1.1, 1.2$, and 1.3. Compare your results with those of Exercise 2, Section 8.2.

w_n and y_n would differ by a term proportional to h^5. Thus the numerical method based on formula (8.11) is said to be a fourth-order method.[5]

In practice, formula (8.11) is usually used only as a *predictor* for the approximate value of y_n. That is, an initial approximation $w_n^{(0)}$ is found from Eq. (8.11) and is then improved by means of a *corrector* formula. We shall present a corrector formula that is frequently used in conjunction with Eq. (8.11). In the derivation of this corrector formula, the integrand in the relation

$$y_n = y_{n-1} + \int_{x_{n-1}}^{x_n} f(x, y(x))\, dx$$

is approximated by the polynomial of degree 3 or less that passes through the 4 points

$$(x_n, f_n),\quad (x_{n-1}, f_{n-1}),\quad (x_{n-2}, f_{n-2}),\quad (x_{n-3}, f_{n-3}).$$

Proceeding as in the derivation of Eq. (8.11) we arrive at the relation

$$w_n = w_{n-1} + \frac{h}{24}(9f_n + 19f_{n-1} - 5f_{n-2} + f_{n-3}), \qquad (8.12)$$

which, like formula (8.11), has a local error proportional to h^5. Notice that w_n appears implicitly in the right-hand member of Eq. (8.12), since

$$f_n = f(x_n, w_n).$$

In using formulas (8.11) and (8.12) an initial approximation $w_n^{(0)}$ to y_n is found from the predictor formula

$$w_n^{(0)} = w_{n-1} + \frac{h}{24}(55f_{n-1} - 59f_{n-2} + 37f_{n-3} - 9f_{n-4}). \qquad (8.13)$$

Then a final approximation w_n is found from the corrector formula

$$w_n = w_{n-1} + \frac{h}{24}(9f_n^{(0)} + 19f_{n-1} - 5f_{n-2} + f_{n-3}), \qquad (8.14)$$

where

$$f_n^{(0)} = f(x_n, w_n^{(0)}).$$

To use this scheme, the starting values w_0, w_1, w_2, and w_3 must be found by another method. The Runge-Kutta method (8.9) is appropriate, since

---

[5] Formula (8.11) is referred to in the literature as the fourth-order Adams-Bashforth method.

the local error in that formula is proportional to h^5, as is the case in both the relations (8.13) and (8.14).

Actually, the corrector formula (8.14) can be iterated to give still another approximation to y_n. An estimate of the error is obtained by comparing successive approximations. One reason why the predictor-corrector methods are sometimes preferred over the Runge-Kutta methods is that the latter provide no such estimate. However, it is not feasible to change the spacing h at some intermediate point in the computations when using the predictor-corrector formulas; this presents no difficulty when using a Runge-Kutta method.

Example   We will illustrate the use of formulas (8.13) and (8.14) for the problem

$$y' = 4x^2 - 2y, \qquad y(0) = 2,$$

with $h = 0.1$. The solution is $y(x) = 2x^2 - 2x + 1 + e^{-2x}$, and we will use $y(0)$, $y(0.1)$, $y(0.2)$, and $y(0.3)$, rounded to five decimal places, as the starting values for w_0, w_1, w_2, and w_3, respectively. The calculation of f_0, f_1, f_2, and f_3 is displayed in the accompanying table.

x	w	$f(x, w)$
0.0	2.00000	$4(0) - 2(2.00000) = -4.00000$
0.1	1.63873	$4(0.1)^2 - 2(1.63873) = -3.23746$
0.2	1.35032	$4(0.2)^2 - 2(1.35032) = -2.54064$
0.3	1.12881	$4(0.3)^2 - 2(1.12881) = -1.89762$

From formula (8.13) we find that

$$w_4^{(0)} = 1.12881 + \frac{0.1}{24} [55(-1.89762) - 59(-2.54064)$$

$$+ 37(-3.23746) - 9(-4.00000)]$$

$$= 0.96973.$$

Then

$$f_4 = f(x_4, w_4^{(0)}) = 4(0.4)^2 - 2(0.96973) = -1.29946$$

and from formula (8.14) we find that

$$w_4 = 1.12281 + \frac{0.1}{24} [9(-1.29946) + 19(-1.89762)$$

$$- 5(-2.54064) + (-3.23746)]$$

$$= 0.96329.$$

A great variety of predictor and corrector formulas can be found from the relation

$$y_n = y_{n-k} + \int_{x_{n-k}}^{x_n} f(x, y(x))\, dx.$$

In the derivations of formulas (8.11) and (8.12), k was chosen to be 1, but other choices are possible. Also, various choices are possible for the degree of the polynomial used to approximate the integrand. The reader is referred to books on numerical analysis for a description of some of these other methods and comparisons of their relative merits.

Exercises for Section 8.4

1. The exact solution values (correct to six significant digits) of the initial value problem

   $$y' = -2x + 2y + 1, \qquad y(0) = 1,$$

   at $x = 0.0, 0.1, 0.2,$ and 0.3 are as indicated in Table 8.3. Use them and the formulas (8.13), (8.14) to find an approximate value for the solution at $x = 0.4$. Use $h = 0.1$.

2. Verify that the solution of the initial value problem

   $$y' = -y^2, \qquad y(0) = 1,$$

   is $y = 1/(x + 1)$. Calculate the exact solution values at $x = 0.0, 0.1, 0.2,$ and 0.3. Then use formulas (8.13), (8.14), with $h = 0.1$, to obtain an approximate value for the solution at $x = 0.4$. Compare this value with the exact value.

3. (a) Find the polynomial P of degree 2 or less that passes through the 3 points

   $$(x_0, f_0), \quad (x_1, f_1), \quad (x_2, f_2).$$

   (b) Using the formula

   $$y_3 = y_2 + \int_{x_2}^{x_3} y'(x)\, dx$$

   $$\approx y_2 + \int_{x_2}^{x_3} P(x)\, dx$$

and the result of Part (a), derive the approximation

$$w_3 = w_2 + \frac{h}{12}(23f_2 - 16f_1 + 5f_0)$$

and thereby the predictor formula

$$w_n = w_{n-1} + \frac{h}{12}(23f_{n-1} - 16f_{n-2} + 5f_{n-3}).$$

4. (a) Find the polynomial Q of degree 2 or less that passes through the 3 points

   $$(x_n, f_n), \quad (x_{n-1}, f_{n-1}), \quad (x_{n-2}, f_{n-2}).$$

   (b) Use the polynomial Q of Part (a) to derive the corrector formula

   $$w_n = w_{n-1} + \frac{h}{12}(5f_n + 8f_{n-1} - f_{n-2}).$$

5. Apply the results of Exercises 3 and 4 to obtain an approximate value for the solution of the initial value problem

   $$y' = -y^2, \qquad y(0) = 1$$

   at $x = 0.3$. Use $h = 0.1$ and calculate the exact solution values at $x = 0.0, 0.1,$ and 0.2. See Exercise 2.

**8.5
SYSTEMS OF
EQUATIONS**

In the previous sections, we have examined some methods for the numerical solution of a single first-order equation. The same approaches can be used to develop techniques for first-order systems, and hence for a single higher-order equation.

Let us consider the first-order system

$$\frac{dx}{dt} = f(t, x, y), \qquad \frac{dy}{dt} = g(t, x, y) \tag{8.15}$$

with initial conditions $x(t_0) = x_0$, $y(t_0) = y_0$. Various approximation schemes can be derived from the Taylor series

$$x(t_n) = x(t_{n-1}) + x'(t_{n-1})h + \frac{1}{2!} x''(t_{n-1})h^2 + \cdots$$

$$y(t_n) = y(t_{n-1}) + y'(t_{n-1})h + \frac{1}{2!} y''(t_{n-1})h^2 + \cdots.$$

The Euler method uses the approximations

$$x(t_n) \approx x(t_{n-1}) + x'(t_{n-1}), \qquad y(t_n) \approx y(t_{n-1}) + y'(t_{n-1})h.$$

Using the differential equations (8.15) to replace x' and y' by $f(x, y)$ and $g(x, y)$, we arrive at the scheme

$$x_n = x_{n-1} + hf(t_{n-1}, x_{n-1}, y_{n-1}),$$
$$y_n = y_{n-1} + hg(t_{n-1}, x_{n-1}, y_{n-1}). \tag{8.16}$$

As a special case, we consider the single second-order equation

$$\frac{d^2x}{dt^2} = F\left(t, x, \frac{dx}{dt}\right). \tag{8.17}$$

Setting $y = \dfrac{dx}{dt}$, we obtain the first-order system

$$\frac{dx}{dt} = y, \qquad \frac{dy}{dt} = F(t, x, y).$$

For this system, the Euler method (8.16) becomes

$$x_n = x_{n-1} + hy_{n-1}, \qquad y_n = y_{n-1} + hF(t_{n-1}, x_{n-1}, y_{n-1}). \tag{8.18}$$

Higher-order Taylor formulas and Runge-Kutta formulas can be developed. The fourth-order Runge-Kutta formulas for the system (8.15) can be written in the form

$$x_n = x_{n-1} + \frac{1}{6}(a_1 + 2a_2 + 2a_3 + a_4),$$

$$y_n = y_{n-1} + \frac{1}{6}(b_1 + 2b_2 + 2b_3 + b_4) \qquad (8.19)$$

where

$$a_1 = hf(t_{n-1}, x_{n-1}, y_{n-1})$$

$$b_1 = hg(t_{n-1}, x_{n-1}, y_{n-1})$$

$$a_2 = hf\left(t_{n-1} + \frac{h}{2}, x_{n-1} + \frac{a_1}{2}, y_{n-1} + \frac{b_1}{2}\right)$$

$$b_2 = hg\left(t_{n-1} + \frac{h}{2}, x_{n-1} + \frac{a_1}{2}, y_{n-1} + \frac{b_1}{2}\right)$$

$$a_3 = hf\left(t_{n-1} + \frac{h}{2}, x_{n-1} + \frac{a_2}{2}, y_{n-1} + \frac{b_2}{2}\right)$$

$$b_3 = hg\left(t_{n-1} + \frac{h}{2}, x_{n-1} + \frac{a_2}{2}, y_{n-1} + \frac{b_2}{2}\right)$$

$$a_4 = hf(t_{n-1} + h, x_{n-1} + a_3, y_{n-1} + b_3)$$

$$b_4 = hg(t_{n-1} + h, x_{n-1} + a_3, y_{n-1} + b_3).$$

In the special case of a second-order equation (8.17), these last formulas become

$$a_1 = hy_{n-1}$$

$$b_1 = hF(t_{n-1}, x_{n-1}, y_{n-1})$$

$$a_2 = h\left(y_{n-1} + \frac{b_1}{2}\right)$$

$$b_2 = hF\left(t_{n-1} + \frac{h}{2}, x_{n-1} + \frac{a_1}{2}, y_{n-1} + \frac{b_1}{2}\right)$$

$$a_3 = h\left(y_{n-1} + \frac{b_2}{2}\right)$$

$$b_3 = hF\left(t_{n-1} + \frac{h}{2}, x_{n-1} + \frac{a_2}{2}, y_{n-1} + \frac{b_2}{2}\right)$$

$$a_4 = h(y_{n-1} + b_3)$$
$$b_4 = hF(t_{n-1} + h, x_{n-1} + a_3, y_{n-1} + b_3).$$

Example   As an example we consider the initial value problem

$$\frac{d^2x}{dt^2} + 0.01\left(\frac{dx}{dt}\right) = 8, \qquad x(0) = x'(0) = 0.$$

The associated first-order system is

$$\frac{dx}{dt} = y, \qquad \frac{dy}{dt} = 8 - 0.01y^2$$

with $x(0) = y(0) = 0$. The Euler scheme (8.18) becomes

$$x_n = x_{n-1} + hy_{n-1}, \qquad y_n = y_{n-1} + h(8 - 0.01y_{n-1}^2).$$

The exact solution, as found by the methods of Section 1.11, is

$$x = 100 \ln(\cosh(\sqrt{2}t/5)), \qquad y = 20\sqrt{2} \tanh(\sqrt{2}t/5).$$

Table 8.4 compares the exact values with those obtained by the Euler method and the fourth-order Runge-Kutta method with a spacing of $h = 0.1$ on the interval $[0,4]$.

A computer program, in BASIC, for the fourth-order Runge-Kutta method concludes this section.

Table 8.4
Euler, Runge-Kutta, and Exact Values
$h = 0.1$

t	*x*-Euler	*y*-Euler	*x*-RK	*y*-RK	*x*-exact	*y*-exact
0.00	0.0000000	0.0000000	0.0000000	0.0000000	0.0000000	0.0000000
0.40	0.4796162	3.1910574	0.6386393	3.1864162	0.6386393	3.1864162
0.80	2.2275443	6.3116044	2.5384483	6.2929648	2.5384484	6.2929649
1.20	5.2041152	9.2872947	5.6526908	9.2475901	5.6526908	9.2475902
1.60	9.3417015	12.0568557	9.9084503	11.9922902	9.9084503	11.9922903
2.00	14.5506951	14.5764780	15.2130731	14.4867395	15.2130731	14.4867396
2.40	20.7267987	16.8212467	21.4614787	16.7090284	21.4614788	16.7090286
2.80	27.7584695	18.7838895	28.5432570	18.6539776	28.5432571	18.6539778
3.20	35.5335519	20.4716660	36.3487478	20.3298991	36.3487480	20.3298993
3.60	43.9445063	21.9023947	44.7736729	21.7547384	44.7736730	21.7547386
4.00	52.8920254	23.1004730	53.7222290	22.9523439	53.7222292	22.9523442

```
100 REM ********************************************************
110 REM
120 REM This program illustrates the fourth-order Runge-Kutta
130 REM method for a particular first-order system of two
140 REM equations. The significance of the variables is as follows:
150 REM
160 REM     T: starting value for the independent variable t
170 REM     X: starting value for dependent variable x
180 REM     Y: starting value for dependent variable y
190 REM     H: spacing for t
200 REM     N1: number of iterations between printed values
210 REM     N2: number of values to be printed
220 REM
230 REM ********************************************************
240 DEF FNA(T,X,Y) = Y
250 DEF FNB(T,X,Y) = 8 - .01 * Y * Y
260 F1$ = "##.##      ###.#####      ###.#####"
270 INPUT "Enter starting values for t, x, and y: "; T, X, Y
280 INPUT "Enter the spacing h: "; H
290 INPUT "Enter the number of iterations between printed values: "; N1
300 INPUT "Enter the number of values to be printed : "; N2
310 PRINT : PRINT "    t              x                y" : PRINT
320 PRINT USING F1$; T; X; Y
330 FOR I = 1 TO N2
340    FOR J = 1 TO N1
350       A1 = H * FNA(T,X,Y)
360       B2 = H * FNB(T,X,Y)
370       A2 = H * FNA(T + .5 * H, X + .5 * A1, Y + .5 * B1)
380       B2 = H * FNB(T + .5 * H, X + .5 * A1, Y + .5 * B1)
390       A3 = H * FNA(T + .5 * H, X + .5 * A2, Y + .5 * B2)
400       B3 = H * FNB(T + .5 * H, X + .5 * A2, Y + .5 * B2)
410       A4 = H * FNA(T + H, X + A3, Y + B3)
420       B4 = H * FNB(T + H, X + A3, Y + B3)
430       X = X + (A1 + 2 * A2 + 2 * A3 + A4) / 6
440       Y = Y + (B1 + 2 * B2 + 2 * B3 + B4) / 6
450       T = T + H
460    NEXT J
470    PRINT USING F1$; T; X; Y
480 NEXT I
490 END
```

Exercises for Section 8.5

1. Given the initial value problem

$$\frac{dx}{dt} = -5x + 4y - 6, \quad \frac{dy}{dt} = -3x + 2y - 2$$

$$x(0) = y(0) = 0,$$

   (a) formulate the Euler formulas (8.16) for the system.

   (b) find the exact solution, using the methods of Chapter 6.

   (c) use a computer program that uses the formulas of Part (a) with $h = 0.1$ to calculate approximate values of x and y on the t-interval $[0,4]$. Compare with the exact values.

   (d) do as in Part (c), but with $h = 0.01$.

2. Do as in Exercise 1 for the initial value problem

$$\frac{dx}{dt} = -x + y, \qquad \frac{dy}{dt} = -x - y$$

$$x(0) = 4, \qquad y(0) = 3.$$

3. Redo Exercise 1, but using the fourth-order Runge-Kutta method.

4. Redo Exercise 2, but using the fourth-order Runge-Kutta method.

5. Given the initial value problem

$$\frac{d^2x}{dt^2} + 2\frac{dx}{dt} + 5x = 0,$$

$$x(0) = 0, \qquad x'(0) = 10,$$

(a) formulate the Euler formulas (8.18) for the problem.
(b) use a computer program that uses the formulas of Part (a) with $h = 0.1$ to calculate approximate values of x and y on the t-interval $[0, 4]$. Compare these values with the exact values.
(c) do as in Part (b), but with $h = 0.01$.

6. Given the initial value problem

$$\frac{d^2x}{dt^2} + 8 \sin x = 0, \quad x(0) = 0.2, \quad x'(0) = 0,$$

do as in Exercise 5, but omit the comparison with exact values.

7. Redo Exercise 5, but using the fourth-order Runge-Kutta method.

8. Redo Exercise 6, but using the fourth-order Runge-Kutta method.

9. Derive the second-order Taylor formulas for the system (8.15).

10. Redo the specified exercise, using the formulas of Exercise 9.
(a) Exercise 1   (b) Exercise 2

11. Write the second-order Taylor formulas of Exercise 9 for the special case of a single second-order equation (8.17).

12. Redo the specified exercise, using the formulas of Exercise 11.
(a) Exercise 5   (b) Exercise 6

Additional Exercises for Chapter 8

1. Find approximate values of the solution of the problem

$$(x + 1)y' = 2y, \qquad y(0) = 1$$

at $x = 0.1, 0.2$ by taking $h = 0.1$ and using
(a) the Euler method.
(b) the second-order Taylor method.

2. Find approximate values of the solution of the problem

$$y' = x^2 + 1/y, \qquad y(1) = 1$$

at $x = 1.1, 1.2$ by taking $h = 0.1$ and using
(a) the Euler method.
(b) the second-order Runge-Kutta method.

3. Use a computer and the fourth-order Runge-Kutta method to obtain approximate values of the solution to the problem of Exercise 1 on the interval $[0, 4]$, with $h = 0.01$. Compare the results with the exact values (found by solving the differential equation).

4. Use a computer and the second-order Runge-Kutta method to obtain approximate values of the solution to the problem of Exercise 2 on the interval $[1, 2]$, with (a) $h = 0.05$, (b) $h = 0.01$.

5. Given the initial value problem

$$\frac{dN}{dt} = N(2 - N), \qquad N(0) = 0.5$$

(see Section 1.9), use a computer to obtain approximate solution values on the interval $[0, 4]$ using $h = 0.01$ and
(a) the Euler method.
(b) the fourth-order Runge-Kutta method.

6. Given the initial value problem

$$\frac{dI}{dt} + 2I = 6, \qquad I(0) = 0,$$

use a computer to obtain approximate solution values on the interval $[0, 4]$ using $h = 0.01$ and (a) the Euler method; (b) the fourth-order Runge-Kutta method. Compare with the exact values.

7. By integrating the equation $y' = f(x, y)$ from x_{n-2} to x_n and using an approximating polynomial through the points (x_{n-1}, f_{n-1}), (x_{n-2}, f_{n-2}), derive the formula

$$w_n = w_{n-2} + 2hf_{n-1}.$$

Although this formula is similar in appearance to that of the Euler method, it is a multistep formula and the local error is proportional to h^3 instead of h^2.

8. By integrating the equation $y' = f(x, y)$ from x_{n-4} to x_n, and approximating the integrand with a polynomial through the appropriate points with abscissae $x_{n-1}, x_{n-2}, x_{n-3}, x_{n-4}$, derive the formula

$$w_n = w_{n-4} + \frac{4h}{3}(2f_{n-1} - f_{n-2} + 2f_{n-3}).$$

9. Given the initial value problem

$$\frac{dx}{dt} = -\frac{x}{r}, \qquad \frac{dy}{dt} = -\frac{y}{r} + 0.5,$$

$$r = \sqrt{x^2 + y^2}, \quad x(0) = 1, \quad y(0) = 0$$

(see Exercise 25 in the set of additional exercises for Chapter 1), find approximate

solution values using the fourth-order Runge-Kutta method and $h = 0.01$ for t in $[0, 2]$.

10. Given the initial value problem

$$\frac{dP}{dt} = P(-2 + H), \qquad \frac{dH}{dt} = H(2 - P),$$

$$P(0) = 1, \qquad H(0) = 1,$$

use a computer to obtain approximate solution values for t in the interval $[0, 4]$ using $h = 0.01$ and (a) the Euler method; (b) the fourth-order Runge-Kutta method.

11. Given the initial value problem

$$\frac{d^2I}{dt^2} + 4I = 6 \sin \omega t, \qquad I(0) = I'(0) = 0,$$

use one of the methods of Section 8.5, with $h = 0.01$, to obtain approximate solution values for t in $[0, 3]$ (a) for $\omega = 1$; (b) for $\omega = 2$.

12. Given the initial value problem

$$\frac{d^2x}{dt^2} + 3\frac{dx}{dt} + 2x = 2t, \qquad x(0) = x'(0) = 0,$$

use one of the methods of Section 8.5 with $h = 0.01$ to obtain approximate solution values for t in $[0, 4]$.

13. Derive the Euler formulas for a first-order system of four equations

$$\frac{dx_i}{dt} = f_i(t, x_1, x_2, x_3, x_4), \qquad i = 1, 2, 3, 4.$$

14. Given the initial value problem

$$x'' = -2x + y, \qquad y'' = x - y,$$

$$x(0) = 1, \quad y(0) = 2, \quad x'(0) = 0, \quad y'(0) = 2$$

convert to a first-order system and write out the Euler formulas derived in the previous exercise.

C H A P T E R

9

Laplace Transforms

9.1

THE LAPLACE TRANSFORM

Let f be a function that is defined on $[0, \infty)$. Associated with f is the improper integral

$$\int_0^\infty e^{-st}f(t)\,dt, \tag{9.1}$$

where s is a real[1] number. It may happen that there is no number s for which the integral converges. Otherwise, there exists a set of real numbers, which we denote by S, such that the integral converges for s in S. In this case we define the function F by means of the relation

$$F(s) = \int_0^\infty e^{-st}f(t)\,dt, \qquad s \text{ in } S.$$

The function F is called the *Laplace transform* of the function f. We write

$$F = \mathscr{L}[f] \qquad \text{and} \qquad F(s) = \mathscr{L}[f](s)$$

to indicate the relations between the functions f and F. (In general, we shall use capital letters for transforms of functions denoted by corresponding lowercase letters. Thus $\mathscr{L}[g] = G$, $\mathscr{L}[h] = H$, and so on.) Actually, we shall be interested only in functions whose transforms exist in an interval

---

[1] In more advanced treatments of the Laplace transform, s is permitted to be a complex number.

of the form (s_0, ∞) for some number s_0. Sufficient conditions that the transform of a function exist on such an interval will be discussed in the next section.

Not every function has a Laplace transform. For instance, if $f(t) = e^{t^2}$, the improper integral (9.1) diverges for all values of s. When s is positive, however, the quantity e^{-st} tends to zero fairly rapidly as t becomes infinite. Consequently many functions do possess Laplace transforms.

Let us now compute the transforms of some specific functions. Starting with the function $f(t) = e^{at}$, where a is a constant, we have

$$\int_0^\infty e^{-st}f(t)\, dt = \int_0^\infty e^{-t(s-a)}\, dt$$

$$= \lim_{T\to\infty}\left[\frac{-1}{s-a}e^{-t(s-a)}\right]_0^T$$

$$= \frac{1}{s-a}$$

if $s > a$. Thus

$$\mathcal{L}[f](s) = \frac{1}{s-a}, \qquad s > a. \tag{9.2}$$

As a second example, let $g(t) = \cos at$. Then

$$\int_0^\infty e^{-st}g(t)\, dt = \int_0^\infty e^{-st}\cos at\, dt$$

$$= \lim_{T\to\infty}\left[\frac{e^{-st}}{s^2+a^2}(a\sin at - s\cos at)\right]_0^T$$

$$= \frac{s}{s^2+a^2}$$

if $s > 0$. Consequently,

$$\mathcal{L}[g](s) = \frac{s}{s^2+a^2}, \qquad s > 0. \tag{9.3}$$

If the functions f and g both possess Laplace transforms for $s > s_0$, then the function $c_1 f + c_2 g$, where c_1 and c_2 are constants, also possesses a transform for $s > s_0$. In fact, from the relation

$$\int_0^\infty e^{-st}[c_1 f(t) + c_2 g(t)]\, dt = c_1\int_0^\infty e^{-st}f(t)\, dt + c_2\int_0^\infty e^{-st}g(t)\, dt,$$

we see that the operator $\mathcal{L}$ is linear:

$$\mathcal{L}[c_1 f + c_2 g](s) = c_1\mathcal{L}[f](s) + c_2\mathcal{L}[g](s), \qquad s > s_0. \tag{9.4}$$

A particularly important property of Laplace transforms comes to light when we consider the transform of the *derivative* of a function f. Let us assume that f and f' are continuous on $[0, \infty)$, and that both functions possess Laplace transforms for $s > s_0$. Using integration by parts, we have

$$\int_0^T e^{-st} f'(t)\, dt = [f(t)e^{-st}]_0^T + s \int_0^T e^{-st} f(t)\, dt.$$

As $T \to \infty$, both integrals tend to finite limits for $s > s_0$. Consequently $f(T)e^{-sT}$ must also tend to a finite limit for $s > s_0$. We shall show that this limit is zero. Given any number s_1, where $s_1 > s_0$, let s_2 be a number such that $s_0 < s_2 < s_1$. Since $f(T)\exp(-s_2 T)$ tends to a finite limit,

$$f(T)\exp(-s_1 T) = f(T)\exp(-s_2 T)\exp[-(s_1 - s_2)T]$$

tends to zero. Therefore

$$\lim_{T \to \infty} [f(t)e^{-st}]_0^T = -f(0)$$

for $s > s_0$, and we have

$$\mathcal{L}[f'](s) = sF(s) - f(0), \qquad s > s_0. \tag{9.5}$$

It is because of this property and its generalization to higher derivatives that Laplace transforms are useful in solving initial value problems for certain types of differential equations. To illustrate, let us consider the simple problem

$$x'(t) + 2x(t) = e^{-t}, \qquad x(0) = 2.$$

Let us assume for the moment that the solution function x and its derivative x' both possess Laplace transforms. For the function $f(t) = e^{-t}$ we have $\mathcal{L}[f](s) = 1/(s+1)$, according to formula (9.2). Transforming both sides of the differential equation, we see that

$$\mathcal{L}[x' + 2x](s) = \frac{1}{s+1}$$

or, by linearity,

$$\mathcal{L}[x'](s) + 2\mathcal{L}[x](s) = \frac{1}{s+1}.$$

Using the property (9.5), we have

$$sX(s) - 2 + 2X(s) = \frac{1}{s+1}.$$

Thus the initial value problem for the function x has been transformed into an algebraic equation for the function X. Solving the last equation for $X(s)$, we have

$$X(s) = \frac{2s + 3}{(s + 1)(s + 2)},$$

or, upon using partial fractions,

$$X(s) = \frac{1}{s + 1} + \frac{1}{s + 2}.$$

Now, from formula (9.2), we recognize that the function

$$x(t) = e^{-t} + e^{-2t}$$

has X as its Laplace transform. It is easy to verify that this function is indeed the solution of the initial value problem.

In applying the method of Laplace transforms to the initial value problem, we went through three main steps. First we transformed a "hard" problem (the initial value problem) into a relatively "easy" problem (the algebraic equation for X). Then we solved the easy problem by finding X. Finally we "inverted"; that is, we found the solution x of the original problem from the solution of the transformed problem. This same procedure is followed in the solution of more complicated initial value problems.

Applications of Laplace transforms to differential equations will be considered later in this chapter. Meanwhile, we shall investigate the properties of Laplace transforms in more detail.

Exercises for Section 9.1

1. Calculate the Laplace transform of the given function f. Determine the values of s for which $F(s)$ exists.
   (a) $f(t) = 1$
   (b) $f(t) = t$
   (c) $f(t) = t^2$
   (d) $f(t) = t^n$, n a positive integer
   (e) $f(t) = \sin at$
   (f) $f(t) = \sinh at$
   (g) $f(t) = \begin{cases} 1, & 0 \le t \le 1 \\ 0, & t > 1 \end{cases}$
   (h) $f(t) = \begin{cases} t, & 0 \le t \le 1 \\ 1, & t > 1 \end{cases}$

   (i) $f(t) = \begin{cases} t, & 0 \le t \le 1 \\ 2 - t, & 1 < t \le 2 \\ 0, & t > 2 \end{cases}$
   (j) $f(t) = \begin{cases} \sin t, & 0 \le t \le \pi \\ 0, & t > \pi \end{cases}$

2. Use formulas (9.2) and (9.3) to find $\mathcal{L}[f](s)$ if f is as given.
   (a) $f(t) = e^{-5t}$
   (b) $f(t) = e^{6t}$
   (c) $f(t) = \cos 5t$
   (d) $f(t) = \cos 2t$
   (e) $f(t) = 3e^{-2t} + 4e^t$
   (f) $f(t) = -3 \cos 2t - 5e^{-3t}$

3. Use formulas (9.2) and (9.3) to find a function whose Laplace transform F is as given.

(a) $F(s) = \dfrac{1}{s+2}$

(b) $F(s) = \dfrac{1}{s-3}$

(c) $F(s) = \dfrac{s}{s^2+4}$

(d) $F(s) = \dfrac{2s}{s^2+9}$

(e) $F(s) = \dfrac{3}{s-2} - \dfrac{5}{s+7}$

(f) $F(s) = \dfrac{3}{s+3} - \dfrac{2s}{s^2+1}$

4. If $f(t) = t^\alpha$, $\alpha > -1$, show that

$$\mathscr{L}[f](s) = \frac{\Gamma(\alpha+1)}{s^{\alpha+1}}, \qquad s > 0,$$

where Γ is the gamma function.

5. Use the result of Exercise 4 to find $\mathscr{L}[f]$ if

(a) $f(t) = \sqrt{t}$.   (b) $f(t) = \dfrac{1}{\sqrt{t}}$.

6. Let f be continuous on $[0, T]$ and periodic with period T. Show that

$$\mathscr{L}[f](s) = \frac{1}{1-e^{-sT}} \int_0^T e^{-st} f(t)\, dt.$$

7. Use the result of Exercise 6 to find the Laplace transform of the given function.

(a) $f(t) = 2t$, $0 \le t < \dfrac{1}{2}$, $f(t) = 2 - 2t$,

$\dfrac{1}{2} \le t < 1$, and $f(t+1) = f(t)$ for $t \ge 0$.

(b) $f(t) = |\sin t|$ for all t.

8. By using Laplace transforms, find the solution of the initial value problem. Verify that your answer is the correct one.

(a) $x'(t) - 2x(t) = 2$, $x(0) = -3$

(b) $x'(t) + 3x(t) = e^{2t}$, $x(0) = -1$

9.2 FUNCTIONS OF EXPONENTIAL ORDER

In the examples of the last section we were able to show that certain functions possessed Laplace transforms by actually carrying out the integration in formula (9.1). In this section we will establish the existence of a Laplace transform for a certain class of functions.

We will have occasion to deal with functions that have discontinuities. A function f is said to be *piecewise continuous* on a closed interval $[a, b]$ if it has at most a finite number of discontinuities at $t_1 < t_2 < \cdots < t_k$, and if the one-sided limits

$$f(t_i-) = \lim_{t \to t_i-} f(t), \qquad f(t_i+) = \lim_{t \to t_i+} f(t)$$

both exist at each such point. (If $t_1 = a$ the right-hand limit $f(a+)$ must exist, and if $t_k = b$ the left-hand limit $f(b-)$ must exist.) In addition, a function is said to be piecewise continuous on an interval of the form $[a, \infty)$ if it is piecewise continuous on every interval of the form $[a, b]$, with $b > a$.

In establishing the existence of the Laplace transform of a function, it is necessary to show that a certain improper integral converges. The following theorem from advanced calculus is useful.

Theorem 9.1 Let f and g be piecewise continuous on the interval $[c, \infty)$. If $|f(t)| \le g(t)$ for $t \ge c$, and if the integral $\int_c^\infty g(t)\, dt$ converges, then the integral $\int_c^\infty f(t)\, dt$ also converges.

In a moment we shall use Theorem 9.1 to establish a set of sufficient conditions for the existence of the Laplace transform of a function. First, however, let us introduce the notation[2]

$$f(t) = O[g(t)],$$

which should be read "$f(t)$ is of the order of $g(t)$." This notation means that there exist positive constants M and N such that

$$|f(t)| \le Mg(t)$$

whenever $t \ge N$. In particular, if $f(t) = O[e^{at}]$, for some constant a, we say that f is of *exponential order*.

We are now ready to prove the following theorem.

Theorem 9.2   Let f be piecewise continuous on the interval $[0, \infty)$, and let $f(t) = O[e^{at}]$, for some constant a. Then the Laplace transform $\mathcal{L}[f](s)$ exists, at least for $s > a$.

Proof   According to the hypotheses of the theorem, there exist positive constants M and t_0 such that $|f(t)| \le Me^{at}$ when $t \ge t_0$. Then $|f(t)e^{-st}| \le Me^{-(s-a)t}$ when $t \ge t_0$. Since the integral $\int_{t_0}^{\infty} Me^{-(s-a)t}\, dt$ converges when $s > a$, the integral $\int_{t_0}^{\infty} e^{-st}f(t)\, dt$ also converges when $s > a$, by Theorem 9.1. Since

$$\int_0^{\infty} e^{-st}f(t)\, dt = \int_0^{t_0} e^{-st}f(t)\, dt + \int_{t_0}^{\infty} e^{-st}f(t)\, dt, \qquad s > a,$$

the Laplace transform $\mathcal{L}[f](s)$ exists for $s > a$.

As an important application of Theorem 9.2, we shall show that if $f(t)$ is of the form

$$t^n e^{at} \cos bt, \qquad t^n e^{at} \sin bt, \tag{9.6}$$

where n is a nonnegative integer, then $\mathcal{L}[f](s)$ exists for $s > a$. We first observe that

$$t^n = O[e^{\varepsilon t}]$$

for every positive number ε. Since $|\cos bt| \le 1$ and $|\sin bt| \le 1$ for all t, we have

$$f(t) = O[e^{(a+\varepsilon)t}].$$

---

[2] The notation $f(t) = o[g(t)]$ also appears in the literature. It means that $f(t)/g(t) \to 0$ as $t \to \infty$.

By Theorem 9.1, $\mathcal{L}[f](s)$ exists for $s > a + \varepsilon$ for every positive number ε. Consequently $\mathcal{L}[f](s)$ exists for $s > a$.

The above result is important in the study of linear differential equations with constant coefficients.

Theorem 9.3   Every solution of the differential equation

$$P(D)x = 0,$$

where $D = d/dt$ and $P(D)$ is a polynomial operator, possesses a Laplace transform. Furthermore, every derivative of every solution possesses a Laplace transform.

Proof   Every solution of the equation is a linear combination of functions of the form (9.6), and therefore every derivative is also a linear combination of functions of this type. By Theorem 9.2, these functions possess Laplace transforms.

We shall give one more result about functions of exponential order.

Theorem 9.4   Let f be piecewise continuous on the interval $[0, \infty)$, and let $f(t) = O[e^{at}]$ for some constant a. Then the function h, where

$$h(t) = \int_0^t f(u) \, du$$

is also of exponential order. If $a > 0$, $h(t) = O[e^{at}]$, and if $a \leq 0$, $h(t) = O[1]$.[3]

Proof   There exist positive constants t_0 and M_1 such that $|f(t)| \leq M_1 e^{at}$ for $t \geq t_0$. Also, there exists a positive constant M_2 such that $|f(t)| \leq M_2$ for $0 \leq t \leq t_0$. Since

$$h(t) = \int_0^{t_0} f(u) \, du + \int_{t_0}^t f(u) \, du$$

for $t > t_0$, we have

$$|h(t)| \leq M_2 \int_0^{t_0} du + M_1 \int_{t_0}^t e^{au} \, du$$

or

$$|h(t)| \leq M_2 t_0 + \frac{M_1}{a} (e^{at} - e^{at_0}).$$

---

[3] The notation $h(t) = O[1]$ means the same thing as $h(t) = O[e^{0t}]$, that is, that h is bounded on $[t_0, \infty)$.

If $a > 0$, then

$$|h(t)| \leq \left(M_2 t_0 + \frac{M_1}{a} \right) e^{at} \qquad \text{for } t \geq t_0$$

and $h(t) = O[e^{at}]$. If $a \leq 0$,

$$|h(t)| \leq M_2 t_0 + 2 \frac{M_1}{a} \qquad \text{for } t \geq t_0,$$

and $h(t) = O[1]$.

Exercises for Section 9.2

1. Suppose that the limit

$$\lim_{t \to \infty} \frac{f(t)}{g(t)}$$

exists (and is finite). Show that $f(t) = O[|g(t)|]$.

2. Show that, as $t \to \infty$,
   (a) $\sin t = O[1]$
   (b) $\dfrac{e^{-t}}{1+t} = O[e^{-t}]$
   (c) $\dfrac{e^{-t}}{1+t} = O\left[\dfrac{1}{t}\right]$
   (d) $t e^t = O[e^{(1+e)t}], \quad \varepsilon > 0$
   (e) $\sinh t = O[e^t]$
   (f) $\dfrac{e^{3t}}{e^t + 1} = O[e^{2t}]$

3. Without actually computing the transform, show that $\mathcal{L}[f](s)$ exists for the indicated values of s.
   (a) $f(t) = \dfrac{1}{1+t}, \quad s > 0$
   (b) $f(t) = \dfrac{e^{at}}{1+t}, \quad s > a$
   (c) $f(t) = \dfrac{\sin t}{t}, \quad s > 0$
   (d) $f(t) = t \ln t, \quad s > 0$
   (e) $f(t) = e^t \sin(t^2), \quad s > 1$

(f) $f(t) = \sqrt{e^t + 1}, \quad s > \dfrac{1}{2}$

4. Let f and g be of exponential order.
   (a) Show that $c_1 f + c_2 g$, where c_1 and c_2 are constants, is of exponential order.
   (b) Show that fg is of exponential order.

5. Let the function b be continuous on $[0, \infty)$ and be of exponential order. Show that every solution of the differential equation

$$x'(t) + ax(t) = b(t),$$

where a is a constant, is of exponential order.

6. If f' exists on $[0, \infty)$ and is of exponential order, show that f is of exponential order.

7. If $\lim_{t \to \infty} f(t)/g(t) = \infty$, show that $f(t) \neq O[g(t)]$.

8. If $f(t) = e^{t^2}$ for all t, show that f is not of exponential order.

9. Let $f(t) = \sin(e^{t^2})$ for all t. Show that f is of exponential order but that f' is not.

10. Let $f(t) = e^{n^2}$ for $n \leq t \leq n + e^{-n^2}$, for every positive integer n, and $f(t) = 0$ elsewhere. Show that f is not of exponential order, but that $\mathcal{L}[f](s)$ exists for $s > 0$.

9.3

PROPERTIES OF LAPLACE TRANSFORMS

In this section we shall develop some of the more useful properties of Laplace transforms. In the formulas listed below, we denote the transforms of f and g by F and G, respectively. For properties (A) through (E) we assume that f and g are piecewise continuous on $[0, \infty)$ and that $f(t) = O[e^{at}]$ and $g(t) = O[e^{bt}]$ for some constants a and b. Then $F(s)$ exists for $s > a$ and $G(s)$ exists for $s > b$.

(A) $\mathscr{L}[c_1 f + c_2 g](s) = c_1 F(s) + c_2 G(s)$, $s > \max(a, b)$.

(B) If $h(t) = e^{ct}f(t)$, then $H(s) = F(s - c)$, $s > a + c$.

(C) If $k(t) = \int_0^t f(u)\,du$, then $K(s) = \dfrac{1}{s}F(s)$, $s > \max(a, 0)$.

(D) If $p_n(t) = t^n f(t)$, then $P_n(s) = (-1)^n \dfrac{d^n F(s)}{ds^n}$, $s > a$.

(E) If $q(t) = \begin{cases} 0, & 0 < t < c, \\ f(t - c), & t > c, \end{cases}$ then $Q(s) = e^{-cs}F(s)$, $s > a$.

(F) Let $f^{(n-1)}(t) = O[e^{at}]$. Let $f, f', \ldots, f^{(n-1)}$ be continuous on $[0, \infty)$ and let $f^{(n)}$ be piecewise continuous on $[0, \infty)$. Then $\mathscr{L}[f^{(n)}](s)$ exists for $s > \max(a, 0)$ and

$$\mathscr{L}[f^{(n)}](s) = s^n F(s) - [s^{n-1}f(0) + s^{n-2}f'(0) + \cdots + f^{(n-1)}(0)].$$

Property (A) follows from Theorem 9.2 and the definition of the Laplace transform.

To prove property (B), we first note that $e^{ct}f(t) = O[e^{(a+c)t}]$. Then we observe that

$$\mathscr{L}[h](s) = \int_0^\infty e^{-(s-c)t}f(t)\,dt = F(s - c).$$

To prove property (C), we use the result of Theorem 9.4, which assures us that the function k is of exponential order. Using integration by parts, and observing that $k'(t) = f(t)$, we have

$$\mathscr{L}[k](s) = \int_0^\infty e^{-st}k(t)\,dt = \left[-\frac{1}{s}e^{-st}k(t) \right]_0^\infty + \frac{1}{s}\int_0^\infty e^{-st}f(t)\,dt.$$

Since $k(0) = 0$, the integrated part vanishes and thus we have $\mathscr{L}[k](s) = F(s)/s$.

Now consider property (D). If we differentiate both sides of the equation

$$F(s) = \int_0^\infty e^{-st}f(t)\,dt, \qquad s > a,$$

with respect to s (the assumptions on f ensure that $F'(s)$ exists and that $F'(s)$ can be obtained by differentiation under the integral sign), we find that

$$F'(s) = -\int_0^\infty e^{-st}tf(t)\,dt = -\mathscr{L}[p_1](s).$$

Repeated differentiation shows that

$$F^{(n)}(s) = (-1)^n \int_0^\infty e^{-st}t^n f(t)\,dt = (-1)^n \mathscr{L}[p_n](s).$$

The verification of property (E) is left as an exercise.

We shall prove property (F) by induction. When $n = 1$, f is assumed to be continuous on $[0, \infty)$. Using integration by parts, we have

$$\int_0^T e^{-st}f'(t)\,dt = [e^{-st}f(t)]_0^T + s\int_0^T e^{-st}f(t)\,dt.$$

Since $f(t) = O[e^{at}]$, it follows that $e^{-sT}f(T) \to 0$ as $T \to \infty$ for $s > a$. Letting $T \to \infty$ in the above relation, we see that the right-hand side has a limit; hence the left-hand side has a limit and

$$\mathscr{L}[f'](s) = sF(s) - f(0), \qquad s > a.$$

Now suppose that property (F) holds for $n = m$, where m is a positive integer. Using the same arguments as in the case $n = 1$, we have

$$\mathscr{L}[f^{(m+1)}](s) = \int_0^\infty e^{-st}f^{(m+1)}(t)\,dt$$

$$= [e^{-st}f^{(m)}(t)]_0^\infty + s\int_0^\infty e^{-st}f^{(m)}(t)\,dt$$

$$= s\mathscr{L}[f^{(m)}](s) - f^{(m)}(0).$$

The assumption that $f^{(m)}(t) = O[e^{at}]$ implies that the function $f^{(m-1)}$ is of exponential order. Consequently we can apply the property, for $n = m$, to find $\mathscr{L}[f^{(m)}](s)$. We have

$$\mathscr{L}[f^{(m+1)}](s) = s[s^m F(s) - s^{m-1}f(0) - \cdots - f^{(m-1)}(0)] - f^{(m)}(0)$$

$$= s^{m+1}F(s) - [s^m f(0) + \cdots + f^{(m)}(0)].$$

Thus if property (F) holds for $n = m$, it also holds for $n = m + 1$. Since it holds for $n = 1$, it holds for every positive integer.

These basic properties of Laplace transforms are frequently useful in finding the transforms of functions. Starting with the relations

$$f(t) = t^n, \qquad F(s) = \frac{n!}{s^{n+1}},$$

$$f(t) = e^{at}, \qquad F(s) = \frac{1}{s-a},$$

$$f(t) = \cos at, \qquad F(s) = \frac{s}{s^2 + a^2},$$ \hfill (9.7)

$$f(t) = \sin at, \qquad F(s) = \frac{a}{s^2 + a^2},$$

we can easily find the transforms of many elementary functions, with the aid of properties (A) through (F).

Example 1   Let $f(t) = \sinh at$. Since

$$f(t) = \frac{1}{2} e^{at} - \frac{1}{2} e^{-at},$$

it follows from property (A) that

$$F(s) = \frac{1}{2} \frac{1}{s-a} - \frac{1}{2} \frac{1}{s+a} = \frac{a}{s^2 - a^2}.$$

Example 2   Let $f(t) = e^{-2t} \cos 3t$. If $g(t) = \cos 3t$, then

$$\mathscr{L}[g](s) = \frac{3}{s^2 + 9},$$

and it follows from property (B) that

$$\mathscr{L}[f](s) = \frac{s+2}{(s+2)^2 + 9} = \frac{s+2}{s^2 + 4s + 13}.$$

Example 3   As an example of the use of property (D), we find the transform of the function $f(t) = t^2 \sin t$. If $g(t) = \sin t$, we have

$$\mathscr{L}[g](s) = \frac{1}{s^2 + 1}$$

and hence

$$\mathscr{L}[f](s) = \frac{d^2}{ds^2}\left(\frac{1}{s^2 + 1}\right) = 2 \frac{3s^2 - 1}{(s^2 + 1)^3}.$$

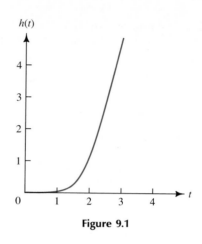

Figure 9.1

Example 4   As a final example, we consider the function

$$f(t) = \begin{cases} 0, & 0 < t < 1, \\ (t-1)^2, & t > 1, \end{cases}$$

whose graph is shown in Fig. 9.1.

If $g(t) = t^2$, then $\mathscr{L}[g](s) = 2/s^3$, and it follows from property (E) that

$$\mathscr{L}[f](s) = e^{-s}\mathscr{L}[g](s) = \frac{2}{s^3} e^{-s}.$$

───────────── **Exercises for Section 9.3** ─────────────

1. Verify property (E).

2. Find the Laplace transform of f, where $f(t)$ is:
   (a) $2e^{-t} - 3\sin 4t$ (b) $\cosh 2t$
   (c) $e^{2t}\sin 3t$ (d) $e^{-t}\cos 2t$
   (e) $e^{-3t}t^4$ (f) $t^3 e^{4t}$
   (g) $t^2\cos t$ (h) $t\sin 2t$

3. Find the Laplace transform of f, where $f(t)$ is as given.
   (a) $e^{2t}\sqrt{t}$ (b) $\displaystyle\int_0^t \sin 2u\, du$
   (c) $\displaystyle\int_0^t x^2 e^x\, dx$ (d) $\displaystyle\int_0^t \cos^2 u\, du$
   (e) $f(t) = \begin{cases} 0, & 0 < t < 2 \\ 1, & t > 2 \end{cases}$
   (f) $f(t) = \begin{cases} 0, & 0 < t < \pi \\ \sin(t-\pi), & t > \pi \end{cases}$

   (g) $f(t) = \begin{cases} 0, & 0 < t < 1 \\ t^2, & t > 1 \end{cases}$
   (h) $f(t) = \begin{cases} 0, & 0 < t < 1 \\ (t-1)e^t, & t > 1 \end{cases}$

4. If $\mathscr{L}[f] = F$, express the transform of the indicated derivative of f in terms of F.
   (a) f'', if $f(0) = 1$ and $f'(0) = 2$
   (b) f'', if $f(0) = -3$ and $f'(0) = 0$
   (c) f''', if $f(0) = 1$, $f'(0) = -1$, and $f''(0) = 5$
   (d) f''', if $f(0) = -2$, $f'(0) = 0$, and $f''(0) = 1$
   (e) $f^{(4)}$, if $f(0) = 2$ and $f'(0) = f''(0) = f'''(0) = 0$
   (f) $f^{(4)}$, if $f(0) = f'(0) = 0$, $f''(0) = 1$, and $f'''(0) = -1$

5. If $g(t) = f(ct)$, where c is a positive constant, show that

$$G(s) = \frac{1}{c} F\left(\frac{s}{c}\right).$$

6. If f is continuous on $[0, \infty)$ and is the derivative of a function g ($f = g'$) of exponential order, show that f possesses a Laplace transform even though f may not be of exponential order. (See, in connection with this exercise, Exercise 9, Section 9.2.)

7. Let f and f' be piecewise continuous on the interval $[0, \infty)$, and be of exponential order. Suppose that f has only a finite number of discontinuities on $(0, \infty)$, at the points $t_1 < t_2 < \cdots < t_k$. Show that

$$\mathcal{L}[f'](s) = sF(s) - f(0+)$$

$$- \sum_{i=1}^{k} e^{-st_i}[f(t_i +) - f(t_i -)].$$

9.4

INVERSE TRANSFORMS

In this section, we shall consider the following problem. Given a function F, what functions, if any, have F as their Laplace transforms? To simplify matters, we shall consider only functions that are piecewise continuous on the interval $[0, \infty)$ and are of exponential order. We first prove the following result.

Theorem 9.5   Let f be a function of the type described above, and let $F(s) = \mathcal{L}[f](s)$. Then

$$\lim_{s \to \infty} F(s) = 0.$$

Proof   There exist positive numbers t_0 and M_1, and a number a, such that $|f(t)| \leq M_1 e^{at}$ for $t \geq t_0$. We write

$$F(s) = \int_0^{t_0} e^{-st} f(t) \, dt + \int_{t_0}^{\infty} e^{-st} f(t) \, dt.$$

Since f is piecewise continuous on the finite interval $[0, t_0]$, there exists a positive number M_2 such that $|f(t)| \leq M_2$ for $0 \leq t \leq t_0$. Then

$$|F(s)| \leq M_2 \int_0^{t_0} e^{-st} \, dt + M_1 \int_{t_0}^{\infty} e^{-t(s-a)} \, dt,$$

so

$$|F(s)| \leq M_2 \frac{1}{s}(1 - e^{-st_0}) + M_1 \frac{1}{s-a} e^{-t_0(s-a)}, \qquad s > a.$$

Letting $s \to \infty$, we see that $F(s) \to 0$.

In view of this result, we can state that unless $F(s)$ tends to zero with increasing s, there exists no function of the type considered which has F as its Laplace transform. For instance, if

$$F(s) = \frac{s(s+1)}{s^2 + 2},$$

no function of the type considered has F as its transform, because $F(s) \to 1 \neq 0$ as $s \to \infty$.

We can also ask if it is possible for two different functions to have the same Laplace transform. A partial answer is given by the following theorem, which we must state without proof. A more general version of this theorem is known as *Lerch's theorem*.

Theorem 9.6   Let f and g be piecewise continuous on the interval $[0, \infty)$, and let $\mathscr{L}[f](s) = \mathscr{L}[g](s)$ for $s > s_0$, for some number s_0. Then at each point t_0 in the interval $[0, \infty)$, where f and g are both continuous, $f(t_0) = g(t_0)$. In particular, if f and g are both continuous on $[0, \infty)$, then $f(t) = g(t)$ for $t \geq 0$.

Let us consider as an example the function

$$F(s) = \frac{1}{s - 2}.$$

If $f(t) = e^{2t}$ for $t \geq 0$, we know that f has F as its transform. Because of Theorem 9.6 we can assert that f is the only *continuous* function that has F as its transform.

More generally, let F be defined on (a, ∞), for some number a, and be such that $F(s) \to 0$ as $s \to \infty$. We may ask whether there exists a function f, continuous on $[0, \infty)$ and of exponential order, which has F as its Laplace transform. We know by Theorem 9.6 that at most one such function can exist. If such a function f does exist, we call it the *inverse transform* of F and write

$$f = \mathscr{L}^{-1}[F], \qquad f(t) = \mathscr{L}^{-1}[F](t).$$

Sufficient conditions that a function F possess an inverse transform can be found in Churchill (1971).

It is possible to find the inverse transforms of a number of functions by using the relations (9.7) and the properties of Laplace transforms that were derived in the last section.

Example 1   Let us consider the function

$$F(s) = \frac{3s}{s^2 + 4s + 5}.$$

By completing the square in the denominator, we can write

$$F(s) = \frac{3s}{(s + 2)^2 + 1} = \frac{3(s + 2)}{(s + 2)^2 + 1} - \frac{6}{(s + 2)^2 + 1}.$$

If $G(s) = s/(s^2 + 1)$ and $H(s) = 1/(s^2 + 1)$, we know that

$$\mathscr{L}^{-1}[G](t) = \cos t, \qquad \mathscr{L}^{-1}[H](t) = \sin t.$$

Using these facts, and properties (A) and (B) of the last section, we see that

$$\mathscr{L}^{-1}[F](t) = e^{-2t}(3 \cos t - 6 \sin t).$$

Example 2   Consider the function

$$F(s) = e^{-2s}\frac{1}{s^2}.$$

If $G(s) = 1/s^2$, then $\mathscr{L}^{-1}[G](t) = t$. By property (E) we have

$$\mathscr{L}^{-1}[F](t) = \begin{cases} 0, & 0 \le t \le 2, \\ t - 2, & t > 2. \end{cases}$$

In cases where F is a rational function, it is often convenient to expand F in a series of partial fractions.

Example 3   Let

$$F(s) = \frac{1}{(s - 2)(s^2 + 1)}.$$

Expansion of F in partial fractions yields the formula

$$F(s) = \frac{1}{5}\frac{1}{s - 2} - \frac{1}{5}\frac{s + 2}{s^2 + 1}.$$

Then we recognize that

$$\mathscr{L}^{-1}[F](t) = \frac{1}{5}(e^{2t} - \cos t - 2 \sin t).$$

Let us now consider the problem of finding the inverse transform of the product FG, where $F = \mathscr{L}[f]$ and $G = \mathscr{L}[g]$. We have

$$F(s)G(s) = \left(\int_0^\infty e^{-sx}f(x)\,dx\right)\left(\int_0^\infty e^{-sy}g(y)\,dy\right)$$

$$= \int_0^\infty \int_0^\infty e^{-s(x+y)}f(x)g(y)\,dx\,dy.$$

The product of the two single integrals can be interpreted as an improper double integral whose region of integration is the first quadrant of a plane in which x and y are rectangular coordinates. Let us now make the change of variables.

$$x = t - u, \qquad u = y,$$

$$y = u, \qquad t = x + y,$$

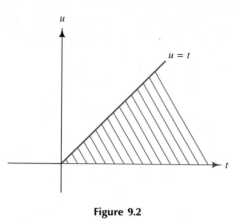

Figure 9.2

from (x, y) to (t, u). The first quadrant of the xy plane corresponds to the region of the tu plane that is described by means of the inequalities $u \geq 0$ and $t - u \geq 0$. This region is shown in Fig. 9.2.

The iterated integral becomes[4]

$$F(s)G(s) = \int_0^\infty \int_0^t e^{-st} f(t - u)g(u) \, du \, dt$$

$$= \int_0^\infty e^{-st} \left[\int_0^t f(t - u)g(u) \, du \right] dt.$$

Therefore

$$\mathcal{L}^{-1}[FG](t) = \int_0^t f(t - u)g(u) \, du.$$

The function defined here is called the *convolution* of the functions f and g, and is denoted by $f * g$. Thus

$$(f * g)(t) = \int_0^t f(t - u)g(u) \, du \tag{9.8}$$

and we may write

$$\mathcal{L}[f * g] = FG.$$

It can be shown (Exercise 3) that $f * g = g * f$, so that

$$(f * g)(t) = (g * f)(t) = \int_0^t f(u)g(t - u) \, du.$$

---

[4] The Jacobian of the transformation of coordinates is unity.

The notion of the convolution of two functions arises in other areas of mathematics, such as probability theory.

Example 4 As an exercise, we shall use the convolution formula (9.8) to find the inverse transform of the function

$$H(s) = \frac{1}{s^2(s^2 + 1)}.$$

Since $H(s) = F(s)G(s)$, where $F(s) = 1/s^2$ and $G(s) = 1/(s^2 + 1)$, we have $h = f * g$. Now $f(t) = t$ and $g(t) = \sin t$, so

$$h(t) = (f * g)(t) = \int_0^t (t - u) \sin u \; du$$

$$= \left[-(t - u) \cos u - \sin u \right]_{u=0}^{u=t}$$

$$= t - \sin t.$$

(This result could also have been obtained by finding the partial fractions expansion of the given function.)

The techniques illustrated here can be used only when the given function F can be expressed in a fairly simple way in terms of functions whose inverse transforms are easily recognizable. For convenience, a short table of functions and their transforms is given at the end of this chapter. A more powerful and direct method for finding inverse transforms exists. This method requires a fairly deep knowledge of complex variables, however, and we cannot discuss it here. We shall be concerned mainly with the solution of initial value problems for differential equations by means of Laplace transforms. The methods at our disposal suffice for the solution of many such problems. In any case, they illustrate the general approach.

--- **Exercises for Section 9.4** ---

1. Find the function f that is continuous on $[0, \infty)$ and has F as its transform if $F(s)$ is:

   (a) $\dfrac{1}{(s + 1)^2}$

   (b) $\dfrac{1}{(s - 2)^2 + 9}$

   (c) $\dfrac{s}{(s + 1)^2 - 4}$

   (d) $\dfrac{1}{s^2 - 3s}$

   (e) $\dfrac{1}{(s + 1)(s + 2)}$

   (f) $\dfrac{1}{s(s + 2)^2}$

2. Find the function f that is continuous on $[0, \infty)$ and has F as its transform if $F(s)$ is:

   (a) $\dfrac{6}{s^2 + s - 2}$

   (b) $\dfrac{3s - 8}{s^2 - 5s + 6}$

   (c) $\dfrac{s^2 + 20s + 9}{(s - 1)^2(s^2 + 9)}$

   (d) $\dfrac{s - 4}{s^2 + 3s + 3}$

   (e) $\dfrac{1}{s^3 - 8}$

   (f) $\dfrac{3s - 2}{s^2 - 2s + 10}$

3. Show that $g * f = f * g$.

4. Find $(f * g)(t)$ if f and g are as given.

   (a) $f(t) = g(t) = t$

   (b) $f(t) = t, \quad g(t) = e^t$

   (c) $f(t) = e^t, \quad g(t) = e^{2t}$

   (d) $f(t) = e^{-t}, \quad g(t) = \cos t$

5. Find the inverse transform of the function whose value at s is:

(a) $e^{-s} \dfrac{1}{(s-2)^2}$

(b) $e^{-2s} \dfrac{1}{s^2 + 9\pi^2}$

(c) $\dfrac{s}{(s-2)^{3/2}(s^2+1)}$

(d) $\dfrac{1}{s^{5/2}(s^2+1)}$

(e) $\dfrac{1}{s+1} F(s)$

(f) $\dfrac{1}{s^2+1} F(s)$

6. Let f be continuous on $[0, \infty)$. Let f' be piecewise continuous on $[0, \infty)$ and be of expo-nential order. Show that

$$\lim_{s \to \infty} sF(s) = f(0).$$

Suggestion: apply Theorem 9.5 to f'.

7. It can be shown that $(f * g) * h = f * (g * h)$. Verify this for the case $f(t) = t$, $g(t) = e^t$, and $h(t) = e^{-t}$.

8. Let f and g be piecewise continuous on $[0, \infty)$. If f and g are both of exponential order, show that $f * g$ is of exponential order.

9.5
APPLICATIONS
TO
DIFFERENTIAL
EQUATIONS

We shall now apply the theory of Laplace transforms to the solution of initial value problems. The method we shall describe applies to those problems where the differential equation, or system of differential equations, is linear and has constant coefficients.

As an illustration, let us consider the problem

$$x''(t) + 4x(t) = 5e^{-t}, \qquad t \geq 0,$$

$$x(0) = 2, \qquad x'(0) = 3.$$

From the theory of linear differential equations we know that this problem possesses a unique solution x on $[0, \infty)$. This solution and its first two derivatives are continuous. Let us assume, for the moment, that x and x' are of exponential order. Then x, x', and x'' possess Laplace transforms on some interval (s_0, ∞). [See property (F), Section 9.3.] If $\mathscr{L}[x] = X$, then

$$\mathscr{L}[x'](s) = sX(s) - x(0) = sX(s) - 2,$$

$$\mathscr{L}[x''](s) = s^2 X(s) - sx(0) - x'(0)$$

$$= s^2 X(s) - 2s - 3.$$

From the differential equation, we see that

$$\mathscr{L}[x'' + 4x](s) = \frac{5}{s+1}$$

or

$$s^2 X(s) - 2s - 3 + 4X(s) = \frac{5}{s+1}.$$

Upon solving for $X(s)$, we find that

$$X(s) = \frac{2s^2 + 5s + 8}{(s+1)(s^2+4)}.$$

The use of partial fractions shows that

$$X(s) = \frac{4}{s^2 + 4} + \frac{s}{s^2 + 4} + \frac{1}{s + 1}.$$

Inverting, we arrive at the formula

$$x(t) = 2 \sin 2t + \cos 2t + e^{-t}.$$

However, we cannot immediately assert that this function is the solution of the initial value problem. For in the derivation, we made the assumption, not yet justified, that the solution and its first derivative were of exponential order.

We shall presently show that our assumptions about the behavior of the solution and its derivatives were correct. But first let us consider the more general problem

$$P(D)x = a_0 x^{(n)} + a_1 x^{(n-1)} + \cdots + a_{n-1} x' + a_n x = b(t),$$
$$x(0) = k_0, \qquad x'(0) = k_1, \ldots, \qquad x^{(n-1)}(0) = k_{n-1}. \tag{9.9}$$

Suppose that b possesses a Laplace transform B. If we "transform" the differential equation formally, taking into account the initial conditions, we arrive at the algebraic equation

$$a_0[s^n X(s) - k_0 s^{n-1} - \cdots - k_{n-1}] + a_1[s^{n-1} X(s) - k_0 s^{n-2} - \cdots - k_{n-2}]$$
$$+ \cdots + a_{n-1}[sX(s) - k_0] + a_n X(s) = B(s)$$

for $X(s)$. This equation can be written as

$$P(s)X(s) = B(s) + Q(s),$$

where Q is a polynomial whose coefficients depend on the constants k_i. Then

$$X(s) = \frac{B(s) + Q(s)}{P(s)}. \tag{9.10}$$

The justification of this procedure can be based on the following theorem.

Theorem 9.7 Let the function b be continuous on $[0, \infty)$ and be of exponential order. Then the solution x of the initial value problem (9.9) is of exponential order, as are its first $n - 1$ derivatives.[5]

---

[5] Actually the nth derivative of x is also continuous and of exponential order, as we can see from the differential equation.

Proof   We know from Theorem 9.3 that the solutions of the associated homogeneous equation, along with their derivatives, are of exponential order. The solution of the problem (9.9) can be expressed in terms of these functions by the method of variation of parameters. The formula for the solution involves an integral. By applying Theorem 9.4, it is easy to see that x has the indicated properties. The details are left as an exercise.

When b is of exponential order, so are the functions $x, x', \ldots, x^{(n-1)}$. Then these functions, along with $x^{(n)}$, possess Laplace transforms, and the transforms of the derivatives can be expressed in terms of $\mathscr{L}[x]$ by the use of property (F). In this case, the derivation of formula (9.10) for the transform of the solution is valid. In particular, the procedure followed in the example at the beginning of this section is valid.

Let us next consider the linear system with constant coefficients,

$$x_i'(t) = \sum_{j=1}^n a_{ij} x_j + b_i(t), \qquad t \ge 0, \quad i = 1, 2, \ldots, n,$$
$$x_i(0) = k_i, \qquad i = 1, 2, \ldots, n. \tag{9.11}$$

Suppose that the components x_i of the solution possess transforms X_i, and that the functions b_i possess transforms B_i. If we formally transform the differential equations, we arrive at the system of algebraic equations

$$sX_i(s) - k_i = \sum_{j=1}^n a_{ij} X_j(s) + B_i(s), \qquad i = 1, 2, \ldots, n \tag{9.12}$$

for the quantities $X_i(s)$. Justification of this procedure can be based on the following theorem, which is the multidimensional analog of Theorem 9.7.

Theorem 9.8   Let each of the functions b_i be continuous on $[0, \infty)$ and be of exponential order. Then the components x_i of the solution of the problem (9.11) are of exponential order.

When the functions b_i are of exponential order, the functions x_i and x_i' therefore possess Laplace transforms, and

$$\mathscr{L}[x_i'] = \mathscr{L}[x_i] - x_i(0).$$

In this case, the derivation of the system (9.12) is valid.

Example 1   Let us consider the problem

$$(D + 3)x + 5y = 2,$$
$$-x + (D - 1)y = 1,$$
$$x(0) = 1, \qquad y(0) = 0.$$

We note that the hypotheses of Theorem 9.8 are satisfied. Let X and Y denote the transforms of x and y, respectively. We transform the system to get

$$sX(s) - 1 + 3X(s) + 5Y(s) = \frac{2}{s},$$

$$-X(s) + sY(s) - Y(s) = \frac{1}{s}.$$

Upon regrouping terms, we have

$$(s + 3)X(s) + 5Y(s) = \frac{2}{s} + 1,$$

$$-X(s) + (s - 1)Y(s) = \frac{1}{s}.$$

Solving for $X(s)$ and $Y(s)$, and using partial fractions, we obtain the formulas

$$X(s) = -\frac{7}{2}\frac{1}{s} + \frac{1}{2}\frac{9(s + 1) + 7}{(s + 1)^2 + 1},$$

$$Y(s) = \frac{5}{2}\frac{1}{s} - \frac{1}{2}\frac{5(s + 1) + 1}{(s + 1)^2 + 1}.$$

Taking inverse transforms, we find that the components of the solution are

$$x(t) = -\frac{7}{2} + \frac{9}{2}e^{-t}\cos t + \frac{7}{2}e^{-t}\sin t,$$

$$y(t) = \frac{5}{2} - \frac{5}{2}e^{-t}\cos t - \frac{1}{2}e^{-t}\sin t.$$

To check, we note that $x(0) = 1$ and $y(0) = 0$.

Example 2   As a final example, let us consider the problem

$$D^2x + y = -2,$$

$$x + D^2y = 0, \tag{9.13}$$

$$x(0) = y(0) = x'(0) = y'(0) = 0.$$

The system is not a first-order system, so Theorem 9.8 does not apply. However, the system can be rewritten as a first-order system for the quantities x, Dx, y, and Dy. Setting

$$x = u_1, \qquad Dx = u_2, \qquad y = u_3, \qquad Dy = u_4,$$

we obtain the first-order system

$$Du_1 = u_2, \qquad Du_2 = -u_3 - 2, \qquad Du_3 = u_4, \qquad Du_4 = -u_1. \quad (9.14)$$

The initial conditions are

$$u_1(0) = u_2(0) = u_3(0) = u_4(0) = 0.$$

The system (9.14) satisfies the hypotheses of Theorem 9.8. It possesses a unique solution that satisfies the initial conditions. The components u_i of this solution are of exponential order. Consequently, the problem (9.13) possesses a unique solution (x, y), and the quantities x, y, Dx, Dy, D^2x, D^2y possess Laplace transforms. Therefore we can apply the method of Laplace transforms directly to the problem (9.13). Transformation of the differential equations yields the relations

$$s^2 X(s) + Y(s) = -\frac{2}{s}, \qquad X(s) + s^2 Y(s) = 0.$$

From these we find that

$$X(s) = \frac{-2s}{s^4 - 1} = \frac{s}{s^2 + 1} - \frac{s}{s^2 - 1}, \qquad Y(s) = \frac{-1}{s^2 + 1} + \frac{1}{s^2 - 1}.$$

Consequently the solution of the problem is

$$x(t) = \cos t - \cosh t, \qquad y(t) = -\sin t + \sinh t.$$

Exercises for Section 9.5

1. Find the solution of the initial value problem by the use of Laplace transforms.

   (a) $x'' + 3x' + 2x = 6e^t$, $\quad x(0) = 2$,
   $x'(0) = -1$

   (b) $x'' + 2x' + x = 4 \sin t$, $\quad x(0) = -2$,
   $x'(0) = 1$

   (c) $x'' + 4x = 8 \sin t$, $\quad x(0) = 0$,
   $x'(0) = 2$

   (d) $x'' + 4x' + 5x = 25t$, $\quad x(0) = -5$,
   $x'(0) = 7$

   (e) $x''' + 2x'' + x' + 2x = 2$, $\quad x(0) = 3$,
   $x'(0) = -2$, $\quad x''(0) = 3$

   (f) $x^{(4)} - 2x'' + x = 8e^t$, $\quad x(0) = 1$,
   $x'(0) = 2$, $\quad x''(0) = 1$, $\quad x'''(0) = 10$

2. Consider the initial value problem

   $$x'' + x = f(t), \qquad t \geq 0, \qquad x(0) = x'(0) = 0,$$

   where

   $$f(t) = \begin{cases} t, & 0 \leq t \leq 1, \\ 1, & t > 1. \end{cases}$$

   (a) Find the solution by means of Laplace transforms.

   (b) Find the solution by using a different method.

3. Find by means of Laplace transforms the solution of the problem

$x'' - x = f(t), \quad t \geq 0, \quad x(0) = 1, \quad x'(0) = 0,$

$f(t) = \begin{cases} 0, & 0 \leq t \leq 1, \\ (t-1), & t > 1. \end{cases}$

4. By using Laplace transforms, express the solution of the problem

$$x'' + x = f(t), \qquad x(0) = 0, \qquad x'(0) = 1$$

as an integral.

5. If x satisfies the given integral equation, determine the Laplace transform of x and then determine x.

(a) $x(t) = 2 + \int_0^t e^{t-u} x(u)\, du$

(b) $x(t) = 1 + t + \int_0^t (t-u)x(u)\, du$

6. Find the solution of the initial value problem by the use of Laplace transforms.
(a) $(D+2)x_1 - 2x_2 = 0,$
$-x_1 + (D+1)x_2 = 2e^t,$
$x_1(0) = 0, \quad x_2(0) = 1$
(b) $(D+1)x_1 + x_2 = 0,$
$-5x_1 + (D-1)x_2 = -4,$
$x_1(0) = 1, \quad x_2(0) = 3$

(c) $(D+2)x_1 + x_2 = e^{-t},$
$-2x_1 + Dx_2 = -e^{-t},$
$x_1(0) = 2, \quad x_2(0) = 0$
(d) $4Dx_1 - (D^2 - D)x_2 = 0,$
$-(D+3)x_1 + x_2 = 0,$
$x_1(0) = 0, \quad x_2(0) = 2, \quad x_2'(0) = -1$
(e) $-4x_1 + (D^2 + D + 4)x_2 = 2,$
$(D+1)x_1 - x_2 = 2,$
$x_1(0) = 2, \quad x_2(0) = 4, \quad x_2'(0) = 2$
(f) $x_1' = -x_1 - 3x_2 + 3x_3, \; x_1(0) = -1$
$x_2' = -4x_2 + 3x_3, \qquad x_2(0) = 1$
$x_3' = -6x_2 + 5x_3, \qquad x_3(0) = 0$

7. (a) Consider the differential equation

$$x''(t) + ax'(t) + bx(t) = h(t),$$

where a and b are constants, h is continuous on $[0, \infty)$, and h is of exponential order. Show that every solution, and the first derivative of every solution, is of exponential order. Suggestion: use the method of variation of parameters and Theorem 9.4.
(b) Generalize the result of part (a) to the nth-order equation $P(D)x = b$. In other words, complete the proof of Theorem 9.7.

9.6 FUNCTIONS WITH DISCONTINUITIES

A simple example of a discontinuous function is the *unit step function u,* where

$$u(t) = \begin{cases} 0, & t < 0 \\ 1, & t \geq 0. \end{cases}$$

More generally, we shall be concerned with step functions of the form u_c, where we define

$$u_c(t) = u(t - c).$$

The graph of u_c is shown in Figure 9.3. In applications, c is positive (we are interested in what happens *after* $t = 0$). In the remainder of this section we assume that the constant c in the symbol u_c is positive unless otherwise specified. The Laplace transform of u_c is

$$\mathscr{L}[u_c](s) = \int_0^\infty e^{-st} u(t-c)\, dt = \int_0^\infty e^{-st}\, dt = \frac{e^{-cs}}{s}.$$

The second integral has lower limit c because $u(t-c)$ is 0 when $t < c$.

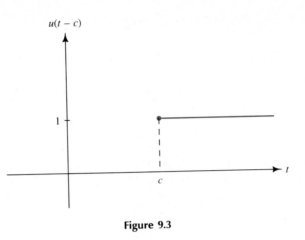

Figure 9.3

We shall be concerned with Laplace transforms of functions of the form

$$f_c(t) \equiv f(t - c)u(t - c).$$

The relationship between f and f_c is indicated in Figure 9.4. The graph of f_c is obtained from that of f by shifting the latter to the right a distance c and then "chopping it off at $t = c$" (since $f_c(t) = 0$ when $t < c$).

There is a simple relationship between the transforms of f and f_c. If the transform of f is denoted by F, this relationship is

$$f(t - c)u(t - c) \leftrightarrow F(s)e^{-cs}. \tag{9.15}$$

To derive this relationship, we first note that

$$\mathcal{L}[f_c](s) = \int_c^\infty e^{-st}f(t - c)u(t - c)\, dt.$$

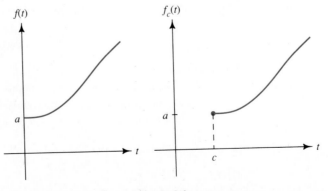

Figure 9.4

Making the change of variable $x = t - c$, we have

$$\mathscr{L}[f_c](s) = \int_0^\infty e^{-s(x+c)}f(x)u(x)\,dx$$

$$= e^{-cs}\int_0^\infty e^{-sx}f(x)\,dx$$

$$= e^{-cs}F(s).$$

The relationship (9.15) can be used to find the inverse transforms of functions that have a factor of the form e^{-cs}. For example, suppose we wish to find the inverse transform of

$$\frac{1}{s^2 + a^2}\,e^{-\pi s}.$$

Recognizing that $1/(s^2 + a^2)$ is the transform of the function

$$f(t) = \frac{1}{a}\sin at,$$

we see from Eq. (9.15) that the desired function is

$$f_\pi(t) = f(t - \pi)u(t - \pi) = \begin{cases} 0, & t < \pi \\ \dfrac{1}{a}\sin a(t - \pi), & t > \pi. \end{cases}$$

In applying formula (9.15), the next two relations are frequently useful. First, if

$$f(t) = \begin{cases} f_1(t), & 0 < t < c \\ f_2(t), & t > c, \end{cases}$$

then

$$f(t) = f_1(t) + [f_2(t) - f_1(t)]u(t - c).$$

Second, if

$$f(t) = f_1(t) + f_2(t)u(t - c),$$

then

$$f(t) = \begin{cases} f_1(t), & 0 < t < c \\ f_1(t) + f_2(t), & t > c. \end{cases}$$

For example, suppose we want to find the Laplace transform of the function f, where

$$f(t) = \begin{cases} t, & 0 < t < 1 \\ 1, & t > 1. \end{cases}$$

Writing $f_1(t) = t$, $f_2(t) = 1$, we have

$$f(t) = f_1(t) + [f_2(t) - f_1(t)]u(t-1) = t + (1-t)u(t-1)$$
$$= t - (t-1)u(t-1)$$

and hence

$$F(s) = \frac{1}{s} - \frac{1}{s^2}\,e^{-s}.$$

Certain problems in electric circuits and mechanics give rise to differential equations of the form

$$P(D)x = f, \qquad D = \frac{d}{dt},$$

where $P(D)$ is a polynomial operator (Chapter 5) and f is a function that may have discontinuities. The discontinuities occur at instants when a switch is suddenly opened or closed, or when there is an abrupt change in the force applied to an object. The formula for f may change from interval to interval. We now concern ourselves with applications of Laplace transforms to problems of this type.

Let us consider the specific initial value problem

$$x' + x = f, \qquad t \geq 0, \tag{9.16}$$

where $x(0) = 0$ and $f(t) = 1$ when $0 \leq t < c$, $f(t) = 0$ when $t \geq c$. Note that f is discontinuous at $t = c$. We may write

$$f(t) = 1 - u(t-c),$$

thus expressing f in terms of function whose transforms are known.

A question that must be answered is what is meant by a solution of this problem. Starting out on the time interval $[0, c)$, we find, by the methods of Chapter 1, a function x_1 that satisfies the differential equation

$$x_1'(t) + x_1(t) = 1. \qquad 0 \leq t < c,$$

and the condition $x_1(0) = 0$. This function (there is only one) is

$$x_1(t) = 1 - e^{-t}, \qquad 0 \leq t < c.$$

We now attempt to extend this "partial solution" to the interval (c, ∞). Accordingly, we seek a function x_2 such that

$$x_2'(t) + x_2(t) = 0, \qquad c \leq t < \infty,$$

and

$$x_2(c) = x_1(c-) = 1 - e^{-c}.$$

This last requirement specifies that the "solution" x, where

$$x(t) = \begin{cases} x_1(t), & 0 \le t < c \\ x_2(t), & t \ge c, \end{cases}$$

will be continuous on the entire interval $[0, \infty)$. We find, by the methods of Chapter 1, that

$$x_2(t) = (e^c - 1)e^{-t}, \qquad t \ge c.$$

Thus we have as our tentative solution to the original problem

$$x(t) = \begin{cases} 1 - e^{-t}, & 0 \le t < c \\ (e^c - 1)e^{-t}, & t \ge c. \end{cases}$$

This function is continuous on $[0, \infty)$ (we chose $x_2(c)$ to insure this property), but the derivative x' does not exist at $t = c$. Thus there is no function that is differentiable throughout the interval $[0, \infty)$ and satisfies the differential equation and initial condition. However, we have found the one function that is continuous, satisfies the initial condition, and satisfies the differential equation except at the point where the discontinuity occurs in the differential equation itself. This function we define to be the solution of the problem.

Let us now consider the general first-order problem

$$x'(t) + a(t)x(t) = f(t), \qquad t \ge 0, \tag{9.17}$$

where $x(0) = x_0$. We assume that the funtion a is continuous on $[0, \infty)$ and that f is piecewise continuous on $[0, \infty)$. Let us denote the discontinuities of f by c_1, c_2, c_3, and so on, where

$$0 < c_1 < c_2 < c_3 \cdots .$$

It can be shown that there is exactly one function, x_1, defined on $[0, c_1)$ that satisfies the differential equation and the condition $x_1(0) = x_0$. It can also be shown that $x_1(c_1 -)$ exists. Similarly, on the next interval $[c_1, c_2)$ there is exactly one function, x_2, that satisfies the differential equation and the condition $x_2(c_1) = x_1(c_1 -)$. Furthermore, $x_2(c_2 -)$ exists. Proceeding along the t-axis, we construct a unique function x that is continuous on $[0, \infty)$, satisfies the initial condition $x(0) = x_0$, and satisfies the differential equation except possibly at the points where f is discontinuous. This function x is defined to be the solution of the initial value problem (9.17). If a is a constant function and if f is of exponential order, it can be shown that x and x' are also of exponential order, and that the method of Laplace transforms applies.

Example 1 Let us solve the previously analyzed problem by the method of Laplace transforms. The problem may be written as

$$x'(t) + x(t) = 1 - u(t - c), \qquad t \geq 0,$$

where $x(0) = 0$. Transforming the differential equation, we have

$$sX(s) + X(s) = \frac{1}{s} - \frac{e^{-cs}}{s}$$

and hence

$$X(s) = \frac{1}{s(s + 1)}(1 - e^{-cs}).$$

Using partial fractions, we find that

$$X(s) = \left(\frac{1}{s} - \frac{1}{s + 1}\right)(1 - e^{-cs}).$$

According to Formulas 1 and 3 in the Table of Transforms (p. 435), the inverse transform of

$$\frac{1}{s} - \frac{1}{s + 1}$$

is $1 - e^{-t}$. Making use of this fact, Eq. (9.15), and the formula for $X(s)$, we see that

$$x(t) = (1 - e^{-t}) - (1 - e^{-(t-c)})u(t - c)$$

or

$$x(t) = \begin{cases} 1 - e^{-t}, & 0 \leq t < c \\ e^{-(t-c)} - e^{-t}, & t \geq c. \end{cases}$$

An obvious advantage of the transform method is that the various subintervals of $[0, \infty)$ need not be considered separately until a formula for the solution has been found.

In the general case of an nth order initial value problem,

$$x^{(n)} + a_1 x^{(n-1)} + \cdots + a_n x = f, \qquad t \geq 0,$$

$$x^{(k)}(0) = x_0^{(k)}, \qquad k = 0, 1, 2, \ldots, n - 1,$$

suppose that the coefficient functions $a_1, a_2, \ldots, a_n$ are continuous and that f is piecewise continuous on the interval $[0, \infty)$. The solution is defined to be the one function x that is continuous along with its first $n - 1$ derivatives on $[0, \infty)$, satisfies the initial conditions, and satisfies the differential equation except possibly at the points where f is discontinuous. If the coefficients a_i are constant and if f is of exponential order, then the solution and its first n derivatives are of exponential order and the method of Laplace transforms is applicable.

Example 2   Let us consider the second-order problem

$$x''(t) + x(t) = t[1 - u(t - c)], \qquad t \geq 0,$$

where $x(0) = x'(0) = 0$. Note that

$$f(t) = t[1 - u(t - c)] = \begin{cases} t, & 0 \leq t < c \\ 0, & t \geq c. \end{cases}$$

In order to make use of Eq. (9.15) to find the transform of f, we write

$$f(t) = t - (t - c)u(t - c) - cu(t - c).$$

Then

$$\mathscr{L}[f](s) = \frac{1}{s^2} - \frac{e^{-cs}}{s^2} - c\frac{e^{-cs}}{s}.$$

Transforming the differential equation, we find that

$$(s^2 + 1)X(s) = \frac{1}{s^2}(1 - e^{-cs}) - c\frac{e^{-cs}}{s}.$$

Now, by partial fractions,

$$\frac{1}{s^2(s^2 + 1)} = \frac{1}{s^2} - \frac{1}{s^2 + 1}, \qquad \frac{1}{s(s^2 + 1)} = \frac{1}{s} - \frac{s}{s^2 + 1},$$

so we have

$$X(s) = \left(\frac{1}{s^2} - \frac{1}{s^2 + 1}\right)(1 - e^{-cs}) - c\left(\frac{1}{s} - \frac{s}{s^2 + 1}\right)e^{-cs}.$$

Using Formulas 2, 5, and 6 in the Table of Transforms, and Eq. (9.15), we see that

$$x(t) = t - \sin t + [(t - c) - \sin(t - c)]u(t - c) - c[1 - \cos(t - c)]u(t - c).$$

By considering the intervals $[0, c)$ and $[c, \infty)$ separately, we may write

$$x(t) = \begin{cases} t - \sin t, & 0 \leq t < c \\ 2(t - c) - \sin t - \sin(t - c) + c\cos(t - c), & t \geq c. \end{cases}$$

Exercises for Section 9.6

1. Write formulas that do not make use of unit step functions to describe each of the following functions.

   (a) $f(t) = u(5 - t)$
   (b) $g(t) = u(t) + 3u(t - 1) - u(t - 2)$
   (c) $h(t) = t^2 u(t - 2)$
   (d) $k(t) = \sin 2\left(t - \dfrac{\pi}{3}\right) u\left(t - \dfrac{\pi}{3}\right)$
   (e) $p(t) = (t - 5)^2[1 - u(t - 5)]$
   (f) $q(t) = e^{3(t-1)}[1 - 2u(t - 1)]$

2. Write formulas that make use of unit step functions to describe each of the following functions.

   (a) $f(t) = \begin{cases} t^2, & 0 \le t < 2 \\ 0, & t \ge 2 \end{cases}$ $t^2 = [0 - t^2] u(t - 2)$
   $t^2 + t^2 \, u(t-2)$
   (b) $g(t) = \begin{cases} 0, & 0 \le t < 4 \\ e^{-2(t-4)}, & t \ge 4 \end{cases}$
   (c) $h(t) = \begin{cases} 0, & 0 \le t < 1 \\ 2, & 1 \le t < 6 \\ 0, & t \ge 6 \end{cases}$ $2 - 2(u(t-1))$
   $+2(u(t-6))$
   (d) $k(t) = \begin{cases} 3 \sin t, & 0 \le t < \pi \\ 0, & t \ge \pi \end{cases}$ $2(1 - u(t-1)) + u(t-6)$

3. Draw the graph of each function and find its Laplace transform by using formula (9.15).

   (a) $4u(t - 3)$
   (b) $e^{t-2}u(t - 2)$
   (c) $e^{6-2t}u(t - 3)$
   (d) $(t - 5)^2 u(t - 5)$
   (e) $\cos 3\left(t - \dfrac{\pi}{2}\right) u\left(t - \dfrac{\pi}{2}\right)$
   (f) $tu(t - 1)$
   (g) $G(t) = \begin{cases} \sin \pi t, & 0 \le t < 1 \\ 0, & t \ge 1 \end{cases}$
   (h) $H(t) = \begin{cases} 4, & 0 \le t < 2 \\ -4, & t \ge 2 \end{cases}$
   (i) $I(t) = \begin{cases} t, & 0 \le t < 1 \\ 2 - t, & 1 \le t < 2 \\ 0, & t \ge 2 \end{cases}$

4. Find the inverse transform of each of the following functions.

   (a) $\dfrac{2}{s} e^{-3s}$
   (b) $\dfrac{5}{s^3} e^{-s}$
   (c) $\dfrac{1}{s + 4} e^{-2s}$
   (d) $\dfrac{2}{(s + 3)^2} e^{-s}$
   (e) $\dfrac{6}{s^2 + 4} e^{-\pi s/2}$
   (f) $\dfrac{s}{s^2 + 9} e^{-\pi s}$
   (g) $\dfrac{6s}{(s^2 + \pi^2)^2} e^{-2s}$
   (h) $\dfrac{4}{s^2(s^2 + 9)} e^{-\pi s/2}$

5. Solve the initial value problem in two ways: (a) by the methods of Chapter 1; (b) by Laplace transforms.

   $x'(t) + 2x(t) = t[1 - u(t - 1)], t \ge 0, x(0) = 0.$

6. Solve the initial value problem by using Laplace transforms.

   (a) $x' + 2x = u(t - 1) - u(t - 5),$
   $x(0) = 1$
   (b) $x' + x = \sin \pi t[1 - u(t - 1)],$
   $x(0) = 0$
   (c) $x' + x = t - 2(t - 1)u(t - 1)$
   $\qquad + (t - 2)u(t - 2), \quad x(0) = 0$
   (d) $x' + x = 1 - u(t - 1) + u(t - 2),$
   $x(0) = 0$

7. Solve the initial value problem by using Laplace transforms.

   (a) $x'' + x = 1 - u(t - 2), \quad x(0) = x'(0) = 0$
   (b) $x'' + 4x = \sin t[1 - u(t - \pi)],$
   $x(0) = x'(0) = 0$
   (c) $x'' + 3x' + 2x = u(t - 1), \quad x(0) = 1,$
   $x'(0) = 0$
   (d) $x'' + 2x' + x = 1 - u(t - 1),$
   $x(0) = x'(0) = 0$

8. Solve the initial value problem by using Laplace transforms.

   (a) $\dfrac{dx}{dt} + x - y = 1 - u(t - 1)$
   $\dfrac{dy}{dt} - x + y = 0, \quad x(0) = y(0) = 0$

(b) $\dfrac{dx}{dt} + 2(x - y) = 1 - u(t - 1)$

$2\dfrac{dy}{dt} + 2(y - x) = 0, \quad x(0) = y(0) = 0$

9. (a) Let f be piecewise continuous on the interval $[0, \infty)$ and periodic with period a (so that $f(t + a) = f(t)$ for $t \geq 0$). Show that

$$\mathcal{L}[f](s) = \frac{1}{1 - e^{-as}} \int_0^a e^{-st} f(t)\, dt.$$

Suggestion: write the integral for the transform of f as an infinite series of integrals over the intervals $[ka, (k + 1)a]$, $k = 0, 1, 2, \ldots$.

(b) Use the result of Part (a) to find the Laplace transform of the function f that is periodic with period 2 and

$$f(t) = \begin{cases} 1, & 0 \leq t < 1 \\ 0, & 1 \leq t < 2. \end{cases}$$

A Table of Transforms

$f(t)$	$F(s)$		$f(t)$	$F(s)$
1. 1	$\dfrac{1}{s}$	9. $t \sin at$		$\dfrac{2as}{(s^2 + a^2)^2}$
2. t^n	$\dfrac{n!}{s^{n+1}}$	10. $t \cos at$		$\dfrac{s^2 - a^2}{(s^2 + a^2)^2}$
3. e^{at}	$\dfrac{1}{s - a}$	11. $t \sinh at$		$\dfrac{2as}{(s^2 - a^2)^2}$
4. $t^n e^{at}$	$\dfrac{n!}{(s - a)^{n+1}}$	12. $t \cosh at$		$\dfrac{s^2 + a^2}{(s^2 - a^2)^2}$
5. $\sin at$	$\dfrac{a}{s^2 + a^2}$	13. $\sin at - at \cos at$		$\dfrac{2a^3}{(s^2 + a^2)^2}$
6. $\cos at$	$\dfrac{s}{s^2 + a^2}$	14. $at \cosh at - \sinh at$		$\dfrac{2a^3}{(s^2 - a^2)^2}$
7. $\sinh at$	$\dfrac{a}{s^2 - a^2}$	15. $e^{ct} f(t)$		$F(s - c)$
8. $\cosh at$	$\dfrac{s}{s^2 - a^2}$	16. $f(t - c) u(t - c)$		$e^{-cs} F(s)$

Additional Exercises for Chapter 9

1. Find the Laplace transform of each given function.

(a) $e^{-2t} \sin 3t$ (b) $t \cos 2t$

(c) $t^4 e^t$ (d) $\int_0^t e^{3x}\, dx$

(e) $|\sin t|$ (f) $t u_1(t)$

2. Find the Laplace transform of the function

f_n, where

$$f_n(t) = e^t \frac{d^n}{dt^n}\left(t^n e^{-t}\right).$$

3. Find the inverse transform of the given function.

(a) $\dfrac{s}{(s - 2)^2}$ (b) $\dfrac{4s}{s^2 + 4s + 8}$

(c) $\dfrac{2s}{(s+1)(s^2+9)}$ (d) $e^{-s}\dfrac{s}{s^2+\pi^2}$

(e) $\dfrac{1}{s^2+\pi^2}(1-e^{-2s})$ (f) $\dfrac{1}{s+2}F(s)$

4. (a) If the Laplace transform of f is F, what is the inverse transform of F^2?

   (b) Find the inverse transform of

$$\frac{1}{(s^2+a^2)^2}.$$

5. If $f(t)=1,\ t\ge 0$, what is $f*f$?

6. Find $f*u_c$.

7. Draw the graph of the function and find its Laplace transform. (See Exercise 9, Section 9.6.)

   (a) $f(t)=\begin{cases}t, & 0\le t<1\\ 2-t, & 1\le t<2,\end{cases}$
   $f(t+2)=f(t),\quad t\ge 0$

   (b) $f(t)=\sin \pi t,\quad 0\le t<1,$
   $f(t+1)=f(t),\quad t\ge 0$

   (c) $f(t)=\begin{cases}1, & 0\le t<1\\ 0, & 1\le t<2,\end{cases}$
   $f(t+2)=f(t),\quad t\ge 0$

   (d) $f(t)=\begin{cases}t-1, & 0\le t<1\\ 0, & 1\le t<2,\end{cases}$
   $f(t+2)=f(t),\quad t\ge 0$

8. Assume that f is continuous on $[0, \infty)$ and of exponential order. Let

$$g(t)=f(t)-f_c(t)+f_{2c}(t)-f_{3c}(t)+\cdots$$
$$+(-1)^n f_{nc}(t)+\cdots$$

for $t\ge 0$.

   (a) Show that the series for $g(t)$ converges for $t\ge 0$.

   (b) Show that

$$G(s)=\frac{F(s)}{1+e^{-cs}}$$

provided that termwise integration of the series is valid.

(c) If f is periodic with period c, how are the graphs of f and g related?

(d) Use the results of Parts (b) and (c) to find the Laplace transform of the function g, where

$$g(t)=\sin \pi t, \qquad 0\le t<1,$$
$$g(t)=0, \qquad 1\le t<2,$$

and $g(t+2)=g(t)$ for $t\ge 0$.

9. (a) Let V be the set of all functions that are piecewise continuous on the interval $[0, \infty)$ and of exponential order. Show that V is a vector space under the usual operations of addition and multiplication by a scalar.

   (b) Show that the Laplace transformation operator is a linear operator on the space V of Part (a).

10. Solve the initial value problem by the method of Laplace transforms.

    (a) $x'+x=t,\quad x(0)=0$
    (b) $x'+x=1-u_1(t),\quad x(0)=0$
    (c) $x'+2x=\sin \pi t,\quad x(0)=0$
    (d) $x'+2x=4,\quad x(0)=3$

11. Solve by the method of Laplace transforms.

    (a) $x''+4x=6e^{-t},\qquad x(0)=x'(0)=0$
    (b) $x''-x=2\sin t,\quad x(0)=2,\quad x'(0)=1$
    (c) $x''+2x'+x=t,\quad x(0)=1,\quad x'(0)=2$
    (d) $x''+3x'+2x=1-u_2(t),$
    $x(0)=x'(0)=0$

12. Solve by the method of Laplace transforms.

    (a) $x''-4y=1\quad x(0)=y(0)=x'(0)$
    $y''-4x=1\qquad =y'(0)=0$
    (b) $x'=x+y+\sin t\quad x(0)=y(0)=0$
    $y'=-6x-4y+\cos t$
    (c) $x'=-2x+y+t[1-u(t-1)]$
    $y'=x-2y,\quad x(0)=y(0)=0$
    (d) $x'=-3x+y,\quad x(0)=y(0)=0$
    $y'=2x-4y+(1-t)[1-u(t-1)]$

CHAPTER

10

Stability
and the
Phase Plane

10.1
BASIC IDEAS

The most general first-order system has the form

$$\frac{dx_1}{dt} = g(t, x_1, x_2, \ldots, x_n)$$

$$\cdots\cdots\cdots\cdots\cdots\cdots\cdots\cdots$$

$$\frac{dx_n}{dt} = g_n(t, x_1, x_2, \ldots, x_n).$$

In this chapter, we restrict our attention to systems of the form

$$\frac{dx_1}{dt} = f_1(x_1, x_2, \ldots, x_n)$$

$$\cdots\cdots\cdots\cdots\cdots\cdots\cdots\cdots \qquad (10.1)$$

$$\frac{dx_n}{dt} = f_n(x_1, x_2, \ldots, x_n)$$

where the functions f_i do not depend on the independent variable t. Such systems are called *autonomous systems.* In vector form, we write

$$\frac{d\mathbf{x}}{dt} = \mathbf{f}(\mathbf{x}). \qquad (10.2)$$

437

One example of such a system, with $n = 3$, is

$$\frac{dx_1}{dt} = 2x_1 - x_2 + x_3 - 6$$

$$\frac{dx_2}{dt} = -x_1 + 3x_2 + 5 \qquad\qquad (10.3)$$

$$\frac{dx_3}{dt} = x_1 + x_2 + x_3 - 2.$$

Another, with $n = 2$, is

$$\frac{dx_1}{dt} = 2x_1(2 - x_2), \qquad \frac{dx_2}{dt} = 3x_2(1 - x_1). \qquad (10.4)$$

If $\mathbf{k} = (k_1, k_2, \ldots, k_n)$ is a constant vector such that $\mathbf{f}(\mathbf{k}) = \mathbf{0}$, then the system (10.2) has the constant solution

$$x_1 = k_1, \qquad x_2 = k_2, \ldots, \qquad x_n = k_n.$$

For instance, any constant solution of the system (10.3) must satisfy

$$2x_1 - x_2 + x_3 - 6 = 0, \quad -x_1 + 3x_2 + 5 = 0, \quad x_1 + x_2 + x_3 - 2 = 0.$$

Upon solving this algebraic system of equations, we find that the only constant solution is

$$x_1 = 2, \qquad x_2 = -1, \qquad x_3 = 1.$$

Any constant solution of the system (10.4) must satisfy

$$2x_1(2 - x_2) = 0, \qquad 3x_2(1 - x_1) = 0.$$

This time we find two constant solutions, $x_1 = 0$, $x_2 = 0$ and $x_1 = 1$, $x_2 = 2$.

A point $\mathbf{k}$ in R^n such that $\mathbf{f}(\mathbf{k}) = \mathbf{0}$ is called a *critical point* for the system (10.2). Any other point is called a *regular point*. A critical point is said to be an *isolated critical point* if there is a positive number δ such that $\mathbf{k}$ is the only critical point in the region

$$\{\mathbf{x}: \|\mathbf{x} - \mathbf{k}\| < \delta\}.$$

The critical points (0,0) and (2,1) of the system (10.4) are both isolated. In fact, if a system has a finite number of critical points, they must all be isolated. However, for the system

$$\frac{dx_1}{dt} = 2x_1 - x_2, \qquad \frac{dx_2}{dt} = -4x_1 + 2x_2,$$

every point on the line $2x_1 - x_2 = 0$ is a critical point, and none of these is isolated. In the important special case of a linear system with constant coefficients,

$$\frac{d\mathbf{x}}{dt} = A\mathbf{x},$$

where A is an $n \times n$ constant matrix, the point $\mathbf{x} = \mathbf{0}$ is a critical point. If A is nonsingular, it is the only critical point. However, if A is singular, there are infinitely many nonisolated critical points; every point in the kernel of A is a critical point.

By a *solution* of the system (10.1) we mean an ordered set of n functions

$$x_1 = \phi_1(t), \qquad x_2 = \phi_2(t), \ldots, \qquad x_n = \phi_n(t)$$

that satisfies the system. The directed curve that is traced out in $x_1 x_2 \cdots x_n$ -space is called a *trajectory* of the system. The curve is directed, with the positive direction being that of increasing t. To illustrate, let us consider the system

$$\frac{dx_1}{dt} = -x_1, \qquad \frac{dx_2}{dt} = -2x_2. \tag{10.5}$$

The solutions of this uncoupled system are

$$x_1 = c_1 e^{-t}, \qquad x_2 = c_2 e^{-2t}.$$

If $c_1 = c_2 = 0$, the constant solution has the critical point $(0, 0)$ as its trajectory. If $c_2 = 0$ and $c_1 > 0$, the trajectory is the positive x_1-axis, directed toward the origin. If $c_2 = 0$ and $c_1 < 0$, the trajectory is the negative x_1-axis, directed toward the origin. The case $c_1 = 0$ is similar; the trajectories consist of the positive x_2-axis and the negative x_2-axis, directed toward the origin. If neither c_1 nor c_2 is zero, we have

$$\frac{x_2}{c_2} = \left(\frac{x_1}{c_1}\right)^2.$$

Each associated trajectory is half a parabola (since $e^{-t} > 0$ for all t) directed toward the origin. Fig. 10.1 shows some of the trajectories.

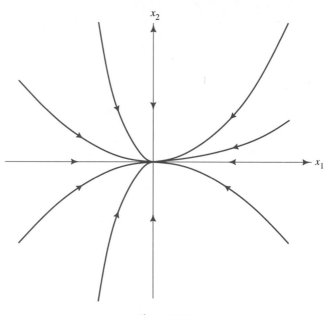

Figure 10.1

In general, different solutions of the system (10.1) may have the same trajectory, but different trajectories cannot have a common point. (See Exercise 9 at the end of this section.)

In case $n = 2$ (as in the previous example), the trajectories are curves in the $x_1 x_2$-plane. This plane is called the *phase plane*. A diagram that shows the pattern of the trajectories in this plane is called *a phase portrait* for the system. As another example, let us consider the system

$$\frac{dx_1}{dt} = -4x_2, \qquad \frac{dx_2}{dt} = x_1. \tag{10.6}$$

The solutions, as found by the method of Section 6.3, are

$$x_1 = -2A \sin(2t + \alpha), \qquad x_2 = A \cos(2t + \alpha),$$

where A and α are arbitrary constants. In addition to the critical point $(0, 0)$, the trajectories are the directed ellipses

$$\frac{x_1^2}{4A^2} + \frac{x_2^2}{A^2} = 1.$$

The directions of the trajectories can be found from the equations (10.6). Since x_2 is increasing when x_1 is positive ($x_2' > 0$ when $x_1 > 0$), the ellipses are oriented counterclockwise. Fig. 10.2 shows some of the trajectories.

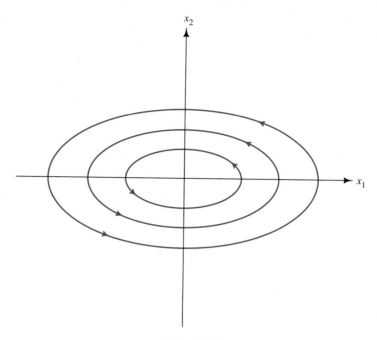

Figure 10.2

We consider one more example,

$$\frac{dx_1}{dt} = -x_1, \qquad \frac{dx_2}{dt} = x_2. \tag{10.7}$$

The solutions

$$x_1 = c_1 e^{-t}, \qquad x_2 = c_2 e^{t}$$

give rise to the trajectories $x_1 x_2 = c_1 c_2$. As in the previous example, the orientations can be found from the differential equations. Some trajectories are shown in Fig. 10.3.

Let us return to the general case (10.1). If $\mathbf{k} = (k_1, k_2, \ldots, k_n)$ is a critical point and $\mathbf{x} = (\phi_1, \phi_2, \ldots, \phi_n)$ is a solution, we say that the solution *approaches* the critical point if

$$\lim_{t \to \infty} \phi_i(t) = k_i$$

for each i. In the example (10.5), the solutions $x_1 = c_1 e^{-t}$, $x_2 = c_2 e^{-2t}$ all approached the critical point $(0, 0)$. In the example (10.6), the solutions did not all approach the critical point $(0, 0)$, but a solution that passed near it did not go far away from it as t increased. In example (10.7), solutions that passed very close to the critical point ultimately went far away from

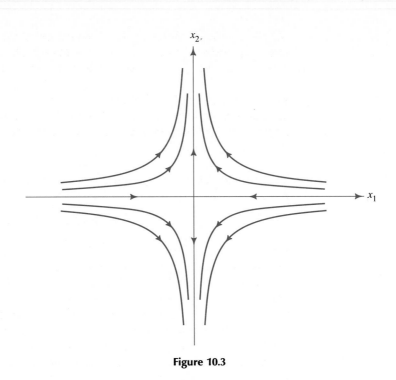

Figure 10.3

it. These examples lead us to formulate some important definitions for the system (10.2).

A critical point **k** for the system (10.2) is said to be *stable* if for every $\epsilon > 0$ there corresponds a $\delta > 0$ such that whenever a solution **x** satisfies

$$\|\mathbf{x}(0) - \mathbf{k}\| < \delta,$$

the solution exists for all $t > 0$ and satisfies

$$\|\mathbf{x}(t) - \mathbf{k}\| < \epsilon$$

for $t \geq 0$. Thus a solution will stay within a distance ϵ of the critical point if it starts out sufficiently close to it. The critical points of the examples (10.5) and (10.6) are stable.

A critical point **k** is said to be *asymptotically stable* if it is stable and if in addition there is a positive number δ_0 such that every solution for which $\|\mathbf{x}(0) - \mathbf{k}\| < \delta_0$ approaches the critical point. The system (10.5) has an asymptotically stable critical point $(0, 0)$. The stable critical point of the system (10.6) is not asymptotically stable.

A critical point that is not stable is said to be *unstable*. The system (10.7) has an unstable critical point $(0, 0)$.

In the case of a linear system $\mathbf{x}' = A\mathbf{x}$ with constant coefficients, we can solve the system and determine from the characteristic values of A what sort of stability the critical point $\mathbf{k} = \mathbf{0}$ possesses. Let us suppose that A is nonsingular, so that the critical point is isolated. Then no characteristic value of A is zero. The facts are as stated in the next theorem.

Theorem 10.1 Let A be nonsingular. The critical point $\mathbf{k} = \mathbf{0}$ of the system $\mathbf{x}' = A\mathbf{x}$ is asymptotically stable if and only if all the characteristic values of A have negative real parts. If even one characteristic value has a positive real part, the critical point is unstable.

Proof Each component of every solution is a linear combination of terms of the form

$$t^n e^{\alpha t}, \quad t^n e^{\alpha t} \cos \beta t, \quad t^n e^{\alpha t} \sin \beta t,$$

where α is the real part of a characteristic value. If $\alpha < 0$ for every characteristic value then every term approaches 0 as $t \to \infty$. It can be shown (Exercise 13) that $\mathbf{0}$ is asymptotically stable in this case. However, if there exists even one characteristic value, α or $\alpha + i\beta$, with nonnegative real part, there exists a solution $\mathbf{x} = \mathbf{k}e^{\alpha t}$ or $\mathbf{x} = \mathbf{k}e^{\alpha t} \cos \beta t$ with $\mathbf{k} \neq \mathbf{0}$, and this solution does not approach $\mathbf{0}$ as $t \to 0$. If $\alpha > 0$, the solution does not stay inside any fixed neighborhood of $\mathbf{0}$, so $\mathbf{0}$ is unstable.

In the case of a nonlinear system, it is often not possible to find explicit formulas for all the solutions. However, in some cases, it is still possible to determine whether a critical point is asymptotically stable or unstable. Nonlinear systems will be investigated in Section 10.3.

─────────────────── **Exercises for Section 10.1** ───────────────────

1. Find the critical points of the system.

   (a) $x_1' = 2x_1 - x_2 - 5$
   $x_2' = x_1 + x_2 - 2$

   (b) $x_1' = -x_1 + 3x_2 - 5$
   $x_2' = 2x_1 + x_2 - 4$

   (c) $x_1' = -2x_1 + 6x_2$
   $x_2' = x_1 - 3x_2$

   (d) $x_1' = 0$
   $x_2' = 0$

   (e) $x_1' = (x_1 + 2)(x_2 + 2)$
   $x_2' = (x_1 - 1)(x_2 - 3)$

   (f) $x_1' = 3x_2$
   $x_2' = \sin(x_1 + x_2)$

   (g) $x_1' = 2x_1 - x_3 + 2$
   $x_2' = x_1 + 2x_2 - x_3$
   $x_3' = x_1 - 2x_2 - 5$

   (h) $x_1' = x_1 - x_2 + x_3$
   $x_2' = x_1 + x_2$
   $x_3' = 2x_1 + x_3$

2. Replace the differential equation with a first-order system, and find the critical points.

   (a) $x'' + \sin x = 0$

   (b) $x'' + x' + \sin x = 0$

   (c) $x'' + 2x' + 3x = 6$

   (d) $x'' + (x')^2 = 4$

   In Exercises 3–8, find all solutions of the system. Determine whether the origin is a stable, asymptotically stable, or unstable critical point. In Exercises 3–6, find the $x_1 x_2$-equations of the trajectories.

3. $x_1' = x_1$
   $x_2' = -2x_2$

4. $x_1' = -9x_2$
   $x_2' = x_1$

5. $x_1' = x_1$
$x_2' = 2x_2$

6. $x_1' = -x_1$
$x_2' = -3x_2$

7. $x_1' = -2x_1 + x_2$
$x_2' = -x_1 - 2x_2$

8. $x_1' = 2x_1 + x_2$
$x_2' = -x_1 + 2x_2$

9. (a) If the functions $x_i = \phi_i(t)$ form a solution of the autonomous system (10.1), show that the functions $\phi_i(t - c)$, where c is a constant, also form a solution, and that the two solutions have the same trajectory.

(b) Show that two distinct trajectories of the system (10.1) cannot meet. (Suppose that $x_i = \phi_i(t)$ and $x_i = \psi_i(t)$ are solutions such that $\phi_i(t_1) = \psi_i(t_2)$. Let $\chi_i(t) = \psi_i(t + t_2 - t_1)$. Show $\chi_i(t_1) = \phi_i(t_1)$.)

10. If the system $x' = f(x)$ has a critical point k, show that the change of variable $y = x - k$ yields a system $y' = g(y)$ with critical point 0.

11. Suppose that all characteristic values of the nonsingular $n \times n$ matrix A have nonpositive real parts. Is the critical point of the system $x' = Ax$ necessarily stable if (a) $n = 2$? (b) $n = 3$? (c) $n = 4$?

12. If the origin is asymptotically stable for the system $x_1' = F_1(x_1, x_2)$, $x_2' = F_2(x_1, x_2)$, show

that it is unstable for the system $x_1' = -F_1(x_1, x_2)$, $x_2' = -F_2(x_1, x_2)$.

13. Suppose that all characteristic values of the $n \times n$ real matrix A have negative real parts, and let γ be the largest real part. Let Φ be the fundamental matrix for the equation $x' = Ax$ for which $\Phi(0) = I$ ($\Phi(t) = e^{tA}$). Then

$$x(t) = \Phi(t)x(0) = e^{\gamma t/2} e^{-\gamma t/2} \Phi(t)x(0).$$

(a) Show that every component of $e^{-\gamma t/2} \Phi(t)$ is bounded on $[0, \infty)$, so that $|e^{-\gamma t/2} \Phi(t)| \le K$ for some constant K.

(b) Show that $|x| \le \|x\| \le \sqrt{n}|x|$.

(c) Given $\epsilon > 0$, suppose that $\|x(0)\| < \epsilon/(K\sqrt{n})$. Show that $\|x(t)\| < \epsilon$ for $t \ge 0$, and hence that 0 is asymptotically stable for the equation $x' = Ax$.

14. Let F_1 and F_2 be continuous in a region D, and let (ϕ_1, ϕ_2) be a solution of the system (10.18). If $\phi_1(t) \to a_1$ and $\phi_2(t) \to a_2$ as $t \to \infty$, and (a_1, a_2) is in D, show that (a_1, a_2) is a critical point. Suggestion: suppose that $F(a_1, a_2) > 0$. Then $F_1(x_1, x_2) \ge k > 0$ for $(x_1 - a_1)^2 + (x_2 - a_2)^2 < \delta^2$. However, $[\phi_1(t) - a_1]^2 + [\phi_2(t) - a_2]^2 < \delta^2$ if $t \ge T$. Hence $\phi_1(t) \to \infty$ as $t \to \infty$, which is false. Show that the assumption $F_1(a_1, a_2) < 0$ also leads to a contradiction. Thus $F_1(a_1, a_2) = 0$. Likewise, $F_2(a_1, a_2) = 0$.

10.2
THE PHASE PLANE

Given a linear system with constant coefficients,

$$\frac{dx}{dt} = Ax, \tag{10.8}$$

where A is an $n \times n$ real matrix, we can attempt to simplify the equations with a change of variable $x = Ky$, where K is a nonsingular $n \times n$ matrix. Equation (10.8) becomes

$$K\frac{dy}{dt} = AKy$$

or

$$\frac{dy}{dt} = By \tag{10.9}$$

where $B = K^{-1}AK$. If A is diagonalizable (Section 4.3), K can be chosen so that $B = \operatorname{diag}(\lambda_1, \lambda_2, \ldots, \lambda_n)$, where the λ_i are the characteristic values of A. Then equation (10.9) becomes

$$\frac{dy_1}{dt} = \lambda_1 y_1, \qquad \frac{dy_2}{dt} = \lambda_2 y_2, \ldots, \qquad \frac{dy_n}{dt} = \lambda_n y_n,$$

and the solutions are

$$y_i = c_i e^{\lambda_i t}$$

for each i. However, we know that not every matrix A is diagonalizable, and whether it is or not the characteristic values may not all be real.

In this chapter we consider the important special case $n = 2$. Then K can be chosen so that $B = K^{-1}AK$ has one of several special forms, called *canonical forms*. If the two characteristic values λ_1 and λ_2 are real and distinct, then K can be chosen (Section 4.3) so that B has the diagonal form

$$\begin{bmatrix} \lambda_1 & 0 \\ 0 & \lambda_2 \end{bmatrix}. \tag{10.10}$$

If $\lambda_1 = \alpha + i\beta$ and $\lambda_2 = \alpha - i\beta$, where $\beta \neq 0$, K can be chosen so that B is

$$\begin{bmatrix} \alpha & \beta \\ -\beta & \alpha \end{bmatrix}. \tag{10.11}$$

However, if $\lambda_1 = \lambda_2$, then B will have one of the forms

$$\begin{bmatrix} \lambda_1 & 0 \\ 0 & \lambda_1 \end{bmatrix}, \quad \begin{bmatrix} \lambda_1 & 1 \\ 0 & \lambda_1 \end{bmatrix}. \tag{10.12}$$

We assume that A is nonsingular, so that the system $\mathbf{x}' = A\mathbf{x}$ has an isolated critical point at $\mathbf{0}$. This implies that zero is not a characteristic value of A.

We first consider the case (10.10) of real distinct characteristic values. The system (10.9) in this case is

$$\frac{dy_1}{dt} = \lambda_1 y_1, \qquad \frac{dy_2}{dt} = \lambda_2 y_2, \tag{10.13}$$

so

$$y_1 = c_1 e^{\lambda_1 t}, \qquad y_2 = c_2 e^{\lambda_2 t}.$$

If $c_2 = 0$, the trajectory is either the positive or negative y_1-axis, depending on whether $c_1 > 0$ or $c_1 < 0$. If $c_1 = 0$, the trajectory is either the positive or negative y_2-axis, depending on whether $c_2 > 0$ or $c_2 < 0$. If neither c_1 nor c_2 is 0, we have

$$\left(\frac{y_1}{c_1}\right)^{\lambda_2} = \left(\frac{y_2}{c_2}\right)^{\lambda_1}$$

or

$$y_2 = c_2 \left(\frac{y_1}{c_1}\right)^{\lambda_2/\lambda_1}.$$

We consider several subcases. First, suppose that λ_1 and λ_2 have the same sign. Then if $\lambda_2/\lambda_1 > 1$, the trajectories are tangent to the y_1-axis; if $\lambda_2/\lambda_1 < 1$, they are tangent to the y_2-axis. Thus all but two trajectories are tangent to the same line at the origin. If λ_1 and λ_2 are both negative, the trajectories approach the origin as $t \to \infty$, and the origin is an asymptotically stable critical point. If λ_1 and λ_2 are both positive, then $|y_1|$ and $|y_2|$ increase as t increases. As $t \to -\infty$, $y_1 \to 0$ and $y_2 \to 0$. The origin is an unstable critical point. All trajectories approach (or leave) the origin in specific directions, but not every direction has an associated trajectory. Such behavior is described by saying that the critical point is an *improper node*. (When every trajectory approaches or leaves with a specific direction, and every direction has an associated trajectory, the critical point is called a *proper node*.)

The case $\lambda_1 = 1$, $\lambda_2 = 2$ is shown in Fig. 10.4. However, we are concerned with the behavior of the original system in the x_1x_2-plane. Consider the system

$$\frac{dx_1}{dt} = -x_1 + 3x_2, \qquad \frac{dx_2}{dt} = -2x_1 + 4x_2, \qquad (10.14)$$

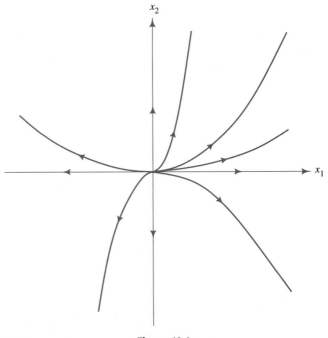

Figure 10.4

whose characteristic values are $\lambda_1 = 1$ and $\lambda_2 = 2$. The trajectories are obtained from those of the system (10.13) by means of reflections, stretchings, and shearings, as was shown in Section 3.9. Let us consider the system (10.14) in more detail. The matrix of the system,

$$\begin{bmatrix} -1 & 3 \\ -2 & 4 \end{bmatrix},$$

has characteristic vectors $\mathbf{k}_1 = (3, 2)$ and $\mathbf{k}_2 = (1, 1)$ associated with the characteristic values λ_1 and λ_2, respectively. The matrix $K = [\mathbf{k}_1, \mathbf{k}_2]$ has the property that $K^{-1}AK = \text{diag}\,(1, 2)$. To see the geometrical significance of the characteristic vectors, recall that

$$\mathbf{x} = K\mathbf{y} = y_1\mathbf{k}_1 + y_2\mathbf{k}_2.$$

If $y_2 = 0$ (on the y_1-axis) then $\mathbf{x} = y_1\mathbf{k}_1$. Thus $\mathbf{k}_1$ has the direction of the line through the origin of the x_1x_2-plane that corresponds to the y_1-axis. In the example, $\mathbf{k}_1 = (3, 2)$, so the line is $2x_1 - 3x_2 = 0$. If $y_1 = 0$ (on the y_2-axis) then $\mathbf{x} = y_2\mathbf{k}_2$. Thus $\mathbf{k}_2$ has the direction of the line through the origin that corresponds to the y_2-axis. Since $\mathbf{k}_2 = (1, 1)$, the line is $x_1 - x_2 = 0$. Some trajectories of the system (10.4) are shown in Fig. 10.5.

The case where $\lambda_1 < 0$ and $\lambda_2 < 0$ is similar, except that all solutions approach the critical point as $t \to \infty$. The critical point is an asymptotically

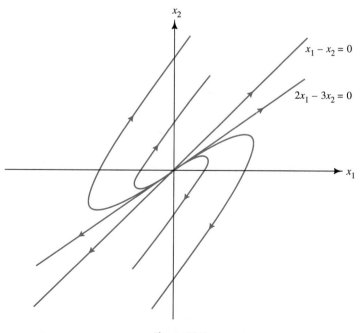

Figure 10.5

stable improper node. An example is provided by the system

$$\frac{dx_1}{dt} = -5x_1 + 2x_2, \qquad \frac{dx_2}{dt} = -4x_1 + x_2.$$

The characteristic values are $\lambda_1 = -1$, $\lambda_2 = -3$, and their associated characteristic vectors are $\mathbf{k}_1 = (1, 2)$ and $\mathbf{k}_2 = (1, 1)$, respectively. The y_1-axis corresponds to the line $2x_1 - x_2 = 0$ and the y_2-axis to $x_1 - x_2 = 0$. Some trajectories are shown in Fig. 10.6.

If λ_1 and λ_2 have opposite signs, then as $t \to \infty$ (or $t \to -\infty$) one variable increases and the other approaches 0. The trajectories consist of the positive and negative y_1- and y_2-axes and the curves

$$y_2 = c_2 \left(\frac{y_1}{c_1}\right)^{\lambda_2/\lambda_1}.$$

If, for example, $\lambda_1 = 1$ and $\lambda_2 = -1$, we have $y_2 = cy_1^{-1}$. The origin is said to be a *saddle point*. It is unstable. An example of such a system is

$$\frac{dx_1}{dt} = -0.5x_1 + 0.5x_2, \qquad \frac{dx_2}{dt} = 1.5x_1 + 0.5x_2.$$

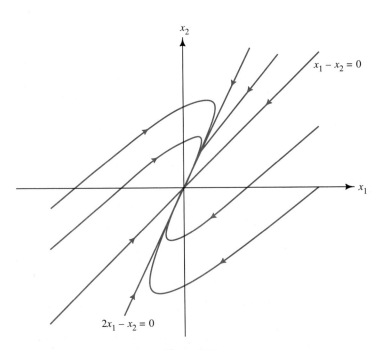

Figure 10.6

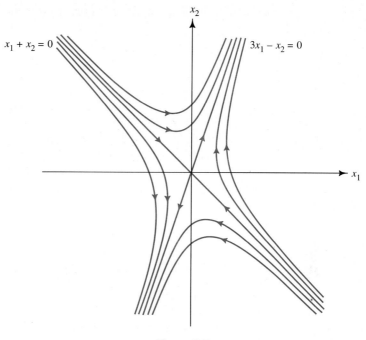

Figure 10.7

The characteristic values $\lambda_1 = 1$ and $\lambda_2 = -1$ have the associated characteristic vectors $\mathbf{k}_1 = (1, 3)$ and $\mathbf{k}_2 = (1, -1)$, respectively. The lines $3x_1 - x_2 = 0$ and $x_1 + x_2 = 0$ correspond to $y_2 = 0$ and $y_1 = 0$, respectively. Some trajectories are shown in Fig. 10.7.

We next consider the case (10.11) of nonreal characteristic values. The corresponding system is

$$\frac{dy_1}{dt} = \alpha y_1 + \beta y_2, \qquad \frac{dy_2}{dt} = -\beta y_1 + \alpha y_2. \qquad (10.15)$$

The solutions and the equations of the trajectories become simpler in polar coordinates. Writing

$$r^2 = y_1^2 + y_2^2, \qquad \tan \theta = \frac{y_2}{y_1},$$

we see upon differentiating with respect to t that

$$2r\frac{dr}{dt} = 2\left(y_1\frac{dy_1}{dt} + y_2\frac{dy_2}{dt}\right), \qquad \sec^2\theta\,\frac{d\theta}{dt} = \left(y_1\frac{dy_2}{dt} - y_2\frac{dy_1}{dt}\right)\Big/ y_1^2.$$

Since $\sec^2 \theta = 1 + \tan^2 \theta = 1 + \dfrac{y_2^2}{y_1^2} = \dfrac{r^2}{y_1^2}$, the second relation simplifies. We find that

$$r\frac{dr}{dt} = y_1 \frac{dy_1}{dt} + y_2 \frac{dy_2}{dt}, \qquad r^2 \frac{d\theta}{dt} = y_1 \frac{dy_2}{dt} - y_2 \frac{dy_1}{dt}. \qquad (10.16)$$

Substituting the expressions for dy_1/dt and dy_2/dt from equations (10.15) into equations (10.16) and simplifying, we find that

$$\frac{dr}{dt} = \alpha r, \qquad \frac{d\theta}{dt} = -\beta.$$

Thus

$$r = c_1 e^{\alpha t}, \qquad \theta = -\beta t + c_2.$$

If $\alpha > 0$, the trajectories spiral away from the origin; if $\alpha < 0$, they spiral toward the origin. No trajectory approaches the origin in a definite direction, unlike the case of a node. The origin here is called a *spiral point*. If $\alpha = 0$, the trajectories are circles, centered at the origin. For the original system $\mathbf{x}' = A\mathbf{x}$, the trajectories would be ellipses or circles, centered at the origin. In this case the origin is called a *center*.

An example of a system with an asymptotically stable spiral point is

$$\frac{dx_1}{dt} = x_1 + 4x_2, \qquad \frac{dx_2}{dt} = -2x_1 - 3x_2.$$

The characteristic values are $\lambda_1 = -1 + 2i$ and $\lambda_2 = -1 - 2i$. Some trajectories are shown in Fig. 10.8.

If the characteristic values are real and equal, there are two subcases. If A is diagonalizable (in which case A is already diagonal), then the system (10.8) is

$$\frac{dx_1}{dt} = \lambda_1 x_1, \qquad \frac{dx_2}{dt} = \lambda_1 x_2.$$

Then $x_1 = c_1 e^{\lambda_1 t}$ and $x_2 = c_2 e^{\lambda_1 t}$, so $c_2 y_1 = c_1 y_2$. Each trajectory is a ray with endpoint at the origin. Each is directed toward the origin if $\lambda_1 < 0$ and away from it if $\lambda_1 > 0$. A trajectory approaches in each possible direction. The origin is said to be a *proper node*. (In the case of an *improper node*, not every direction has a trajectory associated with it.)

In the remaining subcase, where

$$B = \begin{bmatrix} \lambda_1 & 1 \\ 0 & \lambda_1 \end{bmatrix},$$

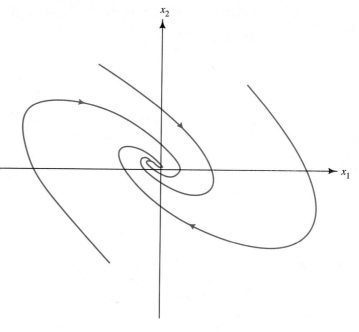

Figure 10.8

the system for y_1 and y_2 is

$$\frac{dy_1}{dt} = \lambda_1 y_1 + y_2, \qquad \frac{dy_2}{dt} = \lambda_1 y_2. \qquad (10.17)$$

We find the solutions

$$y_1 = (c_1 + c_2 t)e^{\lambda_1 t}, \qquad y_2 = c_2 e^{\lambda_1 t}.$$

To investigate the trajectories, we use the differential equations (10.17) to write

$$\frac{dy_1}{dy_2} = \frac{\lambda_1 y_1 + y_2}{\lambda_1 y_2}.$$

This is a homogeneous equation, of the type discussed in Section 1.3. Its solutions are found to be

$$y_1 = \lambda_1 y_2 \ln|c y_2|.$$

Each trajectory is tangent to the y_1-axis. In the $x_1 x_2$-plane, the direction of the single characteristic vector of A is that of the line that corresponds to the y_1-axis under the change of variable. The origin is an improper node.

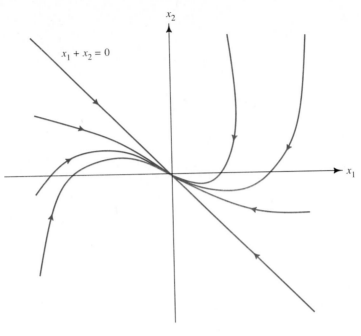

Figure 10.9

An example of such a system is

$$\frac{dx_1}{dt} = -x_1 + x_2, \qquad \frac{dx_2}{dt} = -x_1 - 3x_2.$$

The characteristic values are $\lambda_1 = \lambda_2 = -2$. All characteristic vectors are multiples of $\mathbf{k} = (1, -1)$. The line in the x_1x_2-plane with this direction is $x_1 + x_2 = 0$. Some trajectories are shown in Fig. 10.9.

──────────── **Exercises for Section 10.2** ────────────

In Exercises 1–10, determine whether the origin is asymptotically stable, stable but not asymptotically stable, or unstable. Determine whether it is a proper node, an improper node, a center, a saddle point, or a spiral point. Give the canonical form to which the system corresponds. Find a matrix K such that the change of variable $\mathbf{x} = K\mathbf{y}$ yields a system in canonical form. Find the equations of the lines in the x_1x_2-plane that correspond to the coordinate axes in the y_1y_2-plane. Sketch some trajectories.

1. $x_1' = 3x_1 - 2x_2$
    $x_2' = x_1 + x_2$

2. $x_1' = -5x_1 + 4x_2$
    $x_2' = -2x_1 - x_2$

3. $x_1' = 5x_1 - 3x_2$
    $x_2' = 6x_1 - 4x_2$

4. $x_1' = 5x_1 - 4x_2$
    $x_2' = 8x_1 - 7x_2$

5. $x_1' = -4x_1 + x_2$
$x_2' = -2x_1 - x_2$

6. $x_1' = 2x_2$
$x_2' = -4x_1 + 6x_2$

7. $x_1' = 3x_1 + 3x_2$
$x_2' = -6x_1 - 3x_2$

8. $x_1' = -2x_1 + 2x_2$
$x_2' = -4x_1 + 2x_2$

9. $x_1' = x_1 + x_2$
$x_2' = -4x_1 - 3x_2$

10. $x_1' = -5x_1 + x_2$
$x_2' = -4x_1 - x_2$

In Exercises 11–14, replace the equation by a first-order system and proceed as in Exercises 1–10.

11. $x'' + 2x' + 2x = 0$

12. $x'' + 4x' + 3x = 0$

13. $x'' + 4x = 0$

14. $x'' + 4x' + 4x = 0$

In Exercises 15–18, find the critical points and analyze as in Exercises 1–10. (Make a change of variable $y = x - k$ that places the critical point at the origin of the y_1y_2-plane.)

15. $x_1' = 4x_1 + 6x_2 - 6$
$x_2' = -3x_1 - 5x_2 + 4$

16. $x_1' = x_1 + x_2 - 5$
$x_2' = -2x_1 + 3x_2 - 5$

17. $x_1' = -2x_1 + 4x_2 - 8$
$x_2' = -5x_1 + 2x_2 + 4$

18. $x_1' = x_1 + x_2 - 3$
$x_2' = -x_1 + 3x_2 - 5$

19. Find a first-order system with characteristic values $\lambda_1 = -2, \lambda_2 = -4$ for which the lines $x_1 - x_2 = 0$ and $x_1 + x_2 = 0$ correspond to $y_2 = 0$ and $y_1 = 0$, respectively.

20. Find a first-order system with characteristic values $\lambda_1 = -1, \lambda_2 = 2$, for which the lines $2x_1 + x_2 = 0$ and $x_1 + x_2 = 0$ correspond to $y_2 = 0$ and $y_1 = 0$, respectively.

10.3 NONLINEAR SYSTEMS

We consider two-dimensional autonomous systems of the form

$$\frac{dx_1}{dt} = F_1(x_1, x_2), \qquad \frac{dx_2}{dt} = F_2(x_1, x_2). \qquad (10.18)$$

We assume, without loss of generality, that there is an isolated critical point at $(0, 0)$, so that $F_1(0, 0) = F_2(0, 0) = 0$. (If the critical point is at some other location, say (a_1, a_2), the change of variables $u_1 = x_1 - a_1, u_2 = x_2 - a_2$ puts the critical point at the origin of the u_1u_2-plane.) If the functions F_1 and F_2 can be expanded in Taylor series about $(0, 0)$, then

$$F_1(x_1, x_2) = a_{11}x_1 + a_{12}x_2 + a_{13}x_1^2 + a_{14}x_1x_2 + a_{15}x_2^2 + \cdots$$
$$F_2(x_1, x_2) = a_{21}x_1 + a_{22}x_2 + a_{23}x_1^2 + a_{24}x_1x_2 + a_{25}x_2^2 + \cdots .$$

When $|x_1|$ and $|x_2|$ are small, so that (x_1, x_2) is near the critical point $(0, 0)$, we have the approximations

$$F_1(x_1, x_2) \approx a_{11}x_1 + a_{12}x_2, \qquad F_2(x_1, x_2) \approx a_{21}x_1 + a_{22}x_2.$$

Thus, associated with the nonlinear system (10.18) is the linear system

$$\frac{dx_1}{dt} = a_{11}x_1 + a_{12}x_2, \qquad \frac{dx_2}{dt} = a_{21}x_1 + a_{22}x_2. \qquad (10.19)$$

We shall be interested in determining which properties of the critical point are preserved by this "linearization" process. For instance, if the critical point $(0, 0)$ is asymptotically stable for the linear system (10.19), is it also asymptotically stable for the nonlinear system (10.18)? If it is a node for the linear system, is it also a node for the nonlinear one?

We must first describe the linearization process more carefully. We assume that F_1 and F_2 are such that

$$\frac{dx_1}{dt} = a_{11}x_1 + a_{12}x_2 + f_1(x_1, x_2), \qquad \frac{dx_2}{dt} = a_{21}x_1 + a_{22}x_2 + f_2(x_1, x_2),$$
$$(10.20)$$

where f_1 and f_2 and their first partial derivatives are continuous in a region that contains the origin, and where

$$\lim \frac{f_1(x_1, x_2)}{\sqrt{x_1^2 + x_2^2}} = \lim \frac{f_2(x_1, x_2)}{\sqrt{x_1^2 + x_2^2}} = 0 \qquad (10.21)$$

as $(x_1, x_2) \to 0$. This last condition requires that f_1 and f_2 be "smaller" than the linear terms when (x_1, x_2) is near $(0, 0)$.

Example 1   One such system is

$$\frac{dx_1}{dt} = -x_2 + x_1^3, \qquad \frac{dx_2}{dt} = x_1 - x_2 + 2x_1x_2, \qquad (10.22)$$

with $f_1(x_1, x_2) = x_1^3$ and $f_2(x_1, x_2) = 2x_1x_2$. Using polar coordinates r and θ, where $x_1 = r \cos \theta$ and $x_2 = r \sin \theta$, we see that

$$\frac{f_1(x_1, x_2)}{r} = \frac{r^3 \cos^3 \theta}{r} = r^2 \cos \theta,$$

$$\frac{f_2(x_1, x_2)}{r} = \frac{2r^2 \cos \theta \sin \theta}{r} = 2r \cos \theta \sin \theta.$$

Thus

$$\frac{|f_1(x_1, x_2)|}{r} \le r^2, \qquad \frac{|f_2(x_1, x_2)|}{r} \le 2r,$$

so $f_1(x_1, x_2)/r \to 0$ and $f_2(x_1, x_2)/r \to 0$ as $r \to 0$. Hence the conditions

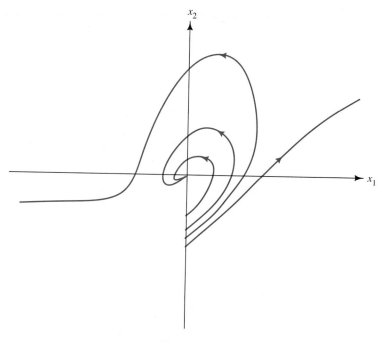

Figure 10.10

(10.21) are satisfied. The associated linear system

$$\frac{dx_1}{dt} = -x_2, \qquad \frac{dx_2}{dt} = x_1 - x_2 \qquad (10.23)$$

has characteristic values $\lambda_1 = \frac{1}{2}(-1 + i\sqrt{3})$, $\lambda_2 = \frac{1}{2}(-1 - i\sqrt{3})$. Since they are nonreal and have negative real parts, the origin is an asymptotically stable spiral point for the linear system. We are concerned with the nature of the critical point for the nonlinear system (10.22). Answers will be provided in the sequel. Some trajectories of the nonlinear system are shown in Fig. 10.10.

Example 2 We consider the system

$$\frac{dx_1}{dt} = -\sin x_1 + 3x_2, \qquad \frac{dx_2}{dt} = -2x_1 + 4x_2 + x_1 x_2. \qquad (10.24)$$

Here we expect that $\sin x_1 \approx x_1$. In fact,

$$\sin x_1 = x_1 - \frac{1}{3!} x_1^3 + \cdots = x_1 + x_1^3 g(x_1),$$

where the function g is continuous. If (x_1, x_2) is restricted to a closed bounded region, there is a constant M such that $|g(x_1)| \leq M$. The system (10.24) becomes

$$\frac{dx_1}{dt} = -x_1 + 3x_2 + f_1(x_1, x_2), \qquad \frac{dx_2}{dt} = -2x_1 + 4x_2 + f_2(x_1, x_2),$$

where

$$f_1(x_1, x_2) = -x_1^3 g(x_1), \qquad f_2(x_1, x_2) = x_1 x_2.$$

Then

$$\frac{|f_1(x_1, x_2)|}{r} = |r^2 \cos^3 \theta g(x_1)| \leq M r^2, \qquad \frac{|f_2(x_1, x_2)|}{r} = |r \cos \theta \sin \theta| \leq r.$$

The conditions (10.21) are again satisfied. The characteristic values of the linear system

$$\frac{dx_1}{dt} = -x_1 + 3x_2, \qquad \frac{dx_2}{dt} = -2x_1 + 4x_2 \tag{10.25}$$

are $\lambda_1 = 1$, $\lambda_2 = 2$. The origin is an unstable improper node for this system. The properties of the nonlinear system will be discussed shortly.

We now state without proof some fundamental theorems that relate the behavior of the critical point of a nonlinear system to that of the associated linear system. (See Coddington and Levinson (1955).)

Theorem 10.2   If the origin is an asymptotically stable critical point for the linear system (10.19), it is also an asymptotically stable critical point for the nonlinear system (10.20). If the origin is unstable for the linear system (10.19), it is also unstable for the nonlinear system (10.20).

On the basis of this theorem, we can conclude that the system (10.22) of Example 1 has an asymptotically stable critical point at the origin, and that the system (10.24) of Example 2 has an unstable critical point at the origin.

Notice that the theorem says nothing about the case when the linear system has a stable critical point that is not asymptotically stable. In this case, the critical point for the nonlinear system may be stable (but not asymptotically stable), asymptotically stable, or unstable, depending on the nonlinear terms.

More information about the nature of the critical point for the nonlinear system is provided by the next two theorems. The case of equal characteristic values is treated separately.

Theorem 10.3   Let f_1 and f_2 satisfy the conditions (10.21).

(a)   If $\lambda_1 = i\beta$ and $\lambda_2 = -i\beta$, $\beta \neq 0$ (in which case the linear system has a center), then the nonlinear system has either a center or a spiral point (which may be asymptotically stable or unstable).

(b)   If λ_1 and λ_2 are real, distinct, and of the same sign, the origin is an improper node for both the linear and nonlinear system.

(c)   If λ_1 and λ_2 are real and of opposite sign, the origin is a saddle point for both the linear and nonlinear system.

(d)   If $\lambda_1 = \alpha + i\beta$, $\lambda_2 = \alpha - i\beta$, $\alpha \neq 0$, $\beta \neq 0$, the origin is a spiral point for both the linear and nonlinear system.

From Theorems 10.2 and 10.3, we conclude that the nonlinear system (10.22) of Example 1 has an asymptotically stable spiral point at the origin. Notice in Figure 10.10, however, that only those solutions that start sufficiently close to the origin spiral toward it. We also conclude that the nonlinear system (10.24) of Example 2 has an unstable improper node at the origin.

The next theorem treats the case $\lambda_1 = \lambda_2$.

Theorem 10.4   Suppose that the characteristic values of the linear system (10.19) are real and equal. If the functions f_1 and f_2 satisfy the conditions (10.21), then the origin may be either a node or a spiral point for the nonlinear system (10.20). However, if there is a positive number ϵ such that

$$\lim \frac{f_1(x_1, x_2)}{r^{1+\epsilon}} = \lim \frac{f_2(x_1, x_2)}{r^{1+\epsilon}} = 0 \qquad (10.26)$$

as $(x_1, x_2) \to 0$, then the origin is a proper node for both the linear and nonlinear systems if the matrix A of the linear system is diagonal, but it is an improper node for both the linear and nonlinear systems if A is not diagonal.

Notice that the conditions (10.26) require that the nonlinear terms be even "smaller" near the origin than the conditions (10.21) do. Functions that satisfy these conditions will be presented in the exercises.

Example 3   We consider a simple pendulum that moves in a medium that resists the motion with a force proportional to the velocity. If θ is the angular displacement from the equilibrium position (see Figure 1.4, Section 1.12), the equation of motion has the form

$$\ddot{\theta} + b\dot{\theta} + a \sin \theta = 0$$

where $b \geq 0$ (if $b = 0$, damping is absent) and $a > 0$. Setting $x_1 = \theta$, $x_2 = \dot{\theta}$, we arrive at the system

$$\frac{dx_1}{dt} = x_2, \qquad \frac{dx_2}{dt} = -a \sin x_1 - bx_2. \qquad (10.27)$$

The critical points are at $(n\pi, 0)$, where n is an integer. We shall investigate the points $(0, 0)$ and $(\pi, 0)$. Because of the periodic nature of the sine function, the points $(n\pi, 0)$ with n even exhibit the same behavior as $(0, 0)$ and those with n odd the same behavior as $(\pi, 0)$. Because of the physical interpretation of $x_1 = \theta$, we should expect the first group of critical points to be stable and the second group to be unstable.

We first consider $(0, 0)$. Treating the sine function as in Example 2, we see that the system (10.27) may be written as

$$\frac{dx_1}{dt} = x_2, \qquad \frac{dx_2}{dt} = -ax_1 - bx_2 + f_2(x_1, x_2)$$

where $f_2(x_1, x_2) = a(x_1 - \sin x_1) = ax_1^3 h(x_1)$, where h is continuous. The same sort of analysis as in Example 2 shows that $f_2/r \to 0$ as $(x_1, x_2) \to (0, 0)$. In fact, $f_2/r^2 \to 0$, so the conditions (10.26), with $\epsilon = 1$, are satisfied. Thus the hypotheses of Theorems 10.2, 10.3, and 10.4 are all satisfied. The associated linear system

$$\frac{dx_1}{dt} = x_2, \qquad \frac{dx_2}{dt} = -ax_1 - bx_2 \qquad (10.28)$$

has the characteristic values

$$\lambda_1 = \frac{1}{2}(-b + \sqrt{b^2 - 4a}), \qquad \lambda_2 = \frac{1}{2}(-b - \sqrt{b^2 - 4a}). \qquad (10.29)$$

If damping is absent ($b = 0$), we have $\lambda_1 = ai$, $\lambda_2 = -ai$. The origin is therefore a center for the linear system (10.28), but Theorem 10.3 does not tell us the exact nature of the critical point for the nonlinear system (10.27). However, further analysis (Exercise 14) yields the information that the origin is also a center for the nonlinear system. If $b > 0$, there are three cases. If $b^2 < 4a$, the characteristic values (10.29) are nonreal and have negative real parts. (This is the underdamped case; the damping coefficient b is small.) According to Theorems 10.2 and 10.3, the origin is an asymptotically stable spiral point for the nonlinear system (10.27). If $b^2 > 4a$, the characteristic values are real, distinct, and negative. (This is the overdamped case.) Hence the origin is an asymptotically stable improper node for the nonlinear system. In the critically damped case, $b^2 = 4a$, the roots are real, equal, and negative. Since the matrix for the linear system (10.28) is not diagonal, the origin is an asymptotically stable improper node for the nonlinear system, according to Theorem 10.4.

Next we consider the critical point $(\pi, 0)$. The change of variable $u_1 = x_1 - \pi$, $u_2 = x_2$ will put the critical point at the origin of the $u_1 u_2$-plane. We find that

$$\frac{du_1}{dt} = u_2, \qquad \frac{du_2}{dt} = -a \sin(u_1 + \pi) - bu_2.$$

Since $\sin(u_1 + \pi) = -\sin u_1$, the system becomes

$$\frac{du_1}{dt} = u_2, \qquad \frac{du_2}{dt} = a \sin u_1 - bu_2.$$

This time, the associated linear system is

$$\frac{du_1}{dt} = u_2, \qquad \frac{du_2}{dt} = au_1 - bu_2,$$

with characteristic values

$$\lambda_1 = \frac{1}{2}(-b + \sqrt{b^2 + 4a}), \qquad \lambda_2 = \frac{1}{2}(-b - \sqrt{b^2 + 4a}).$$

Both values are real. λ_1 is positive, and λ_2 is negative. The origin is unstable and is a saddle point. The critical point $(\pi, 0)$ of the original system has these same properties.

Additional applications are discussed in the next section.

Exercises for Section 10.3

1. Determine whether each function satisfies the condition (10.21).
   (a) $x_1^2 - x_2^3$
   (b) $x_1 x_2$
   (c) $x_1^3 - 3x_2$
   (d) $(x_1^2 + x_2^2)^{3/2}$
   (e) $\sin(x_1 x_2)$
   (f) $\cos x_2 - 1$
   (g) $x_1/\ln(x_1^2 + x_2^2)$
   (h) $e^{x_2} - 1$

2. Determine whether the indicated function satisfies the condition (10.26).
   (a) The function of Exercise 1, Part (a).
   (b) The function of Exercise 1, Part (c).
   (c) The function of Exercise 1, Part (f).
   (d) The function of Exercise 1, Part (g).

In Exercises 3–10, use Theorems 10.2, 10.3, and 10.4 to investigate the nature of the critical point $(0, 0)$.

3. $x_1' = x_1 + 5x_2 - x_1 x_2$
   $x_2' = -x_1 - 3x_2 + x_1^3$

4. $x_1' = -x_2 + x_1^2 + 2x_2^2$
   $x_2' = -5x_1 + 4x_2$

5. $x_1' = x_1 + 2x_2 + (1 - \cos x_2)$
   $x_2' = -x_1 + 4x_2$

6. $x_1' = -x_2 + \sin x_1$
    $x_2' = 8x_1 - 5(e^{x_2} - 1)$

7. $x_1' = x_1 + x_2 \cos x_1$
    $x_2' = 2x_1$

8. $x_1' = -x_1 + 2x_2$
    $x_2' = 2x_1 - x_2 + x_2 e^{x_1}$

9. $x_1' = -2x_1 + x_2$
    $x_2' = -x_1 + (x_1^2 + x_2^2)^{3/4}$

10. $x_1' = -\sin 2x_1$
    $x_2' = -2x_2$

In Exercises 11 and 12, use Theorems 10.2, 10.3, and 10.4 to investigate the nature of the indicated critical point.

11. $x_1' = x_2(2 - x_1 + x_2)$,
    $x_2' = x_1(4 - x_1 - x_2)$;   (3, 1)

12. $x_1' = x_2(1 + x_1 - x_2)$,
    $x_2' = x_1(6x_1 - x_2 - 4)$;   (1, 2)

13. Find all critical points of the system. Investigate each as to type and stability.

$$x_1' = x_1(4 - x_1 - x_2),$$

$$x_2' = x_2(5 - x_1 - 2x_2)$$

14. Show that the critical point $(0, 0)$ is a center for the system

$$x_1' = x_2, \qquad x_2' = -a \sin x_1, \qquad a > 0,$$

which corresponds to the equation $\ddot{\theta} + a \sin \theta = 0$. (Show that the trajectories are

$$a \cos x_1 - \frac{1}{2} x_2^2 = c \; .$$

Show that if $0 < c < a$ these are closed curves. Show that if $|x_1| < \pi/3$, $|x_2| < \sqrt{a}$, there is a closed trajectory through the (x_1, x_2).)

15. Show that the origin is a center for the system that corresponds to the equation

$$\ddot{x} + x^3 = 0.$$

16. Given the system

$$x_1' = x_2 + f_1(x_1, x_2),$$

$$x_2' = -x_1 + f_2(x_1, x_2),$$

where f_1 and f_2 satisfy the conditions (10.21),
(a) show that $(0, 0)$ is a center for the associated linear system.
(b) let

$$f_1 = -x_1(x_1^2 + x_2^2),$$

$$f_2 = -x_2(x_1^2 + x_2^2).$$

Show that f_1 and f_2 satisfy the conditions (10.21). Show that the origin is an asymptotically stable critical point for the nonlinear system. (Use polar coordinates to show that $\dot{r} = -r^3$.)
(c) Let

$$f_1 = x_1(x_1^2 + x_2^2), \qquad f_2 = x_2(x_1^2 + x_2^2).$$

Show that f_1 and f_2 satisfy the conditions (10.21). Show that the origin is an unstable critical point for the nonlinear system.

10.4
COMPETITION
BETWEEN
TWO SPECIES

Let x_1 and x_2 be the populations of two competing species at time t. Then the growth rate $\dfrac{dx_1}{dt} \Big/ x_1$ of the first species should decrease as x_2 increases. If the growth of a species contributes to a decline in its own growth rate, then $\dfrac{dx_1}{dt} \Big/ x_1$ should also decrease as x_1 increases. Similar remarks apply to the growth rate of the second species. A simple model that incorporates these features has the form

$$\frac{dx_1}{dt} = x_1(c_1 - a_1 x_1 - b_1 x_2), \qquad \frac{dx_2}{dt} = x_2(c_2 - a_2 x_1 - b_2 x_2), \quad (10.30)$$

where all constants are positive. Only nonnegative values of x_1 and x_2 are meaningful.

We analyze an example before considering the general case.

Example 1   Suppose that

$$\frac{dx_1}{dt} = x_1(9 - 3x_1 - x_2), \qquad \frac{dx_2}{dt} = x_2(5 - x_1 - x_2). \quad (10.31)$$

The critical points are found to be $(0, 0)$, $(3, 0)$, $(0, 5)$ and $(2, 3)$. We examine them in order. The linear system associated with $(0, 0)$ is $x_1' = 9x_1$, $x_2' = 5x_2$. Since the characteristic values are $\lambda_1 = 9$ and $\lambda_2 = 5$, the origin is an unstable improper node for both the linear system and the nonlinear system (10.31). To investigate the critical point $(3, 0)$ we make the translation $u_1 = x_1 - 3$, $u_2 = x_2$. Then the system (10.31) becomes

$$\frac{du_1}{dt} = -9u_1 - 3u_2 - 3u_1^2 - u_1 u_2, \qquad \frac{du_2}{dt} = 2u_2 - u_1 u_2 - u_2^2. \quad (10.32)$$

The associated linear system, $u_1' = -9u_1 - 3u_2$, $u_2' = 2u_2$ has characteristic values $\lambda_1 = -9$, $\lambda_2 = 2$. Hence $(0, 0)$ is an (unstable) saddle point for the system (10.32), and $(3, 0)$ is a saddle point for the original system (10.31). The critical point $(0, 5)$ is found, in similar fashion, to be an (unstable) saddle point. To investigate the point $(2, 3)$, we make the change of variables $u_1 = x_1 - 2$, $u_2 = x_2 - 3$. The resulting system is

$$\frac{du_1}{dt} = -6u_1 - 2u_2 - 3u_1^2 - u_1 u_2, \qquad \frac{du_2}{dt} = -3u_1 - 3u_2 - u_1 u_2 - u_2^2.$$

The characteristic values of the associated linear system $u_1' = -6u_1 - 2u_2$, $u_2' = -3u_1 - 3u_2$ are $\lambda_1 = \frac{1}{2}(-9 + \sqrt{33})$, $\lambda_2 = \frac{1}{2}(-9 - \sqrt{33})$. Both are negative, so the critical point $(2, 3)$ of the nonlinear system is an asymptotically stable improper node. If, at some time, the populations of the two species are sufficiently near $x_1 = 2$, $x_2 = 3$, they will remain near and approach these values. Graphs of some trajectories are shown in Fig. 10.11.

Let us now consider the general case of the system (10.30). In order to simplify the algebra, we shall write it in the "normalized form"

$$\frac{dx_1}{dt} = hx_1(1 - ax_1 - bx_2), \qquad \frac{dx_2}{dt} = kx_2(1 - cx_1 - dx_2), \quad (10.33)$$

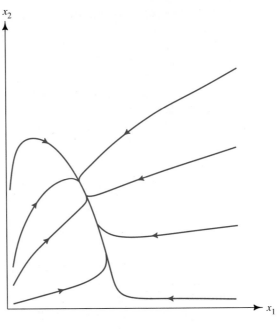

Figure 10.11

where $h = c_1$, $k = c_2$, $a = a_1/c_1$, and so on. All the constants a, b, c, d, h, k are positive. The system always possesses the three critical points $(0, 0)$, $(1/a, 0)$, and $(0, 1/d)$, with nonnegative coordinates. If the two lines

$$\ell_1 : ax_1 + bx_2 = 1; \qquad \ell_2 : cx_1 + dx_2 = 1$$

intersect (but do not coincide), there will be a fourth critical point which may be in the first, second, or fourth quadrant, or on the positive x_1- or x_2-axis. Since x_1 and x_2 represent populations, we are interested only in nonnegative values for the two variables.

If the lines intersect (but do not coincide), the determinant $ad - bc$ is not 0, and the coordinates (p_1, p_2) of the point of intersection are

$$p_1 = \frac{d - b}{ad - bc}, \qquad p_2 = \frac{a - c}{ad - bc}.$$

A necessary and sufficient condition that $p_1 > 0$ and $p_2 > 0$ is that $(a - c)(d - b) > 0$. That is, $a - c$ and $d - b$ must have the same sign. To see this, consider the two cases $ad - bc > 0$ and $ad - bc < 0$. If $ad - bc > 0$, then $p_1 > 0$ and $p_2 > 0$ if and only if $d - b > 0$ and $a - c > 0$. But if $d - b > 0$ and $a - c > 0$ then $d > b$ and $a > c$ so $ad - bc > 0$. The case $ad - bc < 0$ is similar.

Let us now assume that $(a - c)(d - b) > 0$. Then the system has four critical points $(0, 0)$, $(1/a, 0)$, $(0, 1/d)$, and (p_1, p_2), where $p_1 > 0$ and $p_2 > 0$. We investigate them in order.

Associated with the critical point $(0, 0)$ is the linear system $x_1' = hx_1$, $x_2' = kx_2$, with characteristic values $\lambda_1 = h$, $\lambda_2 = k$. Since both are positive, the origin is an unstable proper node for both the linear system and the nonlinear system (10.33).

To investigate the point $(1/a, 0)$ we set $u_1 = x_1 - 1/a$, $u_2 = x_2$. The resulting system is

$$\frac{du_1}{dt} = h\left(u_1 + \frac{1}{a}\right)(-au_1 - bu_2) = -hu_1 - \frac{b}{a}hu_2 + \cdots$$

$$\frac{du_2}{dt} = ku_2\left(1 - cu_1 - \frac{c}{a} - du_2\right) = k\left(1 - \frac{c}{a}\right)u_2 + \cdots$$

where the dots indicate quadratic terms. The associated linear system has the characteristic values $\lambda_1 = -h$, $\lambda_2 = k(1 - c/a)u_2$. The first, λ_1, is always negative. The second, λ_2, is negative if $c > a$ and positive if $c < a$. Thus the critical point $(1/a, 0)$ is an asymptotically stable improper node if $c > a$ and an (unstable) saddle point if $c < a$.

A similar analysis of the critical point $(0, 1/d)$ shows that it is an asymptotically stable improper node if $b > d$ and an (unstable) saddle point if $b < d$.

To investigate the critical point (p_1, p_2), we set $u_1 = x_1 - p_1$, $u_2 = x_2 - p_2$. The resulting system has the form

$$\frac{du_1}{dt} = -ahp_1u_1 - bhp_1u_2 + \cdots, \qquad \frac{du_2}{dt} = -ckp_2u_1 - dkp_2u_2 + \cdots,$$

where the dots represent quadratic terms. The characteristic polynomial of the associated linear system is

$$\lambda^2 + (ahp_1 + dkp_2)\lambda + hkp_1p_2(ad - bc).$$

The characteristic values are

$$\frac{1}{2}\left[-(ahp_1 + dkp_2) \pm \sqrt{D}\right],$$

where

$$D = (ahp_1 + dkp_2)^2 - 4hkp_1p_2(ad - bc) = (ahp_1 - dkp_2)^2 + 4hkp_1p_2bc.$$

The second form for D shows that $D > 0$, which implies that the characteristic values are real. If $ad - bc > 0$, the first form for D shows that $D < (ahp_1 + dkp_2)^2$, which implies that both characteristic values are negative and the critical point is an asymptotically stable improper node. However, if $ad - bc < 0$, $D > (ahp_1 + dkp_2)^2$, the characteristic values have opposite signs, and the critical point is an (unstable) saddle point.

Investigation of the case where $(a - c)(d - b) < 0$ is similar and is left as an exercise (Exercise 16). However, if $(a - c)(b - d) = 0$, one of the critical points has a zero characteristic value. The methods of Section 10.3 do not apply in such a situation. We examine here the case where $a = c$ and $b < d$, leaving the other possibilities to the exercises. (See Exercise 17.)

Suppose that $a = c$, $b < d$ in the system (10.33). Then the lines

$$\ell_1 : 1 - ax_1 - bx_2 = 0, \qquad \ell_2 : 1 - cx_1 - dx_2 = 0$$

separate the first quadrant of the phase plane into three regions, as shown in Fig. 10.12. The arrows in the figure indicate the general directions of the trajectories in the various regions. For example, in Region I, x_1' and x_2' are negative, so the trajectories are directed down and to the left. We must keep in mind that a trajectory with a point in the first quadrant cannot leave it (Exercise 13). A solution $x_1 = \phi_1(t)$, $x_2 = \phi_2(t)$, which has a point $(\phi_1(t_0), \phi_2(t_0))$ in Region I and does not approach $(1/a, 0)$, must enter Region II. (Suppose not. Then ϕ_1 and ϕ_2 are bounded and decreasing, and hence must have limits. According to Exercise 14, Section 10.1, the solution

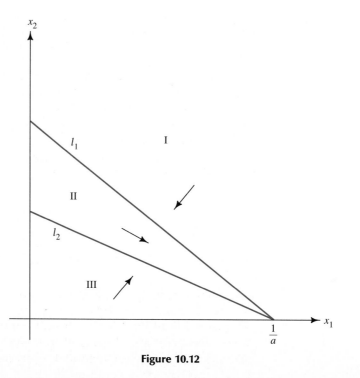

Figure 10.12

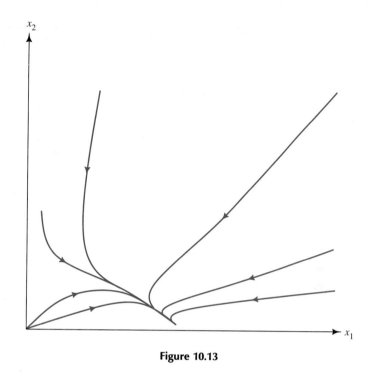

Figure 10.13

must approach a critical point. But there is only the one critical point available.) Likewise, a solution with a point in Region III must énter Region II. A solution with a point in Region II cannot enter Region I or III, and cannot cross the x_1-axis, but its components are monotonic and bounded. Therefore it approaches the critical point $(1/a, 0)$. Thus every trajectory in the first quadrant approaches this critical point which is asymptotically stable *if the domain is restricted to the region* $x_1 \geq 0$, $x_2 \geq 0$. Figure 10.13 illustrates the case where $a = b = c = h = k = 1$, $d = 2$.

We summarize the results, assuming $ad - bc \neq 0$. The case $ad - bc = 0$ is treated in Exercises 14 and 15.

Theorem 10.5   The system

$$\frac{dx_1}{dt} = hx_1(1 - ax_1 - bx_2), \qquad \frac{dx_2}{dt} = kx_2(1 - cx_1 - dx_2)$$

has a critical point (p_1, p_2) in the first quadrant if and only if $(a - c)(d - b) > 0$. If this condition is satisfied, there are two cases.

Case 1. If $ad - bc > 0$, the critical point (p_1, p_2) is an asymptotically stable improper node. The critical point $(0, 0)$ is an unstable improper node. The critical points $(1/a, 0)$ and $(0, 1/d)$ are (unstable) saddle points.

Case 2. If $ad - bc < 0$, the critical point (p_1, p_2) is an (unstable) saddle point. The critical point $(0, 0)$ is an unstable improper node. The critical points $(1/a, 0)$ and $(0, 1/d)$ are asymptotically stable improper nodes.

If $(a - c)(d - b) < 0$, the only critical points with nonnegative coordinates are $(0, 0)$, $(1/a, 0)$, and $(0, 1/d)$. There are again two cases.

Case 3. If $c > a$ and $b < d$, the point $(1/a, 0)$ is an asymptotically stable improper node and the points $(0, 1/d)$ and $(0, 0)$ are (unstable) saddle points.

Case 4. If $c < a$ and $b > d$, the points $(1/a, 0)$ and $(0, 0)$ are (unstable) saddle points and the point $(0, 1/d)$ is an asymptotically stable improper node.

If $(a - c)(d - b) = 0$, there are again two cases.

Case 5. If $a = c$, then every solution that starts in the first quadrant approaches either $(1/a, 0)$ or $(0, 1/d)$, accordingly as $b < d$ or $b > d$.

Case 6. If $b = d$, then every solution that starts in the first quadrant approaches either $(1/a, 0)$ or $(0, 1/d)$, accordingly as $a < c$ or $a > c$.

In case 1, the two species can co-exist at nonzero population levels. In the other cases, one species dies out and the other approaches a limiting value.

Example 2   As an application of the theorem, we reexamine the system

$$\frac{dx_1}{dt} = x_1(9 - 3x_1 - x_2), \qquad \frac{dx_2}{dt} = x_2(5 - x_1 - x_2)$$

of Example 1. We first rewrite it in the normalized form

$$\frac{dx_1}{dt} = 9x_1\left(1 - \frac{1}{3}x_1 - \frac{1}{9}x_2\right), \qquad \frac{dx_2}{dt} = 5x_2\left(1 - \frac{1}{5}x_1 - \frac{1}{5}x_2\right).$$

Then $a = \frac{1}{3}$, $b = \frac{1}{9}$, $c = \frac{1}{5}$, and $d = \frac{1}{5}$, so $a > c$ and $d > b$. Hence there is a critical point in the first quadrant. (Earlier we found this point to be $(2, 3)$.) Since $ad - bc = \frac{2}{45} > 0$, this critical point is an asymptotically stable proper node.

Example 3   As another example, we consider the system

$$\frac{dx_1}{dt} = x_1(3 - x_1 - x_2), \qquad \frac{dx_2}{dt} = x_2(7 - 3x_1 - x_2).$$

The normalized form is

$$\frac{dx_1}{dt} = 3x_1\left(1 - \frac{1}{3}x_1 - \frac{1}{3}x_2\right), \qquad \frac{dx_2}{dt} = 7x_2\left(1 - \frac{3}{7}x_1 - \frac{1}{7}x_2\right),$$

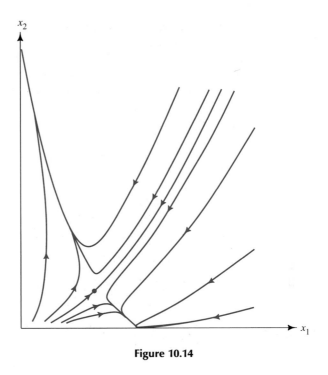

Figure 10.14

with $a = \frac{1}{3}$, $b = \frac{1}{3}$, $c = \frac{3}{7}$, $d = \frac{1}{7}$. Since $a < c$ and $d < b$, there is a critical point in the first quadrant. (This point is (2, 1).) Since $ad - bc = -\frac{8}{21} < 0$, this critical point is an (unstable) saddle point. Some trajectories are displayed in Fig. 10.14.

A similar sort of problem involves two species, one of which is a predator and the other its prey. This problem is referred to in the literature as the predator–prey problem, or the problem of the foxes and the hares, or the problem of the sharks and the fishes.

Let x_1 and x_2 be the populations of the predator and prey, respectively, at time t. Then the growth rate of the predator should be negative in the absence of prey and should increase with the prey population. Our mathematical model in this case is

$$\frac{dx_1}{dt} = x(-c_1 - a_1 x_1 + b_1 x_2), \qquad \frac{dx_2}{dt} = x_2(c_2 - a_2 x_1 - b_2 x_2), \quad (10.34)$$

where all constants are positive. We write these equations in the normalized form

$$\frac{dx_1}{dt} = hx_1(-1 - ax_1 + bx_2), \qquad \frac{dx_2}{dt} = kx_2(1 - cx_1 - dx_2), \quad (10.35)$$

where $h = c_1$, $k = c_2$, $a = a_1/c_1$, and so on. The lines

$$\ell_1 : ax_1 - bx_2 = -1; \qquad \ell_2 : cx_1 + dx_2 = 1$$

intersect at the critical point (p_1, p_2), where

$$p_1 = \frac{b-d}{ad+bc}, \qquad p_2 = \frac{a+c}{ad+bc}.$$

This point is in the first quadrant if and only if $b > d$. Other critical points are $(0, 0)$ and $(0, 1/d)$. We proceed to investigate the nature of these points.

The linear system associated with the critical point $(0, 0)$ is $x_1' = -hx_1$, $x_2' = kx_2$. It is therefore an (unstable) saddle point.

To investigate the point $(0, 1/d)$, we make the change of variables $u_1 = x_1$, $u_2 = x_2 - 1/d$. The resulting system is

$$\frac{du_1}{dt} = h\left(\frac{b}{d} - 1\right)u_1 + \cdots, \qquad \frac{du_2}{dt} = -\frac{kc}{d}u_1 - ku_2 + \cdots,$$

where the dots indicate quadratic terms. The characteristic values of the linear system are

$$\lambda_1 = h\left(\frac{b}{d} - 1\right), \qquad \lambda_2 = -k.$$

If $b < d$, both values are negative and the critical point is an asymptotically stable improper node. If $b > d$, one value is positive and the other negative, so the critical point is an (unstable) saddle point. (We omit the case $b = d$; see, however, Exercise 18.) This critical point corresponds to a state where the predator has died out and the prey population has stabilized at the value $x_2 = 1/d$.

For our examination of the critical point (p_1, p_2), we set $u_1 = x_1 - p_1$, $u_2 - p_2$. The system that results is

$$\frac{du_1}{dt} = -ahp_1u_1 + bhp_1u_2 + \cdots, \qquad \frac{du_2}{dt} = -ckp_2u_1 - dkp_2u_2 + \cdots,$$

where the dots represent nonlinear terms. The characteristic polynomial of the linearized system is

$$\lambda^2 + (ahp_1 + dkp_2)\lambda + hkp_1p_2(ad + bc), \tag{10.36}$$

with zeros

$$\lambda = \frac{1}{2}[-(ahp_1 + dkp_2) \pm \sqrt{D}], \tag{10.37}$$

where

$$D = (ahp_1 + dkp_2)^2 - 4hkp_1p_2(ad + bc) = (ahp_1 - dkp_2)^2 - 4hkp_1p_2bc.$$
(10.38)

From the last expression for D we see that $D < (ahp_1 + dkp_2)^2$, so both characteristic values have negative real parts. The critical point (p_1, p_2) is therefore asymptotically stable. It may be either a spiral point or an improper node, depending on the constants of the system. For instance, if $h = dp_2$ and $k = ap_1$, then $D = -4hkp_1p_2bc < 0$, so the characteristic values are not real and the critical point is a spiral point. However, if $h = 1$, say, and k is sufficiently small, we see from the first expression for D that $D > 0$. In this case both characteristic values are real and the critical point is a node.

We summarize our results as follows:

Theorem 10.6   The predator–prey system

$$\frac{dx_1}{dt} = hx(-1 - ax_1 + bx_2), \qquad \frac{dx_2}{dt} = kx_2(1 - cx_1 - dx_2)$$

has the critical points $(0, 0)$ and $(0, 1/d)$. It also has a critical point (p_1, p_2) in the first quadrant if and only if $b > d$. There are two main cases, depending on whether $b > d$ or $b < d$.

Case 1. If $b > d$, the critical point (p_1, p_2) is asymptotically stable. It may be a spiral point or an improper node, according as the quantity D in formula (10.38) is negative or nonnegative. The critical points $(0, 0)$ and $(0, 1/d)$ are (unstable) saddle points.

Case 2. If $b < d$, the critical point $(0, 1/d)$ is an asymptotically stable improper node, and the point $(0, 0)$ is an (unstable) saddle point.

Example 4   Let us examine the system

$$\frac{dx_1}{dt} = x_1(-1 - 2x_1 + x_2), \qquad \frac{dx_2}{dt} = x_2(7 - x_1 - x_2),$$

which may be written as

$$\frac{dx_1}{dt} = x_1(-1 - 2x_1 + x_2), \qquad \frac{dx_2}{dt} = 7\left(1 - \frac{1}{7}x_1 - \frac{1}{7}x_2\right).$$

Comparing with the standard form (10.35), we see that $h = 7$, $k = 1$, $a = 2$, $b = 1$, $c = \frac{1}{7}$, $d = \frac{1}{7}$. Since $b > d$, there exists a critical point (p_1, p_2) in the first quadrant. This point is found to be $(2, 5)$. The other critical points are

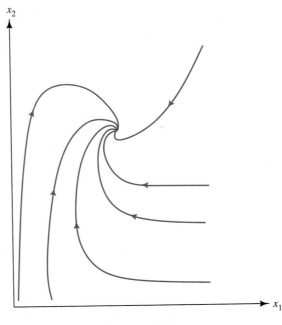

Figure 10.15

$(0, 0)$ and $(0, 7)$. From formula (10.38) we find that

$$D = (ahp_1 - dkp_2)^2 - 4hkp_1p_2bc = (4 - 5)^2 - 4 \cdot 7 \cdot 2 \cdot 5 \cdot \frac{1}{7} = -39 < 0.$$

Hence the point $(2, 5)$ is an asymptotically stable spiral point. The points $(0, 0)$ and $(0, 7)$ are (unstable) saddle points. Some trajectories are shown in Fig. 10.15.

Exercises for Section 10.4

In Exercises 1–12, find the critical points with nonnegative coordinates, and investigate them as to type and stability. Use Theorems 10.5 and 10.6 where applicable.

1. $x_1' = x_1(7 - 2x_1 - x_2),$
    $x_2' = x_2(6 - x_1 - 3x_2)$

2. $x_1' = x_1(5 - 3x_1 - x_2),$
    $x_2' = x_2(5 - x_1 - 2x_2)$

3. $x_1' = x_1(5 - x_1 - x_2),$
    $x_2' = x_2(8 - x_1 - 2x_2)$

4. $x_1' = x_1(4 - x_1 - x_2),$
    $x_2' = x_2(5 - 2x_1 - x_2)$

5. $x_1' = x_1(4 - x_1 - 2x_2),$
    $x_2' = x_2(1 - x_1 - x_2)$

6. $x_1' = x_1(1 - 2x_1 - x_2),$
   $x_2' = x_2(3 - 3x_1 - x_2)$

7. $x_1' = x_1(6 - 3x_1 - 2x_2),$
   $x_2' = x_2(4 - 2x_1 - x_2)$

8. $x_1' = x_1(3 - 3x_1 - x_2),$
   $x_2' = x_2(3 - 6x_1 - x_2)$

9. $x_1' = x_1(-7 - x_1 + 2x_2),$
   $x_2' = x_2(7 - 3x_1 - x_2)$

10. $x_1' = x_1(-7 - x_1 + 3x_2),$
    $x_2' = x_2(8 - 2x_1 - 2x_2)$

11. $x_1' = x_1(-4 - x_1 + 2x_2),$
    $x_2' = x_2(2 - x_1 - 2x_2)$

12. $x_1' = x_1(-5 - 2x_1 + x_2),$
    $x_2' = x_2(2 - x_1 - x_2)$

13. (a)   For the system (10.30), show that every point on the x_1-axis (or the x_2-axis) is contained in a trajectory that is wholly contained in the axis. (Find the solutions for which $x_1 = 0$ or $x_2 = 0$.)
    (b)   Show that a trajectory of the system (10.30) with a point in the first quadrant is wholly contained in the first quadrant. (Make use of Exercise 9, Section 10.1.)

14. If $ad - bc = 0$, then the system (10.33) may be written as

$$x_1' = hx_1(1 - ax_1 - bx_2),$$
$$x_2' = kx_2(1 - a\delta x_1 - b\delta x_2),$$

where all constants are positive. Find the critical points and investigate them as to type and stability. (Consider two cases, $\delta > 1$ and $\delta < 1$. The case $\delta = 1$ is examined in the next exercise.)

15. Consider the system

$$x_1' = hx_1(1 - ax_1 - bx_2),$$
$$x_2' = kx_2(1 - ax_1 - bx_2),$$

where all constants are positive.
(a)   Show that every point on the line $ax_1 + bx_2 = 1$ is a critical point.
(b)   Show that the equations of the trajectories are $y = cx^{k/h}$.
(c)   If $(x_1(0), x_2(0))$ is not on the line of Part (a) and is not $(0, 0)$, show that $(x_1(t), x_2(t))$ approaches a point on the line as $t \to \infty$ (Make use of Exercise 14, Section 10.1.)

16. Prove the statements in Theorem 10.5 for the case $(a - c)(d - b) < 0$.

17. (a)   Suppose that $a = c$ in the system (10.33). Show that every solution that starts in the first quadrant approaches either $(1/a, 0)$ or $(0, 1/d)$, according as $b < d$ or $b > d$. Suggestion: consider the signs of the derivatives in the regions bounded by the coordinate axes and the lines ℓ_1 and ℓ_2.
    (b)   Suppose that $b = d$ in the system (10.33). Show that every solution that starts in the first quadrant approaches either $(1/a, 0)$ or $(0, 1/d)$, according as $a < c$ or $a > c$.

18. Suppose that $b = d$ in the predator–prey problem (10.35). Show that the only critical points with nonnegative coordinates are $(0, 0)$ and $(0, 1/d)$. Show that the linearized system for $(0, 1/d)$ has a zero characteristic value and hence that the theorems of Section 10.3 cannot be applied to determine stability. Some trajectories of the particular system $x_1' = x_1(-1 - x_1 + x_2), x_2' = x_2(1 - x_1 - x_2)$ are shown in Fig. 10.16. Show that every solution of this system for which $(x_1(0), x_2(0))$ is in the first quadrant approaches the critical point $(0, 1)$. (Examine the signs of the derivatives x_1' and x_2' in the regions formed by the lines $-1 - x_1 + x_2 = 0$ and $1 - x_1 - x_2 = 0$. Make use of Exercise 14, Section 10.1.)

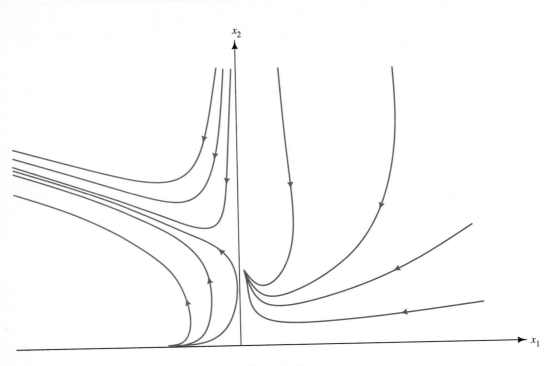

Figure 10.16

19. Consider the predator–prey problem

$$x_1' = x_1(-a + bx_2), \qquad x_2' = x_2(c - dx_1),$$

where the constants are all positive. This model assumes an unlimited food supply for the prey, and that the predator does not interfere with its own growth. Show that there is a critical point $(c/d, a/b)$ in the first quadrant, and that this point is a center for the associated linear system. It can be shown (see Plaat (1971)) that the point is also a center for the nonlinear system.

20. Consider the predator–prey problem

$$x_1' = hx_1(-1 + bx_2),$$
$$x_2' = kx_2(1 - cx_1 - dx_2),$$

where the constants are all positive. Find the critical points and investigate them as to type and stability.

10.5
PERIODIC
SOLUTIONS

A solution of the system

$$\frac{dx_1}{dt} = F_1(x_1, x_2), \qquad \frac{dx_2}{dt} = F_2(x_1, x_2) \qquad (10.39)$$

is said to be periodic with (positive) period T if the solution exists for all t and if $x_1(t + T) = x_1(t)$, $x_2(t + T) = x_2(t)$ for all t. The trajectory of a periodic solution is a closed curve in the phase plane. A constant solution (which corresponds to a critical point) is periodic, and every positive number T is a period for such a solution. However, in what follows, we shall be interested in nonconstant periodic solutions.

In the case of a linear system,

$$\frac{dx_1}{dt} = a_{11}x_1 + a_{12}x_2, \qquad \frac{dx_2}{dt} = a_{21}x_1 + a_{22}x_2,$$

nonconstant periodic solutions exist only in the case when the origin is a center for the system. In this case, all solutions are periodic and their trajectories are ellipses. Thus, for a linear system, either all or none of the (nonconstant) solutions are periodic. For nonlinear systems, the situation is different, as the following example shows.

Example 1   The nonlinear system

$$\frac{dx_1}{dt} = -x_2 + x_1(1 - x_1^2 - x_2^2), \qquad \frac{dx_2}{dt} = x_1 + x_2(1 - x_1^2 - x_2^2)$$

becomes simpler in polar coordinates. We find that the polar equations are

$$\frac{dr}{dt} = r(1 - r^2), \qquad \frac{d\theta}{dt} = 1,$$

with solutions

$$r = \frac{1}{\sqrt{1 + c_1 e^{-2t}}}, \qquad 0 = t + c_2.$$

If $c_1 = 0$, we have the solutions $r = 1$, $\theta = t + c_2$, whose trajectory is the unit circle, directed counterclockwise. When $c_1 < 0$, we see that $r > 1$ and that $r \to 1$ as $t \to \infty$. When $c_1 > 0$, we have $r < 1$ and again $r \to 1$ as $t \to \infty$. Thus the other trajectories spiral toward the circle $r = 1$. The non-linear system possesses only one closed trajectory. The situation is illustrated in Fig. 10.17.

In this example, we were able to show the existence of a periodic solution by actually finding formulas for the solutions. In most cases, we cannot do this. We state without proof two theorems that give sufficient conditions for the existence of a periodic solution. The first is the Poincare–Bendixon theorem.

Theorem 10.7   Let F_1 and F_2 and their first partial derivatives be continuous in a region G. Let D be a closed bounded region contained in G, and suppose that D contains no critical points. If $x_1 = \phi_1(t)$, $x_2 = \phi_2(t)$ is a solution that exists and stays in D for $t > t_0$ for some t_0, then either the solution is periodic or it spirals toward the closed trajectory of a periodic solution. In either case, the system (10.39) has a periodic solution.

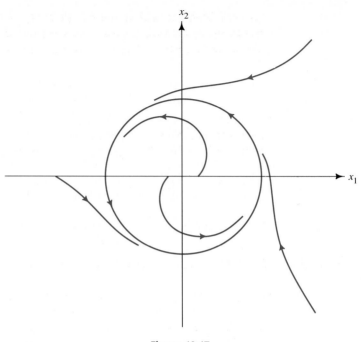

Figure 10.17

The difficulty in applying this theorem is in showing that a solution stays inside the region D. One way to do this is to show that on the boundary of D the vector $\mathbf{F} = F_1\mathbf{i} + F_2\mathbf{j}$ points into D. Then a solution that once enters D can never leave it.

Example 2   We consider the system

$$\frac{dx_1}{dt} = 2x_1 + x_2 - x_1(x_1^2 + x_2^2)^2, \qquad \frac{dx_2}{dt} = -x_1 + 2x_2 - x_2(x_1^2 + x_2^2)^2.$$

This system possesses the single critical point $(0, 0)$. The component of $\mathbf{F}$ in the direction away from the origin is

$$\mathbf{F} \cdot \frac{x_1\mathbf{i} + x_2\mathbf{j}}{r} = \frac{1}{r}\left[2x_1^2 + x_1 x_2 - x_1^2(x_1^2 + x_2^2)^2 - x_1 x_2 \right.$$

$$\left. + 2x_2^2 - x_2^2(x^2 + x_2^2)^2 \right]$$

$$= \frac{1}{r}(2r^2 - r^5) = 2r - r^4.$$

On the circle $r = 1$, this component is positive, while on the circle $r = 2$ it is negative. If we take for D the annular region $1 \le r \le 2$, we see that the vector $\mathbf{F}$ points into D on the boundary. Also, D contains no critical points. Hence the system must possess at least one periodic solution.

Our next theorem deals with certain second order equations of the form

$$\ddot{x} + f(x)\dot{x} + g(x) = 0. \tag{10.40}$$

Such an equation is known as a *Lienard equation*. Lienard, and later Levinson and Smith, established the existence of periodic solutions under certain conditions. In order to state our theorem, we define the functions

$$F(x) = \int_0^x f(s)\,ds, \qquad G(x) = \int_0^x g(s)\,ds,$$

where f and g are the functions in equation (10.40).

Theorem 10.8   Let f and g' be continuous everywhere, and satisfy the following conditions:

(a) f is even and g is odd, with $g(x) > 0$ for $x > 0$.

(b) There exists a positive number a such that $F(x) < 0$ for $0 < x < a$, $F(x) > 0$ for $x > a$, and F is monotonically increasing on (a, ∞).

(c) $F(x) \to \infty$ and $G(x) \to \infty$ as $x \to \infty$.

Then the equation (10.40) possesses a periodic solution whose closed trajectory encloses the origin of the phase plane. There is no other closed trajectory. Every other trajectory (except the point $(0, 0)$) spirals toward the closed trajectory as $t \to \infty$.

Example 3   As an application, we consider the *van der Pol equation*

$$\ddot{x} + \mu(x^2 - 1)\dot{x} + x = 0,$$

where μ is a positive constant. (This equation arises in the study of vacuum tube circuits.) Here $f(x) = \mu(x^2 - 1)$ and $g(x) = x$. Clearly f is even and g is odd, and $g(x) > 0$ for $x > 0$. Since

$$F(x) = \mu\left(\frac{1}{3}x^3 - x\right) = \frac{1}{3}\mu x(x^2 - 3), \qquad G(x) = \frac{1}{2}x^2$$

we see that $F(x) \to \infty$ and $G(x) \to \infty$ as $x \to \infty$. Also, $F(x) < 0$ for $0 < x < \sqrt{3}$ and $F(x) > 0$ for $x > \sqrt{3}$. Since $F'(x) = \mu(x^2 - 1)$, $F'(x) > 0$ for $x > 1$ and hence for $x > \sqrt{3}$. Thus the hypotheses of Theorem 10.8 are satisfied, and we conclude that the equation has a periodic solution.

—————————————— **Exercises for Section 10.5** ——————————————

1. Find formulas for the solutions of the given system, using polar coordinates, and make a diagram showing the behavior of the trajectories.

$$\frac{dx_1}{dt} = x_2 + x_1(x_1^2 + x_2^2 - 4),$$

$$\frac{dx_2}{dt} = -x_1 + x_2(x_1^2 + x_2^2 - 4)$$

2. Do as in Exercise 1 for the system

$$\frac{dx_1}{dt} = x_2 + x_1(r - 1)(r - 2),$$

$$\frac{dx_2}{dt} = -x_1 + x_2(r - 1)(r - 2),$$

where $r = \sqrt{x_1^2 + x_2^2}$.

3. Use Theorem 10.7 to establish the existence of a periodic solution for the system

$$\frac{dx_1}{dt} = 3x_1 + x_2 - x_1 e^{r^2},$$

$$\frac{dx_2}{dt} = -x_1 + 3x_2 - x_2 e^{r^2}.$$

4. Given the system

$$\frac{dx_1}{dt} = x_2 + x_1(1 - x_1^2 - 4x_2^2), \qquad \frac{dx_2}{dt} = -x_1,$$

show that
   (a) the only critical point is $(0, 0)$.
   (b) $\dfrac{dr}{dt} = x_1^2(1 - r^2 - 3\sin^2\theta)$.
   (c) if $r \le \frac{1}{10}$, then $\dot{r} \ge x_1^2(1 - \frac{1}{100} - \frac{3}{100}) \ge 0$.
   (d) if $r \ge 10$, then $\dot{r} \le x_1^2(1 - 100) \le 0$.
   (e) the system has a periodic solution.

5. Proceed as in Exercise 4 to show that the given system has a periodic solution.

$$\frac{dx_1}{dt} = -x_2, \qquad \frac{dx_2}{dt} = x_1 + x_2(5 - 2x_1^2 - x_2^2)$$

In Exercises 6–8, use Theorem 10.8 to show that the equation has a periodic solution.

6. $\ddot{x} + (x^4 - x^2)\dot{x} + x = 0$

7. $\ddot{x} + c(x^{2m} - k^2)\dot{x} + x^{2n-1} = 0$, where c and k are positive constants, and m and n are integers.

8. $\ddot{x} + (\cosh x - 2)\dot{x} + \sinh x = 0$

Additional Exercises for Chapter 10

In Exercises 1–8, determine the type (spiral point, center, etc.) of the critical point $(0, 0)$, and investigate its stability. Determine a matrix K such that the change of variable $\mathbf{x} = K\mathbf{y}$ puts the system in canonical form, and describe the canonical form. Sketch some trajectories, indicating the equations of the lines that correspond to the coordinate axes for the canonical form.

1. $x_1' = -3x_1 + 4x_2$
   $x_2' = -8x_1 + 5x_2$

2. $x_1' = x_1 + 2x_2$
   $x_2' = -4x_1 - 3x_2$

3. $x_1' = -2x_1 + 2x_2$
   $x_2' = -4x_1 + 2x_2$

4. $x_1' = 3x_1 + x_2$
   $x_2' = -x_1 + x_2$

5. $x_1' = -2x_1 - x_2$
   $x_2' = -x_1 - 2x_2$

6. $x_1' = x_1 + 2x_2$
   $x_2' = 2x_1 + x_2$

7. $x_1' = -x_1 + x_2$
   $x_2' = -6x_1 + 4x_2$

8. $x_1' = 3x_1$
    $x_2' = 3x_2$

In Exercises 9–14, determine the type of critical point at $(0, 0)$, and investigate its stability.

9. $x_1' = -2x_1 - x_2 + (x_1^2 + x_2^2)^{3/2}$
    $x_2' = -x_1 - 2x_2$

10. $x_1' = \sin(x_1 + 2x_2)$
    $x_2' = -4x_1 - 3x_2$

11. $x_1' = -x_1 + x_2$
    $x_2' = -6x_1 + 4x_2 - 3x_1x_2^2$

12. $x_1' = 3x_1 - x_2^2$
    $x_2' = 3x_2 + x_1^3$

13. $x_1' = -x_1 + x_2 + 1 - \cos(x_1 + x_2)$
    $x_2' = -6x_1 + 4x_2$

14. $x_1' = 3x_1 + x_2$
    $x_2' = -x_1 + x_2 - (x_1^2 + x_2^2)^{5/8}$

15. Show that the critical point $(0, 0)$ of the linear system

    $$x_1' = a_{11}x_1 + a_{12}x_2, \quad x_2' = a_{21}x_1 + a_{22}x_2$$

    is asymptotically stable if and only if $a_{11} + a_{22} < 0$ and $a_{11}a_{22} > a_{12}a_{21}$.

16. Use the result of Exercise 15 to determine whether the origin is asymptotically stable for the system of
    (a) Exercise 1   (b) Exercise 2
    (c) Exercise 5   (d) Exercise 6.

17. Consider the system $\mathbf{x}' = A\mathbf{x}$, where A is a *singular* 2×2 real matrix. Make diagrams showing the trajectories, for the following three cases:
    (a) $\lambda_1 = 0, \quad \lambda_2 \neq 0$
    (b) $\lambda_1 = \lambda_2 = 0, \quad A \neq 0$
    (c) $A = 0$.

18. Find all the critical points of the given system, and investigate them as to type and stability.

$x_1' = x_1(7 - x_1 - x_2),$
$x_2' = x_2(10 - 2x_1 - x_2)$

19. Do as in Exercise 18 for the system

$x_1' = x_1(2 - x_1 - x_2),$
$x_2' = x_2(5 - 2x_1 - 3x_2).$

20. Consider the predator–prey model

$x_1' = x_1(-ax_1 + bx_2),$
$x_2' = x_2(e - cx_1 - dx_2),$

where the constants are all positive. Find the critical points and investigate them as to type and stability.

21. It can be shown that the system $x_1' = F_1(x_1, x_2)$, $x_2' = F_2(x_1, x_2)$ has no closed trajectory contained entirely in a simply connected region (see Section 1.4) D if

$$\frac{\partial F_1}{\partial x_1} + \frac{\partial F_2}{\partial x_2}$$

is everywhere positive or everywhere negative in D. Use this criterion to show that the given system or equation has no closed trajectory and hence no periodic solution.
    (a) $x_1' = x_1 + x_2^3$
        $x_2' = 3x_2 + x_1^5$
    (b) $x_1' = -2x_1 - x_1^4$
        $x_2' = -3x_2 - x_1^2$
    (c) $\ddot{x} + (x^2 + 1)\dot{x} + x = 0$
    (d) $\ddot{x} + x^2\dot{x} + \dot{x}^3 = 0$

22. (a) Write down a system of differential equations for three competing species, each of which interferes with its own growth as well as that of the others.
    (b) Given three species A, B, and C, suppose that A preys on B and that B preys on C. Write down an appropriate system of differential equations for the populations of the species, assuming that each species interferes with its own growth.

References

General Anton, H. *Elementary Linear Algebra.* 5th ed. New York: Wiley, 1987.

Churchill, R. V. *Operational Mathematics.* 3rd ed. New York: McGraw-Hill, 1971.

Coddington, E. A. *Introduction to Ordinary Differential Equations.* Englewood Cliffs, New Jersey: Prentice-Hall, 1961.

Coddington, E. A., and Levinson, N. *Theory of Ordinary Differential Equations.* New York: McGraw-Hill, 1955.

Conte, S. D., and DeBoor, C. W. *Elementary Numerical Analysis.* 2nd ed. New York: McGraw-Hill, 1972.

Cullen, C. G. *Linear Algebra and Differential Equations.* 2nd ed. Boston: Prindle, Weber and Schmidt, 1991.

Fraleigh, J. B., and Beauregard, R. A. *Linear Algebra.* 2nd ed. Reading, Massachusetts: Addison-Wesley, 1990.

Gaughan, E. D. *Introduction to Analysis.* 3rd ed. Pacific Grove, California: Brooks-Cole, 1987.

Hirsch, M. W., and Smale, S. *Differential Equations, Dynamical Systems, and Linear Algebra.* New York: Academic Press, 1974.

Plaat, O. *Ordinary Differential Equations.* San Francisco: Holden-Day, 1971.

Smiley, M. F. *Algebra of Matrices.* Boston: Allyn and Bacon, 1965.

Strang, G. *Linear Algebra and Its Applications.* 3rd ed. San Diego: Harcourt Brace Jovanovich, 1988.

Applications Chiang, A. *Fundamental Methods of Mathematical Economics.* 3rd ed. New York: McGraw-Hill, 1984.

Edelstein-Keshet, L. *Mathematical Models in Biology.* New York: McGraw-Hill, 1988.

Smith, R. J. *Circuits, Devices and Systems.* 4th ed. New York: Wiley, 1983.

Symon, K. R. *Mechanics.* 3rd ed. Reading, Massachusetts: Addison-Wesley, 1971.

Tables Abramowitz, M., and Stegun, I. A. *A Handbook of Mathematical Functions with Formulas, Graphs, and Mathematical Tables.* New York: Dover, 1964.

Software Koçak, H. *Differential and Difference Equations through Computer Experiments.* 2nd ed. New York: Springer-Verlag, 1989.

Answers to Selected Exercises

Chapter 1

Section 1.1

1. (a) First order, linear (c) First order, nonlinear (e) Second order, nonlinear
   (g) Third order, linear

2. (a) $y = x^2 - 3x + c$ (c) $y = \ln\left|\dfrac{x-4}{x}\right| + c$ (e) $y = -\ln|\cos x| + c_1 x + c_2$
   (g) $y = x^4 - x^3 + c_1 x^2 + c_2 x + c_3$

3. (a) $y = -5$ (c) $y = 2x^2 - 3x - 17$ (e) $y = -x + 3$ (g) $y = -\cos x + 1$

5. (a) $y = ce^{-3x}$ (c) $y = ce^{x/3}$ 9. $g''(-1) = -3,\quad g'''(-1) = 2$

Section 1.2

1. $y = \pm(4x^2 + c)^{1/2},\quad y = -(4x^2 + 5)^{1/2}$ 3. $y = c\sqrt{x^2 + 4},\quad y = 3\sqrt{x^2 + 4}$

5. $y = \dfrac{x+c}{1-cx},\quad y = \dfrac{7x+1}{7-x}$ 7. $y = 1 \pm (e^x + c)^{1/2},\quad y = 1 - (e^x + 8)^{1/2}$

9. $y = (ce^{\sin x} - 1)^{1/3}$ 11. $y = \sin^{-1}(x + c) + 2n\pi$ and $y = -\sin^{-1}(x+c) + (2n+1)\pi$

13. $y = -\ln(c - e^x)$ 15. $y = \exp(ce^{x^2})$ 17. $y = \pm\tan\sqrt{x^2 + c}$

19. $\sin y = c \sin x$ or $y = \sin^{-1}(c \sin x) + 2n\pi,\ y = -\sin^{-1}(c \sin x) + (2n-1)\pi$ 21. (a) $y = -1$

23. (a) $y = 2\tan(2x + c) - 4x + 1$ (c) $y = 3x \pm (12x + c)^{1/2}$

Section 1.3

1. (a) Degree 3   (c) No   (e) No   (g) Degree 1   (i) Degree 2   (k) Degree 0
2. (a) Separable and homogeneous   (c) Not separable, not homogeneous   (e) Homogeneous
    (g) Separable

3. $y = \dfrac{x}{\ln|x| + c}$ and $y = 0$ 5. $y = \pm x(cx^2 + 1)^{1/2}$ 7. $y = x \ln(cx^2 + 1)$

9. $y = \dfrac{1}{2}\left(cx^2 - \dfrac{1}{c}\right)$ 11. $y = x[\sin^{-1} cx + 2n\pi]$, $y = x[-\sin^{-1} cx + (2n - 1)\pi]$, $y = 0$

13. $y = x \tan cx$, $y = 0$ 15. $y = \dfrac{1}{2}\dfrac{c^2 x^2 - 1}{c}$ 18. (a) $x - 1 = (y + 2)(\ln|y + 2| + c)$

21. $y \ln cy = x^3$

Section 1.4

3. $x^2 y^3 - 2xy^2 = c$ 5. Not exact   7. $y = x^2 \pm (2x^4 + c)^{1/2}$ 9. $y = \pm[x^2 \pm (x^4 + c)^{1/2}]^{1/2}$
11. $y = \ln[x \pm (2x^2 + c)^{1/2}]$ 13. $y = [\sin x \pm (\sin^2 x + c)^{1/2}]^{-1}$
17. $\mu(x, y) = y^{-3}$, $y = [x \pm (4x^2 + c)^{1/2}]^{-1}$ 19. $\mu(x, y) = xy^{-2}$, $y = x^{-2}[c \pm (c^2 + x^5)^{1/2}]$
22. (e) $x^3 + cy = 0$ 23. (a) $x^3 - \tan^{-1}\dfrac{x}{y} = c$ (c) $\ln(x^2 + y^2) + 2x^3 = c$

Section 1.5

1. $y = cx^{-2} + x^2$, $y = 3x^{-2} + x^2$ 3. $y = x^2 e^{-x}(3x + c)$
5. $y = e^{x^2}\left(c + \displaystyle\int e^{-x^2}\,dx\right)$, $y = e^{x^2}\left(be^{-a^2} + \displaystyle\int_a^x e^{-t^2}\,dt\right)$ 7. $y = \ln x + c/\ln x$
9. $y = 1 + ce^{-x^2}$ 11. $y = x(x + 1) + cx/(x + 1)$ 15. $y = (cx + xe^x)^{-1}$, $y = 0$
17. $y = x(c + \sin x)^{1/3}$ 19. $y = (1 + ce^{\cos x})^{-2}$, $y = 0$ 21. $y = \ln(3x^2 + cx)$
23. $y = \exp[(x + 1)(x + c)]$ 25. (a) $y = -x + \dfrac{1}{c - x}$ (c) $y = x^2 + \dfrac{3x^2}{c - x^3}$

Section 1.6

1. $y = e^{-x} + k$ 3. $x^2 + y^2 = k$ 5. $y^4 = kx$ 7. $y^3 + 3x^2 y = k$ 9. $x^3 + y^3 = k$
11. $x^2 + 2xy - y^2 = k$ 13. $y = x - 2\tan^{-1} x + k$ 15. $y = \sqrt{3}\,x - 2\ln|2x + \sqrt{3}| + k$
19. $r = k \cos\theta$ 21. $r^2 = k \sin 2\theta$ 23. (a) $y^2 - x^2 = c$ (c) $y = \ln|\sin x| + c$

Section 1.7

1. 4 gm   3. 103.7 gm   5. After 20 yrs   7. 8.281 yrs   9. 4362 yrs ago

Section 1.8

1. $30e^{-2/5} = 20.1$ lb   3. 80 lb   5. 75%   7. $\frac{5}{2}\ln\frac{17}{13} = 0.671$ gal/min

Section 1.9

1. $N(t) = 2000(\frac{5}{4})^{t/2}$ 3. (a) $N(t) = [N_0^{1-\alpha} + (1-\alpha)kt]^{1/(1-\alpha)}$ (b) If $0 < \alpha < 1$, $N(t) \to \infty$

5. $N(t) = 15000\left[1 + 2\exp\left(-\frac{t}{10}\ln\frac{16}{7}\right)\right]^{-1}$ 7. (a) 249 million   (b) 816 million

Section 1.10

1. After $10(\ln 3)/\ln(3/2) = 27.1$ min   3. $u = 60 + \frac{1}{k} - t + \left(140 - \frac{1}{k}\right)e^{-kt}$, $k = \frac{1}{10}\ln(8/7)$

5. Maximum temperature is $85.6238°$. Reaches $21°$ 17.0629 min after removal.

7. (a) $x(t) = y(t) = a/(1 + akt)$ (b) $t = 1/(ka)$

9. $x = a\left[1 + \frac{t}{T}(2^{n-1} - 1)\right]^{-1/(n-1)}$, $n > 1$, $x = a2^{-t/T}$, $n = 1$

Section 1.11

1. $x = c_1 t^3 - t + c_2$ 3. $x = \pm\frac{2}{3c_1}(c_1 t - 1)^{3/2} + c_2$

5. $x = -t - \frac{2}{c_1}\ln|c_1 t - 1| + c_2$, $x = \pm t + c$

7. $x = \frac{t^2}{2} \pm \left\{\frac{t}{2}(t^2 + c_1)^{1/2} + \frac{c_1}{2}\ln|t + (t^2 + c_1)^{1/2}|\right\} + c_2$ 9. $x = \frac{c_1}{8}t^4 - \frac{1}{2c_1}\ln|t| + c_2$

11. $x = \pm[c_1(t + c_2)^2 - c_1^{-1}]^{1/2}$, $x = \pm(\pm 2t + c)^{1/2}$ 13. $x = \pm(2t + c_2)^{1/2} + c_1$, $x = c$

15. $x = \ln[(c_2 e^{c_1 t} - 1)/c_1]$, $x = \ln(t + c)$, $x = c$

17. $x = [t + c_2 \pm ((t + c_2)^2 + 4c_1)^{1/2}]/(2c_1)$, $x = -(t + c)^{-1}$, $x = c$

19. $x = (c_2 e^{c_1 t} + 2)/c_1$, $x = -2t + c$, $x = c$ 21. $x = (t + c_1)\ln(t + c_1) - t + c_2$

Section 1.12

1. $t = 5\sqrt{10}/2 = 7.9$ sec, $v = 80\sqrt{10} = 253$ ft/sec   3. $v = \frac{64}{3}(1 - e^{-3t/2})$, $x = \frac{64}{9}(3t - 2 + 2e^{-3t/2})$

5. $\frac{5}{2}10^4 \ln(1 + \frac{100}{49}) = 27803$ cm, $\frac{50}{7}\arctan(10/7) = 6.86$ sec

7. (a) $h = \frac{2gR^2}{2gR - v_0^2}$ (b) $t_1 = \frac{v_0 R}{2gR - v_0^2} + \frac{2gR^2}{(2gR - v_0^2)^{3/2}}\sin^{-1}\frac{v_0}{(2gR)^{1/2}}$

9. (a) $h = \frac{1}{2}\left(\frac{m}{cg}\right)^{1/2}\tan^{-1}\left[\left(\frac{c}{mg}\right)^{1/2}v_0^2\right]$ (b) $h = \frac{1}{2}\left(\frac{m}{cg}\right)^{1/2}\tanh^{-1}\left[\left(\frac{c}{mg}\right)^{1/2}v_0^2\right]$

Section 1.13

3. $y_0 = 1$, $y_1 = 1 - x + \frac{1}{2}x^2$, $y_2 = 1 - x + \frac{3}{2}x^2 - \frac{2}{3}x^3 + \frac{1}{4}x^4 - \frac{1}{20}x^5$

Additional Exercises

1. $y = x^2(c - \cos x)$ 3. $y = \ln[c - (2x + 1)e^{-2x}]$ 5. $y = x(3 \ln x + 8)^{1/3}$

7. $y = (x^2 + 1)[\ln(x^2 + 1) + c]$ 9. $y = x[(\ln|x| + c)^2 - 1]^{1/3}$ 11. $y^2 + 4xy - 3x^2 = c$

13. $x = t + c_2 + c_1 \int_0^t e^{-s^2}\, ds$ 15. $y = c_1/(1 - c_2 e^{-c_1 x})$ 17. $r = ce^{\theta}$ and $r = ce^{-\theta}$

19. $t = 20 \ln 100 = 92.10$ min   21. $11.11°$ 23. $2x'' = -\dfrac{62.4\pi}{12}[(x + k)^3 - k^3], \quad k = \left(\dfrac{12 \times 64}{62.4\pi}\right)^{1/3}$

25. (b) $x\dfrac{dy}{dx} = y - kr, \quad k = v_1/v_2$ (c) $y = \dfrac{1}{2}x\left[\left(\dfrac{a}{x}\right)^k - \left(\dfrac{a}{x}\right)^{-k}\right]$. If $k > 1$, $y \to \infty$ as $x \to 0$.

───────────── Chapter 2 ─────────────

Section 2.1

1. $x_1 = -1, \; x_2 = 2$ 3. No solution   5. $x_1 = a + \frac{4}{3}, \; x_2 = 3a$

7. $x_1 = 5, \; x_2 = 3, \; x_3 = -2$ 9. $x_1 = 5a + \frac{23}{7}, \; x_2 = a - \frac{22}{7}, \; x_3 = 7a$

11. $x_1 = -2, \; x_2 = 1, \; x_3 = 0$ 13. $x_1 = -2a + \frac{2}{3}, \; x_2 = -a + \frac{1}{3}, \; x_3 = 3a$

15. $x_1 = 3a - \frac{13}{5}, \; x_2 = -2a + \frac{42}{5}, \; x_3 = 2a - \frac{2}{5}, \; x_4 = 5a$

17. The only solution of the original system is $(2, 5)$. However, each ordered pair $(7 - a, a)$ is a solution of the new system. In particular, $(4, 3)$ is a solution of the new system but not of the old system.

19. (a) $3a + b = 0$ (b) $a - b = 0$

Section 2.2

2. Yes. All or some of the equations might be equivalent.   3. $x_1 = x_2 = 0$ 5. $x_1 = x_2 = x_3 = 0$

7. $x_1 = 2a - b, \; x_2 = -a + 4b, \; x_3 = a, \; x_4 = b$ 9. $x_1 = -2a, \; x_2 = 12a, \; x_3 = -3a, \; x_4 = 6a$

11. $x_1 = 2a - b + c, \; x_2 = a, \; x_3 = b, \; x_4 = 3c, \; x_5 = c$ 13. (a) $a = -\frac{2}{3}$ (b) $a = 4$

Section 2.3

1. No such mixture can be obtained.   3. Yes, the ratio is $15:15:1$. $31/28$ lbs

5. $x_1 = 100 - 14c, \; x_2 = 4c, \; x_3 = 50 - c, \; x_4 = 4c, \; 0 \le c \le \frac{50}{7}$ 7. $I_1 = 6/13, \; I_2 = 18/13$

9. $u_{11} = 1/6, \; u_{12} = 11/24, \; u_{21} = 5/24, \; u_{22} = 2/3$

Section 2.4

1. (a) $\begin{bmatrix} 2 & -1 \\ -1 & 3 \end{bmatrix}$ (c) $\begin{bmatrix} 0 & 1 & 2 \\ 1 & 1 & 1 \\ 2 & 0 & -1 \end{bmatrix}$ 2. (a) $\begin{aligned} 3x_1 - 5x_2 &= 0 \\ -x_1 + 2x_2 &= 0 \end{aligned}$ (c) $\begin{aligned} x_1 + x_2 &= 0 \\ 5x_2 - x_3 &= 0 \\ 4x_2 + 2x_3 &= 0 \end{aligned}$

3. (a) $\begin{bmatrix} 6 & -15 \\ 3 & 0 \end{bmatrix}, \quad \begin{bmatrix} -2 & 5 \\ -1 & 0 \end{bmatrix}, \quad \begin{bmatrix} -4 & 10 \\ -2 & 0 \end{bmatrix}, \quad \begin{bmatrix} 0 & 0 \\ 0 & 0 \end{bmatrix}$

4. (a) $\begin{bmatrix} -1 & 2 & 6 \\ -2 & 2 & 7 \end{bmatrix}$,   (c) $\begin{bmatrix} 8 & -6 & -8 \\ -9 & -1 & -1 \end{bmatrix}$ 5. $\begin{bmatrix} -\frac{3}{2} & 1 & 1 \\ 2 & 0 & -\frac{1}{2} \end{bmatrix}$

8. (a) $\begin{bmatrix} 2 \\ 5 \\ 0 \end{bmatrix}$, $\begin{bmatrix} 1 \\ 0 \\ 1 \end{bmatrix}$, $\begin{bmatrix} -3 \\ 0 \\ 4 \end{bmatrix}$, $\begin{bmatrix} 0 \\ 2 \\ 0 \end{bmatrix}$ 9. (a) $[2 \ \ 1 \ \ -3 \ \ 0]$, $[5 \ \ 0 \ \ 0 \ \ 2]$, $[0 \ \ 1 \ \ 4 \ \ 0]$

10. (a) $\begin{bmatrix} 2 & -4 & 3 \\ 3 & 1 & 5 \end{bmatrix}$,   (c) $\begin{bmatrix} 1 & 2 & 1 \\ 0 & 1 & 2 \\ 1 & 5 & 3 \\ 4 & -1 & 4 \end{bmatrix}$ 11. (a) $\begin{bmatrix} -2 & 3 \\ 1 & 0 \\ 4 & 5 \end{bmatrix}$ (c) $\begin{bmatrix} 2 & 3 & 0 & -1 \\ 0 & 1 & 5 & 0 \\ 4 & 3 & 2 & 1 \end{bmatrix}$

Section 2.5

1. $AB = \begin{bmatrix} 0 & 0 \\ -5 & 5 \end{bmatrix}$, $BA = \begin{bmatrix} 3 & -1 \\ -6 & 2 \end{bmatrix}$

3. $AB = \begin{bmatrix} 7 & -1 & 0 \\ -5 & -7 & 5 \\ -4 & 1 & -1 \end{bmatrix}$, $BA = \begin{bmatrix} -3 & 11 & -5 \\ 4 & 2 & 1 \\ -3 & -6 & 0 \end{bmatrix}$

5. $AB = \begin{bmatrix} 7 & 4 & 9 \\ 3 & 1 & 11 \end{bmatrix}$, BA is not defined   7. AB is not defined, $BA = \begin{bmatrix} 0 & 3 \\ 6 & -3 \\ 0 & 0 \end{bmatrix}$

9. Neither AB nor BA is defined.

11. (a) $A = \begin{bmatrix} 3 & 1 \\ 4 & -2 \end{bmatrix}$, $\mathbf{x} = \begin{bmatrix} x_1 \\ x_2 \end{bmatrix}$, $\mathbf{b} = \begin{bmatrix} 7 \\ -3 \end{bmatrix}$

   (c) $A = \begin{bmatrix} 2 & -1 & 1 \\ -1 & 1 & 5 \\ 2 & 1 & 0 \end{bmatrix}$, $\mathbf{x} = \begin{bmatrix} x_1 \\ x_2 \\ x_3 \end{bmatrix}$, $\mathbf{b} = \begin{bmatrix} 4 \\ -2 \\ 3 \end{bmatrix}$

12. (a) $2x_1 - 3x_2 = 2$ (c) $2x_1 - x_2 + 3x_3 = 1$
         $x_1 + 4x_2 = -5$ $3x_2 - 2x_3 = -1$
                                        $x_1 + 4x_3 = 0$

Section 2.6

1. (a) $-8, \sqrt{5}, 3\sqrt{5}$ (c) $-16, \sqrt{14}, \sqrt{26}$ 2. (a) $\cos^{-1} \dfrac{-8}{15}$

3. (a) $1 + 4i, 1 - 4i, \sqrt{6}, \sqrt{5}$ (c) $0, 0, \sqrt{3}, \sqrt{14}$ 4. (a) Yes   (c) No

11. (a) $\pm \dfrac{1}{\sqrt{30}} (2, 1, 5)$ (c) $\pm \dfrac{1}{3\sqrt{5}} (4, -2, -4, 3)$

Section 2.7

1. (a), (c), (d)   3. (d)   5. None (not a square matrix)   7. (a) $\begin{bmatrix} 2 & -2 & 4 \\ 0 & -3 & 2 \\ 0 & 0 & 0 \end{bmatrix}$

8. (a) $\begin{bmatrix} -6 & 0 & -3 \\ 0 & 0 & 9 \\ 8 & 0 & 6 \end{bmatrix}$ 12. (a) $\begin{bmatrix} 2 & 4 \\ -1 & 5 \end{bmatrix}$ (c) $\begin{bmatrix} 6 & 1 & 2 \\ 2 & 0 & 1 \\ 4 & 3 & 1 \end{bmatrix}$

17. (a) $\begin{bmatrix} -2 & \frac{1}{2} \\ \frac{1}{2} & 1 \end{bmatrix}$ (c) $\begin{bmatrix} 3 & -1 & 0 \\ -1 & 1 & \frac{3}{2} \\ 0 & \frac{3}{2} & -4 \end{bmatrix}$

18. (a) $2x_1^2 - 2x_1x_2$ (c) $-2x_1^2 + 4x_2^2 + x_3^2 + 2x_2x_3 + 3x_1x_2$

Section 2.8

1. There are $4! = 24$ permutations.   2. (a) 2   (c) 0   (e) 5   (g) 3
3. (a) -1 (c) -1 (e) -1 7. (a) 22   (c) -17 8. (a) -2 (c) 29
13. $4x^3 + 2$

Section 2.9

5. (a) -10 6. (a) 178   7. (a) -83 8. (a) -324 9. 252   11. det A is 0 or 1

Section 2.10

1. (a) $A_{11} = 0$, $A_{12} = 1$, $A_{21} = -3$, $A_{22} = 2$
   (c) $A_{11} = 2$, $A_{12} = -7$, $A_{13} = 1$, $A_{21} = 4$, $A_{22} = -7$, $A_{23} = -5$, $A_{31} = 2$, $A_{32} = 7$, $A_{33} = 1$
2. (a) 20   (c) 6   3. $(-1)^N d_1 d_2 \cdots d_n,$ $N = \dfrac{n(n-1)}{2}$

Section 2.11

1. $m = n$ and A nonsingular   3. $x_1 = -43/6$, $x_2 = 22/3$ 5. Solutions $x_1 = c - 2$, $x_2 = 3c$
7. $x_1 = -34/37$, $x_2 = -25/37$, $x_3 = 44/37$
9. Solutions $x_1 = -c - 3/2$, $x_2 = 2c$, $x_3 = 5c - 3/2$ 11. $c = 7$; $x_1 = -a$, $x_2 = -3a$, $x_3 = a$
15. This way the rounding errors are multiplied by a numerically smaller quantity.

Section 2.12

1. If $AB = I$ then det $A \cdot$ det $B = 1$ so det $A \neq 0$.

3. $\begin{bmatrix} 4/3 & 1/3 \\ -5/3 & -2/3 \end{bmatrix}$ 5. $\begin{bmatrix} 1/2 & -1/2 \\ -1/2 & 3/2 \end{bmatrix}$ 7. $\begin{bmatrix} 11/3 & -4/3 & -2 \\ 2/3 & -1/3 & 0 \\ -4/3 & 2/3 & 1 \end{bmatrix}$ 9. $\begin{bmatrix} 2/5 & 1/5 & 0 \\ -6/5 & -3/5 & 1 \\ -1 & -1 & 1 \end{bmatrix}$

11. $\begin{bmatrix} 2 & -6 & 1 & -8 \\ 0 & 1 & 0 & 1 \\ 1 & -3 & 1 & -3 \\ 0 & 1 & 0 & 2 \end{bmatrix}$ 15. Any matrix of the form $\begin{bmatrix} 0 & x \\ 1/x & 0 \end{bmatrix}$, $x \neq 0$ is its own inverse.

Additional Exercises

1. (a) $x_1 = -4a - \frac{1}{5}$, $x_2 = -11a + \frac{6}{5}$, $x_3 = 5a$ (c) $x_1 = -\frac{1}{4}$, $x_2 = \frac{3}{4}$, $x_3 = -\frac{1}{2}$

2. (a) $x_1 = 2$, $x_2 = -1$ (c) $x_1 = \frac{11}{4}$, $x_2 = -\frac{9}{4}$, $x_3 = \frac{3}{2}$ 3. $\begin{bmatrix} -50 & 0 & 0 \\ 0 & 25 & 0 \\ 0 & 0 & 5 \end{bmatrix}$

4. (a) -12 5. (a) 18   10. (c)(i) $A = \begin{bmatrix} 2 & 1 \\ 1 & 5 \end{bmatrix} + \begin{bmatrix} 0 & -2 \\ 2 & 0 \end{bmatrix}$

17. Between 2.50 and 11.43 tons

Chapter 3

Section 3.1

3. Yes   5. No   7. No   9. Yes   11. Yes   13. Yes   15. Yes

Section 3.2

1. Line, $x_1 + x_2 = 0$ 3. Plane, $2x_1 + 2x_2 - x_3 = 0$ 5. $4x_1 + 7x_2 + x_3 = 0$
13. V is a subspace of U but $V \neq U$.

Section 3.3

1. (a) Dependent   (c) Dependent   (e) Independent   (g) Dependent   (i) Dependent
4. (b) No. Consider the subset of R^2 with elements $(1, 0), (0, 1), (1, 1)$.   (d) No
7. No   9. Yes   11. (a) Independent   12. (a) Dependent

Section 3.4

1. (a) $W(x) = (b - a)e^{(a+b)x}$; independent   (c) $W(x) = 2$; independent
   (e) $W(x) = -2x^{-6}$; independent   (g) $W(x) = 0$; dependent if m is even, independent if m is odd

Section 3.5

4. (a) No
5. Dimension 4. One basis consists of
$$\begin{bmatrix} 1 & 0 \\ 0 & 0 \end{bmatrix}, \begin{bmatrix} 0 & 1 \\ 0 & 0 \end{bmatrix}, \begin{bmatrix} 0 & 0 \\ 1 & 0 \end{bmatrix}, \begin{bmatrix} 0 & 0 \\ 0 & 1 \end{bmatrix}.$$
7. Dimension 2. Examples of bases are $\{1, x\}, \{x + 1, x - 1\}$.
9. (a) $a = (v_1 - v_2)/3$, $b = (v_1 + 2v_2)/3$
13. (a) Any 2 of the vectors form a basis.   (c) One basis consists of $(1, 3, 0, -2), (3, 1, 2, 0), (1, 0, 1, 0)$.

Section 3.6

1. $\frac{1}{5}(4, -3)$ 3. $\frac{1}{3}(1, 2, -2), \frac{1}{3\sqrt{2}}(4, -1, 1)$ 5. $\frac{1}{\sqrt{10}}(1, 0, 3), \frac{1}{\sqrt{26}}(3, 4, -1)$

7. $\frac{1}{3}(0, 1, 2, 2), \frac{1}{3\sqrt{2}}(3, 2, 1, -2), \frac{1}{3\sqrt{2}}(3, -2, -1, 2)$

9. (a) $(2, -1, 5), (-22, -19, 5)$ is one possibility.   (c) $(1, -1, 0, 1), (2, 1, 3, -1)$ is one possibility.

Section 3.7

1. (a) and (b) are linear, (c) and (d) are not.

3. (a) $(-4, 3)$ (b) $(0, 0)$ (belongs to kernel)   (c) $(-9, 5)$ 5. (b) and (c) belong to the kernel.

7. (a) $(4, 15, 38)$, $\begin{bmatrix} -1 & 2 \\ 0 & 3 \\ 2 & 6 \end{bmatrix}$ (c) $(4, 7)$, $\begin{bmatrix} 1 & 0 & 1 \\ -2 & 3 & 1 \end{bmatrix}$

11. The kernel is the line $2x_1 - x_2 = 0$ and the range is the line $2y_1 + y_2 = 0$.

12. (a) $(15, 12)$, $\begin{bmatrix} 2 & 1 \\ 2 & 0 \end{bmatrix}$ (c) $(1, -7, 2)$, $\begin{bmatrix} 1 & 0 \\ 5 & -3 \\ -2 & 1 \end{bmatrix}$

14. (a) Dimension 2. One basis consists of the first 2 column vectors of A.
    (c) Dimension 2. One basis consists of the first 2 column vectors of A.

Section 3.8

1. (a) $T_2 T_1 \mathbf{x} = BA\mathbf{x}$, $BA = \begin{bmatrix} 5 & 4 \\ 5 & 1 \\ 1 & 2 \end{bmatrix}$ (b) Not defined   (c) Not defined

3. (a) $T_2 T_1 f(x) = \frac{1}{12}(1 - x^4)$, $T_1 T_2 f(x) = \frac{1}{12}(4x - x^4)$, $T_1 T_2 \neq T_2 T_1$
   (b) $(T_1 + T_2)f(x) = \frac{1}{3}$. The range consists of all constant polynomials.

5. (a) $A = \begin{bmatrix} 5 & 2 \\ -2 & 0 \end{bmatrix}$, $B = \begin{bmatrix} 1 & 0 \\ -1 & 4 \end{bmatrix}$, $C = \begin{bmatrix} 0 & 4 \\ 4 & 1 \end{bmatrix}$

7. (a) One basis consists of $(2, 0, -1), (1, 3, 0)$.

Section 3.9

1. (a) $A = G_{12}(2)G_{21}(1)E_1(2) = G_{12}(2)E_1(2)G_{21}(\frac{1}{2})$, etc.   (b) $9y_1 - 4y_2 = 6$
   (c) $55y_1^2 + 15y_2^2 - 58y_1 y_2 = 16$; hyperbola

Section 3.10

1. (a) 0   (c) $x^2 \cos x - (2 + x)\sin x$ 3. (a) and (c) belong to the kernel.

4. (a) $L_1L_2 = D^2 + (2 + x)D + 2x$ (c) $L_1L_2 = D^3 - (x + 1)D^2 + xD - 1$

         $L_2L_1 = D^2 + (2 + x)D + (2x + 1)$ $L_2L_1 = D^3 - (x + 1)D^2 + (x + 1)D - 1$

7. No. Let $L_1 = D$, $L_2 = D + x$, $g(x) = 1$.

Additional Exercises

1. (a) is not a basis,   (b) is a basis.   3. (b) 2   (c) One basis is $\{x^2, x\}$

7. Dimension 2. One basis consists of $(2, -3, 1, 1, 0)$ and $(-1, 0, 1, 0, 1)$.

12. (a) $\dfrac{1}{5}\begin{bmatrix} 5 & -5 \\ 9 & -2 \end{bmatrix}$ (c) The first 3 column vectors form a basis.   18. (a) 2   (c) 2

Chapter 4

Section 4.1

1. $\lambda_1 = -2, (a, -a)$; $\lambda_2 = 3, (a, 4a)$ 3. $\lambda_1 = -1, (a, a)$ 5. $\lambda_1 = i, a(2, 1 + i)$; $\lambda_2 = -i, a(2, 1 - i)$

7. $\lambda_1 = 1, (a, 0, 0)$; $\lambda_2 = 2, (a, 0, -a)$; $\lambda_3 = -2, (-a, 4a, 5a)$ 9. $\lambda_1 = 1, (a, 0, -a)$; $\lambda_2 = -1, (a, a, -a)$

11. $\lambda_1 = 1, (-a - b, a, b)$ 13. $\lambda_1 = 0, (0, a, -a)$; $\lambda_2 = 1 + i, a(1 - i, 1, 1)$; $\lambda_3 = 1 - i, a(1 + i, 1, 1)$

19. $\begin{bmatrix} 0 & 1 & 0 & 0 \\ 0 & 0 & 1 & 0 \\ 0 & 0 & 0 & 1 \\ -7 & -6 & 5 & 2 \end{bmatrix}$

Section 4.2

3. 35.09%   7. (b) Characteristic vectors are $a(1, 1, .7)$. 37.04%

Section 4.3

1. $\lambda_1 = \lambda_2 = 2$ 3. $\lambda_1 = 2$, $\lambda_2 = -1$. $K = \begin{bmatrix} 1 & 1 \\ 2 & -1 \end{bmatrix}$

    Not similar to a diagonal matrix

5. $\lambda_1 = 1$, $m_1 = 1$, $n_1 = 1$ 7. $\lambda_1 = 0$, $m_1 = n_1 = 1$

    $\lambda_2 = -1$, $m_2 = 2$, $n_2 = 1$ $\lambda_2 = 2$, $m_2 = n_2 = 2$

    Not similar to a diagonal matrix   Similar to a diagonal matrix

9. $\lambda_1 = 1$, $m_1 = 3$, $n_1 = 1$ Not similar to a diagonal matrix

11. (a) $K = \begin{bmatrix} 1 & 0 \\ -1 & 1 \end{bmatrix}$ (c) $K = \begin{bmatrix} 1 & 0 \\ -1 & 1 \end{bmatrix}$ 12. (a) $K = \begin{bmatrix} 1 & 0 \\ 0 & \frac{1}{3} \end{bmatrix}$

Section 4.4

1. $K = \dfrac{1}{\sqrt{5}}\begin{bmatrix} 2 & -1 \\ 1 & 2 \end{bmatrix}$, $D = \text{diag}(10, -5)$

3. $K = \dfrac{1}{2}\begin{bmatrix} \sqrt{2 + \sqrt{2}} & -\sqrt{2 - \sqrt{2}} \\ \sqrt{2 - \sqrt{2}} & \sqrt{2 + \sqrt{2}} \end{bmatrix}$, $D = \text{diag}(1 + 2\sqrt{2}, 1 - 2\sqrt{2})$

5. $K = \dfrac{1}{7}\begin{bmatrix} 2 & 6 & 3 \\ -6 & 3 & -2 \\ 3 & 2 & -6 \end{bmatrix}$, $D = \text{diag}(49, 0, 0)$

7. $K = \dfrac{1}{\sqrt{6}}\begin{bmatrix} \sqrt{2} & -\sqrt{3} & 1 \\ \sqrt{2} & \sqrt{3} & 1 \\ \sqrt{2} & 0 & -2 \end{bmatrix}$, $D = \text{diag}(0, 0, 6)$

13. (a) $7x_1^2 - 2x_2^2 + 12x_1x_2$; not positive-definite   (c) $3x_1^2 - x_2^2 + 4x_1x_2$; not positive-definite
    (e) $4x_1^2 + 36x_2^2 + 9x_3^2 - 36x_2x_3 + 12x_1x_3 - 24x_1x_2$; not positive-definite
    (g) $x_1^2 + x_2^2 + 4x_3^2 - 4x_2x_3 - 4x_1x_3 + 2x_1x_2$; not positive-definite

Section 4.5

1. (a) $\begin{bmatrix} 9 & 1 \\ -2 & 5 \end{bmatrix}$ 2. (a) $\begin{bmatrix} 1 & 4 \\ 4 & -11 \end{bmatrix}$ (c) $\begin{bmatrix} 128 & 256 \\ 64 & 128 \end{bmatrix}$

3. (a) $\dfrac{1}{3}\begin{bmatrix} 2 & 1 \\ 1 & -1 \end{bmatrix}$ (c) Singular

Additional Exercises

1. (a) $\lambda_1 = 1$, $a(2, 5)$; $\lambda_2 = -2$, $a(1, 1)$; $K = \begin{bmatrix} 2 & 1 \\ 5 & 1 \end{bmatrix}$

    (c) $\lambda_1 = 1$, $(2a + b, a, b)$; $\lambda_2 = 2$, $(2a, a, a)$; $K = \begin{bmatrix} 1 & 2 & 2 \\ 0 & 1 & 1 \\ 1 & 0 & 1 \end{bmatrix}$

2. (a) $\lambda_1 = 2i$, $\lambda_2 = -2i$, $K = \begin{bmatrix} 1 & 0 \\ 1 & 1 \end{bmatrix}$ (c) $\lambda_1 = \lambda_2 = 3$, $K = \begin{bmatrix} 1 & 0 \\ 2 & 1 \end{bmatrix}$

15. (b) For Ex. 1(c), $m(\lambda) = (\lambda - 1)(\lambda - 2)$.

17. (a) $\sqrt{A} = \begin{bmatrix} -2\sqrt{2} + 3 & -3\sqrt{2} + 3 \\ 2\sqrt{2} - 2 & 3\sqrt{2} - 2 \end{bmatrix}$ (c) $e^A = \begin{bmatrix} -2e^2 + 3 & -3e^2 + 3 \\ 2e^2 - 2 & 3e^2 - 2 \end{bmatrix}$

Chapter 5

Section 5.1

1. $y = 0$ 3. (a) $y = 2\sin x$ (c) $y = 0$ 5. It is the zero function.
9. (b) $xy'' - (x + 1)y' + y = 0$

Section 5.2

1. (a) $(D^2 - 3D + 2)y = 0$ (c) $(D^3 - 3D^2 - D + 1)y = 0$
2. (a) $(D - 1)(D + 2)y = 0$ (c) $(D - 1)^2(D + 2)y = 0$
3. (a) $(D^2 - D - 2)y = 0$ (c) $(D^3 - 4D^2 + 4D)y = 0$ 5. $y = c_1 e^{2x} + c_2 e^{3x}$
7. $y = c_1 e^x + c_2 e^{-x} + c_3 e^{-5x}$ 9. Not possible

Section 5.3

3. (a) $\cos 3x + i \sin 3x$ (c) $e^{2x}(\cos 3x - i \sin 3x)$ (e) $\frac{1}{2}(e^{2ix} + e^{-2ix})$ (g) $(1/2i)(e^{ix} - e^{-ix})$
7. $e^{-x} \cos 2x$, $e^{-x} \sin 2x$ 8. (a) $\cos 3x$, $\sin 3x$ (c) $e^{2x} \cos x$, $e^{2x} \sin x$
9. $2 \cos 2x - 4 \sin 2x$, $4 \cos 2x + 2 \sin 2x$ 11. $\cos(2 \ln x)$, $\sin(2 \ln x)$

Section 5.4

1. $y = c_1 e^{-2x} + c_2 e^{3x}$ 3. $y = c_1 + c_2 e^{-2x}$ 5. $y = c_1 + c_2 e^x + c_3 e^{-4x}$ 7. $y = (c_1 + c_2 x)e^{-x}$
9. $y = (c_1 + c_2 x + c_3 x^2)e^{2x}$ 11. $y = c_1 + c_2 x + c_3 e^{-x}$ 13. $y = c_1 \cos 3x + c_2 \sin 3x$
15. $y = e^{3x}(c_1 \cos 2x + c_2 \sin 2x)$ 17. $y = (c_1 + c_2 x)\cos x + (c_3 + c_4 x)\sin x$
19. $y = (c_1 + c_2 x)e^{2x} + c_3 \cos \sqrt{2} x + c_4 \sin \sqrt{2} x$ 21. $y = -3e^x + 2e^{3x}$
23. $y = \cos 2x - 2 \sin 2x$ 25. $y = 3 - e^{-x}$
29. (a) $y'' + 4y' + 4y = 0$ (c) $y'' + 4y = 0$ (e) $(D^2 + 9)^2 y = 0$
30. (a) All solutions approach 0.   (c) Not all solutions approach 0.

Section 5.5

1. $y = c_1 x^2 + c_2 x^{-1}$ 3. $y = c_1 + c_2 x^{1/3}$ 5. $y = (c_1 + c_2 \ln x)x^{1/2}$
7. $y = c_1 + c_2 \ln x + c_3 x$ 9. $y = c_1 \cos(2 \ln x) + c_2 \sin(2 \ln x)$
11. $y = c_1 x + c_2 \cos(\ln x) + c_3 \sin(\ln x)$ 13. $y = 4x^{-1} - 3x^{-2}$ 15. $y = \cos(2 \ln x) + 2 \sin(2 \ln x)$
17. $y = cx$ 21. (a) $y = [c_1 + c_2 \ln(x - 3)](x - 3)^{-1}$
22. (a) All zeros must have positive real parts.

Section 5.6

1. $y = c_1 e^x + c_2 e^{-x} + \sin 2x$ 3. (a) $y = ce^{2x} + 2e^{5x}$ 7. (a) $y = ce^{-3x} - \frac{2}{5}e^{2x} + 4e^{-x}$

Section 5.7

1. $y = c_1 e^x + c_2 e^{-3x} + e^{2x}$, $y = 3e^x + e^{-3x} + e^{2x}$ 3. $y = c_1 e^{-x} + c_2 e^{-2x} + (6x - 5)e^x$
5. $y = c_1 e^{-x} + c_2 e^{-2x} - \cos 2x + 3 \sin 2x$, $y = -\cos 2x + 3 \sin 2x$ 7. $y = c_1 e^{3x} + c_2 e^{-2x} - \frac{1}{3}$
9. $y = (c_1 + c_2 x)e^x + c_3 e^{-x} + \frac{2}{5}(\cos 2x - 2 \sin 2x)$ 11. $y = c_1 e^{-x} + c_2 e^{-2x} - 5xe^{-2x}$

13. $y = c_1e^{-x} + c_2e^{2x} + (x^2 + 2x/3)e^{-x}$ 15. $y = c_1e^{-x} + c_2e^{-3x} + (x^3 - 3x^2/2 + 3x/2)e^{-x}$

17. $y = c_1 + c_2e^{-x} + x^3 - 3x^2 + 6x$ 19. $y = e^{-x}(c_1 \cos 2x + c_2 \sin 2x) + xe^{-x} \sin 2x$

22. (a) $y = c_1e^x + c_2e^{2x} + e^{-x}$ (c) $y = c_1e^{-x} + c_2e^{2x} - 3 \cos x - \sin x$

   (e) $y = (c_1 + c_2x)e^x + c_3e^{-x} + 3e^{2x}$

24. (a) $y = c_1e^{2x} + c_2e^{-x} + 2xe^{2x}$ (c) $y = c_1 \cos x + c_2 \sin x + 2x \sin x$

25. $y = c_1x^{-2} + c_2x^3 + x^4$ 27. $y = c_1x + c_2x^3 - 2$ 29. $y = c_1x + c_2x^{-2} + 2x \ln x$

Section 5.8

1. $y = c_1e^x + c_2e^{-x} - 1 + e^x \ln(1 + e^{-x}) - e^{-x} \ln(1 + e^x)$ 3. $y = (c_1 + c_2x)e^{-x} + x^2e^{-x}(2 \ln|x| - 3)$

5. $y = c_1 \cos(x/2) + c_2 \sin(x/2) + \cos(x/2)\ln|\cos(x/2)| + \frac{1}{2}x \sin(x/2)$

7. $y = e^{-x}(c_1 \cos x + c_2 \sin x - 4 + 2 \sin x \ln|\sec x + \tan x|)$

9. $y = (c_1 + c_2x + c_3x^2)e^x - 2xe^x \ln|x|$ 13. $y = c_1x + c_2xe^{1/x}$ 15. $y = c_1e^x + c_2x^{1/2}e^x + xe^x$

Section 5.9

1. $x = \frac{2}{3} \cos 8t$, $P = \pi/4$, $f = 4/\pi$ 3. $k = 10\pi^2$ lbs/ft   5. $x = \frac{8}{9}e^{-4t} - \frac{2}{9}e^{-16t}$; overdamped

7. $c = (50g - 25\pi^2)^{1/2} = 15.597$ newton-sec/m   11. $c < 4$ 13. (a) $a = 2$ (b) $x(t) \to 2$

15. $x = F_0/k$

Section 5.10

1. (a) $I = 2(1 - e^{-10t})$ (b) $4(1 - e^{-10t})$ and $4e^{-10t}$

3. $e^{-2t}(c_1 \cos 4t + c_2 \sin 4t)$, $\frac{1}{17}(4 \cos 4t + 16 \sin 4t)$ 5. $I = -3\sqrt{10} \sin \sqrt{10}\,t$

Additional Exercises

1. $y = c_1e^{-3x} \cos x + c_2e^{-3x} \sin x$ 3. $y = c_1x^{-3} + c_2x^{-3} \ln x$

5. $y = (c_1 + c_2x)e^{2x} - \frac{3}{4} \sin 2x + 2x + 2$ 7. $y = (c_1 + c_2x + c_3x^2)e^{-x} + \frac{2}{3}x^3e^{-x}$

9. $y = c_1 \cos x + c_2 \sin x - x \cos x + (\sin x)\ln|\sin x|$ 11. $y = c_1x^2 + c_2x^{-1} + (x + 2 + 2/x)e^{-x}$

13. $y = c_1' \cos x + c_2 \sin x + c_3 \cos \sqrt{2}\,x + c_4 \sin \sqrt{2}\,x$ 15. $k = 384\pi^2$ dynes/cm

17. (a) $x = \frac{5}{4} \sin 2t$ (b) $x = \frac{5}{13}(\cos \sqrt{3}\,t + 2\sqrt{3} \sin \sqrt{3}\,t)$ (c) $x = \frac{5}{6}(\cos \sqrt{2}\,t + 2 \sin \sqrt{2}\,t)$

   The values for α are the undamped natural angular frequency, the damped natural angular frequency, and the value for which the amplitude of the steady-state solution is largest.

19. (a) $I = e^{-5t}(c_1 \cos 10t + c_2 \sin 10t) + A \cos(\alpha t - \beta)$, $\alpha = 120\pi$, $A = 250\alpha/[(125 - \alpha^2)^2 + 100\alpha^2]^{1/2}$,

$$\tan \beta = 10\alpha/(125 - \alpha^2), \quad c_1 = -A \cos \beta, \quad c_2 = -A\left(\frac{1}{2} \cos \beta + \frac{\alpha}{10} \sin \beta\right)$$

   (b) $C = 2.5(120\pi)^{-2} = 1.759 \times 10^{-5}$ farads

21. (a) All solutions approach 0.   (c) Not all solutions approach 0.

Chapter 6

Section 6.1

3. Solve the first equation for x_1, then the second for x_2, and then the third for x_3.

5. (a) $x_1' = -k_1x_1$, $\quad x_2' = -k_2x_1' - k_3x_2$

   (b) $x_1(t) = ae^{-k_1t}$, $\quad x_2(t) = \left(b - \dfrac{ak_1k_2}{k_3 - k_1}\right)e^{-k_3t} + \dfrac{ak_1k_2}{k_3 - k_1}e^{-k_1t}$

7. $m_1\ddot{x}_1 = km_1m_2/(x_1 - x_2)^2$, $\quad m_2\ddot{x}_2 = -km_1m_2/(x_1 - x_2)^2$

Section 6.2

1. $x_1' = 4x_1 + \cos t + e^t$, $\quad x_1(t_0) = k_1$,
  $x_2' = 3x_1 + e^t$ $\quad x_2(t_0) = k_2$

3. $u_1' = u_3 + e^t$ $\qquad\qquad u_1 = x_1$, $\quad x_1(t_0) = k_1$
  $u_2' = u_3$ $\qquad\qquad\qquad u_2 = x_2$, $\quad x_2(t_0) = k_2$
  $u_3' = u_1 + u_2 + u_3 + \sin t$ $\quad u_3 = x_2'$, $\quad x_2'(t_0) = k_3$

5. $u_1' = u_2$, $\qquad\qquad u_1 = x_1$, $\quad x_1(t_0) = k_1$
  $u_2' = u_3$, $\qquad\qquad u_2 = x_1'$, $\quad x_1'(t_0) = k_2$
  $u_3' = u_2 - u_1u_5 + \sin t$, $\quad u_3 = x_1''$, $\quad x_1''(t_0) = k_3$
  $u_4' = u_5$, $\qquad\qquad u_4 = x_2$, $\quad x_2(t_0) = k_4$
  $u_5' = u_3 - u_4u_2 - \cos t$, $\quad u_5 = x_2'$, $\quad x_2'(t_0) = k_5$

7. $u_1' = u_2$,
  $u_2' = -\dfrac{k_1}{m_1}u_1 + \dfrac{c}{m_1}(u_4 - u_2)$,
  $u_3' = u_4$,
  $u_4' = -\dfrac{k_2}{m_2}u_3 - \dfrac{c}{m_2}(u_4 - u_2)$

9. $u_1' = u_2$, $\quad u_2' = tu_2 - u_1^2 + \sin t$
11. $u_1' = u_2$, $\quad u_2' = u_3$, $\quad u_3' = u_3 - u_1 + e^t$

13. $u_1' = u_2$, $\quad u_2' = -\dfrac{g}{L}\sin u_1$

Section 6.3

1. $x_1 = -2c_1e^{-t} - 3c_2e^{-2t} - 3e^{-3t}$, $\quad x_1 = -6e^{-2t} - 3e^{-3t}$
  $x_2 = c_1e^{-t} + c_2e^{-2t} + 2e^{-3t}$, $\quad x_2 = 2e^{-2t} + 2e^{-3t}$

3. $x_1 = c_1e^{-t} + 3e^{-2t}$
  $x_2 = -2c_1te^{-t} + c_2e^{-t} + 12e^{-2t}$

5. $x_1 = 5c_1\cos t + 5c_2\sin t + e^{-t}$
  $x_2 = (c_1 + 2c_2)\cos t + (-2c_1 + c_2)\sin t + c_3e^{-2t} + 2e^{-t}$

7. $x_1 = c_1e^{-t} + 4c_2e^{-2t} + c_3\cos t + c_4\sin t$, $\quad x_1 = e^{-t} + \sin t$
  $x_2 = c_1e^{-t} + 5c_2e^{-2t}$, $\quad x_2 = e^{-t}$

9. $x_1 = 5c_1e^{-t} - 2c_2e^{-2t} + c_3e^t$, $\quad x_1 = -2e^{-2t} - e^t$
  $x_2 = 2c_1e^{-t}$, $\qquad\qquad x_2 = 0$
  $x_3 = -2c_1e^{-t} + 3c_2e^{-2t}$, $\quad x_3 = 3e^{-2t}$

11. $x_1 = 2c_1 + c_2e^{-t} + c_3e^t + t^2$
  $x_2 = c_1 + c_2e^{-t} - t$
  $x_3 = -2c_1 - 2c_2e^{-t} - c_3e^t$

13. $x_1 = 7c_1e^t + c_2\cos 2t + c_3\sin 2t$
  $x_2 = 2c_1e^t + c_2\cos 2t + c_3\sin 2t$
  $x_3 = 3c_1e^t + (c_2 + c_3)\cos 2t + (-c_2 + c_3)\sin 2t$

15. $x_1 = c_1e^{-2t} + c_2e^{-t} + \cos t$
  $x_2 = c_1e^{-2t} + c_3e^{-t}$
  $x_3 = c_1e^{-2t} + \sin t$

17. $x_1 = 5(e^{-t/50} - e^{-3t/50})$, $\quad x_2 = 10(e^{-t/50} + e^{-3t/50})$

Section 6.4

1. (a) $x_1' = 2x_1 - x_2 + e^t$ (c) $x_1' = 2x_1 - x_2 + e^{2t}$
      $x_2' = 3x_1 + 3e^{-2t}$ $x_2' = x_2 + x_3$
                                         $x_3' = 3x_1 + 2x_2 + e^t$

2. (a) $A(t) = \begin{bmatrix} -2 & 1 \\ -2 & -1 \end{bmatrix}$, $\quad b(t) = \begin{bmatrix} \cos 2t \\ -2 \sin 2t \end{bmatrix}$ (c) $A(t) = \begin{bmatrix} 2 & 1 & -1 \\ 1 & -1 & 0 \\ 0 & 1 & 2 \end{bmatrix}$, $\quad b(t) = \begin{bmatrix} 2e^{-t} \\ -e^{-t} \\ 0 \end{bmatrix}$

3. (a) $A = \begin{bmatrix} 0 & 1 \\ -a_2 & -a_1 \end{bmatrix}$, $\quad b = \begin{bmatrix} 0 \\ f \end{bmatrix}$

Section 6.5

1. (a) Independent   (c) Dependent   2. (a) Dependent   (c) Dependent
3. (a) Independent   (c) Independent   5. (a) Yes   (b) No

7. $x(t) = 2 \begin{bmatrix} \cos 2t \\ \sin 2t \end{bmatrix} - 3 \begin{bmatrix} \sin 2t \\ -\cos 2t \end{bmatrix}$

Section 6.6

1. $x(t) = c_1 e^{2t} \begin{bmatrix} 1 \\ 1 \end{bmatrix} + c_2 e^{-3t} \begin{bmatrix} 4 \\ -1 \end{bmatrix}$, $\quad x(t) = 2e^{2t} \begin{bmatrix} 1 \\ 1 \end{bmatrix} - e^{-3t} \begin{bmatrix} 4 \\ -1 \end{bmatrix}$

3. $x(t) = c_1 \left(\begin{bmatrix} 1 \\ -1 \end{bmatrix} \cos 2t - \begin{bmatrix} 1 \\ 0 \end{bmatrix} \sin 2t \right) + c_2 \left(\begin{bmatrix} 1 \\ 0 \end{bmatrix} \cos 2t + \begin{bmatrix} 1 \\ -1 \end{bmatrix} \sin 2t \right),$

   $x(t) = \begin{bmatrix} 1 \\ 3 \end{bmatrix} \cos 2t + \begin{bmatrix} 7 \\ -4 \end{bmatrix} \sin 2t$

5. $x(t) = c_1 \begin{bmatrix} 1 \\ 0 \end{bmatrix} e^{-2t} + c_2 \begin{bmatrix} 0 \\ 1 \end{bmatrix} e^{-2t}$ 7. $x(t) = c_1 \begin{bmatrix} 1 \\ 0 \\ 2 \end{bmatrix} + c_2 \begin{bmatrix} 1 \\ 1 \\ -1 \end{bmatrix} e^{-t} + c_3 \begin{bmatrix} 1 \\ 0 \\ -1 \end{bmatrix} e^{-3t}$

9. $x(t) = c_1 \begin{bmatrix} 1 \\ -1 \\ -1 \end{bmatrix} + c_2 \begin{bmatrix} 0 \\ 1 \\ 0 \end{bmatrix} e^{-2t} + c_3 \begin{bmatrix} 3 \\ 0 \\ -1 \end{bmatrix} e^{-2t}$

11. $x(t) = c_1 \begin{bmatrix} 0 \\ 1 \\ -1 \end{bmatrix} e^{2t} + c_2 e^{2t} \left(\begin{bmatrix} 5 \\ -2 \\ 5 \end{bmatrix} \cos t + \begin{bmatrix} 0 \\ 1 \\ 0 \end{bmatrix} \sin t \right) + c_3 e^{2t} \left(\begin{bmatrix} 0 \\ -1 \\ 0 \end{bmatrix} \cos t + \begin{bmatrix} 5 \\ -2 \\ 5 \end{bmatrix} \sin t \right)$

Section 6.7

1. $x_1 = \begin{bmatrix} 1 \\ 2 \end{bmatrix} e^t$, $x_2 = \begin{bmatrix} 0 \\ 1 \end{bmatrix} e^t + \begin{bmatrix} 1 \\ 2 \end{bmatrix} te^t$ 3. $x_1 = \begin{bmatrix} 1 \\ -1 \end{bmatrix} e^{3t}$, $x_2 = \begin{bmatrix} 1 \\ 0 \end{bmatrix} e^{3t} + \begin{bmatrix} 2 \\ -2 \end{bmatrix} te^{3t}$

5. $\mathbf{x}_1 = \begin{bmatrix} 1 \\ -2 \end{bmatrix} e^{-t}, \quad \mathbf{x}_2 = \begin{bmatrix} 0 \\ 1 \end{bmatrix} e^{-t} + \begin{bmatrix} 2 \\ -4 \end{bmatrix} te^{-t}$

7. $\mathbf{x}_1 = \begin{bmatrix} 1 \\ 0 \\ 1 \end{bmatrix} e^t, \quad \mathbf{x}_2 = \begin{bmatrix} 0 \\ 1 \\ 0 \end{bmatrix} e^t + \begin{bmatrix} 1 \\ 0 \\ 1 \end{bmatrix} te^t, \quad \mathbf{x}_3 = \begin{bmatrix} 0 \\ 1 \\ -1 \end{bmatrix} e^{2t}$

9. $\mathbf{x}_1 = \begin{bmatrix} 1 \\ 0 \\ 1 \end{bmatrix} e^{-t}, \quad \mathbf{x}_2 = \begin{bmatrix} 0 \\ 2 \\ -3 \end{bmatrix} e^{-t}, \quad \mathbf{x}_3 = \begin{bmatrix} 0 \\ 0 \\ 1 \end{bmatrix} e^{-t} + \begin{bmatrix} 2 \\ 4 \\ -4 \end{bmatrix} te^{-t}$ 11. $u_1' = u_2, \quad u_2' = -u_1 - 2u_2$

12. (a) No   (c) Yes; $p(\lambda) = \lambda^4 + \lambda^3 + 3\lambda^2 + \lambda + 1$

Section 6.8

7. $\begin{bmatrix} 1 & 1 & 2 \\ 0 & 1 & 4 \\ 0 & 0 & 1 \end{bmatrix}$

Section 6.9

3. (a) $\begin{bmatrix} e^{4t} & 0 \\ 0 & e^{-t} \end{bmatrix}$ 8. (a) $\mathbf{x}(t) = \begin{bmatrix} 3e^t \\ 5e^{2t} \end{bmatrix}$ 9. (a) $\mathbf{x}(t) = \begin{bmatrix} 3\exp(t - t_0) \\ 5\exp 2(t - t_0) \end{bmatrix}$

Section 6.10

1. $e^{tA} = \frac{1}{2}(A + I)e^t - \frac{1}{2}(A - I)e^{-t} = \frac{1}{2}e^t \begin{bmatrix} 1 & 1 \\ 1 & 1 \end{bmatrix} - \frac{1}{2}e^{-t} \begin{bmatrix} -1 & 1 \\ 1 & -1 \end{bmatrix}, \quad \mathbf{x}(t) = e^t \begin{bmatrix} 2 \\ 2 \end{bmatrix} + e^{-t} \begin{bmatrix} 1 \\ -1 \end{bmatrix}$

3. $e^{tA} = [I + t(A - I)]e^t = e^t \begin{bmatrix} 1 & 0 \\ 0 & 1 \end{bmatrix} + te^t \begin{bmatrix} -1 & -1 \\ 1 & 1 \end{bmatrix}, \quad \mathbf{x}(t) = e^t \begin{bmatrix} 2 \\ -3 \end{bmatrix} + te^t \begin{bmatrix} 1 \\ -1 \end{bmatrix}$

5. $e^{tA} = (\cos 3t)I + \frac{1}{3}(\sin 3t)A = (\cos 3t) \begin{bmatrix} 1 & 0 \\ 0 & 1 \end{bmatrix} + \frac{1}{3}(\sin 3t) \begin{bmatrix} 1 & -5 \\ 2 & -1 \end{bmatrix}, \quad \mathbf{x}(t) = (\cos 3t) \begin{bmatrix} 3 \\ 0 \end{bmatrix} + (\sin 3t) \begin{bmatrix} 1 \\ 2 \end{bmatrix}$

7. $e^{tA} = -\frac{1}{2}e^t(A^2 - A - 2I) + \frac{1}{6}e^{-t}(A^2 - 3A + 2I) + \frac{1}{3}e^{2t}(A^2 - I)$

$= -\frac{1}{2}e^t \begin{bmatrix} -2 & 0 & 2 \\ 2 & 0 & -2 \\ 0 & 0 & 0 \end{bmatrix} + \frac{1}{6}e^{-t} \begin{bmatrix} -2 & -2 & 2 \\ 6 & 6 & -6 \\ -2 & -2 & 2 \end{bmatrix} + \frac{1}{3}e^{2t} \begin{bmatrix} 1 & 1 & 2 \\ 0 & 0 & 0 \\ 1 & 1 & 2 \end{bmatrix}$

$\mathbf{x}(t) = e^t \begin{bmatrix} 6 \\ -6 \\ 0 \end{bmatrix} + e^{-t} \begin{bmatrix} -2 \\ 6 \\ -2 \end{bmatrix} + e^{2t} \begin{bmatrix} -1 \\ 0 \\ -1 \end{bmatrix}$

9. $e^{tA} = e^t[I + t(A - I) + \frac{1}{2}t^2(A - I)^2]$

$= e^t \begin{bmatrix} 1 & 0 & 0 \\ 0 & 1 & 0 \\ 0 & 0 & 1 \end{bmatrix} + te^t \begin{bmatrix} 2 & -2 & 1 \\ 2 & -2 & 1 \\ -4 & 4 & 0 \end{bmatrix} + t^2e^t \begin{bmatrix} -2 & 2 & 0 \\ -2 & 2 & 0 \\ 0 & 0 & 0 \end{bmatrix}, \quad \mathbf{x}(t) = e^t \begin{bmatrix} 1 \\ 1 \\ 0 \end{bmatrix}$

11. $e^{tA} = \frac{1}{4}(A - 2I)^2 + \frac{1}{4}e^{2t}A(4I - A)[I + t(A - 2I)]$

$$= \frac{1}{4}\begin{bmatrix} 6 & 2 & -2 \\ 0 & 0 & 0 \\ 6 & 2 & -2 \end{bmatrix} + \frac{1}{4}e^{2t}\begin{bmatrix} -2 & -2 & 2 \\ 0 & 4 & 0 \\ -6 & -2 & 6 \end{bmatrix}, \quad x(t) = e^{2t}\begin{bmatrix} 0 \\ 2 \\ 2 \end{bmatrix}$$

14. $y = -2e^x + 5xe^x$

Section 6.11

1. $\mathbf{u}_p(t) = 5te^t \begin{bmatrix} 1 \\ 1 \end{bmatrix} + \frac{1}{2}(e^t - e^{-t})\begin{bmatrix} -3 \\ 9 \end{bmatrix}$ 3. $\mathbf{u}_p(t) = \begin{bmatrix} 2 - t + 2t^2 - 2e^{-t} \\ -3 + 3t - 2t^2 + 3e^{-t} \end{bmatrix}$

5. $\mathbf{u}_p(t) = \begin{bmatrix} -3\cos t - \sin t + 3e^t \\ 3\cos t + 3\sin t - 3e^t \end{bmatrix}$ 7. $\mathbf{u}_p(t) = e^t \begin{bmatrix} 0 \\ t \\ t \end{bmatrix}$ 10. (a) $\mathbf{u}_p(t) = (e^{-t} - e^t)\begin{bmatrix} 1 \\ 1 \end{bmatrix}$

Section 6.12

1. (b) In the equation $P(r) = ar^4 + br^2 + c = 0$, $b^2 - 4ac = (m_2 k_1 - m_1 k_2)^2 + m_2 k_2(m_2 k_2 + 2m_1 k_2 + 2m_2 k_1)$, which is positive. Hence $-b \pm (b^2 - 4ac)^{1/2}$ is negative.

3. (a) $m_1 \ddot{x}_1 = -k_1 + c(\dot{x}_2 - \dot{x}_1), \quad m_2 \ddot{x}_2 = -k_2 x_2 - c(\dot{x}_2 - \dot{x}_1)$

   (b) $x_1 = c_1 \cos 2t + c_2 \sin 2t + (c_3 + c_4 t)e^{-2t}, \quad x_2 = c_1 \cos 2t + c_2 \sin 2t - (c_3 + c_4 t)e^{-2t}$

5. (a) $v_0(2h/g)^{1/2}$ (b) $y = -\frac{1}{2}gx^2/v_0^2 + h$

7. (a) $(v_0^2/g)\sin 2\alpha$ (b) $y = -\dfrac{g}{2(v_0 \cos \alpha)^2} x^2 + (\tan \alpha)x$

9. $m\ddot{x} = -kx(x^2 + y^2)^{-3/2}, \quad m\ddot{y} = -ky(x^2 + y^2)^{-3/2}$

Section 6.13

1. $v = 17{,}470$ mph   9. Hyperbola $(\varepsilon = 1.04)$

Section 6.14

1. $LI_1'' + R_1 I_1' + \dfrac{1}{C}(I_1 - I_2) = 0, \quad R_2 I_2' + \dfrac{1}{C}(I_2 - I_1) = E'(t), \quad I_1(0) = I_1'(0) = 0, \quad I_2(0) = E(0)/R_2$

3. $L(I_1' - I_2') + R_1 I_1 = 0, \quad L(I_2' - I_1') + R_2 I_2 = E, \quad I_1(0) = E/(R_1 + R_2),$

   $I_1(t) = 2e^{-t/6}, \quad I_2(t) = 6 - 4e^{-t/6}$

5. $L_1 I_1' + (R_1 + R_2)I_1 - R_2 I_2 = 0, \quad R_2 I_2' - R_2 I_1' + \left(\dfrac{1}{C_1} + \dfrac{1}{C_2}\right)I_2 - \dfrac{1}{C_2} I_3 = 0$

   $R_3 I_3' + \dfrac{1}{C_2}(I_3 - I_2) = E'(t), \quad I_1(0) = I_2(0) = 0, \quad I_3(0) = E(0)/R_3$

Additional Exercises

1. $x_1 = c_1 \cos 3t + c_2 \sin 3t, \quad x_2 = (2c_1 + 3c_2)\cos 3t + (-3c_1 + 2c_2)\sin 3t$

3. $x_1 = 3c_1 e^t - c_2 e^{-t}, \quad x_2 = c_1 e^t + c_2 e^{-t} + c_3 e^{-2t}$ 5. $\mathbf{x} = c_1 e^{2t}\begin{bmatrix} 2 \\ 1 \end{bmatrix} + c_2 e^{-t}\begin{bmatrix} 1 \\ 1 \end{bmatrix}$

7. $x_1 = c_1 \cos 2t + c_2 \sin 2t + \cos t - 2 \sin t, \quad x_2 = -(c_1 + c_2)\cos 2t + (c_1 - c_2)\sin 2t + 4 \sin t$

9. $x_1 = e^{-2t}(2c_1 \cos t + 2c_2 \sin t) + \cos t + \sin t, \quad x_2 = e^{-2t}[(-c_1 + c_2)\cos t - (c_1 + c_2)\sin t] + \cos t$

11. $x_1 = (c_1 + c_2 t)e^{-t} + c_3 e^t, \quad x_2 = (c_1 + c_2 t)e^{-t}, \quad x_3 = c_2 e^{-t} + 3c_3 e^t$

13. $\mathbf{x} = c_1\left(\begin{bmatrix} 1 \\ -1 \\ 1 \end{bmatrix}\cos t + \begin{bmatrix} 1 \\ 0 \\ 0 \end{bmatrix}\sin t\right) + c_2\left(\begin{bmatrix} -1 \\ 0 \\ 0 \end{bmatrix}\cos t + \begin{bmatrix} 1 \\ -1 \\ 1 \end{bmatrix}\sin t\right) + c_3 e^{-2t}\begin{bmatrix} 1 \\ -1 \\ 0 \end{bmatrix}$

15. $\mathbf{x} = c_1\begin{bmatrix} 1 \\ 0 \\ 1 \end{bmatrix}e^{-t} + c_2 e^{-2t}\left(\begin{bmatrix} 1 \\ -1 \\ 1 \end{bmatrix}\cos t - \begin{bmatrix} 1 \\ 0 \\ 0 \end{bmatrix}\sin t\right) + c_3 e^{-2t}\left(\begin{bmatrix} 1 \\ 0 \\ 0 \end{bmatrix}\cos t + \begin{bmatrix} 1 \\ -1 \\ 1 \end{bmatrix}\sin t\right)$

18. (a) Yes   (c) No; $P(\lambda) = \lambda^4 - 5\lambda^3 - 73\lambda^2 + 908\lambda + 225$

20. (a) All solutions approach $(2, 1)$.   (c) All solutions approach $(-\frac{15}{4}, -\frac{7}{4}, 0)$.

21. $u_1' = u_2, \quad u_2' = -\dfrac{k_1}{m_1} u_1 + \dfrac{c}{m_1}(u_4 - u_2), \quad u_3' = u_4, \quad u_4' = -\dfrac{k_2}{m_2} u_3 - \dfrac{c}{m_2}(u_4 - u_2)$

23. $I_1 = e^{-5t/4}\left(3 \cos \dfrac{5\sqrt{7}}{4} t + \dfrac{9\sqrt{7}}{7} \sin \dfrac{5\sqrt{7}}{4} t\right), \quad I_2 = e^{-5t/4}\left(3 \cos \dfrac{5\sqrt{7}}{4} t - \dfrac{3\sqrt{7}}{7} \sin \dfrac{5\sqrt{7}}{4} t\right)$

25. $mx'' = -k_1(x - a\theta) - k_2(x + a\theta), \quad mr^2\theta'' = k_1 a(x - a\theta) - k_2 a(x + a\theta)$

Chapter 7

Section 7.1

1. $(0, 2)$ 3. $(1, 3)$ 5. All x 7. $(-1, 1)$

9. $f(x) + g(x) = \displaystyle\sum_{n=0}^{\infty} (n + 2)(x - 2)^n, \quad |x - 2| < 1$ $f(x)g(x) = \dfrac{1}{2}\displaystyle\sum_{n=0}^{\infty} (n + 1)(n + 2)(x - 2)^n, \quad |x - 2| < 1$

11. $f(x) + g(x) = \displaystyle\sum_{n=0}^{\infty} \dfrac{n^2 + 2n + 2}{n + 1} x^n, \quad |x| < 1$ $f(x)g(x) = \displaystyle\sum_{n=0}^{\infty} \left(\sum_{k=0}^{n} \dfrac{n - k + 1}{k + 1}\right)x^n, \quad |x| < 1$

13. $\displaystyle\sum_{n=0}^{\infty} (n + 2)(n + 1)x^n$ 15. $\displaystyle\sum_{n=1}^{\infty} (n^2 - 2n + 3)x^n$

17. $f'(x) = \displaystyle\sum_{n=1}^{\infty} \dfrac{(-1)^n n x^{n-1}}{2^n(n + 1)}, \quad f''(x) = \displaystyle\sum_{n=2}^{\infty} \dfrac{(-1)^n n(n - 1)x^{n-2}}{2^n(n + 1)}, \quad |x| < 2$

Section 7.2

1. $3 + 13(x - 2) + 9(x - 2)^2 + 2(x - 2)^3$ 3. $\displaystyle\sum_{n=1}^{\infty} \dfrac{(-1)^{n+1}}{n}(x - 1)^n$

5. $1 + \frac{1}{2}x + \displaystyle\sum_{n=2}^{\infty} (-1)^{n+1} \dfrac{1 \cdot 3 \cdot 5 \cdots (2n - 3)}{2^n n!} x^n$ 7. $\displaystyle\sum_{n=0}^{\infty} (-1)^n\left(\dfrac{x}{4}\right)^n, \quad |x| < 4$

9. $\dfrac{2}{3} \displaystyle\sum_{n=0}^{\infty} [1 - (-\tfrac{1}{2})^{n+1}]x^n$, $|x| < 1$ 11. $\displaystyle\sum_{n=0}^{\infty} (n+1)x^n$, $|x| < 1$ 13. $\displaystyle\sum_{n=0}^{\infty} \dfrac{x^{2n+1}}{2n+1}$, $|x| < 1$

15. $x\dfrac{d}{dx}(xe^x) = (x^2 + x)e^x$ 17. $\dfrac{d}{dx}\left[x\displaystyle\int_0^x \dfrac{1}{1-t}\,dt\right] = \dfrac{x}{1-x} - \ln(1-x)$

Section 7.3

1. (a) $x = -2, 0, 1$ (c) None

3. $y = A_0\left[1 + \displaystyle\sum_{m=1}^{\infty} \dfrac{x^{2m}}{2^m 1 \cdot 3 \cdot 5 \cdots (2m-1)}\right] + A_1 \displaystyle\sum_{m=1}^{\infty} \dfrac{x^{2m-1}}{2^{2m-2}(m-1)!}$,   for all x

5. $y = A_0\left[1 + \displaystyle\sum_{m=1}^{\infty} \dfrac{(-1)^m 2^m m! x^{2m}}{1 \cdot 3 \cdot 5 \cdots (2m-1)}\right] + A_1 \displaystyle\sum_{m=1}^{\infty} (-1)^{m+1} \dfrac{1 \cdot 3 \cdot 5 \cdots (2m-1)}{2^{m-1}(m-1)!} x^{2m-1}$, $|x| < \sqrt{2}$

7. $y = A_0 \displaystyle\sum_{m=0}^{\infty} \dfrac{x^{3m}}{3^m m!} + A_1 \displaystyle\sum_{m=0}^{\infty} \dfrac{x^{3m+1}}{1 \cdot 4 \cdot 7 \cdots (3m+1)}$,   for all x

9. $y = A_0(1 + \tfrac{1}{2}x^2 - \tfrac{1}{6}x^3 + \tfrac{1}{8}x^4 + \cdots) + A_1(x + \tfrac{1}{6}x^3 - \tfrac{1}{12}x^4 + \cdots)$, $|x| < 1$

11. $y = A_0 \displaystyle\sum_{m=0}^{\infty} \dfrac{(-1)^m (x-1)^{2m}}{2^m m!} + A_1 \displaystyle\sum_{m=1}^{\infty} \dfrac{(-1)^{m+1}(x-1)^{2m-1}}{1 \cdot 3 \cdot 5 \cdots (2m-1)}$,   for all x

13. $y = A_0\left[1 - 3(x-2)^2 + (x-2)^4 + 3\displaystyle\sum_{m=3}^{\infty} \dfrac{(x-2)^{2m}}{(2m-3)(2m-1)}\right] + A_1[(x-2) - (x-2)^3]$, $|x-2| < 1$

Section 7.4

1. (a) $0, -1$ (c) $-\tfrac{1}{2}$

3. $y = c_1 x\left[1 + \displaystyle\sum_{n=1}^{\infty} \dfrac{x^n}{n!5 \cdot 7 \cdot 9 \cdots (2n+3)}\right] + c_2 x^{-1/2}\left[1 - x - \displaystyle\sum_{n=2}^{\infty} \dfrac{x^n}{n!1 \cdot 3 \cdot 5 \cdots (2n-3)}\right]$

5. $y = c_1 x^{1/3}\left[1 + \displaystyle\sum_{n=1}^{\infty} \dfrac{(-1)^n x^n}{n!4 \cdot 7 \cdot 10 \cdots (3n+1)}\right] + c_2\left[1 + \displaystyle\sum_{n=1}^{\infty} \dfrac{(-1)^n x^n}{n!2 \cdot 5 \cdot 8 \cdots (3n-1)}\right]$

7. $y = c_1 x^{1/3} \displaystyle\sum_{m=0}^{\infty} 4\dfrac{(-1)^m x^{2m}}{2^m m!(6m+4)} + c_2 x^{-1}$

9. $y = c_1 x^{-1} \displaystyle\sum_{n=0}^{\infty} \dfrac{(n+1)!x^n}{1 \cdot 3 \cdot 5 \cdots (2n+1)} + c_2 x^{-3/2} \displaystyle\sum_{n=0}^{\infty} \dfrac{1 \cdot 3 \cdot 5 \cdots (2n+1)}{4^n n!} x^n$

11. $y = c_1(x-1)^{1/3}[1 - \tfrac{1}{5}(x-1)] + c_2(x-1)^{-1/3}\left[1 + \displaystyle\sum_{n=1}^{\infty} \dfrac{(-5)(-2)1 \cdot 4 \cdots (3n-8)}{3^n n!1 \cdot 4 \cdot 7 \cdots (3n-2)}(x-1)^n\right]$

Section 7.5

1. $y(x) = c_1 y_1(x) + c_2\left[y_1(x)\ln x - x^2 \displaystyle\sum_{n=1}^{\infty} nx^n\right]$, $y_1(x) = x^2 \displaystyle\sum_{n=0}^{\infty} (n+1)x^n$

3. $y(x) = c_1 y_1(x) + c_2\left[y_1(x)\ln x - 2x \displaystyle\sum_{n=1}^{\infty} \dfrac{\phi(n)}{(n!)^2} x^n\right]$, $y_1(x) = x \displaystyle\sum_{n=0}^{\infty} \dfrac{x^n}{(n!)^2}$

5. $y(x) = c_1 y_1(x) + c_2\left[y_1(x)\ln x - \displaystyle\sum_{m=1}^{\infty} \dfrac{\phi(m)}{2^m(m!)^2} x^{2m}\right]$, $y_1(x) = \displaystyle\sum_{m=0}^{\infty} \dfrac{x^{2m}}{2^m(m!)^2}$

7. $y(x) = c_1 y_1(x) + c_2 \left[y_1(x)\ln x - 2x^{-2} \sum_{n=1}^{\infty} \frac{\phi(n)}{(n!)^2} x^n \right]$, $\quad y_1(x) = x^{-2} \sum_{n=0}^{\infty} \frac{x^n}{(n!)^2}$

9. $y(x) = c_1 y_1(x) + c_2 \left[y_1(x)\ln x - 2x^2 + \sum_{n=2}^{\infty} \frac{(-1)^{n+1} x^{n+1}}{n(n-1)} \right]$, $\quad y_1(x) = x + x^2$

Section 7.6

The general solution is $y = c_1 y_1 + c_2 y_2$, where y_1 and y_2 are as given.

1. $y_1(x) = x \sum_{n=0}^{\infty} \frac{x^n}{n!} = xe^x$, $\quad y_2(x) = y_1(x)\ln x + 1 - \sum_{n=2}^{\infty} \frac{\phi(n-1)}{(n-1)!} x^n$

3. $y_1(x) = x^{-1}(1 - 3x^2 + 3x^4 - x^6)$, $\quad y_2(x) = x^{-2} \left[1 + \sum_{m=1}^{\infty} \frac{(-7)(-5)(-3) \cdots (2m-9)}{1 \cdot 3 \cdot 5 \cdots (2m-1)} x^{2m} \right]$

5. $y_1(x) = x^2 \sum_{n=0}^{\infty} \frac{x^n}{n!(n+1)!}$, $\quad y_2(x) = y_1(x)\ln x + x - x^2 - x \sum_{n=2}^{\infty} \frac{\phi(n-1) + \phi(n)}{(n-1)!n!} x^n$

7. $y_1(x) = 1 + 2 \sum_{n=1}^{\infty} \frac{x^n}{(n+2)!}$, $\quad y_2(x) = x^{-2}(1+x)$

9. $y_1(x) = \sum_{n=0}^{\infty} \frac{x^{n+1}}{n!(n+1)!}$, $\quad y_2(x) = y_1(x)\ln x + 1 - x - \sum_{n=2}^{\infty} \frac{\phi(n) + \phi(n-1)}{(n-1)!n!} x^n$

11. $y_1(x) = \sum_{n=0}^{\infty} \frac{2^{n+1} x^{n+1}}{n!(n+2)!}$, $\quad y_2(x) = -2y_1(x)\ln x + x^{-1} \left[1 - 2x + x^2 + \sum_{n=3}^{\infty} \frac{2^n[\phi(n) + \phi(n-2) - 1]}{(n-2)!n!} x^n \right]$

13. $y_1(x) = x \sum_{n=0}^{\infty} x^n = \frac{x}{1-x}$, $\quad y_2(x) = x^{-3}(1 + x + x^2 + x^3)$

Section 7.7

1. (a) 0, ∞ (neither regular)   (c) 2 (regular), ∞ (not regular)   (e) ∞ (regular), 0 (regular)

3. $y(x) = c_1 \sum_{m=0}^{\infty} \frac{x^{-2m}}{2^m m!} + c_2 \sum_{m=1}^{\infty} \frac{x^{-2m+1}}{1 \cdot 3 \cdot 5 \cdots (2m-1)}$

5. $y(x) = c_1 x^{-1} \sum_{n=0}^{\infty} (-1)^n \frac{2n+3}{3n!} x^{-n} + c_2 x^{-1/2} \left[1 + \sum_{n=1}^{\infty} \frac{(-1)^n 2^n (n+1)}{1 \cdot 3 \cdot 5 \cdots (2n-1)} x^{-n} \right]$

7. $y(x) = c_1 \sum_{m=0}^{\infty} (m+1)x^{-2m} + c_2 \left[y_1(x)\ln x + \frac{1}{2} \sum_{m=1}^{\infty} mx^{-2m} \right]$

Section 7.9

1. (a) $\sqrt{\pi}/2$ (c) $-2\sqrt{\pi}$ 4. (a) 0.990   (c) 0.100

6. (a) $J_3(x) = \frac{8 - x^2}{x^2} J_1(x) - \frac{4}{x} J_0(x)$ (b) $J_2'(x) = \left(1 - \frac{4}{x^2} \right) J_1(x) + \frac{2}{x} J_0(x)$

10. (a) $y = x^{1/2}[c_1 J_{1/4}(x^2/2) + c_2 J_{-1/4}(x^2/2)]$ (c) $y = c_1 x^{-1/3} J_{1/3}(\frac{1}{2}x^2) + c_2 x^{-1/3} J_{-1/3}(\frac{1}{2}x^2)$
    (e) $y = c_1 x^{1/2} J_0(\frac{1}{2}x^2) + c_2 x^{1/2} Y_0(\frac{1}{2}x^2)$

Additional Exercises

1. $y(x) = A_0\left[1 + \displaystyle\sum_{m=1}^{\infty} \frac{x^{3m}}{3\cdot 6\cdot 9\cdots (3m)}\right] + A_1\left[x + \displaystyle\sum_{m=1}^{\infty} \frac{x^{3m+1}}{4\cdot 7\cdot 10\cdots (3m+1)}\right]$

3. $y = A_0\left[1 + \displaystyle\sum_{m=1}^{\infty} (-1)^m \frac{3\cdot 5\cdot 7\cdots (2m+1)}{(2m)!} x^{2m}\right] + A_1\left[x + \displaystyle\sum_{m=2}^{\infty} (-1)^{m+1} \frac{4\cdot 6\cdot 8\cdots (2m)}{(2m-1)!} x^{2m-1}\right]$

5. $y(x) = c_1 x^{1/2} \displaystyle\sum_{n=0}^{\infty} \frac{x^n}{2^n n!} + c_2 x^{-1}\left[1 - x - \displaystyle\sum_{n=2}^{\infty} \frac{x^n}{1\cdot 3\cdot 5\cdots (2n-3)}\right]$

7. $y_1(x) = \displaystyle\sum_{n=0}^{\infty} \frac{x^n}{(n!)^2}$, $y_2(x) = y_1(x)\ln x - 2\displaystyle\sum_{n=1}^{\infty} \frac{\phi(n)}{(n!)^2} x^n$

9. $y_1(x) = x^{-1}\displaystyle\sum_{n=0}^{\infty} \frac{x^n}{n!(n+1)!}$, $y_2(x) = y_1(x)\ln x + x^{-2}\left[1 - x - \displaystyle\sum_{n=2}^{\infty} \frac{\phi(n-1) + \phi(n)}{(n-1)!n!} x^n\right]$

11. $y_1(x) = x^{-2}\displaystyle\sum_{n=0}^{\infty} \frac{x^n}{n!} = x^{-2}e^x$, $y_2(x) = y_1(x)\ln x - x^{-2}\displaystyle\sum_{n=1}^{\infty} \frac{\phi(n)}{n!} x^n$

13. $y_1(x) = x\left[1 + \dfrac{1}{6}\displaystyle\sum_{n=1}^{\infty} \frac{(n+2)(n+3)}{n!} x^n\right]$

$y_2(x) = 3y_1(x) + 1 - 2x + \dfrac{1}{2}\displaystyle\sum_{n=2}^{\infty} \frac{(n+1)(n+2)}{(n-1)!}\left[-\dfrac{3}{2} + \dfrac{1}{n+1} + \dfrac{1}{n+2} - \phi(n-1)\right]x^n$

15. $y_1(x) = 1 - \tfrac{1}{2}x + \tfrac{1}{10}x^2$, $y_2(x) = x^{-3}(1 - 5x + 10x^2)$

17. All values of a except $a = 1$; $y_1(x) = \dfrac{1}{1-x}$, $y_2(x) = \dfrac{x^{1-a}}{1-x}$

18. (a) $y = x^{-1/2}[c_1 J_{1/2}(2x) + c_2 J_{-1/2}(2x)] = x^{-1}(c_1' \sin 2x + c_2' \cos 2x)$

(c) $y = x^{-1}[c_1 J_1(x^2) + c_2 Y_1(x^2)]$

21. $y = 1 + \displaystyle\sum_{k=1}^{n} \frac{(0^2 - n^2)(1^2 - n^2)\cdots[(k-1)^2 - n^2]}{k!1\cdot 3\cdot 5\cdots 2(k-1)}(x-1)^k$

--- Chapter 8 ---

Section 8.1

1.

x	Euler	Exact
0.1	1.60000	1.63873
0.2	1.28400	1.35032
0.3	1.04320	1.12881

3.

x	h = 0.1	h = 0.05
0.1	0.00000	0.00250
0.2	0.01000	0.01500
0.3	0.03001	0.03755

6. (a)

x	Euler	Modified Euler	Exact
0.1	1.60000	1.64200	1.63873
0.2	1.28400	1.35604	1.35032

Section 8.2

1. $w_1 = 1.64000$, $w_2 = 1.35240$, $w_3 = 1.13137$

3. $w_1 = 0.005000$, $w_2 = 0.020008$, $w_3 = 0.045128$

5. $w_1 = 1.32000$, $w_2 = 1.68840$

7. $w_1 = 1.32133$, $w_2 = 1.69165$

Section 8.3

1. $w_1 = 1.64200$, $w_2 = 1.35604$, $w_3 = 1.13635$ 3. $w_1 = 0.00500$, $w_2 = 0.02001$, $w_3 = 0.04511$
5. $w_1 = 1.63860$, $w_2 = 1.35009$

Section 8.4

1. $w_4^{(0)} = 2.55153$, $w_4 = 2.61057$ 5. $w_3^{(0)} = 0.76876$, $w_3 = 0.76930$

Section 8.5

1. (a) $x_n = x_{n-1} + h(-5x_{n-1} + 4y_{n-1} - 6)$, $y_n = h(-3x_{n-1} + 2y_{n-1} - 2)$
   (b) $x = -10e^{-t} + 8e^{-2t} + 2$, $y = -10e^{-t} + 6e^{-2t} + 4$

(c), (d)

t	x(h=0.1)	y(h=0.1)	x(h=0.01)	y(h=0.01)	x-exact	y-exact
0.00	0.000000	0.000000	0.000000	0.000000	0.000000	0.000000
0.50	-1.283460	0.061180	-1.136703	0.134957	-1.122271	0.141970
1.00	-0.627791	1.157461	-0.599367	1.135394	-0.596112	1.133217
1.50	0.222564	2.152195	0.171850	2.075258	0.166995	2.067421
2.00	0.876467	2.853409	0.800907	2.765731	0.793172	2.756541
2.50	1.312325	3.304769	1.240655	3.227845	1.233054	3.219578
3.00	1.585992	3.583516	1.528251	3.523586	1.521959	3.517002
3.50	1.752930	3.752118	1.710095	3.708396	1.705321	3.703497
4.00	1.853255	3.852989	1.822969	3.822350	1.819527	3.818856

3. (a) $x_n = x_{n-1} + \frac{1}{6}(a_1 + 2a_2 + 2a_3 + a_4)$, $y_n = y_{n-1} + \frac{1}{6}(b_1 + 2b_2 + 2b_3 + b_4)$
   $a_1 = h(-5x_{n-1} + 4y_{n-1} - 6)$, $b_1 = h(-3x_{n-1} + 2y_{n-1} - 2)$
   $a_2 = h(-5x_{n-1} - \frac{5}{2}a_1 + 4y_{n-1} + 2b_1 - 6)$, $b_2 = h(-3x_{n-1} - \frac{3}{2}a_1 + 2y_{n-1} + b_1 - 2)$
   $a_3 = h(-5x_{n-1} - \frac{5}{2}a_2 + 4y_{n-1} + 2b_2 - 6)$, $b_3 = h(-3x_{n-1} - \frac{3}{2}a_2 + 2y_{n-1} + b_2 - 2)$
   $a_4 = h(-5x_{n-1} - 5a_3 + 4y_{n-1} + 4b_3 - 6)$, $b_4 = h(-3x_{n-1} - 3a_3 + 2y_{n-1} + 2b_3 - 2)$

(c), (d)

t	x(h=0.1)	y(h=0.1)	x(h=0.01)	y(h=0.01)	x-exact	y-exact
0.00	0.000000	0.000000	0.000000	0.000000	0.000000	0.000000
0.50	-1.122227	0.142002	-1.122271	0.141970	-1.122271	0.141970
1.00	-0.596081	1.133240	-0.596112	1.133217	-0.596112	1.133217
1.50	0.167011	2.067432	0.166995	2.067421	0.166995	2.067421
2.00	0.793179	2.756545	0.793172	2.756541	0.793172	2.756541
2.50	1.233056	3.219579	1.233054	3.219578	1.233054	3.219578
3.00	1.521960	3.517002	1.521959	3.517002	1.521959	3.517002
3.50	1.705321	3.703497	1.705321	3.703497	1.705321	3.703497
4.00	1.819527	3.818856	1.819527	3.818856	1.819527	3.818856

5. (a) $x_n = x_{n-1} + hy_{n-1}$, $y_n = y_{n-1} + h(-5x_{n-1} - 2y_{n-1})$
   (b), (c) $x = 5e^{-t} \sin 2t$, $y = e^{-t}(10 \cos 2t - 5 \sin 2t)$

t	x(h=0.1)	y(h=0.1)	x(h=0.01)	y(h=0.01)	x-exact	y-exact
0.00	0.000000	10.000000	0.000000	10.000000	0.000000	10.000000
0.50	2.958100	0.102800	2.587898	0.663072	2.551890	0.725209
1.00	1.810890	-4.374121	1.682636	-3.304642	1.672559	-3.203478
1.50	-0.203934	-2.723362	0.127261	-2.396367	0.157441	-2.366412
2.00	-0.928346	0.273633	-0.545849	-0.323566	-0.512110	-0.372500
2.50	-0.477828	1.375883	-0.402450	0.684846	-0.393566	0.626411
3.00	0.119395	0.720876	-0.057754	0.566160	-0.069556	0.547597
3.50	0.285106	-0.169181	0.112795	0.112271	0.099196	0.128462
4.00	0.121560	-0.423426	0.094914	-0.138506	0.090604	-0.117253

7.

t	x(h=0.1)	y(h=0.1)	x(h=0.01)	y(h=0.01)	x-exact	y-exact
0.00	0.000000	10.000000	0.000000	10.000000	0.000000	10.000000
0.50	2.551967	0.725131	2.551890	0.725209	2.551890	0.725209
1.00	1.672609	-3.203685	1.672559	-3.203478	1.672559	-3.203478
1.50	0.157405	-2.366530	0.157441	-2.366413	0.157441	-2.366412
2.00	-0.512178	-0.372450	-0.512110	-0.372500	-0.512110	-0.372500
2.50	-0.393600	0.626523	-0.393566	0.626411	-0.393566	0.626411
3.00	-0.069545	0.547658	-0.069556	0.547597	-0.069556	0.547597
3.50	0.099222	0.128451	0.099196	0.128462	0.099196	0.128462
4.00	0.090618	-0.117291	0.090604	-0.117253	0.090604	-0.117253

9. $x_n = x_{n-1} + hf + \frac{1}{2}h^2(f_t + f_x f + f_y g)$, $y_n = y_{n-1} + hg + \frac{1}{2}h^2(g_t + g_x f + g_y g)$ where f, g, f_t, g_t, f_x, f_y, g_x, g_y are evaluated at $(t_{n-1}, x_{n-1}, y_{n-1})$.

10. (a) $x_n = x_{n-1} + h(-5x_{n-1} + 4y_{n-1} - 6) + \frac{1}{2}h^2(13x_{n-1} - 12y_{n-1} + 22)$
    $y_n = y_{n-1} + h(-3x_{n-1} + 2y_{n-1} - 2) + \frac{1}{2}h^2(9x_{n-1} - 18y_{n-1} + 14)$

11. $x_n = x_{n-1} + hy_{n-1} + \frac{1}{2}h^2 F$, $y_n = y_{n-1} + hF + \frac{1}{2}(F_t + y_{n-1}F_x + F_y F)$
    where F, F_t, F_x, F_y are evaluated at $(t_{n-1}, x_{n-1}, y_{n-1})$.

12. (a) $x_n = x_{n-1} + hy_{n-1} + \frac{1}{2}h^2(-5x_{n-1} - 2y_{n-1})$
    $y_n = y_{n-1} + h(-5x_{n-1} - 2y_{n-1}) + \frac{1}{2}h^2(10x_{n-1} - y_{n-1})$

Additional Exercises

1. (a) $w_1 = 1.200000$, $w_2 = 1.418182$ (b) $w_1 = 1.210000$, $w_2 = 1.440000$
3. Exact values are given by $y = (x + 1)^2$. The fourth-order Runge-Kutta method gives values correct to 6 decimal places.

5.

x	y-Euler	y-RK
0.00	0.500000	0.500000
0.50	0.949306	0.950734
1.00	1.422083	1.422469
1.50	1.741242	1.740097
2.00	1.897151	1.895830
2.50	1.961254	1.960373
3.00	1.985710	1.985237
3.50	1.994772	1.994544
4.00	1.998093	1.997989

9.

t	x	y
0.00	1.000000	0.000000
0.50	0.507907	0.175351
1.00	0.103803	0.144371
1.50	0.002559	0.003442
2.00	0.002400	0.005369

13. $x_{i,n} = x_{1,n-1} + hf_i(t_{n-1}, x_{1,n-1}, x_{2,n-1}, x_{3,n-1}, x_{4,n-1}),$ $i = 1, 2, 3, 4$

Chapter 9

Section 9.1

1. (a) $\dfrac{1}{s}$ (c) $\dfrac{2}{s^3}$ (e) $\dfrac{a}{s^2 + a^2}$ (g) $\dfrac{1 - e^{-s}}{s}$ (i) $\dfrac{1 - 2e^{-s} + e^{-2s}}{s^2}$

2. (a) $\dfrac{1}{s+5}$ (c) $\dfrac{s}{s^2 + 25}$ (e) $\dfrac{3}{s+2} + \dfrac{4}{s+1}$ 3. (a) e^{-2t} (c) $\cos 2t$ (e) $3e^{2t} - 5e^{-7t}$

5. (a) $\pi^{1/2} 2^{-1} s^{-3/2}$ 7. (a) $\dfrac{2}{s^2(1 - e^{-s})}(e^{-s} - 2e^{-s/2} + 1)$ 8. (a) $x(t) = -2e^{2t} - 1$

Section 9.3

2. (a) $\dfrac{2}{s+1} - \dfrac{12}{s^2 + 16}$ (c) $\dfrac{3}{(s-2)^2 + 9}$ (e) $\dfrac{24}{(s+3)^5}$ (g) $\dfrac{2s^3 - 6s}{(s^2 + 1)^3}$

3. (a) $\dfrac{\sqrt{\pi}}{2(s-2)^{3/2}}$ (c) $\dfrac{2}{s(s-1)^3}$ (e) $\dfrac{1}{s}e^{-2s}$ (g) $\left(\dfrac{2}{s^3} + \dfrac{2}{s^2} + \dfrac{1}{s}\right)e^{-s}$

4. (a) $s^2 F(s) - s - 2$ (c) $s^3 F(s) - s^2 + s - 5$ (e) $s^4 F(s) - 2s^3$

Section 9.4

1. (a) te^{-t} (c) $e^{-t}\cosh 2t$ (e) $e^{-t} - e^{-2t}$

2. (a) $2e^t - 2e^{-2t}$ (c) $\tfrac{8}{5}e^t + 3te^t - \tfrac{8}{5}\cos 3t - \tfrac{6}{5}\sin 3t$

  (e) $\dfrac{1}{12}e^{2t} - \dfrac{1}{12}e^{-t}\cos\sqrt{3}\,t - \dfrac{\sqrt{3}}{9}e^{-t}\sin\sqrt{3}\,t$

4. (a) $\dfrac{t^3}{6}$ (c) $e^{2t} - e^t$

5. (a) $f(t) = \begin{cases} 0, & 0 \le t \le 1, \\ (t-1)e^{2(t-1)}, & t > 1 \end{cases}$ (c) $\left(\dfrac{1}{\sqrt{\pi}}\right)\displaystyle\int_0^t \sqrt{u}\,e^{2u}\cos(t-u)\,du$ (e) $\displaystyle\int_0^t f(t-u)e^{-u}\,du$

Section 9.5

1. (a) $x = e^{-2t} + e^t$ (c) $x = -\frac{1}{3}\sin 2t + \frac{8}{3}\sin t$ (e) $x = e^{-2t} + \cos t + 1$

3. $x = \begin{cases} \cosh t, & 0 \le t \le 1, \\ \cosh t + (1 - t) + \sinh(t - 1), & t > 1 \end{cases}$ 5. (a) $x = 1 + e^{2t}$

6. (a) $x_1 = e^t - \frac{1}{3}e^{-3t} - \frac{2}{3}$, $x_2 = \frac{3}{2}e^t + \frac{1}{6}e^{-3t} - \frac{2}{3}$
   (c) $x_1 = e^{-t}(2\cos t - \sin t)$, $x_2 = e^{-t}(1 - \cos t + 3\sin t)$
   (e) $x_1 = -\frac{2}{5}e^{-t}\cos 2t + \frac{4}{5}e^{-t}\sin 2t + 2t + \frac{12}{5}$, $x_2 = \frac{8}{5}e^{-t}\cos 2t + \frac{4}{5}e^{-t}\sin 2t + 2t + \frac{12}{5}$

Section 9.6

1. (a) $f(t) = \begin{cases} 1, & t \le 5 \\ 0, & t > 5 \end{cases}$ (c) $h(t) = \begin{cases} 0, & t < 2 \\ t^2, & t \ge 2 \end{cases}$ (e) $p(t) = \begin{cases} (t - 5)^2, & t < 5 \\ 0, & t \ge 5 \end{cases}$

2. (a) $f(t) = t^2[1 - u(t - 2)]$ (c) $h(t) = 2[u(t - 1) - u(t - 6)]$

3. (a) $\dfrac{4}{s}e^{-3s}$ (c) $\dfrac{1}{s + 2}e^{-3s}$ (e) $\dfrac{s}{s^2 + 9}e^{-\pi s/2}$ (g) $\dfrac{\pi}{s^2 + \pi^2}(1 + e^{-s})$ (i) $\dfrac{1}{s} + \dfrac{2}{s}e^{-s} + \dfrac{1}{s^2}e^{-2s}$

4. (a) $2u(t - 3)$ (c) $e^{-4(t - 2)}u(t - 2)$ (e) $3\sin 2\left(t - \dfrac{\pi}{2}\right)u\left(t - \dfrac{\pi}{2}\right)$

   (g) $\dfrac{3}{\pi}(t - 2)\sin \pi(t - 2)u(t - 2)$

5. $y(t) = -\frac{1}{4} + \frac{1}{2}t + \frac{1}{4}e^{-2t} + \left[\dfrac{1}{4} - \dfrac{t}{2} + \dfrac{1}{4}e^{-2(t-1)}\right]u(t - 1)$

6. (a) $y(t) = e^{-2t} + \frac{1}{2}[u(t - 1) - u(t - 5)] - \frac{1}{2}e^{-2(t-1)}u(t - 1) + \frac{1}{2}e^{-2(t-5)}u(t - 5)$
   (c) $y(t) = -1 + t + e^{-t} - 2[-2 + t + e^{-(t-1)}]u(t - 1) + [-3 + t + e^{-(t-2)}]u(t - 2)$

7. (a) $y(t) = 1 - \cos t - [1 - \cos(t - 2)]u(t - 2)$
   (c) $y(t) = 2e^{-t} - e^{-2t} + [\frac{1}{2} - e^{-(t-1)} + \frac{1}{2}e^{-2(t-1)}]u(t - 1)$

8. (a) $x = \begin{cases} \frac{1}{4} + \frac{1}{2}t - \frac{1}{4}e^{-2t}, & 0 < t < 1 \\ \frac{1}{2} + \frac{1}{4}(e^2 - 1)e^{-2t}, & t > 1 \end{cases}$, $y = \begin{cases} -\frac{1}{4} + \frac{1}{2}t + \frac{1}{4}e^{-2t}, & 0 < t < 1 \\ \frac{1}{2} + \frac{1}{4}(1 - e^2)e^{-2t}, & t > 1 \end{cases}$

9. (b) $\dfrac{1}{s}\dfrac{1 - e^{-s}}{1 - e^{-2s}} = \dfrac{1}{s(1 + e^{-s})}$

Additional Exercises

1. (a) $\dfrac{3}{s^2 + 4s + 13}$ (c) $\dfrac{24}{(s - 1)^5}$ (e) $\dfrac{1}{1 + s^2}\dfrac{1 + e^{-\pi s}}{1 - e^{-\pi s}}$

3. (a) $(1 + 2t)e^{2t}$ (c) $\frac{1}{5}(-e^{-t} + \cos 3t + 3\sin 3t)$

   (e) $\dfrac{1}{\pi}[\sin \pi t - \sin \pi(t - 2)u(t - 2)] = \dfrac{1}{\pi}\sin \pi t[1 - u(t - 2)]$

5. If $f(t) = 1$ then $(f * f)(t) = t$ 7. (a) $\dfrac{1 - 2e^{-s} + e^{-2s}}{s^2(1 - e^{-2s})}$ (c) $\dfrac{1 - e^{-s}}{s(1 - e^{-2s})}$

10. (a) $x(t) = -1 + t + e^{-t}$ (c) $x(t) = \dfrac{1}{\pi^2 + 4}(\pi e^{-2t} - \pi \cos \pi t + 2\sin \pi t)$

11. (a) $x(t) = \frac{3}{5}(2e^{-t} - 2\cos 2t + \sin 2t)$ (c) $x(t) = -2 + t + 3e^{-t} + 4te^{-t}$

12. (a) $x(t) = y(t) = -\frac{1}{4} + \frac{1}{8}e^{2t} + \frac{1}{8}e^{-2t}$

(c) $x = \begin{cases} -\frac{5}{9} + \frac{2}{3}t + \frac{1}{2}e^{-t} + \frac{1}{18}e^{-3t}, & 0 < t < 1 \\ \frac{1}{2}e^{-t} + \frac{1}{18}(1 + 2e^3)e^{-3t}, & t > 1 \end{cases}$, $y = \begin{cases} -\frac{4}{9} + \frac{1}{3}t + \frac{1}{2}e^{-t} - \frac{1}{18}e^{-3t}, & 0 < t < 1 \\ \frac{1}{2}e^{-t} - \frac{1}{18}(1 + 2e^3)e^{-3t}, & t > 1 \end{cases}$

Chapter 10

Section 10.1

1. (a) $(\frac{7}{3}, -\frac{1}{3})$ (c) All points on $x_1 - 3x_2 = 0$ (e) $(-2, 3), (1, -2)$ (g) None
2. (a) $(\pm n\pi, 0)$ (c) $(2, 0)$ 3. $x_1 = c_1 e^t, x_2 = c_2 e^{-2t}$; unstable; $x_1^2 x_2 = c_1^2 c_2$
5. $x_1 = c_1 e^t, x_2 = c_2 e^{2t}$; unstable; $c_1^2 x_2 = c_2 x_1^2$
7. $x_1 = Ae^{-2t} \cos(t + \alpha), x_2 = -Ae^{-2t} \sin(t + \alpha)$; asymptotically stable

Section 10.2

1. Unstable, spiral point; $y_1' = 2y_1 + y_2, y_2' = -y_1 + 2y_2$; $K = \begin{bmatrix} 1 & 1 \\ 1 & 0 \end{bmatrix}$; $x_1 - x_2 = 0, x_2 = 0$

3. Unstable, saddle point; $y_1' = 2y_1, y_2' = -y$; $K = \begin{bmatrix} 1 & 1 \\ 1 & 2 \end{bmatrix}$; $x_1 - x_2 = 0, 2x_1 - x_2 = 0$

5. Asymptotically stable, improper node; $y_1' = -2y_1, y_2' = -3y_2$; $K = \begin{bmatrix} 1 & 1 \\ 2 & 1 \end{bmatrix}$; $2x_1 - x_2 = 0, x_1 - x_2 = 0$

7. Stable, center; $y_1' = 3y_2, y_2' = -3y_1$; $K = \begin{bmatrix} -1 & 1 \\ 1 & 0 \end{bmatrix}$; $x_1 + x_2 = 0, x_2 = 0$

9. Asymptotically stable, improper node; $y_1' = -y_1 + y_2, y_2' = -y_2$; $K = \begin{bmatrix} 1 & 0 \\ -2 & 1 \end{bmatrix}$; $2x_1 + x_2 = 0, x_1 = 0$

11. Asymptotically stable, spiral point; $y_1' = -y_1 + y_2, y_2' = -y_1 - y_2$
13. Stable, center; $y_2' = 2y_2, y_2' = -2y_1$ 15. Critical point $(3, -1)$ is unstable, a saddle point
17. Critical point $(2, 3)$ is stable, a center   19. $x_1' = -3x_1 + x_2, x_2' = x_1 - 3x_2$

Section 10.3

1. (a) Yes   (c) No   (e) Yes   (g) Yes   2. (a) Yes   (c) Yes
3. Asymptotically stable, spiral point   5. Unstable, improper node   7. Unstable, saddle point
9. Asymptotically stable, improper node   11. Asymptotically stable, spiral point
13. $(0, 0)$ is an unstable improper node; $(0, \frac{5}{2})$ is an unstable saddle point; $(4, 0)$ is an unstable saddle point; $(3, 1)$ is an asymptotically stable improper node.

Section 10.4

1. $(0, 0)$ and $(3, 1)$ are unstable; $(\frac{7}{2}, 0)$ and $(0, 2)$ are asymptotically stable.
3. $(0, 0), (5, 0)$, and $(0, 4)$ are unstable; $(2, 3)$ is asymptotically stable.

5. (0, 0) and (0, 1) are unstable; (4, 0) is asymptotically stable.   7.   All solutions approach (0, 4).
9. (0, 0) and (0, 7) are unstable; (1, 4) is asymptotically stable.
11. (0, 0) is unstable; (0, 1) is asymptotically stable.

Section 10.5

1. $r = \dfrac{2}{\sqrt{1 + c_1 e^{8t}}}$, $\theta = -t + c_2$. Trajectories spiral away from $r = 2$.

Additional Exercises

1. Unstable, spiral point. $K = \begin{bmatrix} 1 & 0 \\ 1 & 1 \end{bmatrix}$; $y_1' = y_1 + 4y_2$, $y_2' = -4y_1 + y_2$; $x_1 - x_2 = 0$, $x_1 = 0$

3. Stable, center. $K = \begin{bmatrix} 1 & 0 \\ 1 & 1 \end{bmatrix}$; $y_1' = 2y_2$, $y_2' = -2y_2$; $x_1 - x_2 = 0$, $x_1 = 0$

5. Asymptotically stable, improper node. $K = \begin{bmatrix} 1 & 1 \\ -1 & 1 \end{bmatrix}$; $y_1' = -y_1$, $y_2' = -3y_2$;

    $x_1 + x_2 = 0$, $x_1 - x_2 = 0$

7. Unstable, improper node. $K = \begin{bmatrix} 1 & 1 \\ 2 & 3 \end{bmatrix}$; $y_1' = y_1$, $y_2' = 2y_2$; $2x_1 - x_2 = 0$, $3x_1 - x_2 = 0$

9. Asymptotically stable, improper node   11.   Unstable, improper node

13. Unstable, improper node

19. (0, 0) is an unstable, improper node; (1, 1) is an aymptotically stable spiral point; (2, 0) and $(0, \tfrac{5}{3})$ are unstable saddle points.

Index

Numbers in parentheses indicate exercises

A

Abel's formula, 207(7)
Adams–Bashforth method, 395
Additive inverse, 131
Algebraic function, 333
Amplitude, 252
Analytic function, 339
Angular frequency, 250
Augmented matrix, 65
Autonomous equation, 62(26)
Autonomous system, 437
Auxiliary polynomial, 210

B

BASIC, programs in, 385, 392, 401
Basis
　definition of, 146
　orthogonal, 151
　orthonormal, 151
　standard, 148
Bernoulli equation, 26
Bessel function
　of first kind, 375
　modified, 380(19)
　of second kind, 375
Bessel's equation, 374
Binomial series, 339

C

C^n, 95, 132, 148
Canonical form, 445
Cauchy–Euler equation, 152, 347, 360
Cayley–Hamilton theorem, 198
Center, 450
Characteristic equation, 180
Characteristic polynomial, 180

Characteristic value
　complex, 180, 187
　definition of, 178
　and differential equations, 288–96, 306–12
　multiplicity of, 187
Characteristic vector, 178
Chebyshev's polynomial, 380(21)
Coefficient matrix, 65
Coefficients
　of an algebraic system, 65, 82
　of a differential equation, 203
Cofactor, 112, 124
Column vector, 84
Companion matrix, 182(19)
Complementary function, 239
Complete set of vectors, 185, 193
Complex exponential function, 215
Complex vector function, 282
Conservation of mass, 36
Consistent system, 65
Constants of a system, 65
Convolution, 420
Corrector, 395
Cramer's rule 118
Critical point
　asymptotically stable, 442
　definition of, 438
　isolated, 438
　stable, 442
　unstable, 442

D

Damped harmonic motion, 252
Damping constant, 51
Dashpot, 252

DeMoivre's formula, 217(6)
Determinant
　definition of, 104
　derivative of, 106
　properties of, 107–9
　of a triangular matrix, 110
Differential equation
　definition of, 1
　linear, 4
　order of, 2
　solution of, 2
Differential operator, 170
Dimension, 146
Domain, 155

E

Eccentricity, 325
Eigenvalue. *See* Characteristic value
Eigenvector. *See* Characteristic vector
Electric circuits, 79, 257–60, 327–28
Element of a matrix, 65, 83, 279
Elementary row operation, 71, 115, 119, 148, 167
Elimination method, 272
Equation of motion, 48, 250
Equivalent systems, 66, 268
Escape velocity, 54
Euler method, 382, 398
Exact equation, 17
Exponential matrix
　definition of, 301
　properties of, 302–4
Exponential order, 410
Exponents at a singular point, 353

F
Field, 132
First-order method, 382
First-order system
    definition of, 267
    linear, 271
    reduction to, 267–70
Fourth-order system, 391
Frobenius, method of, 349
Fundamental matrix, 285, 304
Fundamental set of solutions,
    284, 313

G
Gamma function, 371
Gauss reduction, 117
General solution, 2, 267
Geometric series, 340
Gram–Schmidt procedure, 153
Gravity, 48, 318

H
Half-life, 32
Hermite's equation, 371(7)
Homogeneous equation, 13,
    172, 203
Homogeneous function, 13
Homogeneous system of
    equations, 65, 75, 119, 271

I
Identity matrix, 98
Identity transformation,
    159(10), 169
Image, 155
Indicial equation, 353
Initial conditions, 6, 267
Initial value problem, 6, 267
Inner product. *See* Scalar
    product
Integral curve, 5
Integrating factor, 21
Inverse
    of a linear transformation, 162
    of a matrix, 122
Inverse transform, 418

J
Jordan form, 191, 192(20), 294

K
Kepler's laws, 326(2), 327(3)
Kernel, 156
Kirchoff's laws, 80, 258, 327
Kronecker delta, 98

L
Laguerre's equation, 371(6)
Laplace transforms
    definition of, 405
    of periodic functions, 435(9)
    properties of, 405, 413
    table of, 435
Lattice point, 81
Legendre polynomial, 369
Legendre's equation, 368–69
Leibniz's rule, 19
Lerch's theorem, 418
Lienard equation, 475
Limiting velocity, 51
Linear combination, 93(20), 143
Linear dependence, 137
Linear differential equation
    definition of, 4, 23
    first-order, 23
Linear independence. *See*
    Linear dependence
Linear mapping, 156
Linear operator, 156, 406
Linear space. *See* Vector space
Linear transformation
    definition of, 156
    domain of, 155
    elementary, 164
    inverse of, 162
    kernel of, 156
    one-to-one, 162
    onto, 162
    properties of, 160–62
    range of, 165
    rank of, 175(19)
Linearization, 454
Lipschitz condition, 56
Local error, 385, 387

M
Maclaurin series, 339
Matrix
    definition of, 83
    diagonal, 98
    diagonalizable, 185
    element of, 83
    elementary, 164
    fundamental, 285, 304
    identity, 98
    inverse of, 122
    lower triangular, 99, 110
    nonsingular, 105
    norm of, 297
    order of, 83
    orthogonal, 126(16), 174(11), 195

product, 89
    rank of, 174(17)
    scalar, 99
    singular, 105
    size of, 83
    skew-symmetric, 100
    symmetric, 100, 192–96
    transpose of, 100
    upper triangular, 99, 110
    zero, 97
Matrix function, 279
Mechanical systems, 317–20,
    322–26
Minimal polynomial, 201(12)
Minor, 112, 222
Mixing problems, 35, 278(16),
    (17), 279(18)
Modulus, 217(5)
Multi-step method, 382, 393

N
Newton's inverse square law, 52
Newton's law of cooling, 39
Newton's law of motion, 48
Node, 446, 450
Nonhomogeneous differential
    equation, 203, 229
Nonhomogeneous system, 65,
    313
Nonsingular matrix, 105
Norm, 95, 155(10), 297
Null space. *See* Kernel

O
Order
    of a differential equation, 2
    of a numerical method, 382
    of a square matrix, 83
Ordinary point, 341
Orthogonal basis, 151
Orthogonal complement, 137(11)
Orthogonal functions, 155(11)
Orthogonal trajectories, 29
Orthogonal transformation,
    174(10)
Orthogonal vectors, 96
Orthonormal basis, 151
Orthonormal set of vectors, 96

P
Partial differential equation, 2
Pendulum, 55(11), 457
Permutation
    definition of, 104

inversion of, 104
    parity of, 104
Permutation matrix, 127(7)
Phase plane, 440
Phase portrait, 440
Picard's method, 57
Piecewise continuous function, 409
Poincare–Bendixon theorem, 473
Polar coordinates, 31(17), 319, 324
Polynomial operator, 208, 272
Population growth, 37, 460
Positive-definite form, 195
Power series, 334
Predator–prey problem, 467
Predictor, 395
Principal leading minor, 222

Q
Quadratic form, 100, 195–96

R
R^n, 85, 129, 148
Radioactive carbon, 34
Range, 156
Rank, 174(17), 175(19)
Rate of reaction, 40, 41
Recurrence relation, 344, 369
Regular point, 438
Regular singular point, 348
Resonance, 255
Riccati equation, 28(24), 262(24)
Rodrigues' formula, 370
Routh–Hurwitz criterion, 222, 296, 330(21)
Row vector, 84
Row-echelon form, 72, 119, 175(17)
Runge–Kutta methods, 390–91, 399

S
Saddle point, 448
Scalar, 132
Scalar product, 95, 96, 155(12)
Schwarz inequality, 97(9)
Second-order method, 387, 390–91
Separable equation, 8
Similar matrices, 184
Simple harmonic motion, 252
Simply connected region, 19
Single-step method, 382
Singular matrix, 105
Singular point, 341
Size of a matrix, 83
Solution
    complex, 213, 218
    definition of, 2, 63
    general, 2
    periodic, 472
    steady-state, 257(16), 260
    transient, 260
    trivial, 75, 203, 282
Spiral point, 450
Spring constant, 249
Stability
    asymptotic, 442
    definition of, 442
Step size, 381
Submatrix, 112
Subspace, 134
Successive approximations, 57
Superposition principle, 231
System of differential equations, 263, 437

T
Taylor series, 339
Taylor's formula, 339, 382, 386
Trajectory, 439
Transpose of a matrix, 100

Trial solution, 234
Trivial solution, 75, 203, 282

U
Uniqueness of solutions, 57, 270
Unit step function, 427
Units
    electrical, 258
    mechanical, 48

V
Van der Pol equation, 475
Vandermonde's determinant, 145
Variation of parameters, 241–46, 314, 316
Vector
    column, 84
    component of, 84
    geometric, 93
    length of, 95–96
    in R^n, 84
    row, 84
    unit, 96
Vector function, 280
Vector space
    basis for, 146
    complex, 132
    definition of, 130
    dimension of, 146
    real, 131

W
Weight, 48
Wronskian, 143

Z
Zero element, 130, 131
Zero matrix, 97
Zero transformation, 159(9)
Zero vector, 85